Trends and Applications of Heat Treatment

Trends and Applications of Heat Treatment

Edited by **Edgar Miller**

NY RESEARCH PRESS

New York

Published by NY Research Press,
23 West, 55th Street, Suite 816,
New York, NY 10019, USA
www.nyresearchpress.com

Trends and Applications of Heat Treatment
Edited by Edgar Miller

International Standard Book Number: 978-1-63238-455-3 (Hardback)

Printed in China.

Contents

Preface

The trends and novel applications of heat treatment are discussed in this profound book. Heat treatment and surface engineering are important factors in the design and manufacture of strategic components in a vast range of market sectors and industries, inclusive of air, sea and land transportation, energy production, mining, defence, and agriculture. This book presents an overview of the latest advancements in the applications of thermal and thermo-chemical processing to alter the microstructure and properties of a vast range of engineering substances. The book presents two different functional areas of heat treatment. The first area deals with conventional heat treatment, with the processing of bearing rings, wrought and cast steels, aluminium alloys, basics of thermo-chemical treatment, details of carbonitriding and a design for cooling units. The second area demonstrates the usage of unconventional thermal routes for the production of substances like nanoparticles, titanium dioxide, carbides of vanadium, battery substances and cinder mortars. A mixture of conventional and new functions, analysis of several processes employing heating, quenching and thermal diffusion, makes the book beneficial for a vast audience of scientists and engineers from academics and industries.

The information contained in this book is the result of intensive hard work done by researchers in this field. All due efforts have been made to make this book serve as a complete guiding source for students and researchers. The topics in this book have been comprehensively explained to help readers understand the growing trends in the field.

I would like to thank the entire group of writers who made sincere efforts in this book and my family who supported me in my efforts of working on this book. I take this opportunity to thank all those who have been a guiding force throughout my life.

Editor

Chapter 1

A Review on the Heat Treatment of Al-Si-Cu/Mg Casting Alloys

A.M.A. Mohamed and F.H. Samuel

Additional information is available at the end of the chapter

1. Introduction

The trend of the automotive industry goes toward the construction of high-powered, comfortable, economical, ecological and safe vehicles. A few Al alloys containing Cu and a few containing Mg and Si are heat treatable in the cast condition due to the precipitation strengthening mechanisms. Two of the major families of heat treatable aluminum alloys containing magnesium and silicon are the 6xxx series in wrought aluminum alloys, and 3xx series in casting aluminum alloys. Al-Si-Cu/Mg alloys are well studied and there exists a lot of publication about the effect of alloying elements and solidification rate on the microstructure formation [1-3]. The influence of heat treatment on the mechanical properties including hardness and tensile strengths is also well studied, while the influence on plastic deformation behavior and elongation to fracture is less studied.

Although the benefit of heat treatment is undisputed, there exist several challenges for heat treatment operators, including market expectations of higher performance and reliability, lower production costs and energy use, as well as concern over environmental impacts. The heat treatment of age hardenable aluminum alloys involves solutionizing the alloys, quenching, and then either aging at room temperature (natural aging) or at an elevated temperature (artificial aging). The enhancement in mechanical properties after thermal treatment has largely been attributed to the formation of non-equilibrium precipitates within primary dendrites during aging and the changes occurring in Si particles characteristics from the solution treatment. The age hardening response depends on the fraction size, distribution and coherency of precipitates formed. Al-Si-Cu-Mg alloys and Al-Si-Mg alloys generally have a high age hardening response, while Al-Si-Cu alloys have a slow and low age hardening response.

2. Solidification process

During the solidification from a melt, chemical thermodynamics and kinetics are generally considered in terms of the enthalpy and Gibbs free energy changes, the solidification path, composition changes, and phase transformations etc.. Chemical thermodynamics describes the most stable phases at equilibrium conditions (i.e. temperature, pressure, compositions etc.) relating to only the initial and final states of a system. Accordingly, the solidification rate in a metallurgical system can be estimated by the enthalpy (H) and heat capacity (Cp), and how these thermodynamic properties reflect the system thermal state and heat energy requirements. Chemical equilibrium is controlled by the Gibbs free energy (G) of the system which is minimized for equilibrium conditions. In contrast, the dynamic system transformation between initial and final states controlled by chemical kinetics, indicates the path and phase changes of a chemical reaction in a system when the limited atomic movement (i.e. in solids, low temperatures, etc.) becomes dominant in a short process time. Hence the solidification rate under a real time condition will be greatly influenced by the nucleation efficiency and the atom diffusion between phases [4].

The solidification rate determines the coarseness of the microstructure including the fraction, size and distribution of intermetallic phases and the segregation profiles of solute in the α-Al phase. Large and brittle intermetallic phases form during a slow solidification, which may initiate or link fracture, decreasing elongation to fracture. Additional, the defect size such as pore size, is also controlled to some extent by the solidification rate. The influence of defects on the elongation to fracture depends on their size, shape, distribution and fraction. Dendrite arms with smaller radius may remelt into the molten into the molten liquid along with the decreasing total interfacial energy. The Ostwald-ripening effect on the formation of dendrite arm spacing (DAS) is determined by local solidification time, allowing smaller particles to grow and merge into the larger ones due to the reduced total surface energy in the system. DSA, which is proportional to (average cooling rate)$^{-n}$ where n =1/2 and 1/3 for the primary and secondary dendrites respectively, generally ranging from 10 to 150 mm and which are controlled mainly by the solidification rate [5]. To gain an optimum property of an alloy, the DAS therefore must be minimized and distributed homogeneously.

The major phases in as-cast microstructure of Al-Si alloys are large size grains and primary α-Al, acicular eutectic Si, coarse primary Si, and also other harmful intermetallic phases such as needle like β-Al_5FeSi, with uncontrolled and unevenly distributed porosities etc. [6]. Table 1 summarizes the sequence of phase precipitation in hypoeutectic Al-Si alloys [7]. Al in the eutectic has been reported to have mainly the same crystallographic features as the primary α-Al dendrites in unmodified alloys [8]. Figure 1.a indicates a basic structure of hypoeutectic Al-Si alloys consisting of grains (sizes at 1~10 mm in general), dendrites (typical DAS - 10~150 μm), and eutectic Si which can be in acicular shapes as long as 2 mm or round particle as small as 1 μm. The acicular Si might be chemically modified to a fibrous morphology by using effective modifiers.

Heterogeneous nucleation should be the major approach to refine the grains, which nucleate on some of the foreign nuclei sites and grow slowly in the melt. Effective grain refiners, such as $TiAl_3$ and TiB_2, must match their lattice perfect coherently to the Al matrix with their lattice coherencies (Figure 1.b). In contrast, particles with a poor lattice matching have little influence on increasing the nucleation of grains (Figure 1.c), resulting in an unrefined grain structure [9]. Typical examples of the microstructure of unmodified, Sr-modified, and Sb-modified alloys are shown in Figure 2.

Temperature (°C)	Phases precipitated	Suffix
650	Primary $Al_{15}(Mn, Fe)_3Si_2$ (sludge)	Pre-dendrite
600	Aluminum dendrites and ($Al_{15}(Mn, Fe)_3Si_2$) and /or Al_5FeSi	Dendritic Post-dendritic Pre-eutectic
550	Eutectic Al + Si and Al_5FeSi Mg_2Si	Eutectic Co-eutectic
500	$CuAl_2$ and more complex phases	Post-eutectic

Table 1. Sequence of phase precipitation in hypoeutectic Al-Si alloys [7]

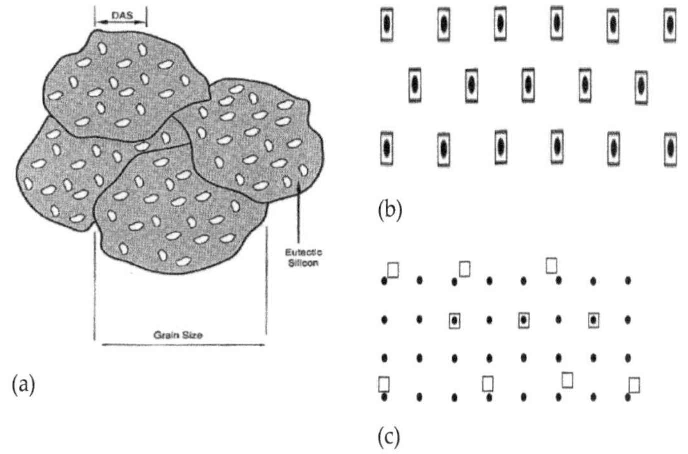

Figure 1. Schematics of a) Three essential elements (grains, Al dendrites, DAS, and eutectic Si in a basic hypoeutectic Al-Si microstructure; b) Perfect grain refiner particles (squares) with one to one lattice matching to Al atoms (points); c) Poor lattice matching [9].

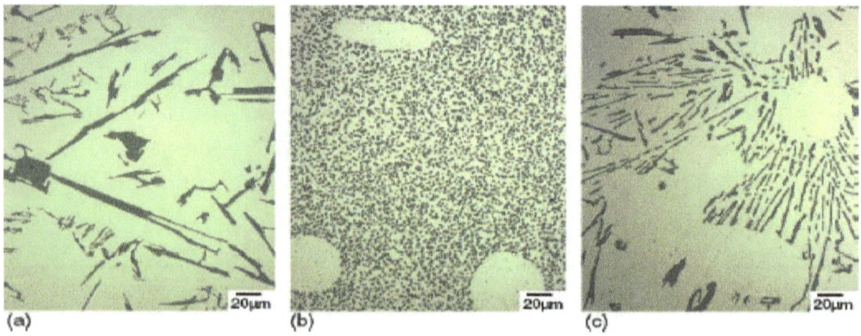

Figure 2. Comparison of the silicon morphology in: (a) unmodified; (b) Sr-modified (300 ppm Sr); and (c) Sb-modified (2400 ppm Sb), hypoeutectic aluminum-silicon alloys [10].

Copper forms an intermetallic phase with Al that precipitates during solidification either as blocky $CuAl_2$ or as alternating lamellae of α-Al + $CuAl_2$ [11]. During solidification, in the presence of iron, other copper containing phases form, such as Cu_2FeAl_7 or Q-$Al_5Cu_2Mg_8Si_6$ [12]. The $CuAl_2$ phase can be blocky shape or finely dispersed α-Al and $CuAl_2$ particles within the interdendritic regions, as shown in Figure 3. The presence of nucleation sites, such as $FeSiAl_5$ platelets or high cooling rates during solidification can result in fine $CuAl_2$ particles [11]. The blocky $CuAl_2$ phase particles are difficult to dissolve during solid solution heat treatment, unlike the fine $CuAl_2$ phase particles that can dissolve within 2 hrs solid solution heat treatment [13]. Magnesium is present as Mg_2Si in Al-Si-Mg alloys if Mg is not in solution. Mg can also form a true quaternary compound $Cu_2Mg_8Si_6Al_5$ with other alloy elements in Al-319 alloy. In the absence of Cu, high Fe and Mg result in the appearance of π-$FeMg_3Si_6Al_8$. The π phase is difficult to dissolve during solid solution heat treatment [8].

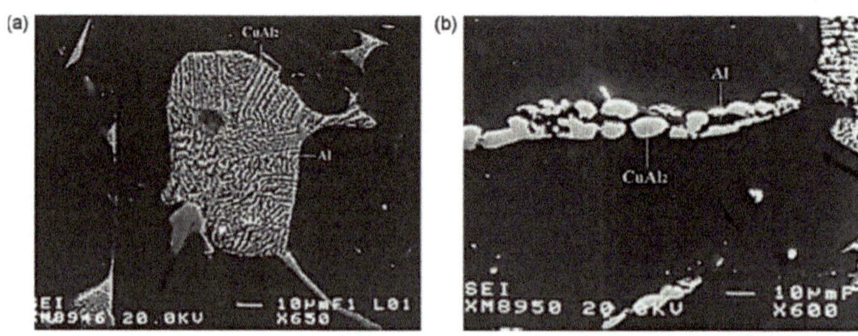

Figure 3. Cu-rich phases in as-cast 319 alloy: (a) Eutectic Al_2Cu and (b) blocky Al_2Cu [14].

A comparative study of the mechanical properties of Al-Si-Cu-Mg alloys was carried out by Cáceres *et al.* [15] to investigate the effects of Si, Cu, Mg, Fe, and Mn, as well as solidification rate. The authors observed that increasing the Cu and Mg content generally resulted in an increase in strength and a decrease in ductility, whereas an increased Fe content (at an Fe/Mn ratio of 0.5) dramatically lowered the ductility and strength of low-Si alloys. They also reported that the Cu + Mg content of the alloys determines the precipitation strengthening and the volume fraction of the Cu-rich and Mg-rich intermetallics obtained.

Yi [16] adopt the enhanced solid diffusion coefficient of Cu in his model. The diffusion coefficient of Cu in α-Al phase is increased by 4-fold. The presence of Si-phase also has great influence on the diffusion of Cu in the matrix. It is assumed that the diffusion coefficient of Cu increases by 20-fold due to the presence of Si-phase. The distribution of Mg and Si across the dendrite arm spacing also changes due to the increase of Cu diffusion in the matrix. This is attributed to the change of solidification path.

3. Heat treatment of cast al alloys

Heat-treatment is of major importance since it is commonly used to alter the mechanical properties of cast aluminum alloys. Heat-treatment improves the strength of aluminum alloys through a process known as precipitation-hardening which occurs during the heating and cooling of an aluminum alloy and in which precipitates are formed in the aluminum matrix. The improvement in the mechanical properties of Al alloys as a result of heat treatment depends upon the change in solubility of the alloying constituents with temperature. Figure 5 shows the major steps of the heat treatment which are normally used to improve the mechanical properties of aluminum. The alloy should first be solution treated at a temperature just below the eutectic temperature for long enough to allow solutionizing of the second phase. Then it should be quenched to room temperature. Finally it should be heated to a lower temperature to allow precipitation. Table 2 details a few of the more commonly applied heat treatments.

The T6 heat treatment is illustrated in Figure 4 for an Al-Si-Cu alloy as an example. The evolution of the microstructure is shown; from (1) atoms in solid solution at the solution treatment temperature, through (2) a supersaturated solid solution at room temperature after quench, to (3) precipitates formed at the artificial ageing temperature.

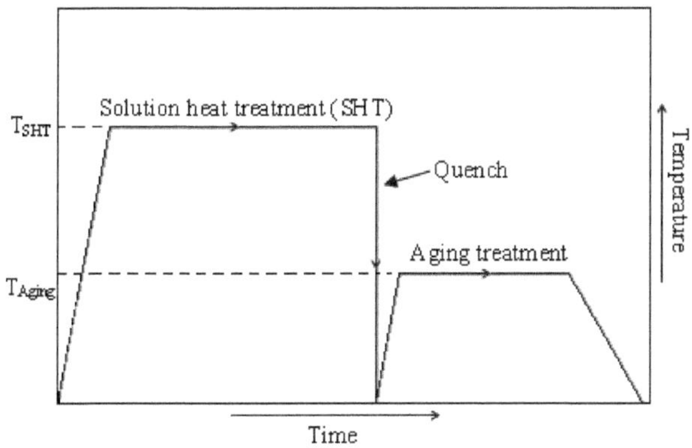

Figure 4. Diagram showing the three steps for precipitation hardening.

Treatment	Solution	Quench	Aging
T4	Yes	Yes	Room Temperature only
T5	No	No	Elevated Temperatures
T6	Yes	Yes	Elevated (to yield increased strength)
T7	Yes	Yes	Elevated (to yield dimensional stability)

Table 2. Common aluminum heat treatment designations

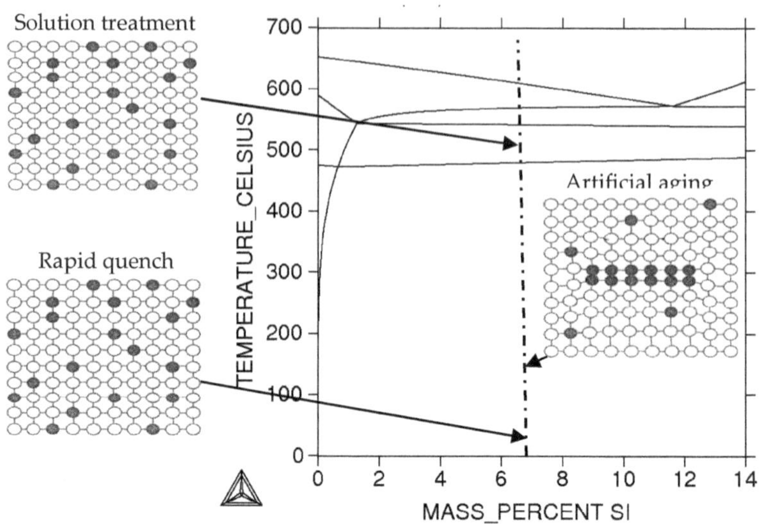

Figure 5. The T6 heat treatment process [17].

3.1. Solution heat treatment

Solution heat treatment must be applied for a sufficient length of time to obtain a homogeneous supersaturated structure, followed by the application of quenching with the aim of maintaining the supersaturated structure at ambient temperature. In Al-Si-Cu-Mg alloys, The solution treatment fulfils three roles: [18,19]

i. Homogenization of as-cast structure.
ii. Dissolution of certain intermetallic phases such as Al_2Cu and Mg_2Si.
iii. Change of the morphology of eutectic silicon.

The segregation of solute elements resulting from dendritic solidification may have an adverse effect on mechanical properties. The time required for homogenization is determined by the solution temperature and by the dendrite arm spacing. Hardening alloying elements such as Cu and Mg display significant solid solubility in heat-treatable aluminum alloys at the solidus temperature; this solubility decreases noticeably as the temperature decreases.

The changes in the size and morphology of the silicon phase have a significant influence on the mechanical properties of the alloy. It has been proposed that the granulation or spheroidization process of silicon particles through heat treatment takes place in two stages: (i) fragmentation or dissolution of the eutectic silicon branches and (ii) spheroidization of the separated branches [20]. During solution treatment, the particles undergo changes in size and in shape. In the initial stages, the unmodified silicon particles undergo necking and separate into segments, which retain their original morphology. As a result of the separation, the average particle size decreases and the fragmented segments are eventually spheroidized. The spheroidization and the coarsening of eutectic Si can occur concurrently during the second stage.

The solution treatment process needs to be optimized because too short a solution treatment time means that not all alloying elements added will be dissolved and made available for precipitation hardening, while too long a solution treatment means using more energy than is necessary. The solution heat treatment may be carried out in either a single step or in multiple steps. Single-step solution treatment is normally limited to about 495°C, in view of the fact that higher temperatures lead to higher thermal stresses induced during quenching and the risk of the incipient melting of the Cu-rich phases [21-23]. This incipient melting tends to lower the mechanical properties of the casting. Solution treatment at temperatures of 495°C or less, however, is not capable of maximizing the dissolution of the copper-rich phases, nor is it able to modify the silicon particle morphology sufficiently. In Al-Si-Cu-Mg alloys having a low magnesium content (0.5 wt.%), Ouellet *et al.* [24] used a solution temperature of 500°C because, at 505°C, fusion of low melting point phases can occur; Wang *et al.* [25], on the other hand, reported that, for a similar alloy with a solution temperature of 520°C, mechanical properties increase without any observable localized melting.

Based on conventional solution treatment rules, the solution temperature of Al-Si-Cu-Mg alloys is restricted to 495°C, in order to avoid incipient melting of the copper-rich phase [26,27]. The time at the nominal solution treatment temperature must be long enough to homogenize the alloy and to ensure a satisfactory degree of precipitate solution. In alloys containing high levels of copper, complete dissolution of the Al_2Cu phase is not usually possible. The solution time must then be chosen carefully to allow for the maximum dissolution of this intermetallic phase, bearing in mind nevertheless, that solutions treating the alloy for long times are expensive and may not be necessary to obtain the required alloy strength. Moreover, the coarsening of the microstructural constituents and the possible formation of secondary porosity which result after prolonged annealing at such temperatures can have a deleterious effect on the mechanical properties [28].

Studies by Gauthier *et al.* [19] on the solution heat treatment of 319 alloy over a temperature range of 480°C to 540°C, for solution times of up to 24 hours, showed that the best combination of tensile strength and ductility was obtained when the as-cast material was solution heat-treated at 515°C for 8 to 16 hours, followed by quenching in warm water at 60°C. A higher solution temperature was seen to result in the partial melting of the copper phase, the formation of a structureless form of the phase and related porosity upon quenching, with a consequent deterioration of the tensile properties. A two-stage solution heat treatment suggested by Sokolowski *et al.* [29] is reported to reduce the amount of the copper-rich phase in the 319 alloys significantly, giving rise to better homogenization prior to aging and improving mechanical properties. Also, Crowell *et al.* [30] stated that the blocky Cu phase in Al-Si-Cu alloys dissolves with increasing solution time at the recommended solution temperature of 495°C; also the rate of dissolution increases with Sr concentration.

A two-step solution treatment, namely, conventional solution treatment followed by a high-temperature solution treatment, as suggested by Sokolowski *et al.*, [31, 32] is reported to reduce the amount of the copper-rich phase in 319 alloys significantly, thereby giving rise to better homogenization prior to aging and thus also to improvements in the mechanical properties. The holding time for the first stage and the solution temperature of the second stage are both significant parameters. Sokolowski *et al.* [32] studied the improvement in 319 aluminum alloy casting durability by means of high temperature solution treatment. Their results showed that a two-step solution treatment of 495ºC/2h followed by 515ºC/4h produced the optimum combination of strength and ductility compared to the traditional single-step solution treatment of 495ºC/8h.

Dissolve the micro-segregation of Mg and Si elements to form a supersaturated solid solution in the primary Al matrix in order to enable the formation of a large number of strengthening precipitates during subsequent natural and artificial ageing processes. Homogenize the casting, and attain a globular morphology of the eutectic Si phase to impart improved ductility and fracture toughness to the component. Reduce micro-segregation of other alloying elements in the primary Al matrix.

3.2. Quenching

Following solution heat treatment, quenching is the next important step in the heat-treatment cycle. The objectives of quenching are to suppress precipitation during quenching; to retain the maximum amount of the precipitation hardening elements in solution to form a supersaturated solid solution at low temperatures; and to trap as many vacancies as possible within the atomic lattice [33,34].

The quench rate is especially critical in the temperature range between 450 °C and 200 °C for most Al-Si casting alloys where precipitates form rapidly due to a high level of supersaturation and a high diffusion rate. At higher temperatures the supersaturation is too low and at lower temperatures the diffusion rate is too low for precipitation to be critical. 4°C/s is a limiting quench rate above which the yield strength increases slowly with further increase in quench rate [35-37].

Faster rates of quenching retain a higher vacancy concentration enabling higher mobility of the elements in the primary Al phase during ageing. An optimum rate of quenching is necessary to maximize retained vacancy concentration and minimize part distortion after quenching. A slow rate of quenching would reduce residual stresses and distortion in the components, however, it causes detrimental effects such as precipitation during quenching, localized over-ageing, reduction in grain boundaries, increase tendencies for corrosion and result in a reduced response to ageing treatment [38,39].

The best combination of strength and ductility is achieved from a rapid quenching. Cooling rates should be selected to obtain the desired microstructure and to reduce the duration time over certain critical temperature ranges during quenching in the regions where diffusion of smaller atoms can lead to the precipitation of potential defects [40]. The effectiveness of the quench is dependent upon the quench media (which controls the quench rate) and the quench interval. The media used for quenching aluminum alloys include water, brine solution and polymer solution [41-43]. Water used to be the dominant quenchant for aluminum alloys, but water quenching most often causes distortion, cracking, and residual stress problems [44,45]. It has been reported that the water temperature affects the properties of the cast aluminum alloy A356 subjected to T6 heat treatment once the water exceeds 60-70°C, with UTS and YS being significantly more sensitive than ductility. Detailed TEM investigations on A356 alloy, reported elsewhere [46], revealed that, at the peak-aged condition and with a water quench at 25°C, the α-Al matrix consists of a large number of needle-shaped and coherent β''-Mg2Si precipitates. The size of the precipitates is approximately 3 to 4 nm in diameter and 10 to 20 in length. With a water quench at 60°C, they observed how the density of the precipitates decreases and the size of the precipitates increases slightly; at the same time a significant number of fine Si precipitates resulting from precipitation of excess Si could be observed in the α-Al matrix.

With a slow quenching in air, very different precipitation features are normally evidenced. By air quenching, the material remains at high temperatures for a longer period, which enhances the diffusion of silicon and magnesium. Besides a high density of fine β''-Mg2Si precipitates, the α-Al matrix also contained a large number of areas with coarse rods β'-Mg2Si grouped parallel to each other [46]. While the first precipitates have an average size approximately 2 to 3 nm in diameter and around 40 nm in length, the latter show an average size ~15 nm in diameter and 300 nm in length.

3.3. Aging

Age-hardening has been recognized as one of the most important methods for strengthening aluminum alloys, which involves strengthening the alloys by coherent precipitates which are capable of being sheared by dislocations [47]. By controlling the aging time and temperature, a wide variety of mechanical properties may be obtained; tensile strengths can be increased, residual stresses can be reduced, and the microstructure can be stabilized. The precipitation process can occur at room temperature or may be accelerated by artificial aging at temperatures ranging from 90 to 260°C.

After solution treatment and quench the matrix has a high supersaturation of solute atoms and vacancies. Clusters of atoms form rapidly from the supersaturated matrix and evolve into GP zones. Metastable coherent or semi-coherent precipitates form either from the GP zones or from the supersaturated matrix when the GP zones have dissolved. The precipitates grow by diffusion of atoms from the supersaturated solid solution to the precipitates. The precipitates continue to grow in accordance with Ostwald ripening when the supersaturation is lost. The length of each step in the sequence depends on the thermal history, the alloy composition and the artificial ageing temperature.

The phenomenon of precipitation was originally discovered by Ardel in 1906 [48]. He found that the hardness of aluminum alloys which contained magnesium, copper, and other trace elements increased with time at room temperature, which was later explained by precipitation hardening. Over the years, much research was carried out to understand the aging kinetics of T4 and T6 heat treatments and to study the effects of underaging, peak-aging, and overaging on hardness [48-50], ultimate tensile strength, crack propagation behavior [51], and the cyclic stress-strain response of cast aluminum-silicon alloys [52].

The precipitation sequence for an Al-Si-Cu alloy, such as 319, is based upon the formation of Al_2Cu-based precipitates. The Al_2Cu precipitation sequence is generally described as follows: [53-55]

$$\alpha_{ss} \rightarrow \text{GP Zones} \rightarrow \theta'' \rightarrow \theta' \rightarrow \theta \ (Al_2Cu)$$

The sequence begins with the decomposition of the solid solution and the clustering of Cu atoms; the clustering then leads to the formation of coherent, disk-shaped GP zones. At room temperature aging conditions, GP zones arise homogeneously; these zones manifest as two-dimensional, copper-rich disks with diameters of approximately 3-5 nm. As time increases, these GP zones increase in number while remaining approximately constant in size. With regard to the Al-Cu alloys, as the aging temperature is increased above 100°C, the GP zones dissolve and are replaced by the θ'' precipitate. This precipitate is a three-dimensional disk-shaped plate having an ordered tetragonal arrangement of Al and Cu atoms; θ'' also appears to nucleate uniformly in the matrix, and is coherent with the matrix in binary Al-Cu alloys. The high degree of coherency causes extensive coherency-strain fields to arise [56], giving peak strength to the material at this time.

As aging proceeds, the θ'' starts to dissolve, and θ' begins to form by nucleating on dislocations and/or cell walls [54,55]; θ' also has a plate-like shape and is composed of Al and Cu atoms in an ordered tetragonal structure; θ' loses coherency with the matrix, however, as it grows. Thus, since the long-range coherency-strain fields do not arise, a decrease in strength properties may be observed, while continued aging causes the equilibrium θ (Al_2Cu) precipitate to occur. Tetragonal in shape, the θ phase is completely incoherent with the matrix; this fact, combined with its relatively large size and coarse distribution, reduces the strength properties significantly [56].

Increases in Cu were found mainly to reduce ductility and change the morphology of the Cu-containing phases [57]. The strength of an age-hardenable alloy is governed by the interaction of moving dislocations and precipitates. The obstacles in precipitation-hardened alloys which hinder the motion of dislocations may be either the strain field around the GP zones resulting from their coherency with the matrix, or the zones and precipitates themselves, or both. The dislocations are then forced to cut through them or go around them forming loops. The preceding thus implies clearly that there are three sources for age hardening: strain field hardening, chemical hardening and dispersion hardening. Gloria *et al.* [58] investigated the dimensional changes occurring during the heat treatment of an automotive 319 alloy by means of T6 and T7 tempers involving solution treatment, quenching and artificial aging. They observed that increasing the solution temperature has the greatest influence in the dimensional change of samples due to dissolution of the Al-Cu(θ) eutectic phase. By increasing the aging temperatures, however, expansion is produced as a result of the transformation of the metastable phases into equilibrium phases.

Shivkumar *et al.* [59] have studied the parameters which control the tensile properties of A356 alloy in the T6 temper. The improvement in the alloy strength has been attributed to the precipitation of negligible phases from a supersaturated matrix. The sequence of precipitation in Al-Si-Mg alloys, see Figure 5, can be described as follows:

i. Precipitation of GP zones, (needles about 10 nm long);
ii. Intermediate phase β''-Mg_2Si, (homogeneous precipitation);
iii. Intermetallic phase β'-Mg_2Si, (heterogeneous precipitation);
iv. Equilibrium phase β-Mg_2Si, FCC structure (a= 0.639), rod or plate-shaped.

The maximum alloy strength (peak-aging) is achieved just before the precipitation of the incoherent β-platelets. Apelian *et al.* [60] studied the aging behaviour of Al-Si-Mg alloys and observed that the precipitation of very fine β'-Mg_2Si during aging leads to a pronounced improvement in strength properties. Both aging time and temperature determine the final properties, see Figures 6 and 7. Their study also established that increasing the aging temperature by 10°C is equivalent to increasing the aging time by a factor of two. The effect of natural ageing on precipitation can be justified as follows. A high concentration of quenched in vacancies enhances the rate of solute clustering in the early stages of natural ageing, and this clustering of solute leads to a reduced supersaturation of solute in the matrix. The solute clusters have a fine distribution within the matrix, and if they were to act as successful nuclei for the formation of β'' during subsequent artificial ageing, a fine precipitate distribution would result. Evidently, this is not the case, a reason being that many of the clusters are below the critical size for stability at artificial ageing temperatures. Furthermore, a lower solute supersaturation is expected to reduce the kinetics of precipitation. Thus, during artificial ageing, the dissolution of unstable clusters increase the solute concentration, while larger clusters that are stable remove solute by growing into GP zones that become nucleation sites for β''. Therefore, the solute supersaturation is maintained at a relatively low level during artificial; ageing and the density of the b'' is much lower than that occurring in alloys without natural ageing.

The precipitation sequence for Al-Si-Cu-Mg alloys is similar, but more complex, as the Q'' phase and the θ' phase may also form. Cu can increase the fraction of the β'' phase formed, but it can also form the Q'' phase, which has a lower strength contribution compared to the β′′ phase. The β′′ phase is therefore preferred, rather than the Q′′ phase. It is however not clearly stated when the Q′′ phase forms at the expense of the β′′ phase in cast alloys. For wrought alloys it has been shown that the fraction of the Q′′ phase increases with natural ageing and artificial ageing time and temperature [64-66].

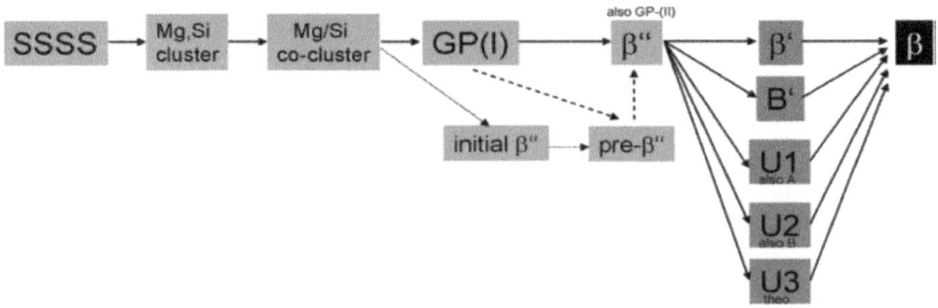

Figure 6. Sequence of phases found during age hardening of Al-Mg-Si alloys [60-62]. Supersaturated solid solution (SSSS) decomposes as Mg and Si atoms are attracted first to themselves (cluster) then to each other to form precipitates GP(I), sometimes also called initial-β′′. GP(I) zones either further evolve directly to a phase β′′ and then to a number of other metastable phases labelled β′, B′, U1, U2 (another one, U3, has been postulated theoretically), or first form an intermediate phase called pre-β′′.

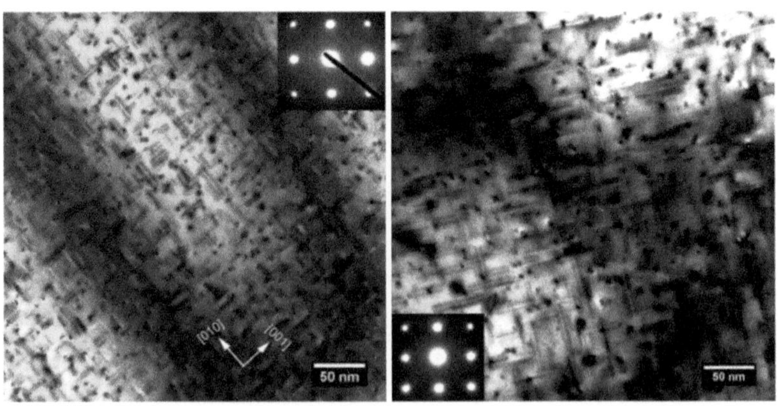

Figure 7. TEM images of Al-Si-Mg alloy subjected to 2 different heat treatments. (a) solutionising and quenching, immediate aging at 180°C for 540 min, (b) solutionising and quenching, natural ageing for 10,000 min at 20°C, aging at 180°C for 540 min [63].

The precipitation of metastable Mg-rich phases depends on the Mg-to-Si ratio. The excess of Si in solid solution can significantly alter the kinetics of precipitation and the phase composition. In other words, equilibrium phases are enriched in Mg and metastable phases are enriched in Si. Silicon precipitates are observed if stable phases are formed [67,68].

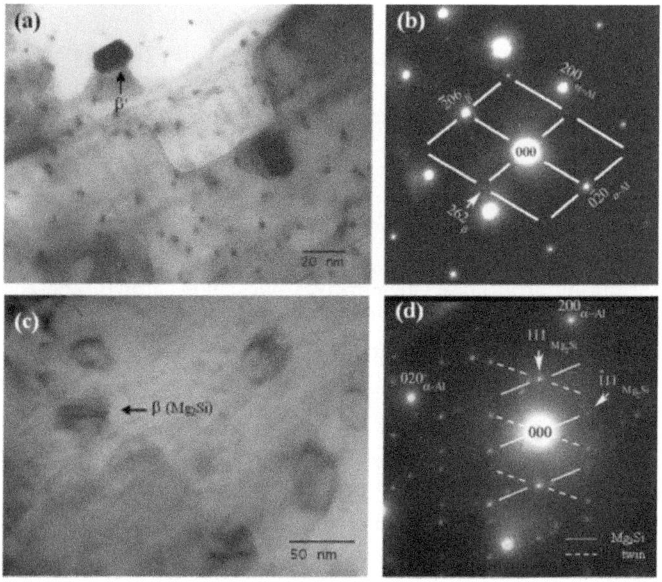

Figure 8. TEM-BF micrographs show precipitated phases in association with T6 over-ageing period; (a) 100 hours, (c) 300 hours. (b) and (d) are corresponding SADPs from (Mg2Si) particles indicated by the arrows in (a) and (c), respectively.

4. Modelling of the heat treatment process

Designing an alloy and a heat treatment process for a material that meets specified requirements for a certain component can be facilitated by the use of models. Development of models can also help in the search for new alloys as knowledge is gained about the influence of a specific part of the microstructure on the alloy properties. The first model where the yield strength is coupled to the evolution of the microstructure during artificial ageing was developed by Shercliff and Ashby in 1990 [69]. They defined their model as a mathematical relation between the process variables (e.g. alloy composition, heat treatment temperature and time), and the mechanical response of the alloy (e.g. yield strength, hardness), based on physical principles (e.g. thermodynamics, kinetics of precipitation, strengthening mechanisms etc.).

More refined models have been developed since then for prediction of yield strength [70,71] and elongation to fracture [72] after artificial ageing. To be able to model the tensile strength after heat treatment, the evolution of the microstructure has to be modelled from casting to artificial ageing. Empirical equations as the Hollomon's [73] and the Ludwigson's [74] and equations where the parameters are coupled to the microstructure as in the KM strain hardening theory can be used to describe the plastic deformation behavior. The KM strain hardening theory has already been successfully used to couple the plastic deformation behavior to the microstructure for heat treatable wrought alloys and Al-Si-Mg casting alloys.

The Scheil equation is a simple model giving fair results for segregation profiles and fraction of particles formed during solidification for aluminum alloys. The Scheil equation assumes no diffusion in the solid and complete diffusion in the liquid [75]. The correctness of the predictions of the Scheil segregation model depends on the diffusivity of the alloying elements in the α-Al phase.

From the as-cast microstructure the time needed for dissolution and homogenization can be modelled. The model developed by Rometsch et al. [76], which handles solution treatment of Al-Si-Mg alloys, is an example of a simple, but efficient model. The evolution of the microstructure during artificial ageing involves nucleation, growth and coarsening. Two main approaches are used; precipitates having an average radius or precipitates having a size distribution. For the case of precipitates which a size distribution, coupled nucleation, growth and coarsening can be calculated, while for an average radius growth is sequentially followed by coarsening.

The strength of an alloy derives from the ability of obstacles, such as precipitates and atoms in solid solution, to hinder the motion of mobile dislocations. The strength contributions from atoms in solid solution and from shearable and non-shearable precipitates change during ageing, while contributions from lattice, dislocations and grain boundaries are constant. Small and not too hard precipitates are normally sheared by moving dislocations, see Figure 8.a. When the precipitates are larger and harder the moving dislocations pass the precipitates by bowing, leaving a dislocation ring around the precipitate, see Figure 8.b. The strength of the precipitates increases with size as long as it is sheared by dislocations. When dislocations pass the precipitates by looping, the alloy strength decreases with increasing radius of the precipitates. Figure 8.c shows the different strength contributions to the total yield strength for different ageing times.

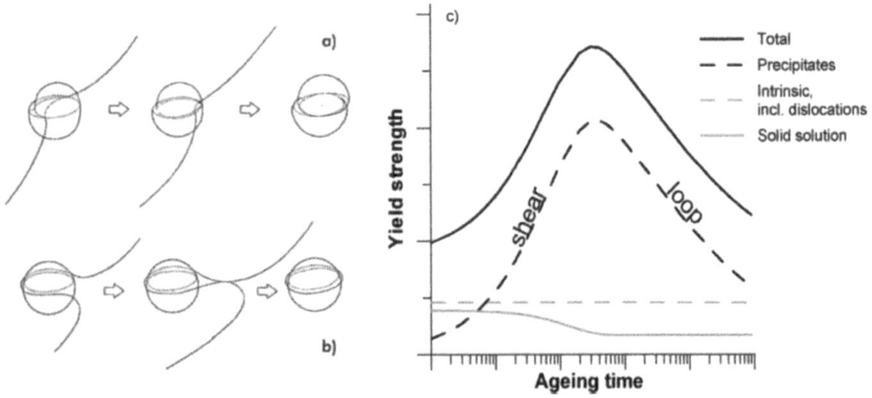

Figure 9. Dislocations passing a precipitate by a) shearing and b) looping (Orowan mechanism) [77] c) Illustrates the different strength contributions to the total yield strength

5. Conculsions

Although many previous investigations into the thermal processing of Al-Si-Cu/Mg casting alloys have been carried out, most focus on a single aspect of the overall process and a comprehensive experimental study considering all heat treatment stages is still required. This review shows that it is of vital importance to take the whole heat treatment process into consideration in order to achieve the optimal mechanical properties of an alloy. It is not sufficient to consider only the solution treatment and the artificial ageing parameters. Furthermore, the development of process models for the prediction of microstructure and mechanical property changes in aluminum alloys has focused on wrought alloys, while casting alloys that contain more complex microstructures have been overlooked and the evolution of the solution treated microstructure and its influence on subsequent ageing behaviour has not been incorporated into the models.

Author details

A.M.A. Mohamed
Materials Technology Unit (MTU), Qatar University, Doha, Qatar

Department of Metallurgical and Materials Engineering,
Faculty of Petroleum and Mining Engineering, Suez Canal University, Suez, Egypt,

F.H. Samuel
Université du Québec à Chicoutimi, Chicoutimi, Québec, Canada

6. Acknowledgement

The authors would like to thank Professor Mariam Al-Maadeed, Head of Materials Technology Unit, MTU, at Qatar University for her help and support.

7. References

[1] Mohamed A.M.A, Samuel F.H., Alkahtani S. Influence of Mg and Solution Heat Treatment on the Occurrence of Incipient Melting in Al-Si-Cu-Mg Cast Alloys. Journal of Materials Science and Engineering A 2012; A543 22-34.
[2] Mohamed A.M.A, Samuel A.M., Samuel F.H., Doty H.W. Influence of Additives on the Microstructure and Tensile Properties of Near-Eutectic Al-10.8%Si Alloys. Materials and Design 2009;30(10) 3943-3957.
[3] Djurdjevic M., Stockwell T., Sokolowski J., 1999. The Effect of Strontium on the Microstructure of the Aluminium-Silicon and Aluminium-Copper Eutectics in the 319 Aluminium Alloy. Int. J. Cast Metal. Res. 1999;12 67–73.
[4] Stefanescu D. Casting, ASM International 1988, vol.15.
[5] Pekguleryuz M.O. Aluminum physical metallurgy. 2007.
[6] ASTM. (2006). Aluminum and Magnesium Alloys Annual (Volume 02.02). ASTM standards

[7] Samuel A.M., Samuel F.H. Modification of Iron Intermetallics by Magnesium and Strontium in Al-Si Alloys. International Journal of Cast Metals Research 1997;10 147-157.
[8] Meyers C.W., Hinton K.H., Chou J.S. Towards the Optimization of Heat Treatment in Aluminum Alloys. Materials Sci Forum 1992;72 102-104.
[9] Gruzleski, J. E., & Closset, B. M. (1990). The treatment of liquid Al-Si alloys. AFS inc.
[10] Dahle A., Nogita K., McDonald S., Dinnis C., Lu L. Eutectic Modification and Microstructure Development in Al-Si Alloys. Materials Science and Engineering A, 2005;413-414 243-248.
[11] Mondolfo L.F. Alumnium Alloys: Structure and Properties. Butter Worths 1976.
[12] Li Z., Samuel A.M. AFS Transactions. 2003;100 114-
[13] Adibhatla S. Master's Thesis, IMS, University of Connecticut 2003.
[14] Li Z., Samuel A.M., Samuel F.H., Ravindran C., Valtierra S. 2003. Effect of Alloying Elements on the Segregation and Dissolution of $CuAl_2$ Phase in Al–Si–Cu 319 Alloys. J. Mater. Sci. 2003;38 1203–1218.
[15] Cáceres C.H., Svensson I., Taylor J. Strength-Ductility Behavior of Al-Si-Cu-Mg Casting Alloys in T6 Temper. International Journal of Cast Metals Research, 2003;15 721-726.
[16] Yi F. Computer Simulation of Solidification and Solution Treatment of Multiphase, Multicomponent Alloys. Master's Thesis, IMS, University of Connecticut 2005.
[17] Sundman B., Jansson B., Andersson J.O. Calphad 9, 1985; 153-190.
[18] J. Barresi, M.J. Kerr, H. Wang, M.J. Couper, "Effect of Magnesium, Iron, and Cooling Rate on Mechanical Properties of Al-7Si-Mg Foundry Alloys", *AFS Transactions*, 2000, 563-70.
[19] Gauthier J., Louchez P., Samuel F.H. Heat Treatment of 319.2 Al Automotive Alloy: Part 1, Solution Heat Treatment. Cast Metals 1995;8(1) 91-106.
[20] [20] F. Paray, J. Gruzleski, "Modification - A Parameter to Consider in the Heat Treatment of Al-Si Alloys", *Cast Metals*, 1993, vol. 5(4), pp. 187-198.
[21] J. Gauthier, P.R. Louchez and F.H. Samuel, Cast Metals (1994) 91-114.
[22] A.M.A Mohamed, A.M. Samuel, F.H. Samuel, H.W. Doty, Materials and Design, 30 (2009) 3943-3957.
[23] N. Crowell, S. Shivkumar, AFS Transactions 103 (1995) 721-726.
[24] P. Ouellet, F.H. Samuel, J. Mater. Sci. 34 (1999) 4671- 4697.
[25] P.S. Wang, S.L. Lee, J.C. Lin, J. Mater. Res .15 (2000) 2027-2035.
[26] P. Ouellet, F.H. Samuel, "Effect of Mg on the Ageing Behavior of Al-Si-Cu 319 Type Aluminum Casting Alloys " *Journal of Materials Science*, 1999, vol. 34(19), pp. 4671-4697.
[27] G. Wang, X. Bain, W. Wang, J. Zhang, "Influence of Cu and Minor Elements on Solution Treatment of Al-Si-Mg-Cu Cast Alloys", *Materials Letters*, 2003, vol. 57, pp. 4083-4087.
[28] L. Lasa, J. Ibabe, "Characterization of the Dissolution of the Al_2Cu Phase in Two Al-Si-Cu-Mg Casting Alloys Using Calorimetry", *Materials Characterization*, 2002, vol. 48, pp. 371-378.
[29] J.H. Sokolowski, X.C. Sun, G. Byczynski, D.O. Northwood, D.E. Penord, R. Thomas, A. Esseltine, "A Metallurgical Study of the Heat Treatment of Aluminum Alloy 319 (Al-6Si-3.5Cu) Castings", *Journal of Materials Processing Technology*, 1995, vol. 53(1-2), pp. 385-392.
[30] N. Crowell, S. Shivkumar, "Solution Treatment Effects in Cast Al-Si-Cu Alloys", *AFS Transactions*, 1995, vol. 103, pp. 721-726.
[31] J.H. Sokolowski, X-C. Sun, G. Byczynski, D.E. Penrod, R. Thomas, A. Esseltine, Journal of Materials Processing Technology, 53(1995) 385-392.

[32] J.H. Sokolowski, M.B. Djurdjevic, C.A. Kierkus, D.O. Northwood, Journal of Materials Processing Technology, 109 (2001) 174-180.
[33] D. Apelian, S. Shivkumar, G. Sigworth, "Fundamental Aspects of Heat Treatment of Cast Al-Si-Mg Alloys", *AFS Transactions*, 1989, vol. 97, pp. 727-742.
[34] D.L. Zhang and L. Zheng, "The Quench Sensitivity of Cast Al-7 Wt Pct Si-0.4 Wt Pct Mg Alloy", *Metallurgical and Materials Transactions A*, 1996, vol. 27A, pp. 3983-3991.
[35] S. Seifeddine, G. Timelli, I.L. Svensson, Int. Foundry Res. 59 (2007) 2-10.
[36] D.L. Zhang, L. Zheng, Metall. Mater. Trans. A 27 (1996) 3983-3991.
[37] P.A. Rometsch, G.B. Schaffer, Int. J. Cast Metal. Res. 12 (2000) 431-439.
[38] lW. Martin, *"Precipitation hardening* (2nd ed.}", *Butterworth-Heinemann,* Oxford, UK(1998).
[39] J.T. Staley, "Quench factor analysis of aluminum alloys", *Material Science and Technology*,3 (1987), 923-935.
[40] D. Emadi, "Optimal Heat Treatment of A356.2 Alloy", *Light Metals,* The Minerals, Metals, and Materials Society, Warrendale, PA, 2003, pp. 983-989.
[41] T. Croucher, B. Butler, "Polymer Quenching of Aluminum Castings", *26th National SAMPE Symposium*, 1981, pp. 527-535.
[42] G.E. Totten and D.S. Mackenzie, "Aluminum Quenching Technology: A Review", *Materials Science Forum*, 2000, vols 331-337, pp. 589-594.
[43] A.V. Sverdlin, G.E. Totten, G.M. Vebster, "Polyalkyleneglycol Base Quenching Media for Heat Treatment of Aluminum Alloys" *Metallovedenie Termicheskaya Obrabotka Metallov*, 1996, vol. 6, pp. 17-19.
[44] O.G. Senatorova, "Low Distortion Quenching of Aluminum Alloys in Polymer Medium", *Materials Science Forum*, 2002, vols 396-402, pp. 1659-1664.
[45] H. Beitz, "None-Combustible Water-Based Quenchants in Forging Shops for Automotive Parts- Latest Development", *The 1st International Automotive Heat Treating Conference*, Puerto Vallerta, Mexico, 1998, pp. 106-109.
[46] Zhang D.L. & Zheng L. (1996). Quench sensitivity of cast Al-7 wt pct Si-0.4 wt pct Mg Alloy, *Metallurgical and Materials Transaction A*, Vol.27, No.12, pp. 3983-3991, ISSN 1073-5623
[47] H.M. Kandil, "Recent Development in Age Hardening Behaviour of Aluminum Alloys-A Review Article", *In Heat Treating: Proceeding of the 21st Conference,* Indianapolis, Indiana: ASM International, 2001, pp. 343-351.
[48] A.J. Ardel, "Precipitation Hardening", *Metallurgical Transactions A*, 1985, vol. 16A, pp. 2132-65.
[49] A. Zanada, G. Riontino, "A Comparative Study of the Precipitation Sequences in Two AlSi7Mg Casting Alloys and their Composites Reinforced by 20% Al_2O_3 Discontinuous Fibers", *Materials Science Forum*, 2000, vols 331-337, pp. 229-234.
[50] S. Kumai, "Hardness Characteristics in Aged Particulate", *Scripta Metallurgica et Materialia*, 1992, vol. 27, pp. 107-110.
[51] R.X. Li, "Age-Hardening Behavior of Cast Al-Si Base Alloy", *Materials Letters*, 2004, vol. 58, pp. 2096-2101.
[52] D.S. Jiang, L.H. Chen, T.S. Liu, "Effect of Aging on the Crack Propagation Behaviour of A356", *Materials Transactions*, JIM, 2000, vol. 41(4), pp. 499-506.

[53] S.W. Han, "Effects of Solidification Structure and Aging Condition on Cyclic Stress-Strain Response in Al-7%Si-0.4%Mg Cast Alloys", *Materials Science and Engineering A*, 2002, vol. 337, pp. 170-178.
[54] A.D. Porter, K.E. Easterling, *Phase Transformations in Metals and Alloys*, Van Nostrand Reinhold, Berkshire, England, 1981.
[55] J.D. Verhoeven, *Fundamentals of Physical Metallurgy*, J. Wiley and Sons, New York, N.Y., 1975.
[56] W.F. Smith, *Structure and Properties of Engineering Alloys*, McGraw-Hill, New York, N.Y., 1981.
[57] J.E. Hatch, *Aluminum: Properties and Physical Metallurgy*, American Society for Metals, Metals Park, OH, 1984, pp. 50-51.
[58] F.H. Samuel, A.M. Samuel, H. Liu, "Effect of Magnesium Content on the Aging Behaviour of Water-Chilled Al-Si-Cu-Mg-Fe-Mn (380) Alloy Castings", *Journal of Materials Science*, 1995, vol. 30, pp. 1-10.
[59] A.M. Samuel, J. Gauthier, F.H. Samuel, "Microstructural Aspects of the Dissolution and Melting of Al_2Cu Phase in Al-Si Alloys during Solution Heat Treatment", *Metallurgical and Materials Transactions A*, 1996, vol. 27A, pp. 1785-1798.
[60] D. Gloria, F. Hernandez, S. Valtierra, "Dimensional Changes During Heat Treating of an Automotive 319 Alloy", *20th ASM Heat Treating Society Conference Proceedings*, 9-12 October 2000, St. Louis, MO, ASM International Materials Park, OH, 2000, pp. 674-679.
[61] S. Shivkumar, C. Keller, D. Apelian, "Aging Behaviour in Cast Al-Si-Mg Alloys", *AFS Transactions*, 1990, vol. 98, pp. 905-911.
[62] D. Apelian, S. Shivkumar, G. Sigworth, "Fundamental Aspects of Heat Treatment of Cast Al-Si-Mg Alloys", *AFS Transactions*, 1989, vol. 97, pp. 727-742.
[63] I. Polmear, *Light Alloys*, Butterworth-Heinemann, Amsterdam 2006 (4th ed.).
[64] M. A. van Huis, J. H. Chen, M. H. F. Sluiter, H. W. Zandbergen, *Acta Mater*. 2007, *55*, 2183.
[65] R. Vissers, M. A. van Huis, J. Jansen, H. W. Zandbergen, C. D. Marioara, S. J. Andersen, *Acta Mater*. 2007, *55*, 3815.
[66] X. Wang, S. Esmaeili, D.J. Lloyd, Metall. Mater. Trans. A 37 (2006) 2691-2699.
[67] D.G. Eskin, J. Mater. Sci. 38 (2003) 279-290.
[68] X. Wang, W.J. Poole, S. Esmaeili, D.J. Lloyd, J.D. Embury, Metall.Mater. Trans. A34(2003) 2913-2924.
[69] Maruyama N., Uemori R., Hashimoto N., Saga M., Kikuchi M. Scr. Mater. 1997;36 89-
[70] Zhen L., Fei W.D., Kang S.B., Kim H.W., J. Mater. Sci. 1997;32 1895
[71] H.R. Shercliff, M.F. Ashby, Acta Metall. Mater. 38 (1990) 1789-1802. [42] S. Esmaeili, D.J. Lloyd, W.J. Poole, Acta Mater. 51 (2003) 2243-2257.
[72] O.R. Myhr, O. Grong, S.J. Andersen, Acta Mater. 49 (2001) 65-75.
[73] J.D. Robson, Acta Mater. 52 (2004) 4669-4676.
[74] G. Liu, J. Sun, C.W. Nan, K.H. Chen, Acta Mater. 2005;53 3459-3468.
[75] J.H. Hollomon, Trans. AIME 162 (1945) 268-290.
[76] D.C. Ludwigson, Metall. Trans. 2 (1971) 2825-2828.
[77] D.M. Stefanescu, Science and Engineering of Casting Solidification, Kluwer Academic /Plenum Publisher, New York, 2002.
[78] P.A. Rometsch, L. Arnberg, D.L. Zhang, Int. J. Cast Metal. Res. 12 (1999) 1-8.
[79] V. Gerold, in: F.R.N. Nabarro (Ed.) Dislocations in solids, North-Holland, 1979, 222.

Chapter 2

Regenerative Heat Treatment of Low Alloy Cast Steel

Grzegorz Golański

Additional information is available at the end of the chapter

1. Introduction

Cylinder and valve chamber castings of large power steam turbines are usually made of low alloy Cr - Mo - V and Cr - Mo cast steels. Forming of the microstructure and mechanical properties of cast steels takes place through heat treatment, thus far mostly consisting of normalizing and tempering. As a result of such a treatment the cast steels of diverse wall thickness reveal microstructures from ferritic – pearlitic to bainitic – ferritic with various ferrite, pearlite and bainite amount.

Operation of the cast steels under creep conditions contributes to the occurrence of deformations, fractures and changes in the microstructure, decreasing their functional properties. The resistance to crack expressed by impact energy falls drastically. The value of impact energy of test pieces taken from cast steels after long-term service was considerably below the required level of 27J, frequently reaching the value of 6 - 10J (Fig. 1).

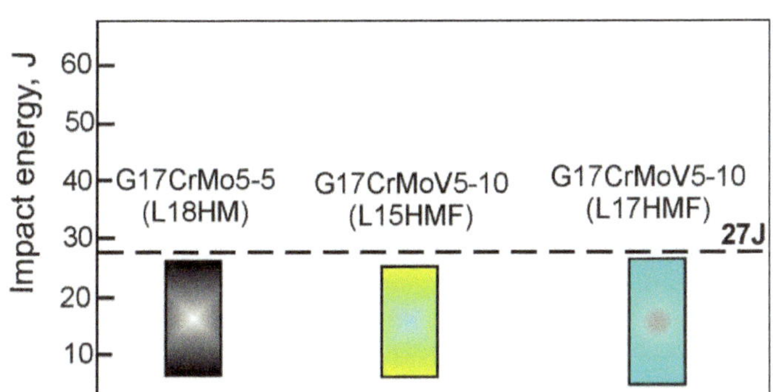

Figure 1. Impact energy of turbine cylinder cast steel in the post-operating condition

Along with the fall of impact energy there is also a growth of nil ductility transition (NDT) temperature, frequently rising above 50 ÷ 60 °C. Large decrease in crack resistance is usually accompanied by a slight decrease in the strength properties (Fig. 2).

Unfavourable changes in mechanical properties of the castings are related to the changes in microstructure which occur during long term service at elevated temperatures, first and foremost to:

- the preferential precipitation of carbides on grain boundaries, as well as changes in morphology and dispersion of precipitates;
- segregation of phosphorus and other trace elements to grain boundaries and near boundary areas; disintegration of pearlite or/and bainite areas.

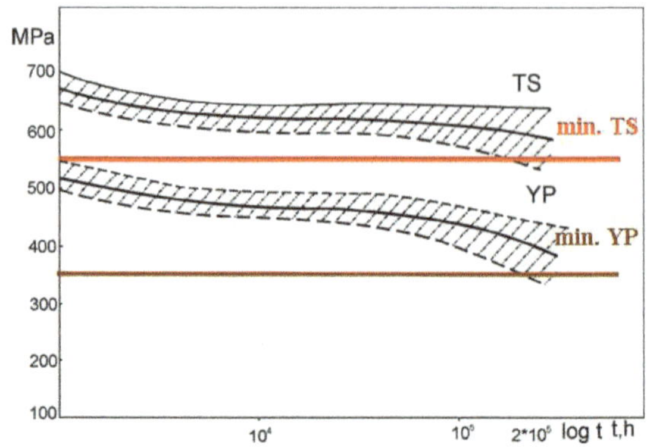

Figure 2. Changes in tensile strength (TS) and yield strength (YS) depending on the time of service

Lowering of impact energy as a result of long-term service depends largely on the as received microstructure of a cast steel. The impact energy decrease is the smallest in the case of tempered bainite microstructure or bainitic – ferritic microstructure, with ferrite amount not higher than 5% (Fig. 3). High impact energy of quenched and tempered cast steel, considerably higher than 100J, guarantees that during long-term service of steel castings with low phosphorus volume fraction (≤ 0.015% P), the impact energy will not fall below the minimum required value of 27J.

Similar tendency has been noticed in new low-alloy bainitic 7CrWVMoNb9 – 6 (P23) steel. Impact energy in the case of this cast steel, whose microstructure is of tempered bainite in the as-received condition, after around 10 years of operation at the temperature of 555 °C and pressure 4.2MPa, was on the level of 70 – 80 J/cm^2.

Degradation of the microstructure of castings and the related gradual decreasing of mechanical properties, however, do not limit the possibility of their further operation, especially as in most of the examined castings there were no irreversible creep changes observed. One of the conditions for extending the time of safe operation for cast steels above the calculated service time is running the process of revitalization of the castings.

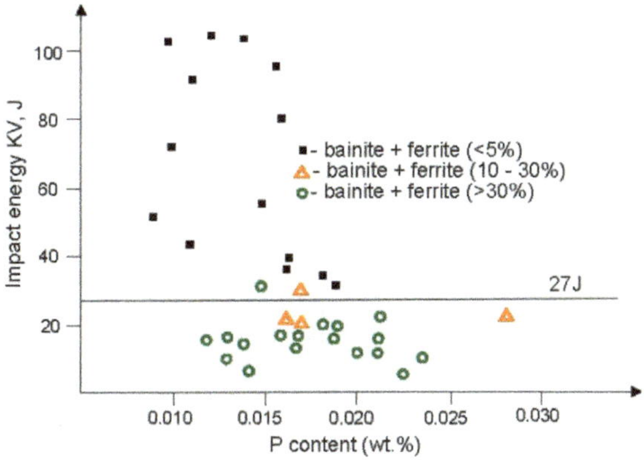

Figure 3. Influence of phosphorus amount and microstructure on impact energy KV of the Cr – Mo – V cast steel after long-term operation at the temperature of 535 °C

The revitalization process consists in heat treatment of turbine cylinders in order to regenerate the structure to the extent which allows the improvement of plastic properties (increase in impact energy, decrease in the nil ductility transition temperature). Regenerative heat treatment of castings applied in industry so far consists in normalizing/full annealing from the austenitizing temperature with the following high-temperature tempering/under annealing. The ferritic – pearlitic or ferritic – bainitic structure, obtained as a result of the above-mentioned heat treatment, provides the required impact energy of KV > 27J, however, with the strength properties being comparable to those after operation.

Modern hardening plants applying aqueous solutions of polymers as a cooling agent make the cooling of massive castings possible at programmed rate which provides an optimum structure throughout the whole casting section.

Regenerative heat treatment at costs not exceeding 40% of a new casting's price, allows obtaining functional properties (yield strength, impact energy, NDT temperature) comparable to the properties of new castings. Regenerated cylinder is fit for further operation at least for another 100 000 hours.

In order to achieve an improvement in mechanical properties of cast steels after long-term operation, the following changes in the degraded microstructure are necessary:

- grain refinement – leads to an increase in crack resistance, decreases the NDT temperature and raises yield strength;
- eliminating irreversible brittleness caused by phosphorus segregation to grain boundaries and interphase boundary: matrix/carbides;
- removal of the needle shaped ferrite (Widmannstätten's ferrite);
- removal of pearlite precipitated on ferrite grain boundaries;
- dissolving of carbides in austenite, especially the carbides precipitated on grain boundaries, in order to obtain the required strength properties in the regenerated microstructure (hardness and tensile strength).

The research aim: The aim of the performed research was to determine the influence of regenerative heat treatment on the microstructure and properties of Cr – Mo – V cast steel with its microstructure degraded by long-term service and mechanical properties being lower than the minimum ones expected in the new castings.

2. Material for research

The material for study was Cr – Mo – V low-alloy **L21HMF** cast steel (designation according to Polish Standards) with its chemical composition given in Table 1. Test pieces for investigation were taken from an inner cylinder of a steam turbine serviced for around 186 000 hours at the temperature of 540 °C and pressure of 13.5MPa.

C	Mn	Si	P	S	Cr	Mo	V
0.19	0.74	0.30	0.017	0.014	1.05	0.56	0.28

Table 1. Chemical composition of the L21HMF cast steel, % mass.

2.1. Microstructure and properties of the examined cast steel after service

In the post-operating condition the L21HMF cast steel was characterized by a degraded ferritic-pearlitic microstructure (Fig. 4). The dominant phase in the microstructure after operation was quasi-polygonal ferrite. The size of ferrite grain in the cast steel was diverse and ranged from 88.4 to 31.2μm, which corresponds to the grain size grade: 4 ÷ 7, according to ASTM standard scale.

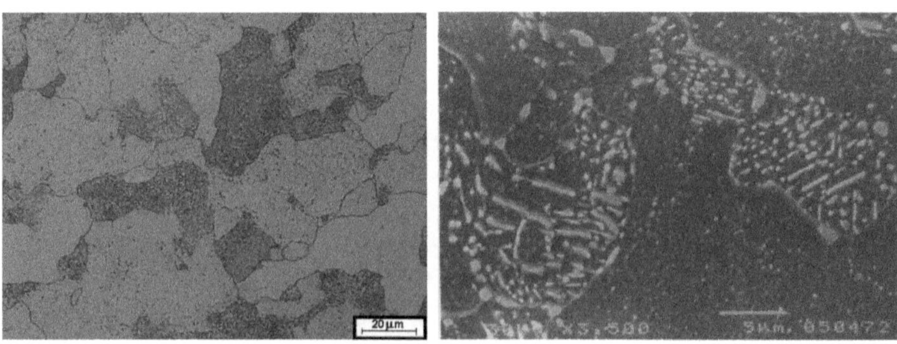

Figure 4. The microstructure of L21HMF cast steel after service

Long-term service of Cr – Mo – V cast steel contributed to the changes in microstructure, including:

- preferential carbide precipitation of $M_{23}C_6$ carbides on ferrite grain boundaries. In some areas the number of carbides precipitated on boundaries was so large that they formed the so-called "continuous grid" of precipitates;

- the process of degradation of pearlite grains consisting in fragmentation, spheroidization and coagulation of pearlitic carbides. Performed identifications have revealed the occurrence of the M_3C and M_7C_3 type of precipitations in those areas (Fig. 5);

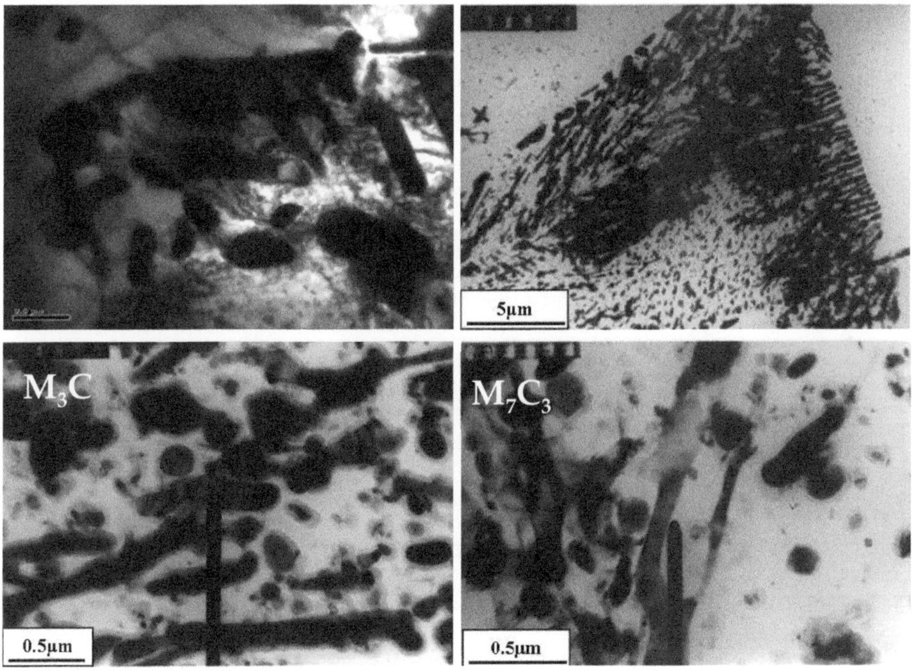

Figure 5. Morphology and type of carbides in pearlite grain

- precipitation of compound carbide complexes called „H – carbides". The compound complexes of precipitates are created by MC and M_2C carbides, where the MC carbide is a "horizontal" precipitation, while M_2C carbides are precipitations of "vertical" type (Fig. 6). This sort of compound precipitations is defined as „H – carbide". During long-term operation the MC carbide is enriched in molybdenum as a result of diffusion. The growth of molybdenum concentration in the interphase areas of MC/matrix makes it possible for the "needle-shaped" precipitations of M_2C (rich in molybdenum) to nucleate on the interphase boundary: MC carbide/ferrite. These processes run more intensely in the border areas of grains, which results in the occurrence of precipitation free zones. The appearance of such zones may be the cause of slow reduction of the strength properties, the yield strength in particular, during long-term operation. The occurrence of this type of complexes results in a decay of fine-dispersion MC carbides which may lead to the fall of creep resistance in the serviced materials. A similar phenomenon can be seen at present in the new high-chromium cast steels for power industry, where the Z phase is being formed and developed at the expense of fine dispersion precipitates of the MX type, which causes a drastic drop of creep resistance of these cast steels.

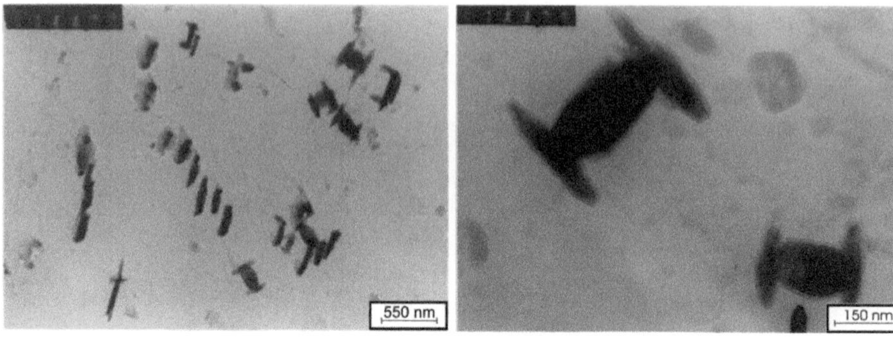

Figure 6. Precipitation of „H - carbide" type in the cast steel

Mechanical properties of L21HMF cast steel after long-term service are shown in Table 2.

Material	TS MPa	YS MPa	El. %	KV J	HV30	DBTT °C	Microstructure
L21HMF	545	305	26	10	156	65	ferritic-pearlitic
Requirements of PN *	500 ÷ 670	min. 320	min. 20	min. 27	140 ÷ 197**	—	—

*- PN - 89/ H - 83157 , ** - hardness according to Brinell,

Table 2. Mechanical properties and microstructure of the L21HMF cast steels after service

Tensile strength and elongation of the examined cast steel after service were higher than the minimum values required for new castings, while the value of yield strength was lower than the minimum required by 15MPa. Hardness of the investigated cast steel after operation amounted to 156HV30.

A significant feature of the material proving its strain capacity, apart from elongation determined in the static test of tension, is the value of impact energy. Knowledge of this factor gives the possibility of assuming the right temperature for the hydraulic pressure tests used in industrial practice, as well as the right conditions of start-ups and shut-downs of a boiler, adjusted to the material state after long-term service. After operation the examined cast steel was characterized by low impact energy amounting to 10J, and the cracking of samples occurred through the transcrystalline fissile mechanism (typical for brittle fractures) with little energy absorbed due to the limited plastic strain preceding the decohesion. Fissile cracking requires little energy supply which is necessary for crack propagation, hence the low impact energy of the cast steel after service (Fig. 7). Low impact energy of the examined materials is related to the nil ductility temperature (brittle temperature). The fracture appearance transition temperature determined for the examined cast steel amounted to 65 °C.

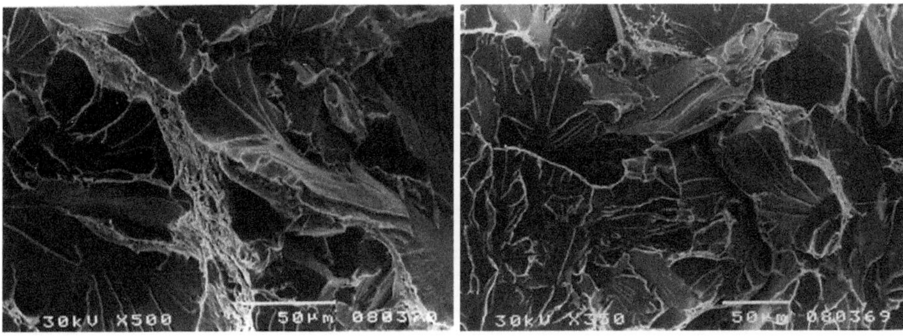

Figure 7. Transcrystalline ductile fracture with areas of microductility and secondary cracks

2.2. Influence of austenitizing parameters on the size of prior austenite grain

Influence of austenitization parameters on the prior austenite grain size has been described in a quantitative way using chosen stereological and statistical parameters, such as: mean diameter and mean area of grain, and also the coefficient of variation of grain size ν was calculated. The ν coefficient is characterized by the inhomogeneity of grain sizes: the more heterogeneous grains in terms of size within the casting, the higher the values of variation coefficient. The tests were run for the austenitizing temperature range of - 910 ÷ 970 °C with the „measurement step" – 15 °C and times of holding at the austenitizing temperature: 3 and 5 hours.

The character of austenite grain distributions was determined using the λ - Kolmogorov test of goodness of fit with normal distribution for logarithmed values (Fig. 8). The assumed significance level was α = 0.01, with its limiting statistics value amounting to 1.63. Selected logarithm-normal layouts of mean diameters and mean surface areas of former austenite grains for austenitization option of 925 °C and holding time 3 hours, are shown in Fig. 8. Obtained results of the tests are presented in Table 3 and 4 and graphically shown in Fig. 9 ÷ 11.

Heat treatment parameters, °C/h	Amount n	Min. diameter of grain, µm	Max. diameter of grain, µm	Diameter of grain, µm	Standard deviation	λ_{emp}	λ_α=0.01	$\frac{\lambda_{emp}}{\lambda_\alpha = 0.01}$
910/3	976	2	29	11.34	6.36	1.54	1.63	0.945
925/3	969	2	30	9.84	5.34	1.36	1.63	0.834
940/3	954	2	31	10.16	6.22	1.56	1.63	0.957
955/3	964	2	38	14.08	9.34	1.36	1.63	0.834
970/3	2024	2	297	22.14	17.73	2.26	1.63	1.387
910/5	946	2	27	9.36	5.70	1.35	1.63	0.828
925/5	915	2	28	11.00	7.01	1.55	1.63	0.951
940/5	959	2	32	9.05	5.41	1.33	1.63	0.816
955/5	937	2	39	12.02	8.37	1.18	1.63	0.724
970/5	2034	2	324	23.67	18.31	1.43	1.63	0.877

Table 3. The results of measurements and calculations of the size of prior austenite grains for the cast steel

Heat treatment parameters, °C/h	Amount n	Min. area of grain, μm²	Max. area of grain, μm²	Mean area of grain, μm²	Standard deviation	λ_{emp}	λ_{α}=0.01	$\frac{\lambda_{emp}}{\lambda_{\alpha}=0.01}$
910/3	976	4	695	142.67	204.58	1.74	1.63	1.067
925/3	969	3	741	104.27	173.60	1.27	1.63	0.779
940/3	954	3	810	121.34	172.77	1.40	1.63	0.859
955/3	964	3	1302	187.41	379.31	1.59	1.63	0.975
970/3	2024	3	95722	812.62	2138.56	1.99	1.63	1.221
910/5	946	3	608	101.07	162.93	1.07	1.63	0.656
925/5	915	2	741	144.72	248.95	1.61	1.63	0.988
940/5	959	3	842	94.03	147.10	1.00	1.63	0.613
955/5	937	2	1533	391.66	423.09	0.98	1.63	0.601
970/5	2034	3	102305	915.33	2408.43	1.53	1.63	0.939

Table 4. The results of measurements and calculations of the size of prior austenite grains for the cast steel

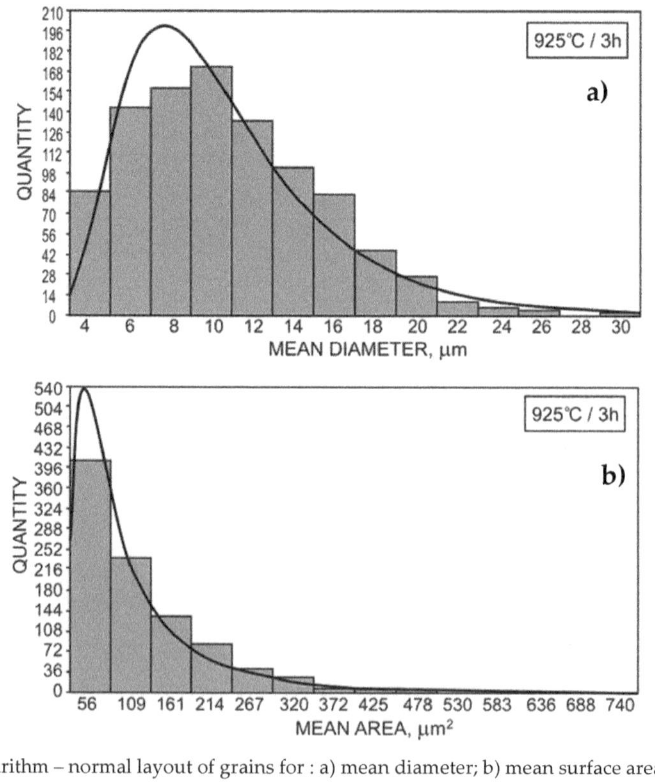

Figure 8. Logarithm – normal layout of grains for : a) mean diameter; b) mean surface area

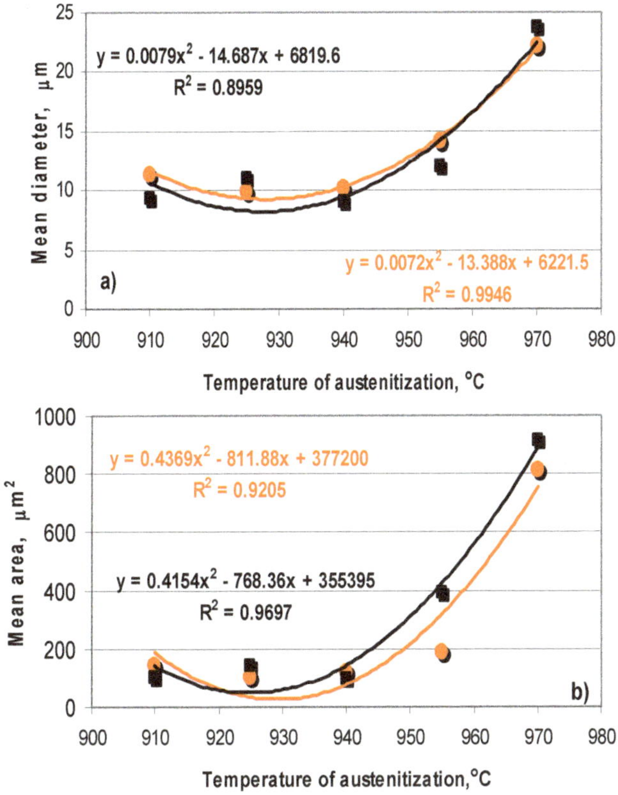

Figure 9. Influence of austenitization temperature on: a) mean grain diameter; b) mean surface area depending on the time of holding (3hrs – orange, 5hrs – black) of L21HMF cast steel

The mean diameters of grains and their mean areas change continuously and reveal log-normal layouts on the significance level of $\alpha = 0.01$ ($\lambda_{emp.} / \lambda_{0.01} < 1$). The exceptions were the treatment variants as follows: austenitization temperature of 910 °C and time - 3h for the mean area (fulfilled for lower significance level of $\alpha = 0.001$) and the temperature of 970 °C and time - 3h for both: mean diameter and mean area of prior austenite grain. Within the range of austenitization temperatures: 910 ÷ 940 °C for holding times: 3 and 5 hrs, the mean diameters and areas of prior austenite grain do not reveal any considerable differences. (Fig. 9). The values of the mean diameters and areas for this range of austenitization amounted to: 9.05 ÷ 11.34 µm and 94.03 ÷ 144.72 µm², respectively. The above-mentioned measurements can be confirmed by the calculated values of v coefficient, specifying heterogeneity in terms of grain sizes, which are the lowest for this range of austenitizing temperatures (Fig. 10), and by the distributions of frequency of cumulated grains (Fig. 11).

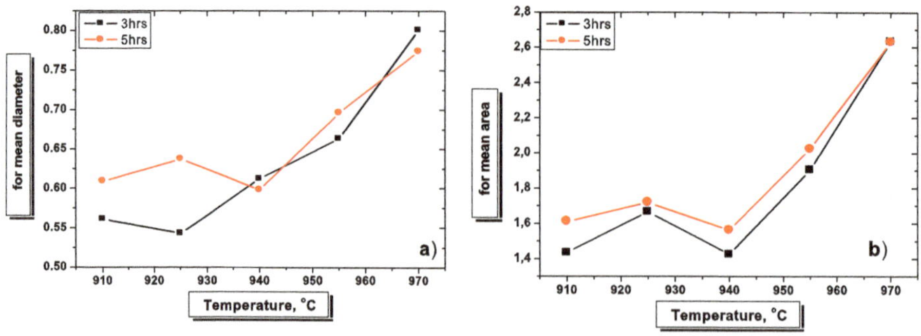

Figure 10. Interrelation between the heterogeneity factor (ν) of former austenite grain size in the cast steel and the austenitizing parameters for: a) mean diameter; b) mean grain area

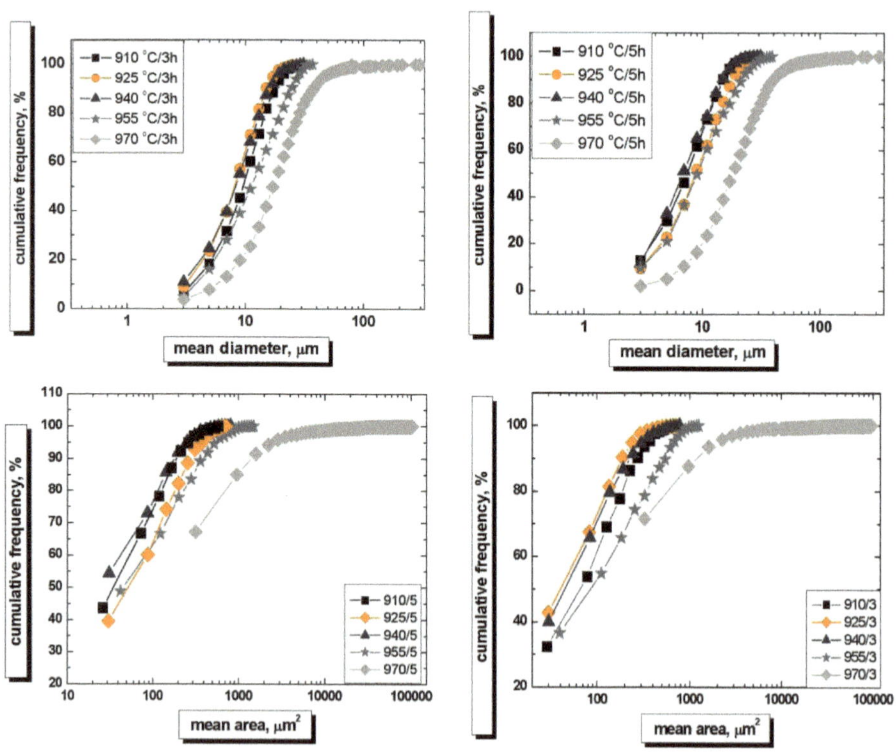

Figure 11. Distributions of frequency of cumulated grains

At the temperature of 970 °C the grain growth was observed – mean diameter increased over two times, while the mean area about five times in comparison with the temperature range of 910 ÷ 940 °C.

2.3. Determining the influence of cooling rate on the microstructure and properties

In order to determine the influence of cooling rates, allowing proper selection of parameters for the regenerative heat treatment, the TTT curves were plotted for L21HMF cast steel. On the basis of results achieved by means of dilatometric tests, a graph was drawn up, as shown in Fig 12. It illustrates the influence of cooling rate in the temperature range of 800 ÷ 500 °C on the structure and hardness of the investigated cast steel.

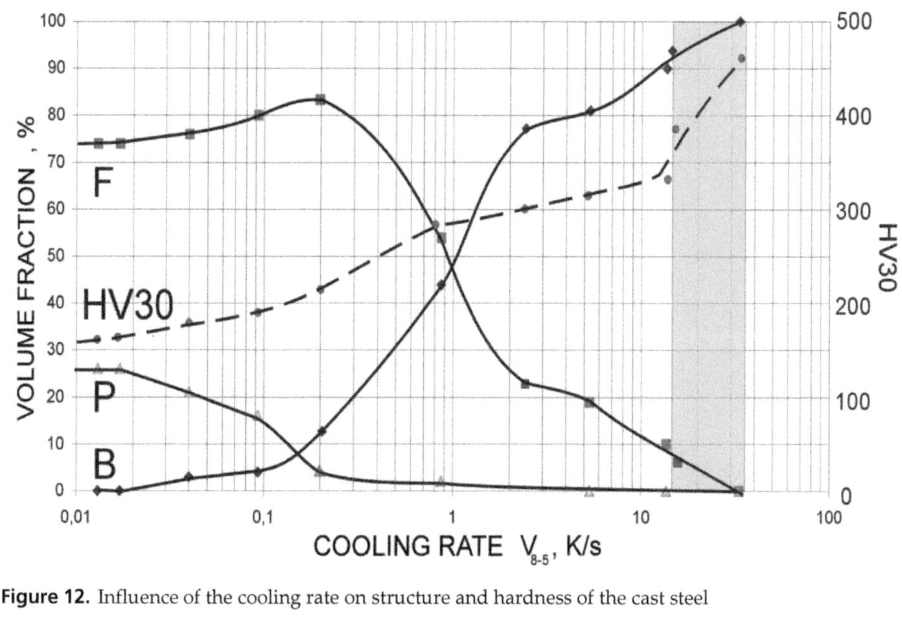

Figure 12. Influence of the cooling rate on structure and hardness of the cast steel

Analysis of the curves presented in Fig. 10 allows to state that in the case of L21HMF cast steel, whose chemical composition is given in Table 1, austenite cooled at 0.004 K/s $\leq v_{8-5} \leq$ 0.017 K/s gets transformed into ferrite and pearlite. The rate of cooling for austenite: 0.023 < $v_{8-5} \leq$ 0.869 K/s makes it possible to obtain ferritic – pearlitic – bainitic structures. Whereas after cooling of the cast steel at the range of 0.869 K/s < $v_{8-5} \leq$ 14.630 K/s bainitic – ferritic structures were obtained, with an increasing bainite volume fraction as the cooling rate increased. Bainitic structure with around 6% volume fraction was received for the cooling rate of $v_{8-5} \geq$ 14.630 K/s.

2.4. Influence of heat treatment on the microstructure and properties of L21HMF cast steel

The L21HMF cast steel was subject to heat treatment consisting in three-hour austenitizing of test pieces at the temperature of 910 °C and the following cooling at the rate corresponding to the processes of: bainitic hardening, normalizing and full annealing. The test pieces, bainite-hardened and normalized, were then tempered in the temperature range of 690 ÷ 730 °C and 690 ÷ 720 °C, respectively. While the test pieces cooled slowly from the austenitizing temperature (fully annealed), were subject to (α + γ) annealing (under annealing) at the temperatures of 780 ÷ 860 °C. Examples of microstructure of the examined cast steel after heat treatment are illustrated in Fig. 13.

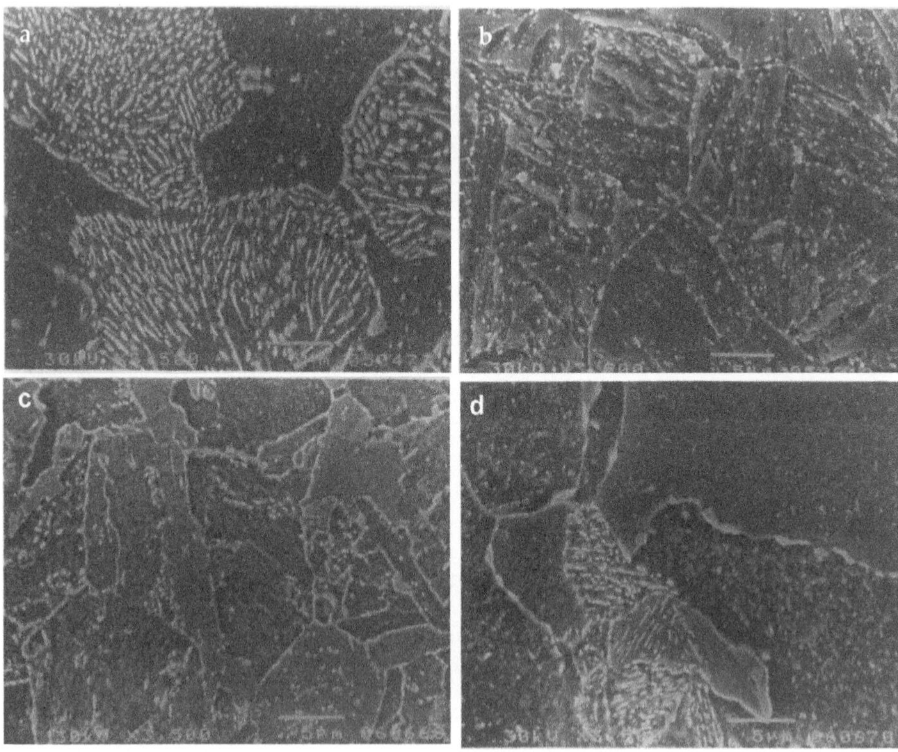

Figure 13. Microstructure of cast steel after: a) service; b) bainitic hardening and tempering; b) normalizing and tempering; c) full annealing and tempering

Bainitic hardening made it possible to obtain bainitic – ferritic microstructure in Cr – Mo – V cast steel. The amount of ferrite in the microstructure did not exceed 6%. In the tempered microstructure there were numerous precipitations of carbides on the lath boundaries, as well as inside and on the boundaries of prior austenite grain. Matrix after heat treatment was characterized by high dislocation density, however, some sparse polygonized areas were observed as well - showing lower density of dislocations (Fig. 14). Presence of the polygonized areas in the cast steel after heat treatment can be caused by the difference in chemical composition of particular grains resulting from a dendritic micro segregation or from the lack of austenite homogeneity during heat treatment. Differences of chemical composition may cause local decrease in the temperature of recrystallization.

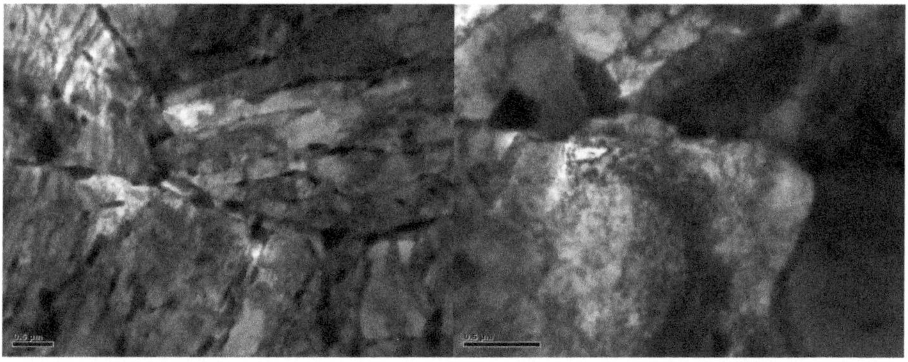

Figure 14. Microstructure of the L21HMF cast steel after heat treatment (bainitic hardening and high-temperature tempering)

Normalizing and tempering allowed obtaining tempered bainitic – ferritic structure with around 20% of ferrite in the Cr – Mo – V cast steel.

The observed microstructures after bainitic hardening and normalizing, apart from ferrite amount in the structure, differed in bainite morphology as well. After bainitic hardening only the "needle-shaped" form of bainite was observed, and it was morphologically similar to martensite, which indicates lower bainite presence in the structure (it can also be proved by the characteristic arrangement of carbides illustrated in Fig. 14). After normalization, however, the Cr – Mo – V cast steel microstructure showed the "feathery" bainite form, which indicates the presence of upper bainite. Apart from the "feathery" bainite also some single areas of „needle-shaped" bainite could be seen.

The identifications of precipitates performed by means of the extraction carbon replicas revealed in the investigated cast steel after heat treatment (in the tempered bainitic and bainitic – ferritic structure) the occurrence of the following carbide types: MC, M_3C, M_7C_3 and $M_{23}C_6$.

The study of mechanical properties at room temperature has shown that the structure of high-temperature tempered bainite provides the combination of high strength properties and impact energy. Tensile strength and yield strength after tempering exceeded the minimum requirements considerably, and similarly, the impact energy was several times higher than the required minimum of 27J for the new castings (Table 5, Fig. 15). Tempering of L21HMF cast steel with bainitic structure at the temperatures which are 10 and 20 °C higher than the maximum tempering temperature recommended by the standard, i.e. at 720 and 730 oC, caused an increase in impact energy by 8 and 35%, respectively, with the hardness decrease by 2 ÷ 5% in comparison with the tempering temperature of 710 °C (Fig. 15).

Therefore, it can be concluded that for the cast steels of bainitic microstructure it is possible to apply higher temperatures of tempering compared to the ones recommended by the standards, without concern that the strength properties can go down below the required minimum. Apart form obtaining high impact energy with the required strength properties maintained, it also allows to achieve the microstructure of higher thermodynamic stability, which can guarantee slower process of its degradation.

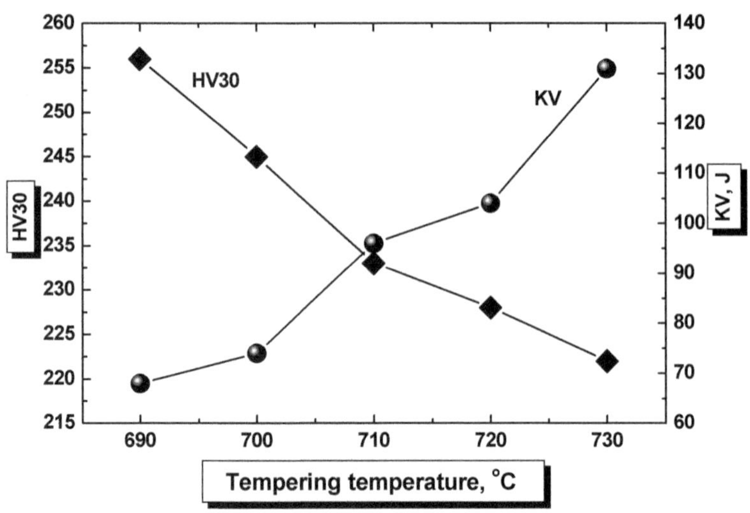

Figure 15. Influence of the tempering temperature on hardness and impact energy of the L21HMF cast steel with bainitic structure

The cast steel of tempered mixed (bainitic – ferritic) microstructure was characterized by the strength properties on a similar level as the cast steels with bainitic microstructure. However, the crack resistance of those cast steels was almost two times as low compared to that of cast steels with bainitic microstructure (Table 6).

High impact energy of the cast steel with the microstructure of tempered bainite is a consequence of large total amount of grain boundaries (boundaries of bainite packets) and high ductility of the tempered microstructure of lower bainite. Whilst, lower impact energy of the cast steel with mixed bainitic – ferritic structure results from the presence of ferrite in the microstructure, which favours the fissile cracking, and from the presence of upper bainite, characterized by greater brittleness than lower bainite.

Full annealing allows to obtain ferritic – pearlitic microstructure for the examined cast steel grade (Fig. 13d), with pearlite located mostly on ferrite grain boundaries. In pearlite the processes of fragmentation and spheroidization of carbides could be observed. The ferritic – pearlitic microstructure obtained as a result of repeated cooling from the austenitizing temperature was morphologically similar to the microstructure after long-term service.

Heat treatment parameters	TS MPa	YS MPa	El. %	KV J	HV30	Microstructure
after service	545	305	26	10	156	ferritic-pearlitic
bainitic hardening + 720 °C/4h	728	620	18	104	228	bainitic
normalizing + 720 °C/4h	721	594	17	62	220	bainitic-20%ferritic
full annealing + 720 °C/4h	558	336	27	26	153	ferritic-20%pearlitic
full annealing + 800 °C/4h	552	316	31	42	162	ferritic-20%pearlitic
full annealing + 820 °C/4h	550	324	28	42	164	ferritic-20%pearlitic
*PN requirements	500 ÷ 670	min. 320	min. 20	min. 27	140 ** ÷ 197	—

*- PN - 89/ H - 83157 ; ** - hardness according to Brinell

Table 5. Microstructure and properties of the L21HMF cast steel after heat treatment

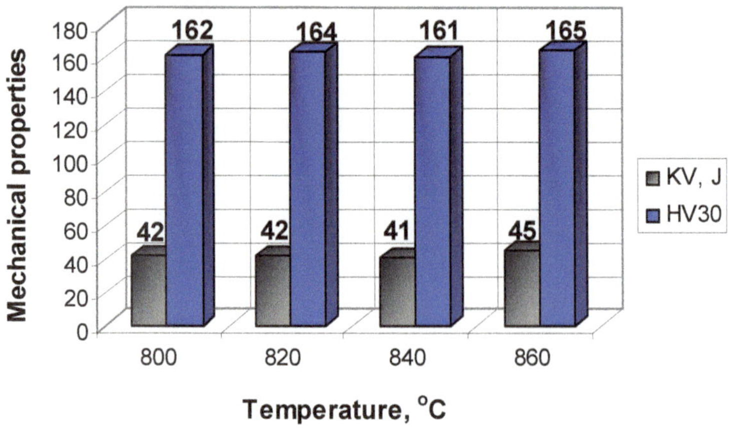

Figure 16. Change in the values of hardness and impact energy of the cast steel depending on the temperature of (α + γ) annealing

For the L21HMF cast steel of ferritic – pearlitic microstructure it is required to apply (α + γ) annealing (under annealing) instead of tempering which did not always provide the required impact energy. Applying under annealing causes: dissolution of carbides precipitated on grain boundaries during slow cooling from the temperature of austenitization, decrease of phosphorus segregation on ferrite grain boundaries and further reduction of austenite grain size. This allows to obtain the required strength properties and impact energy KV on the level ~ 40J. The influence of (α + γ) annealing temperature on the value of impact energy and hardness is presented in Fig.16.

The performed heat treatment, apart from the changes in microstructure and properties of the examined cast steels, also caused a change in the mechanism of cracking (Fig. 17). In the cast steel of high-temperature tempered bainite structure, on the entire surface under the fracture, there was a transcrystalline ductile fracture initiated by fine-dispersion precipitates of carbides and sulfide inclusions (Fig. 17a). The characteristic feature of plastic cracking is its ability to absorb significant amounts of energy connected with plastic deformations preceding the decohesion. The cast steel of bainitic – ferritic structure was subject to decohesion through mixed mechanism. Directly under the notch, at a depth of about 1.0 ÷ 1.5 mm, cracking proceeded in plastic manner through transcrystalline ductile mechanism. Below the area of plastic strain, fissile cracking could be observed, running through a transcrystalline fissile mechanism with micro fields of ductile character.

The cast steel with regenerated ferritic – pearlitic structure, obtained as a result of slow cooling and under annealing, was cracking through a mechanism similar to decohesion of the cast steel after service, i.e. transcrystalline fissile mechanism with micro fields of ductile character (Fig. 17b).

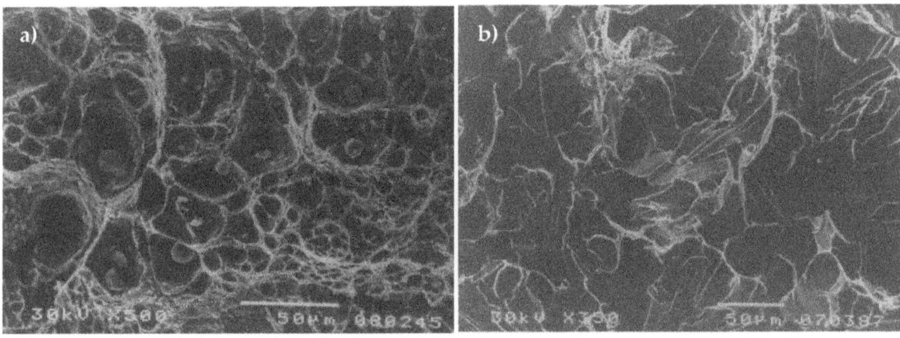

Figure 17. Cracking mechanism of cast steel: a) transcrystalline ductile for tempered bainitic microstructure; b) transcrystalline fissile for ferritic – pearlitic microstructure

3. Summary

The research performed on the L21HMF cast steel, taken from a steam turbine cylinder serviced for around 186 000 hours at the temperature of 540 °C, has revealed that long-term service contributed to: the processes of recovery and polygonization of ferrite grains, preferential precipitation of $M_{23}C_6$ carbides on grain boundaries and formation of „H – carbide" complexes near the boundary areas of ferrite grains. During long-term operation the strength properties were decreasing slowly – yet faster in the case of yield strength than tensile strength, and the impact energy decreased drastically below the required minimum level of 27J.

Changes in the microstructure and properties of the long-term serviced cast steel do not eliminate the possibilities of their further safe operation. Extending the safe operation time beyond the calculative time of 100 000 hours (with the target up to 200 ÷ 250 000 hours) is possible thanks to regenerative heat treatment.

Performed research has proved that applying bainitic hardening instead of normalizing/full annealing, thus far applied in the castings, allows to achieve the best combination of high strength properties and very high impact energy. Moreover, the bainitic microstructure makes it possible to apply high temperatures of tempering, amounting to 710 ÷ 730 °C. This allows increasing the stability of microstructure of long-term serviced cast steels without concern for reduction in the strength properties below the required minimum. High impact energy KV > 100J of the cast steel with high-tempered bainite structure guarantees that after long-term operation the impact energy will not drop below the minimum required level of 27J.

Applying normalizing for the castings allows to obtain bainitic – ferritic microstructure, which is characterized by similar strength properties as the cast steel with tempered bainitic microstructure, with the impact energy, however, being almost two times as low. What seems evident here, is the negative influence of ferrite in the microstructure on the impact strength.

The ferritic – pearlitic microstructure, obtained as a result of slow cooling of the castings from the austenitizing temperature (full annealing), allows to obtain the strength properties comparable to those after service and impact energy on the level of 40J. After the process of full annealing it is recommended to apply the ($\alpha + \gamma$) annealing (under annealing) instead of the process of tempering, which makes it possible to obtain the required impact energy.

Author details

Grzegorz Golański
Institute of Materials Engineering, Czestochowa University of Technology, Poland

4. References

[1] Zieliński A., Dobrzański J., Krztoń H., *Structural changes in low alloy cast steel Cr- Mo – V after long time creep service*, JAMME, 25, 2007, 33

[2] Stachura S., *Changes of structure and mechanical properties in steels and cast steels utilised in increased temperatures*, Energetyka, 2, 1999, 109

[3] Balyts'kyi O.I., Ripei I.V., Protsakh Kh. A., *Degradation of the cast elements of steam turbines of thermal power plants made of 20KhMFL steel in the course of long – term operation*, Materials Science, 41, 3, 2005, 423

[4] Dobosiewicz J., *Influence of operating conditions on the changes in mechanical properties of steam turbine cylinders*, Energetyka, 1992, 1, 27

[5] Stachura S., Kupczk J., Gucwa M., *Optimization of structure and properties of Cr – Mo and Cr – Mo – V cast steel intended for use at increased temperature*, Foundry Review, 54, 5, 2004, 402

[6] Stachura S., Trzeszczyński J., *The choice of the regenerative thermal treatment of the Cr – Mo and Cr – Mo –V cast steel*, Inżynieria Materiałowa, 6, 1997, 227

[7] Lu Z., Faulkner R.G., Flewitt P. E. J., *The role of irradiation – induced phosphorus segregation in the DBTT temperature in ferritic steel*, Mater. Sc. Eng., A 437, 2006, 306

[8] Islam M. A.,. Knott J. F, Bowen P., *Kinetics of phosphorus segregation and its effect on low temperature fracture behaviour in 2.25Cr - 1Mo pressure vessel steel*, Materials Sc. and Techn., 21, 1, 2005, 76

[9] Molinie E., Piques R., Pineau A., *Behaviour of a 1Cr – 1Mo – 0.25V steel after long – term exposure – I. Charpy impact toughness and creep properties*, Fatique Fract. Eng. Mater. Struct., 14, 5, 1991, 531

[10] Stachura S., Golański G., *Metallographic and mechanical properties of steel and cast steel after long term service at elevated temperatures*, Report BZ – 202 – 1/01 unpublished researched

[11] Golański G., Stachura S., Kupczyk J., Kucharska – Gajda B., *Heat treatment of cast steel using normalization and intercritical annealing*, Arch. Foundry Eng., 7, 2007, 123

[12] Komai N., Masuyama F., Igarashi M., *10 – year experience with T23(2.25Cr – 1.6W) and T122 (12Cr – 0.4Mo – 2W) in a power boiler*, Transations of the ASME, 127, 2005, 190
[13] Trzeszczyński J., Grzesiczek E.,. Brunne W, *Effectiveness of solutions extending operation time of long operated cast steel elements of steam turbines and steam pipelines*, Energetyka, 3, 2006, 179
[14] Rehmus – Forc A., *Change of structure after revitalization cylinders of a steam turbine*, Inżynieria Materiałowa , 2006, 3, 265
[15] Golański G., *Influence tempering temperature on mechanical properties of cast steels*, Archives of Foundry Eng., 8, 4, 2008, 47
[16] Dobosiewicz J., *Reasons for regenerating steam turbine cylinders*, Energetyka, 1996, 11, 634
[17] Łukowski J., *Modern installations for quenching with the application of aqueous polimer solutions*, Przegląd Odlewnictwa, 51, 2001, 2, 57
[18] Bakalarski A., Nowarski A., Sus J., Wyszyński K., *Theoretical investigation and experimental application of aqueous polymer solution for quenching products of low-alloy cast steel*, Przegląd Odlewnictwa, 51, 2001, 2, 61
[19] Golański G., Stachura S., Gajda B., Kupczyk J., *Influence of the cooling rate on structure and mechanical properties of L21HMF cast steel after regenerative heat treatment*, Archives of Foundary, 6, 21, 2006, 143
[20] Zieliński A., Dobrzański J., Golański G., *Estimation of residual life of L17HMF cast steel elements after long – term service*, J. Achievements in Materials and Manufacturing Eng.., 34, 2, 2009, 137
[21] Golański G., *Microstructure and mechanical properties of G17CrMoV5 – 10 cast steel after regenerative heat treatmen*t, J. Pressure Vessel Techn., 132, 2010, 064503-1
[22] Golański G., Stachura S., Kupczyk J., Kucharska – Gajda B., *Optimisation of regenerative heat treatment of G21CrMoV4 – 6 cast steel*, Archives Mater. Sc. Eng., 28, 6, 2007, 341
[23] Murphy M.C., Branch G. D., *The microstructure, creep and creep – rupture properties of Cr – Mo – V steam turbine castings*, JISI, 1969, 1347
[24] Senior B. A., *A critical review of precipitation behavior in 1Cr – Mo – V rotor steels*, Mater. Sc. Eng. A, 1988, 103, 263
[25] Williams K. R., Wilshire B., *Microstructural instability of 0.5Cr – 0.5Mo-0.25V creep – resistant steel during service at elevated temperatures*, Mater. Sci. Eng., 1981, 47, 151
[26] Danielsen H. K., Hald J., *Behaviour of Z phase in 9 – 12 %Cr steels*, Energy, 1, 1, 2006, 49
[27] Golański G., *Mechanical Properties of G17CrMoV5 – 10 Cast Steel after Regenerative Heat Treatment*, Solid State Phenom. 147 – 149, 2009, 732
[28] Ryś J., *Stereology of materials*, FOTOBIT Publ., Cracow, 1995
[29] Taylor J.P., Blondeau R., *The respective roles of the packet size and the lath width on toughness*, Metall. Trans., 7A, 1976, 891
[30] H. Kotilainen, *The micromechanisms of cleavage fracture and their relationship to fracture toughness in a bainitie Low Alloy Steel*, VTT, Espoo, Finland, 1980

[31] Bhadeshia H. K. D. H., *Bainite in steels*, 2nd edition, The University Press Cambridge, Cambridge, UK, 2001
[32] PN - 89/ H – 83157 Cast steels for elevated temperature applications. Grades.

Chapter 3

Design of Cooling Units for Heat Treatment

Michal Pohanka and Petr Kotrbáček

Additional information is available at the end of the chapter

1. Introduction

Microstructure and nature of grains, grain size and composition determine the overall mechanical behavior of steel. Heat treatment provides an efficient way to manipulate the properties of steel by controlling the cooling rate. The way of heat treatment depends on many aspects. One of the most important parameter is the amount of production. Another important parameter is the size of products. We focus here on large production such as interstand [1] and run-out table cooling of hot rolled strip, run-out table cooling of sheets and plates, cooling of long products at the exit from a rolling mill, cooling of rails, tubes and special profiles [2], continuous hardening and heat treatment lines for steel strips. Such a treatment is called in-line heat treatment of materials and has become frequently used by hot rolling plants. This method achieves the required material structure without the necessity of reheating. In-line heat treatment is characterized by running of hot material through the cooling section. However, many of discussed topics can be applied on smaller production as well.

The design procedure of cooling sections for obtaining the demanded structure and mechanical properties is iterative research involving several important steps. We begin with the Continuous Cooling Transformation (CCT) diagram for the selected material. Numerical simulation of cooling follows to find appropriate cooling intensity and its duration. Knowing the desired cooling intensity new cooling section is designed and tested under the laboratory conditions [3]. From the laboratory experiments boundary conditions are obtained and tested using a numerical model. When the best solution is found it is tested on the real sample and the result structure is studied. In most cases the process must be repeated as the CCT diagram is aimed at a different size of the sample and the cooling rate in the designed section is not constant.

2. Strategy of design

When preparing a design of the cooling system for the continuous heat treatment, we should know the optimum cooling regime for the material and the product. Any continuous heat treatment process needs to vary the cooling intensity with time. Moreover, the practice has shown that the results obtained by using small samples (usually for the CCT diagram) are usually different from the results achieved when using real products of a large cross-section because it is not possible to achieve the identical temperature regimes in the whole volume due to the low diffusivity. We cannot expect same behavior on product as on the small sample.

Next considerations focus on technical means that can be used to achieve the demanded temperature mode. There are varieties of technical methods for hot steel cooling; one of them uses the spray cooling. The cooling section should ensure reaching a required temperature history in the cooled piece prescribed by the metallurgists. The nozzles applied allow controlling the cooling over a wide range. The cooling intensity of groups of the nozzles must be measured and then the results obtained can be used in a numerical model of the temperature field in the cooled material.

2.1. Leidenfrost effect and its impact

It should be understand that intensity of cooling strongly depends on the surface temperature. So called Leidenfrost effect can be observed above certain temperature. During this effect a liquid, which is near significantly hotter object than the liquid's boiling point, produces vapor layer which insulates the liquid from the hot object and keeps out that liquid from rapid boiling. This is because of the fact that at temperatures above the Leidenfrost point, the part of the water, which is near hot surface, vaporizes immediately on contact with the hot plate and the generated gas keeps out the rest of the liquid water, preventing any further direct contact between the liquid water and the hot plate. The temperature at which the Leidenfrost effect begins to occur is not easy to predict. It depends on many aspects. Ones of them are velocity and size of droplets. As a rough estimate, the Leidenfrost point might occur for quite low temperatures such as 200 °C. On the other hand for high water velocity the Leidenfrost point can be even above 1000 °C. Fig. 1 shows measured heat transfer coefficients (HTC) for water-air mist nozzle. Graph shows three measurements for same nozzle using varying water and air parameters. Three regimes can be found. The first one is for low temperatures when the HTC is relatively high and decreasing slowly. This part is below Leidenfrost point. From certain temperature HTC decreases rapidly. This is a transient regime in which some droplets are above Leidenfrost point and some are below that point. For the last regime HTC is relatively low and is constant or may be increasing due to the increasing radiation with the increasing surface temperature. Designed cooling section should work in the first regime for low temperatures or in the third regime with almost constant HTC. It is strongly recommended to avoid the second transient regime as the product surface temperature is not usually at uniform temperature. Due to the strong dependency of HTC on surface temperature non homogeneous cooling is achieved and causes distortion of the product.

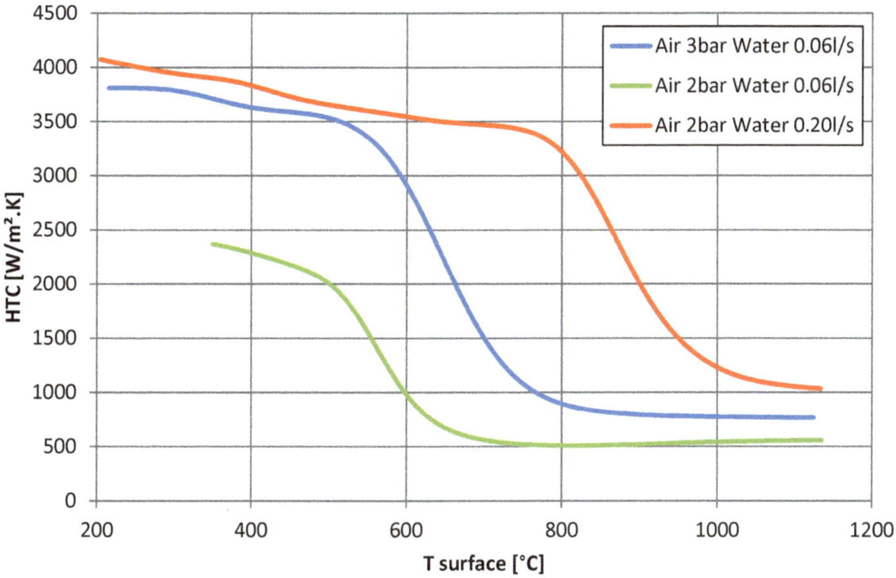

Figure 1. Moving Leidenfrost point for water-air mist nozzle and various water and air conditions.

2.2. Nozzle types and controllability

Nozzle produces usually one of three typical sprays: flat-jet, full-cone, and solid-jet (see Fig. 2). However, other shapes can be found such as hollow-cone, square, spiral etc. An important parameter is controllability of the cooling section and intensity of cooling. The water-air mist nozzles can be used for a soft cooling and a wide controllability range (see Fig. 3). The HTC can vary from several hundreds of $W/m^2.K$ up to several thousand of $W/m^2.K$. Water-air mist nozzles are not the cheapest ones and also the pressurized air is expensive in terms of power consumption. Water-only nozzles can often provide a lower cost solution. Small full-cone nozzles with high pressure and bigger distance from surface can provide also very soft cooling. On the other hand with high pressure flat-jet or solid-jet nozzles at small distances HTC over 50000 $W/m^2.K$ can be obtained even for high surface temperatures (see Fig. 4). This results in enormous heat flux above 50 MW/m^2. The distance of the nozzle form surface is very important in this case for flat-jet nozzles because for 100 mm the HTC can be 50000 $W/m^2.K$ but for 1000 mm it can be similar to water-air mist nozzle. As conclusion we can say that for soft cooling air-mist or full-cone water only nozzles can be used and for hard cooling flat-jet nozzles with small distances are used. In some cases solid-jet nozzles are used for hard cooling but there are often two major problems: large amount of water generates a water layer on the product and the spray spot is small which causes non homogenous cooling. On the other hand, clogging is not big problem for solid-jet nozzles.

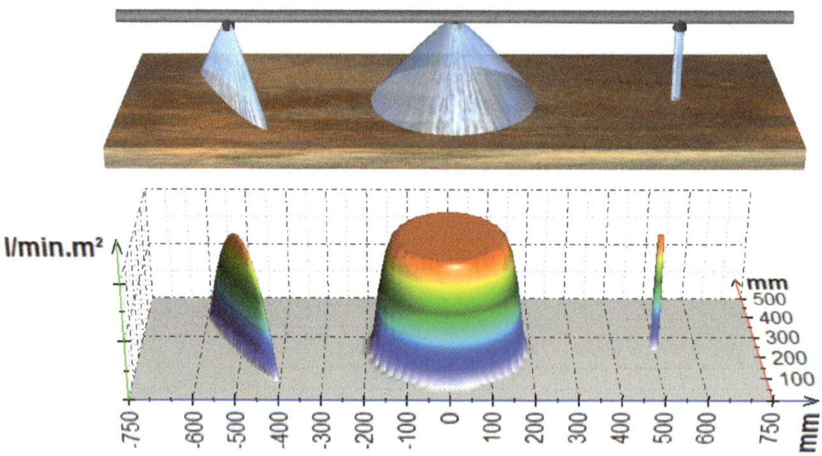

Figure 2. Flat-jet, full-cone, and solid-jet nozzles with computed water distribution on flat surface

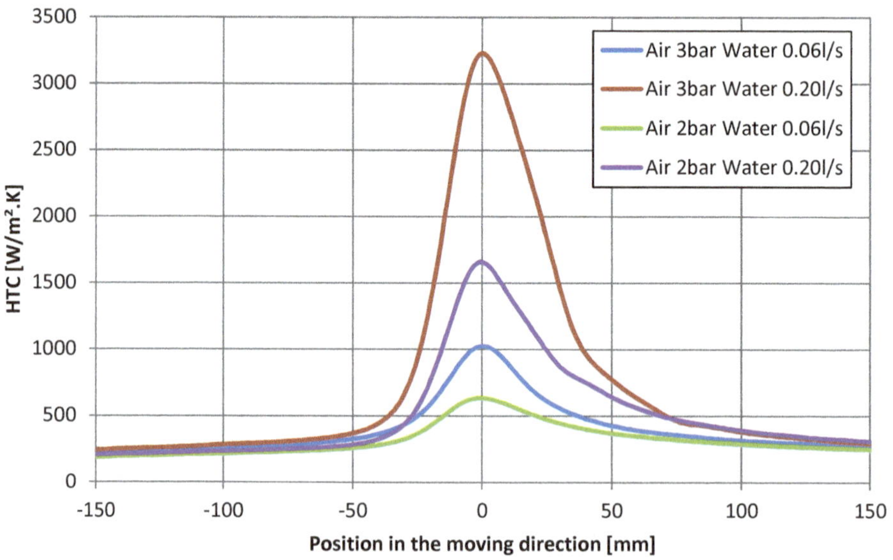

Figure 3. Controllability of water-air mist nozzle for surface temperatures 1000 °C

2.3. Influence of product velocity on heat transfer coefficient

Three measurements are compared when the only different parameter is the casting speed. The first experiment was stationary with no movement, the second experiment used a velocity of 2 m/s and the last experiment was done for a velocity of 5 m/s. These three experiments used the identical water-air mist nozzle, and the same pressure settings were

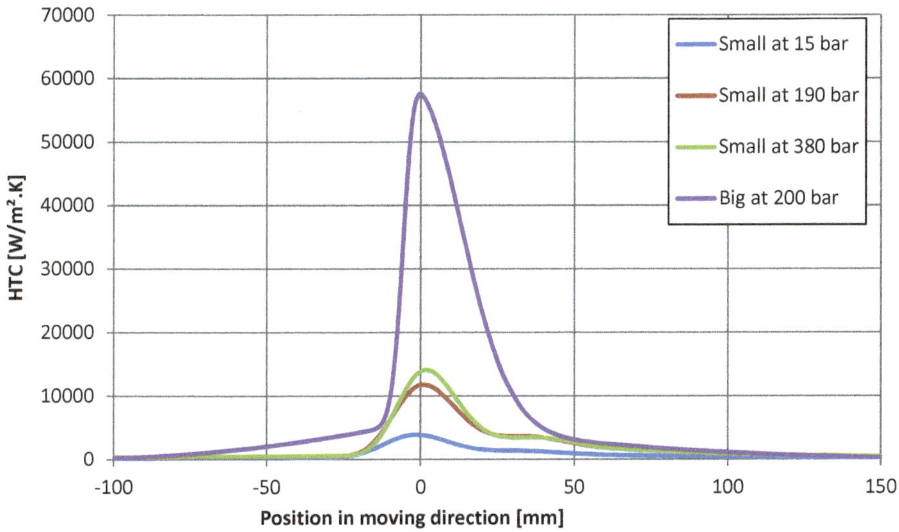

Figure 4. Distribution of HTC under spray for high pressure flat-jet nozzles for surface temperature 1000 °C.

used in all experiments - water pressure of 2 bar, air pressure of 2 bar. Fig. 5 shows the distribution of heat transfer coefficient in experimental group with a variation of velocity. HTC for the stationary case (not possible in mill) is symmetrical and the peak is narrow. The cooling intensity decreases with the increasing velocity. Heat transfer coefficient distribution is more non-symmetrical when product speed increases. The observed effect is caused by the flow on the surface and different vapor forming conditions in front and behind the impinging jet.

3. Cooling intensity and numerical models

In order to design a cooling section, knowledge of the cooling intensity is required for a group of nozzles and nozzle headers. Exact knowledge of the heat transfer coefficient as a function of spray parameters and surface temperature is the key problem for any design work. The cooling intensity is a function of several parameters, mainly, nozzle types, chosen pressures and flow rates, surface temperature of a material, and velocity of a material movement whilst under spray. There is no function available which describes cooling intensity using all the mentioned parameters. This is the reason why real measurement is absolutely necessary.

3.1. Experimental procedure

During the in-line heat treatment the product is moving so our testing sample should be also moving through the cooling section. Fig. 6 shows schematically a suitable experiment used for obtaining boundary conditions for a numerical simulation. The hot sample is moving at

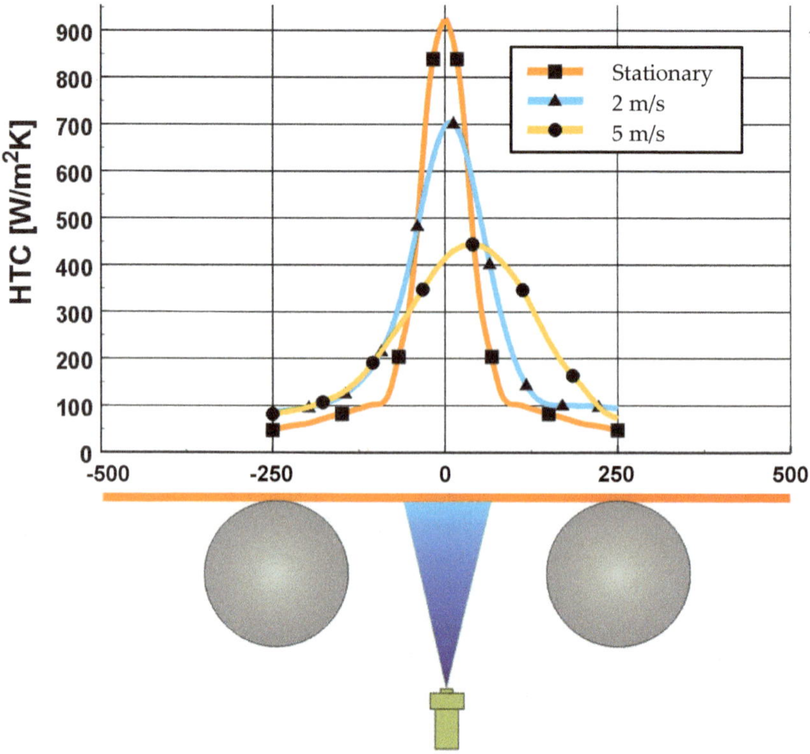

Figure 5. Influence of velocity at heat transfer coefficient.

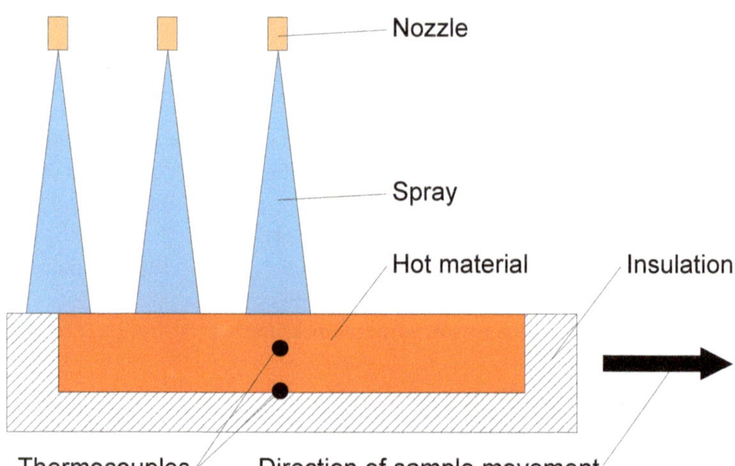

Figure 6. Moving sample with embedded thermocouples cooled down by water spray

prescribed velocity which is similar to real conditions. The sample passes under spray which cools down the hot sample. For a simple shape like plate it is recommended to insulate all surfaces excluding the one on which the cooling intensity is investigated. One or more thermocouples are embedded in the sample and measure temperature during the experiment. The installed thermocouples should not disturb the cooled surface. This is the reason why they should be installed inside the sample, not on the investigated surface. In principle when all surfaces are insulated except the one which is investigated one thermocouple is enough. However, this thermocouple should be as close to the investigated surface as possible. Otherwise the resolution of description of boundary conditions will be degraded. After the measurement an inverse algorithm is used to compute boundary conditions on the investigated surface from the measured temperature history inside the sample. An example of recorded temperature history is shown in Fig. 7.

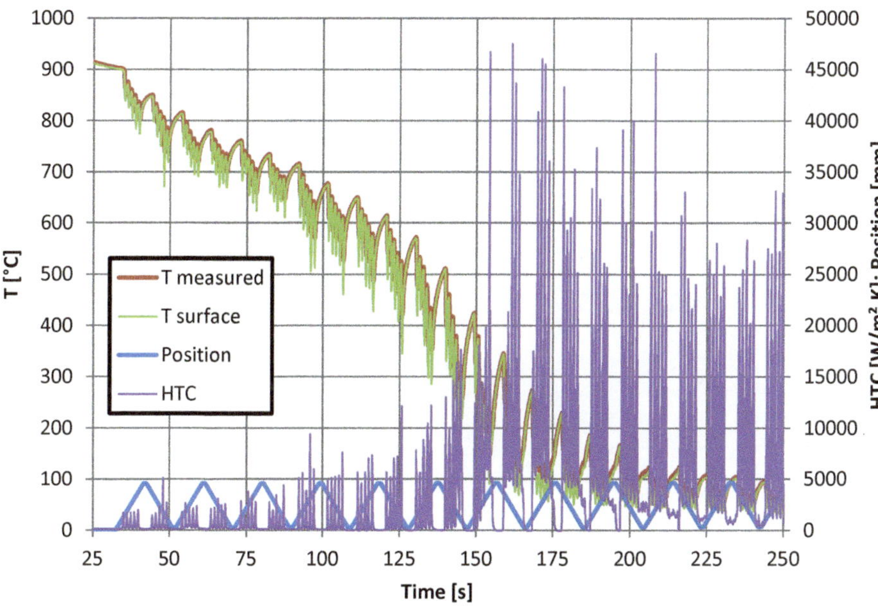

Figure 7. Example of recorded temperature history by one thermocouple inside the sample, computed surface temperature above the installed thermocouple, recorded position of the thermocouple in the cooling section, and computed heat transfer coefficient. Cooling section equipped with five rows of flat-jet nozzles.

Heat transfer test bench presented in [4] is designed so that it enables progression of samples up to the weight of 50 kg with infinitely adjustable speed from 0.1 to 6 m/s (see Fig. 7). On the supporting frame there is a carriage moving, on which the sample under examination with embedded temperature sensors and measuring system is fixed (see Fig. 9). The carriage's progression is provided by a hauling rope through a drive pulley and a motor with a gearbox. The motor is power supplied by a frequency converter with the possibility

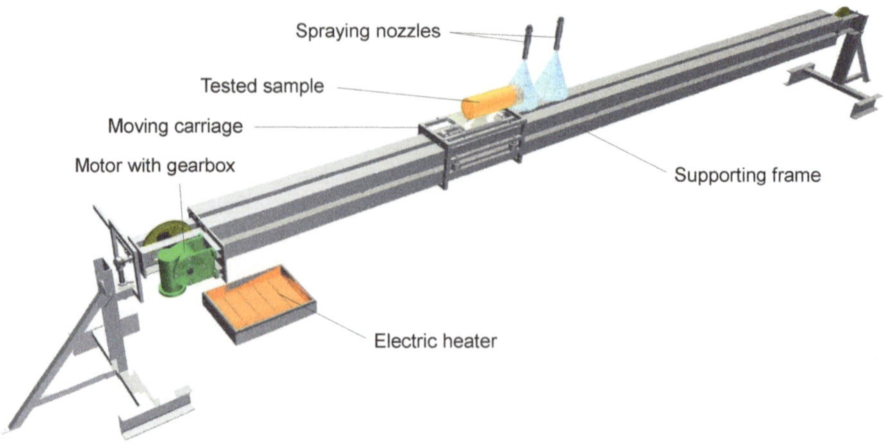

Figure 8. Heat transfer test bench

Figure 9. Examples of boundary conditions measurements on steel plate, rail, and pipe.

of a smooth change of speed. The direction of the carriage can be reversed and passages repeated in a requiring number. The whole cycle is programmed and controlled through the superior PC. There is a spraying section in the central sector where arbitrary jets configuration can be arranged when distribution of heat transfer coefficients or heat fluxes must be measured. The sample is equipped with thermocouples connected to the data logger. The thermocouples are calibrated before use and the results of calibration are used to eliminate dynamic error in measurement of highly transient thermal processes. Before the actual experiment the carriage with the sample is positioned to the electric heater and it is heated to the required temperature inside the furnace. After the temperature in the sample is stabilized, the heating device is removed, the stand is turned to spraying position, the pump for the water gets going and the carriage's runs through the cooling section. The position of the cooled surface can be horizontal with spraying upper or bottom surfaces or vertical. Signals from the sensors are read by the data logger which moves together along with the sample. At the same time, the signal indicating the actual carriage's position is recorded as well. After performing the required number of passes through cooling zone, data are exported from data logger's internal memory into the computer for further processing.

3.2. Inverse computation of boundary conditions and numerical models

Information from temperature histories in a particular depth under the investigated surface are used as entry parameters for the thermal conduction's inverse task. Inverse task outputs are surface temperature histories, heat flows, and heat transfer coefficients (HTC) as function of time and position. Most often, in mathematical models the boundary condition of the 3'd type is used where heat flow is specified by the HTC value and the cooling water temperature.

If the boundary conditions of a solid must be determined from transient temperature measurements at one or more interior locations, it is an inverse heat conduction problem (IHCP) during which the dispersed impulse on boundary must be found. The IHCP is much more difficult to solve than the direct problem. Such problems are extremely sensitive to measurement errors. There are number of procedures that have been advanced for the solution of ill-posed problems in general. Tikhonov has introduced the regularization method [5] to reduce the sensitivity of ill-posed problems to measurement errors. The mathematical techniques for solving sets of ill-conditioned algebraic equations, called single-value decomposition techniques, can also be used for the IHCP [6]. There were extremely varied approaches to the IHCP. These included the use of Duhamel's theorem (or convolution integral) which is restricted to linear problems [7]. Numerical procedures such as finite differences [8][9][10] and finite elements [11] were also employed, due to their inherent ability to treat non-linear problems. Exact solution techniques were proposed by Burggaf [12], Imber and Khan [13], Langford [14], and others. Some techniques used Laplace transforms but these are limited to linear cases [15]. Combined approach is also described in [16]. The improvement in artificial intelligence has brought new approaches, such as genetic algorithm [17] and neural networks [18][19][20].

All the mentioned algorithms need a precise mathematical model of the tested sample for computing the direct heat conduction problem. Analytical methods may be used, in certain cases, for exact mathematical solutions of conduction problems. These solutions have been obtained for many simplified geometries and boundary conditions and are well documented in the literature [21][22][23]. However, more often than not, geometries and boundary conditions preclude such a solution. In these cases, the best alternative is the one using a numerical technique. For situations where no analytical solution is available, the numerical method can be used. Nowadays there are several methods that enable us to solve numerically the governing equations of heat transfer problems. These include: the finite difference method (FDM), finite volume method (FVM), finite element method (FEM), boundary element method (BEM), and others. For one-dimensional model with constant material properties there exists nice similarity. All of the FDM, FVM, and FEM with tent weighting function equations can be put in a similar form:

$$\frac{d}{dt}\left(\beta T_1 + \gamma T_2\right) = \frac{-\alpha}{\Delta x^2}\left(T_1 - T_2\right) + \frac{q_1(t)}{\rho c \Delta x}, \tag{1}$$

$$\frac{d}{dt}\left(\gamma T_{j-1} + 2\beta T_j + \gamma T_{j+1}\right) = \frac{-\alpha}{\Delta x^2}\left(T_j - T_{j+1}\right) + \frac{\alpha}{\Delta x^2}\left(T_{j-1} + T_j\right), \tag{2}$$

$$\frac{d}{dt}\left(\gamma T_{N-1} + \beta T_N\right) = \frac{-\alpha}{\Delta x^2}\left(T_{N-1} - T_N\right) + \frac{q_N(t)}{\rho c \Delta x}, \tag{3}$$

where β and γ have the values listed in Tab. 1. Equations (1–3) are restricted to temperature-independent thermal properties but the concepts can be extended to T-variable cases. In general for multidimensional models and temperature dependent material properties the simplest equations are obtained for FDM while the complexity of equation for FVM and FEM is several times higher.

	β	γ	$\beta + \gamma$
FDM	1/2	0	1/2
FVM	3/8	1/8	1/2
FEM	2/6	1/6	1/2

Table 1. Values of the β and γ of Eq. (1–3)

3.3. Phase change implementation

Physical processes, such as solid/liquid and solid state transformations, involve phase changes. The numerical treatment of this non-linear phenomenon involves many problems. Methods for solving the phase change usually use a total enthalpy H, an apparent specific heat coefficient c_A, or a heat source $\dot{q}$.

The nature of a solidification phase change can take many forms. The classification is based on the matter in the phase change region. The most common cases follow:

a. *Distinct*: The phase change region consists of solid and liquid phases separated by a smooth continuous front – freezing of water or rapid solidification of pure metal.
b. *Alloy*: The phase change region has a crystalline structure consisting of grains and solid/liquid interface has a complex shape – most metal alloys.
c. *Continuous*: The liquid and solid phases are fully dispersed throughout the phase change region and there is no distinct interface between the solid and liquid phase – polymers or glasses.

In a distinct phase change, the state is characterized by the position of the interface. In such cases the class of the so called *front tracking* methods is usually used. However, in cases b) and c) the models use the phase fraction.

The phase change process can be described by a single enthalpy equation

$$\frac{\partial H}{\partial t} + \nabla \cdot \left(g_d H_d s_d + g_l H_l s_l \right) = \nabla \cdot \left(k \nabla T \right) \tag{4}$$

where g is the phase volume fraction, s is the phase velocity, and subscript d and l refer to solid and liquid phases (or structure A and structure B), respectively [24]. The k is (in this case) a mixture conductivity defined as

$$k = g_d k_d + g_l k_l \tag{5}$$

and H is the mixture enthalpy

$$H = g_s \int_{Tref}^{T} \rho_d c_d dT + g_l \int_{Tref}^{T} \rho_l c_l dT + \rho_l c_l L \tag{6}$$

where $Tref$ is an arbitrary reference temperature. To overcome the problem of the non-linear (discontinuity) coefficient of a specific heat a non-linear source term is used. The term $\partial H / \partial t$ can be expanded as

$$\frac{\partial H}{\partial t} = c_{vol} \frac{\partial T}{\partial t} + \delta H \frac{\partial g_l}{\partial t} \tag{7}$$

Neglecting convection effects in Eq. (4) and substituting Eq. (7) results in

$$c_{vol} \frac{\partial T}{\partial t} = \nabla \cdot \left(k \nabla T \right) + \dot{q} \tag{8}$$

where

$$\dot{q} = -\delta H \frac{\partial g_l}{\partial t} \tag{9}$$

Eq. (4) is non-linear and it contains two related but unknown variables H and T. It is convenient to reformulate this equation in terms of a single unknown variable with non–linear latent heat. Song [25] and [26] Comini uses so called Apparent heat capacity. The apparent specific heat can be defined as

$$c_A = \frac{dH}{dT} = c_{vol} + \delta H \frac{dg_l}{dT} \tag{10}$$

where

$$\begin{aligned} c_{vol} &= g_d \rho_d c_d + g_l \rho_l c_l \\ \delta H &= \int_{Tref}^{T} \left(\rho_l c_l - \rho_d c_d\right) dT + \rho_l L. \end{aligned} \tag{11}$$

Neglecting convection effects and substituting into Eq. (4) yields apparent heat capacity equation

$$c_A \frac{\partial T}{\partial t} = \nabla \cdot \left(k \nabla T\right). \tag{12}$$

Another approach is total enthalpy. From Eq. (6) it can be written

$$\nabla T = \nabla H / c_{vol} - \delta H \nabla g_l / c_{vol} . \tag{13}$$

Substitution in Eq. (4) will result in a total enthalpy equation

$$\frac{\partial H}{\partial t} = \nabla \cdot \left(\frac{k}{c_{vol}} \nabla H \right) + \nabla \cdot \left(\frac{k}{c_{vol}} \delta H \nabla g_l \right). \tag{14}$$

3.4. Sequential identification inverse method

For measurements where installed thermocouple inside the investigated body disturbs also surface temperature as it is very close to investigated surface HTC must be computed directly by an inverse method. Classical and very efficient sequential estimation proposed by Beck [27], which computes heat flux instead of HTC, has several limitations. Thus new sequential identification method was developed by Pohanka to solve such inverse problems. The basic principle of time-dependent boundary conditions determination (heat flux, HTC, and surface temperature) from measured transient temperature history is based on cooling (or heating) of heated (or cold) sample with thermocouple installed inside (see Figure 6). Let us assume one-dimensional inverse problem with 3D model involving installed thermocouple for simplicity:

- Known dimensions of the sample.
- Known thermal temperature-dependent material properties of the sample.
- Known temperature profile at the beginning of the cooling (usually constant).
- All surfaces are insulated except the cooled one.
- HTC is not dependent on position.

The sample is heated before starting the measurement. Cooling is applied on one surface and temperature response inside the sample is recorded. Time-dependent boundary conditions are computed using inverse technique from the measured temperature history (see Fig. 7). Cooling of more surfaces can also be investigated when more thermocouples is used.

This new proposed approach computes step by step (time step) heat transfer coefficients (HTC) on the investigated surface using measured temperature history inside the cooled or heated solid body. However, this method can be very easily changed to compute any kind of boundary conditions, e. g. heat flux. The method uses sequential estimation of the time varying boundary conditions and uses future time steps data to stabilize the ill-posed inverse problem [28]. To determine the unknown surface HTC at the current time t_m, the measured temperature responses T_m^* are compared with the computed temperature T_m from the forward solver using n future times steps

$$SSE = \sum_{i=m+1}^{m+n} \left(T_i^* - T_i\right)^2 . \tag{15}$$

Any forward solver can be used e. g. finite volume method described by Patankar [29]. The computational model should include drilled hole, whole internal structure of the embedded thermocouple, and temperature dependent material properties.

At time zero homogeneous temperature is in the sample and thereby zero heat flux and thereby zero HTC on all surfaces is assumed. Otherwise there cannot be homogeneous temperature. This can be done e. g. by heating in furnace after enough long time. If the initial temperature is not homogeneous some modification of the algorithm is necessary for the first time step.

The algorithm starts at time index zero when the HTC is equal to zero (see Fig. 10). The algorithm uses forward solver and it computes temperature response at thermocouple position for linearly changing (increasing or decreasing) HTC (see HTC1 and T1 computed in Fig. 10) over few time steps. These time steps are called future time steps n and five of them are used in Fig. 10–Fig. 11. Determination of minimum number of necessary future time steps to stabilize sequential algorithm is described in [28]. The computed and measured temperatures histories are compared using Eq. (15) the same one as for sequential Beck approach. The slope of linearly changing HTC defined as

$$v = \frac{\partial h}{\partial t} \tag{16}$$

should be changed until the minimum of SSE function in Eq. (15) is found. Such a minimum says that the computed temperature history matches the measured temperature history the best for used linearly changing HTC during n future time steps.

When the best slope of HTC is found the forward solver is used to compute temperature field in the next time step using the computed boundary conditions. The algorithm is repeated for next time steps until the end of recorded temperature history is reached (see Fig. 11). For k

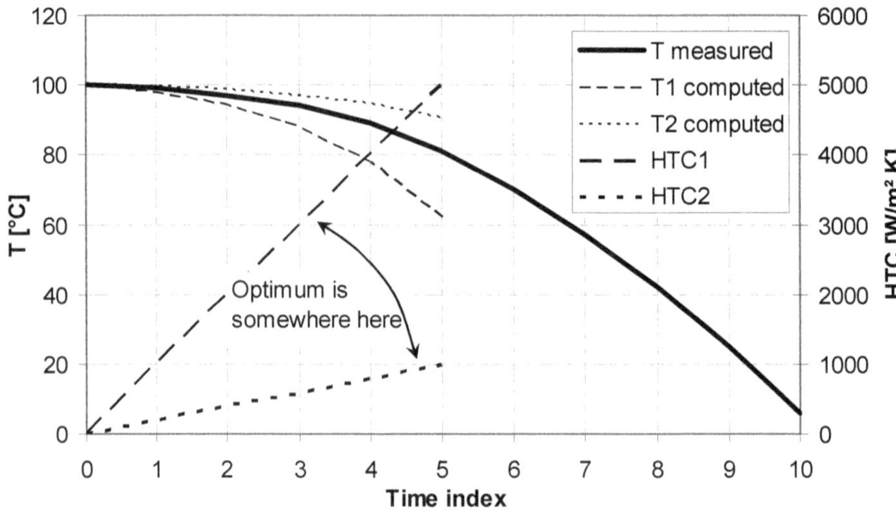

Figure 10. Measured temperature history and two computed temperature histories using two different slopes of HTC for n future time steps.

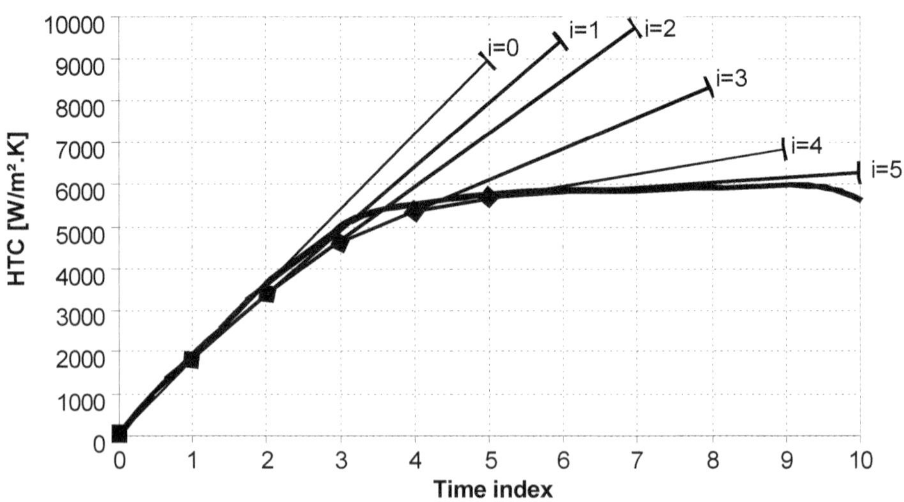

Figure 11. Real HTC and six optimum linearly changing HTC.

measured time steps only $k-n$ time steps can be computed owing to the use of future data. This method works perfectly when real HTC is almost linear in time. When the slope is abruptly changing the computed HTC curve is slightly smoother than the real one; the more future time steps are used the smoother is the computed curve of HTC (bigger difference between computed and real HTC) but the sequential identification inverse algorithm is more stable.

The SSE function described by Eq. (15) has only one minimum and is dependent only on one variable – slope of HTC (see Eq. 16). Even more the function is very close to parabolic function near the searched minimum because it is sum of square of temperature differences. Brent's optimization method [30], which uses inverse parabolic interpolation, is perfect candidate for finding the minimum of the SSE function in Eq. (15).

Brent's optimization method is based on parabolic interpolation and golden section. The searched minimum must be between two given points 1 and 2 (see Fig. 12). Convergence to a minimum is gained by inverse parabolic interpolation. Function values of the SSE function are computed only in few points. A parabola (dashed line) is drawn through the three original points 1, 3, 2 on the SSE function (solid line). The function is evaluated at the parabola's minimum, 4, which replaces point 1. A new parabola (dotted line) is drawn through points 3, 4, 2. The algorithm is repeated until the minimum with desired accuracy is found. If the three points are collinear the golden section [30] is used instead of parabolic interpolation.

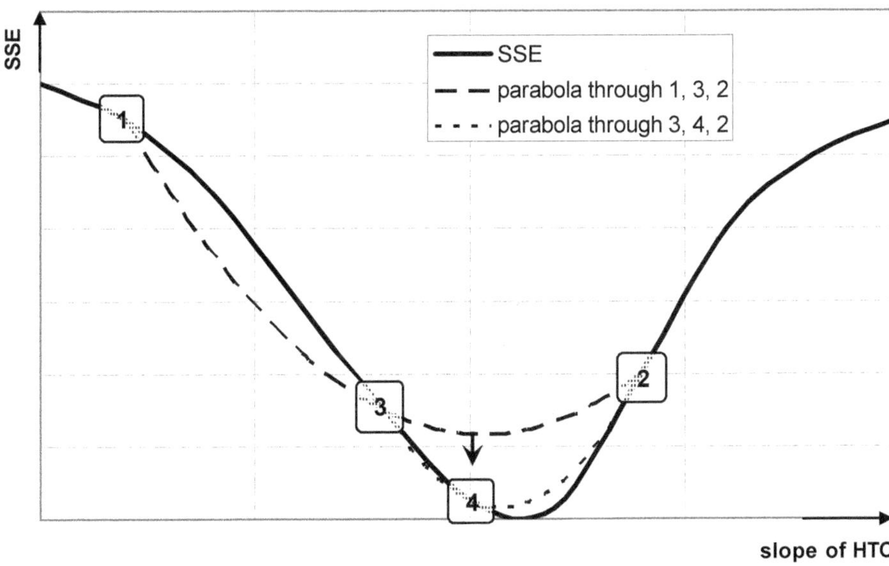

Figure 12. Convergence to a minimum by inverse parabolic interpolation.

3.5. Evaluation of boundary conditions

To demonstrate the procedure a real measurement is used. The cooling section consists of five rows of flat-jet nozzles. The heated sample passes repeatedly under the spraying nozzles. Several thermocouples in one row, which is perpendicular to the sample movement direction, were installed in the sample to be able to investigate also the cooling homogeneity across the sample. For simplicity we focus now on only one thermocouple, however, it is easy to do the procedure for all thermocouples. Temperature record from such measurement is shown in Figure 7. Recorded position (zigzag line) of the thermocouple is shown as well and demonstrates the repeated passes through cooling section. Using the inverse method boundary conditions were computed: surface temperature and HTC. All the shown lines are function of time; however, for numerical simulation we need HTC as function of position and surface temperature. We start with surface temperature from the measurement. See Figure 13 with shown surface temperature drawn using green line as function of position. The green lines represents surface temperatures through which the plate pass during experiment. In the place where the green line is shown we also know HTC from the measurement. HTC values are shown using the color scale. HTC values between green lines are interpolated. This chart shows HTC distribution as a function of surface temperature and position and is the key point for accurate numerical simulation. HTC values above the most top green line are extrapolated and are not accurate as there are no data available from measurement. We should avoid usage of these values during numerical simulation.

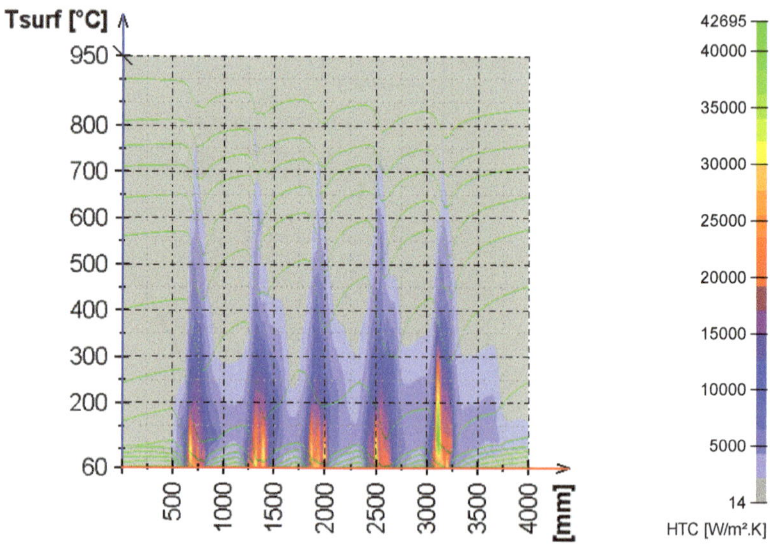

Figure 13. Prepared boundary conditions for numerical simulation from measurement shown in Fig. 7. Chart shows HTC as function dependent on position in cooling section in the direction of sample movement and on surface temperature of the cooled sample.

3.6. Numerical simulation

Having prepared boundary conditions we can do numerical simulation of cooling of products of various material properties and of various thicknesses. By repeating the boundary conditions we can simulate long cooling section with more rows of cooling nozzles. An example of such simulation is shown in Fig. 14 and is drawn in CCT. You can see computed temperature at the surface and at the center of the material. It is obvious that the cooling rate for the surface temperature is higher than in the center. The results are drawn in CCT diagram, however, you should not that the cooling rate is far away from constant. This is very important because the CCT diagram is only informative and the final structure has to be verified by experimental measurement. There are three major reasons why cooling rate is not constant. One is caused by passing product under separate row of nozzles. The passes are obvious from T surface curve in Fig. 14. You can see drops of temperature when the product is passing under spray followed by reheating due to the internal capacity of the heat in the product. The second reason is mentioned Leidenfrost point. You can see low cooling rate in the center up to 20 s as the surface temperature is above Leidenfrost point and after that cooling rate is increasing and reaching maximum which is almost triple in comparison to value above Leidenfrost point. Decreasing of cooling rate is followed as the surface temperature is getting closer to the temperature of water. The third reason is low diffusivity for big products. The product cannot be cooled down at the same cooling rate on the surface as in the center. The lower is the diffusivity and the bigger is the product the bigger difference is between the cooling rate on the surface and in the center.

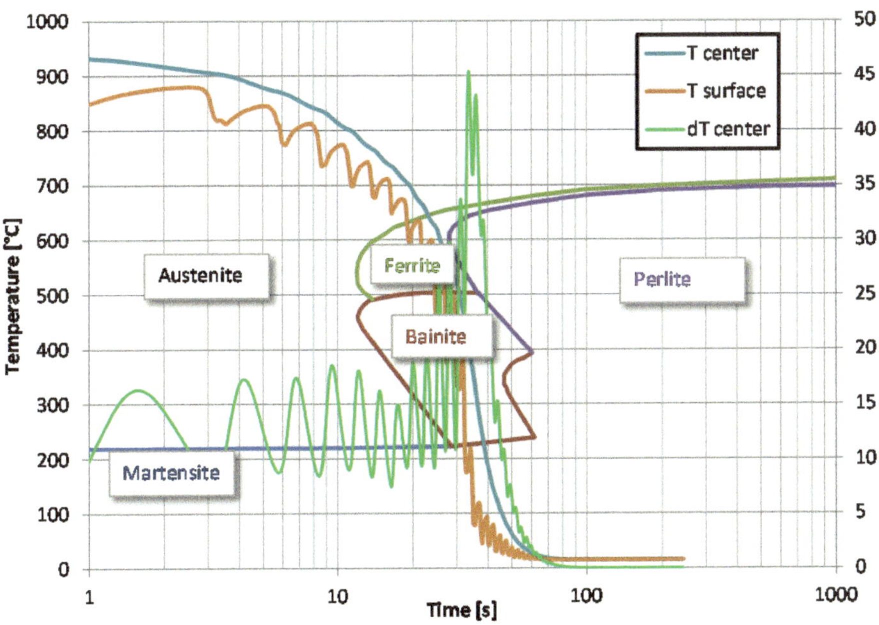

Figure 14. Simulation of cooling in the CCT diagram.

4. Verification conducted at pilot test bench

It is important to understand that cooling rate in cooling section in industrial application is far away from constant value using which CCT diagrams are obtained. Verification functionality of a newly designed cooling system prior to its plant implementation is essential. The design obtained by using the numerical model must be verified and fine-tuned by further full-scale cooling tests. Pieces of tube, rail, wire or plate of real dimensions with implemented thermocouples are tested in the designed cooling section. The length of a laboratory test bench shown in Fig. 8 and Fig. 9 is limited, hence the sample must be accelerated prior entering the cooling section, to a velocity normally used in a plant, and after pass through cooling section, the direction of movement is reversed. In this way, the sample moves several times under the cooling sections. This cooling process is controlled by computer to simulate running under the long cooling section used normally in the plant. Nozzles, pressures, and header configurations are tested. The design of the cooling and the pressures used are modified until the demanded temperature regime and final structure is obtained. The full-scale material samples are then cut for the tests of material properties and structure.

When heat treatment is performed on larger product such as rail, mainly its head, it is not possible to achieve same cooling rate at surface and in the center of rail head. The cooling rate near surface are much faster and even more reheating can appear and can cause very different material properties (see Fig. 15). As the rail head passed under the spray the surface temperature dropped fast and was followed by reheating due to the heat stored inside the head. The reheating caused lower hardness near the surface as shown in Fig. 14. The center of the head is harder because no reheating occurred in the bigger depth. To avoid this problem the cooling section should be modified. One solution is to use more row of nozzles with smaller row pitch and also nozzles with lower HTC. This can be achieved by smaller pressure or smaller nozzles. Also replacement of flat-jet nozzle by full-cone nozzles can be considered. The Leidenfrost temperature should be also considered. We should be above Leidenfrost temperature or below but definitely not near to avoid big different cooling rates for small changes in surface temperature.

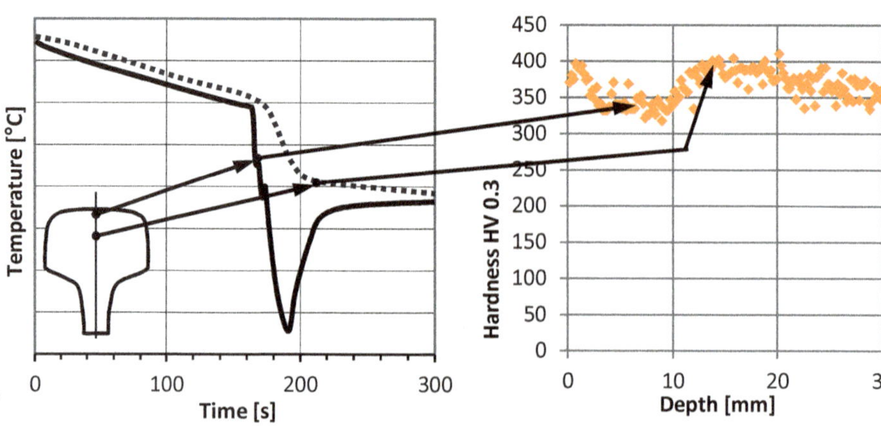

Figure 15. Measured temperature histories in a rail head in two depths and measured micro-hardness in rail head after heat treatment.

5. Concluding remarks

The design of cooling sections used for in-line heat treatment for hot rolling plants is very extensive work. It utilizes laboratory measurement, numerical modeling, inverse computations, and also pilot mill tests. The first step is the search of the best cooling regime for steels for which this is not yet known. The second step is to obtain a selection of technical means in order to guarantee obtaining the prescribed cooling rates. Nozzle configurations and cooling parameters are selected and controllability of the cooling section is checked. The final step of the design is a laboratory test using a full size sample simulating plant cooling.

Design based on laboratory measurement therefore minimizes the amount of expensive experimentation performed directly on the plant. Elimination of potential errors and enabling adjustment of control models in the plant is possible after the cooling process is tested in laboratory conditions.

Author details

Michal Pohanka and Petr Kotrbáček
Brno University of Technology, Heat transfer and Fluid Flow Laboratory, Brno, Czech Republic

6. References

[1] Raudenský, M.; Horský, J.; Hajduk, D.; Čecho, L. Interstand Cooling - Design, Control and Experience. *Journal of Metallurgical and Mining Industry*, 2010, No. 2, Vol. 3, pp. 193-202.

[2] Hnízdil, M.; Raudenský, M.; Horký, J.; Kotrbáček P.; Pohanka M. In-Line Heat Treatment and Hot Rolling. *In proc. AMPT 2010*, Paris, France, October 2010, pp. 563-568.

[3] Horký, J.; Raudenský, M.; Kotrbáček, P. Experimental study of long product cooling in hot rolling. *Journal of Materials Processing Technology*, August 1998, Vol. 80-81, pp. 337-340.

[4] Pohanka, M.; Bellerová, H.; Raudenský, M.; Experimental Technique for Heat Transfer Measurements on Fast Moving Sprayed Surfaces. *Journal of ASTM International*, Sept. 2010, Vol. 1523, pp. 3-15.

[5] Tikhonov, A. N.; Arsenin, V. Y. *Solution of Ill-Posed Problems*. Washington, D.C.: Winston, 1977. ISBN 0470991240.

[6] Mandrel, J. Use of the singular value decomposition in regression analysis. *Am. Stat.*, 1982, Vol. 36, pp. 15-24.

[7] Stloz, G. Jr. Numerical solutions to an inverse problem of heat conduction for simple shapes. *Int. J. Heat Transfer*, 1960, Vol. 82, pp. 20-26.

[8] Smith, G. D. *Numerical solution of partial differential equations*. UK: Oxford University Press, 1978. ISBN 0-198-59650-2.

[9] Beck, J. V. Nonlinear estimation applied to the nonlinear heat conduction problem. *Int. J. Heat and Mass Transfer*, 1970, Vol. 13, pp. 703-716.

[10] Beck. J. V.; Litkouhi, B.; St. Clair, C. R. Jr. Efficient sequential solution of the nonlinear inverse heat conduction problem. *J. Numerical Heat Transfer*, 1982, Vol. 5, pp. 275-286.

[11] Bass, B. R. Applications of the finite element to the inverse heat conduction problem using Beck's second method. *J. Eng. Ind.*, 1980, Vol. 102, pp. 168-176.

[12] Burggraf, O. R. An exact solution of the inverse problem in heat conduction theory and applications. *Int. J. Heat Transfer*, 1964, Vol. 86C, pp. 373-382.

[13] Imber, M.; Khan, J. Prediction of transient temperature distributions with embedded thermocouples. *Al AA J.*, 1972, Vol. 10, pp. 784-789.

[14] Lengford, D. New analytic solutions of the one-dimensional heat equation for temperature and heat flow rate both prescribed at the same fixed boundary (with applications to the phase change problem). *Q. Appl. Math.*, 1967, Vol. 24, pp. 315-322.

[15] Grysa, K.; Cialkowski, M. J.; Kaminski, H. An inverse temperature filed problem of the theory of thermal stresses. *Nucl. Eng. Des.*, 1981, Vol. 64, pp. 169-184.

[16] Sláma, L.; Raudenský, M.; Horský, J.; Březina, T.; Krejsa, J. *Evaluation of quenching test of rotating roll with unknown time constant of sensor using genetic algorithm*. Int. Conf. Mendel, Brno, 1996.

[17] Raudenský, M.; Pohanka, M.; Horský, J. Combined inverse heat conduction method for highly transient processes. In *Advanced computational methods in heat transfer VII*, Halkidiki: WIT Press, 2002, pp. 35–42. ISBN 1-85312-9062.

[18] Dumek, V.; Grove, T.; Raudenský, M.; Krejsa, J. Novel approaches to the IHCP: Neural networks. In *Int. symposium on inverse problems - Inverse problems in Engineering Mechanics*, Paris, 1994, pp. 411-416.

[19] Krejsa, J.; Sláma, L.; Horský, J.; Raudenský, M.; Pátiková, B. The comparison of traditional and non-classical methods solving the inverse heat conduction problem. In *Int. Conf. Advanced Computational Methods in Heat Transfer*, Udine, July 1996, pp. 451-460.

[20] Pohanka, M.; Raudenský, M.; Horský, J. Attainment of more precise parameters of a mathematical model for cooling flat and cylindrical hot surfaces by nozzles. In *Advanced computational methods in heat transfer VI*. Madrid: WIT Press, 2000, pp. 627-635. ISBN 1-85312-818-X.

[21] Incropera, F. P.; DeWitt, D. P. *Fundamentals of Heat and Mass Transfer*. 4th ed. New York: Wiley, 1996. ISBN 0-471-30460-3.

[22] Kakac, S.; Yener, Y. *Heat Conduction*. New York: Hemisphere Publishing, 1985.

[23] Poulikakos, D. *Conduction Heat Transfer*. Englewood Cliffs, NJ: Prentice-Hall, 1994.

[24] Voller, V. R.; Swaminathan, C. R. Fixed grid techniques for phase change problems: A review. *Int. J. for Numerical Methods in Engineering*, 1990, Vol. 30, pp. 875-898.

[25] Song, R.; Dhatt, G.; Cheikh, A. B. Thermo-mechanical finite element model of casting systems. *Int. J. for Numerical Methods in Engineering*, 1990, Vol. 30, pp. 579-599.

[26] Comini, G.; Giudice, S. D.; Saro, O. A conservative alghorithm for multidimensional conduction phase change. *Int. J. for Numerical Methods in Engineering*, 1990, Vol. 30, pp. 697-709.

[27] Beck, J. V.; Blackwell, B.; Charles, R. C. Inverse Heat Conduction: *Ill-posed Problems*. New York: Wiley, 1985. ISBN 0-471-08319-4.

[28] Pohanka, M. Limitation of thermal inverse algorithm and boundary conditions reconstruction for very fast changes on boundary. In *Engineering mechanics 2007*. Svratka (Czech Republic), 2007, pp. 229–230. ISBN 978-80-87012-06-2.

[29] Patankar, S. V. *Numerical Heat Transfer and Fluid Flow*. Hemisphere Publishing Corporation, 1980. ISBN 0-891-16522-3.

[30] William, H. P.; Saul A. T.; William, T. V.; Brian, P. F. *Numerical Recipes in C*. 2nd ed. 1997. ISBN 0-521-43108-5.

Chapter 4

Deformation Reduction of Bearing Rings by Modification of Heat Treating

Anton Panda, Jozef Jurko and Iveta Pandová

Additional information is available at the end of the chapter

1. Introduction

Heat treatment of bearing rings implies the risk of deformations caused by internal tension. In order to eliminate internal tension, hardening is followed by tempering. In general, tempering will remove the tension. However, this tempering is not sufficient for the so-called thin-walled rings AX series bearing rings to eliminate the tension. This article discusses how to effectively eliminate occurrence of tensions in thin-walled bearing rings made from 100Cr6 by optimising their heat treatment. Results have been verified by experiments.

Manufacture of roller bearings (figure 1) is a challenging production process. Even though its specific manufacturing operations are widely known and established, some operations in bearing manufacture must be performed within narrow tolerances ranging from only a few micrometres to comply with requirements of tolerance analysis done before the parts are manufactured to ensure that clients receive a quality product that influences safety of plant operation, therefore safety of people. The manufacturing comprises a number of operations needed to produce rings, rolling elements and cages. It includes hammering of forgings at the beginning, turning, heat treatment, cutting, forming, grinding, washing of parts, their description, assembling bearing components and packaging. A number of preventive, intra-operational and final inspections and dimensional, chemical, metallurgic, endurance and other tests are carried out during the manufacturing process [ZVL & ZKL, 1996; ZVL, 2008]. Customers assemble these products in common applications with standard requirements on bearings, but sometimes they have special requirements either for the bearing as a whole, or for any of its parts. In this case it is not sufficient to implement only standard methods and working practices; they have to be modified or optimised. One of such requirements was a request from one of great important American and Deutschland company to produce bearings needed to place a rotor for an axial piston hydroelectric generating set with an inclined plate for one of its tractors. This application has its particulars. It was necessary to modify the outer ring, inner ring and rolling elements. Mentioned company is considered to be a "Mercedes" in the field of production of tractors, harvesters and similar equipment. The demands on parts are therefore accordingly high. This article will not discuss all the modifications necessary to adjust the technological process of production in order to meet customer's requirements, although they are all interrelated, influence each other and represent a desired outcome of the targeted modification. We will deal only with the parameter that can be affected by heat treatment of the material in order to avoid unwanted deformations in the material and comply with a stricter requirement on ovality for inner rings.

Papers of authors [Jech, 1983; Panda et al., 2011; Vasilko, 1998] address processes of heat treatment of bearing components with related checks and tests (dimensional, chemical, metrological and endurance tests) and technological aspects of production of bearing rings are discussed in [ZVL & ZKL, 1996; ZVL, 2008].

The goal of the work was to modify the tempering process to eliminate or preclude undesirable deformations of the material and to ensure compliance with the stricter requirements for ovality of inner rings (figure 2).

Figure 1. Tapered roller bearing

Figure 2. Inner ring of Tapered roller bearing

2. Ovality measuring on production plant

Technologically, ovality results from non-symmetrical distribution of internal tensions before hardening and uneven heating and cooling. Ovality is a certain type of circularity deviation.

Since 1950s, devices using very accurate rotational tables or spindles are used in modern industry to measure circularity. The measuring base in this method is the axis of the component to be measured. The measuring device ensures very high accuracy, often better than 0.1 μm. However, it requires time-consuming preparations including centring and alignment of the component, therefore it is intended rather for laboratory measurements. These measuring devices are rarely used in production, also due to their low measuring capacity.

Ovality is checked at the output from the heat treatment line. A special diameter gauge (Figure 3) that included a dial deviation meter was used to measure ovality (table 1). Ovality was measured on three rings so that an ovality deviation was recorded after each rotation by 10°. The following procedure was applied. An bearing ring was inserted into the gauge. The ring was rotated manually in 10° increments. A value was read from the deviation meter after each rotation. If the deviation was higher than 0.2 mm, the ring being measured was discarded. It did not meet quality criteria and could not be passed on for subsequent hard machining.

A high deviation in ovality may indicate uneven heating or cooling during heat treatment. Ovality measurement was made on three inner rings selected randomly from different batches. Ring No. 1 shows acceptable ovality up to 0.2 mm in maximum. Subsequent grinding operations can reduce or remove this deviation from circularity. Ring No. 2 shows ovality over 0.2 mm, so it is not suitable for further machining. It must be discarded. The lowest ovality value can be seen in ring No.3. For ring No. 2, it is necessary to identify the cause of such a significant deviation in cylindricality. The whole batch the ring No. 2 comes from should be checked.

Ring rotation (°)		0	10	20	30	40	50	60	70	80	90	100
Ovality values measured for the rings	Ring no. 1	0.02	0.04	0.09	0.13	0.16	0.17	0.17	0.18	0.19	0.2	0.19
	Ring No. 2	0.21	0.2	0.19	0.18	0.17	0.16	0.13	0.09	0.04	0.03	0.03
	Ring No. 3	0.09	0.05	0.02	0	0.02	0.04	0.08	0.09	0.1	0.11	0.12

Table 1. Measured values of ovality

110	120	130	140	150	160	170	180	190	200	210	220	230
0.18	0.17	0.17	0.16	0.13	0.09	0.04	0.02	0.03	0.04	0.09	0.13	0.16
0.04	0.09	0.13	0.16	0.17	0.18	0.19	0.21	0.2	0.19	0.18	0.17	0.17
0.13	0.14	0.15	0.15	0.14	0.13	0.12	0.11	0.1	0.09	0.05	0.02	0.04

Table 2. Measured values of ovality - continuation

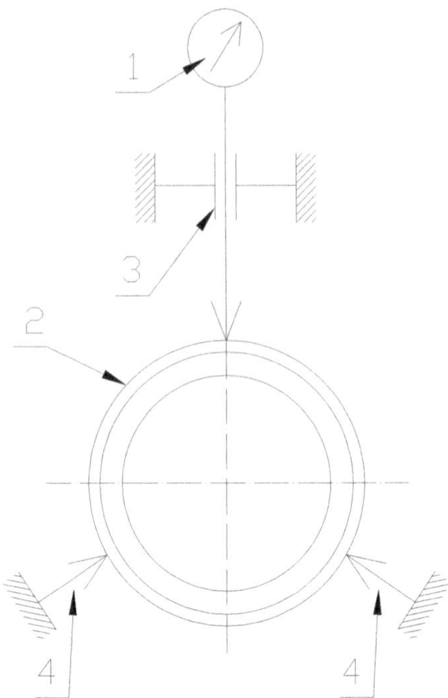

Figure 3. The principle of measuring ovality of outer shape of the outer ring of a roller bearing. 1 - dial deviation meter, 2 - outer ring, 3 - moving measuring contact line, 4 - supporting parts

240	250	260	270	280	290	300	310	320	330	340	350	360
0.17	0.17	0.18	0.19	0.2	0.19	0.18	0.17	0.17	0.16	0.13	0.09	0.05
0.16	0.13	0.09	0.05	0.02	0.04	0.09	0.13	0.16	0.18	0.19	0.2	0.21
0.08	0.1	0.1	0.11	0.12	0.13	0.14	0.15	0.15	0.14	0.13	0.12	0.1

Table 3. Measured values of ovality - continuation

3. Customer's requirements

In principle, the customer required a bearing that would withstand higher axial loads without any major run-in and with reduced ring ovality due to the placement of the rotor in an axial piston hydroelectric generating set. Table 4 shows a comparison of basic modified parameters between the standard design (Standard) and the design required by the customer (Special). As mentioned before, this paper will discuss only the issue of reducing ovality in the inner ring. Finding a solution for such task is even more difficult because the bearing is a thin-walled bearing (AX series) that is much more sensitive to material deformations and ring ovality compared to other bearings with a more favourable ratio of ring thickness and width.

Bearing part and parameter	32010AX Standard	32010AX Special
Inner ring		
- Raceway run-out to the hole [mm]	0,007	0,005
- Support face run-out to the main face [mm]	0,007	0,005
- Raceway roughness [μm]	0,20	0,15
- Support face roughness [μm]	0,20	0,07
- Waviness [μm]	0,35	0,25
- Ovality [mm]	0,006	0,003

Table 4. Comparison of basic parameters of 32010AX bearings between "Standard" and "Special" designs

4. Experimental section

Material deformations and ring ovality are caused by internal tension generated during machining and heat treatment operations. To process a bearing ring by turning, it has to be fixed at three points. The fixing is done pneumatically. A deformation may occur due to poor fixing, or due to a failure to follow technological conditions, when more material is removed. When rings are grinded after heat treatment, similar undesired deformations occur, if technological conditions are not followed. Major deformations occur even during the heat treatment itself, i.e. when the rings are hardened, due to uneven heating and cooling. Deformation that appeared after heat treatment are then reproduced at subsequent grinding, worsening this effect even more.

The customer accepted only 0.003 mm of stricter ovality for the outer ring after grinding, compared to the standard prescribed value of 0.006 mm (see Table 4), which tightens the requirements by 50%. To achieve this final ovality of rings after grinding , then the ovality of rings after hardening can be no more than 0.1 mm, which is also a stricter value compared to the standard requirement of 0.2 mm.

Ovality is defined as the difference in diameters measured in one plane perpendicular to each other. This means that, for example, the maximum diameter D_{max} is measured first – i.e. the maximum value is found when the ring is turned, then the ring is rotated by 90 ° and the second, minimum diameter D_{min} is measured. The outer shape of the ring should be close to a circle, but in fact, the outer ring is elliptical in shape. Our aim is to keep this ovality as small as possible.

$$\text{Ovality} = D_{max} - D_{min}$$

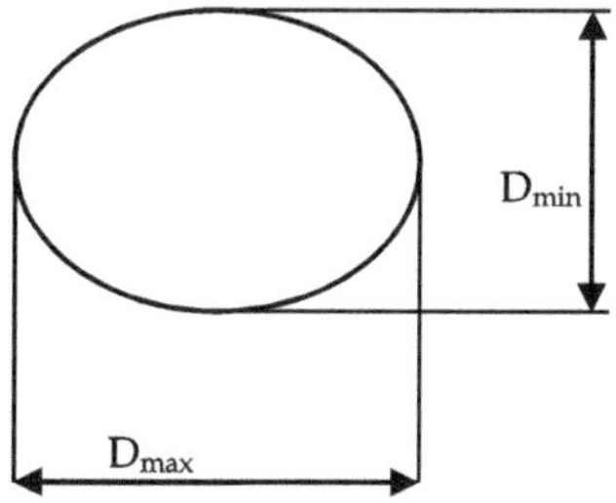

Figure 4. Definition of ovality

4.1. Hardening and tempering of bearing components

Steel 100Cr6 with the following chemical composition (values in % by weight)[2] was used to manufacture bearing rings and rolling elements: C=0.9-1.1; Mn = 0.3-0.5; Si = 0.15-0.35; Cr = 1.3-1.65; P=max 0.027; S=max 0.03; Ni=max 0.3; Cu=max 0.25; Ni+Cu=max 0.5. Desired mechanical properties of roller bearings are obtained by hardening and tempering their components at low temperature. The required hardness of bearing components is achieved by hardening and tempering is used to reduce internal tension and fragility of the hardened bearing steel.

The resulting mechanical properties are then determined primarily by the microstructural state, distribution of internal tensions before hardening, uniformity of heating to austenitizing temperature, austenitizing conditions and cooling down from the austenitizing temperature.

The method of heating affects resultant oxidation and surface decarburization. Local overheating and imperfect soaking must be avoided during heating, because they lead to cracks formed during cooling. Austenitizing conditions, i.e. austenitizing temperature and dwell at the austenitizing temperature, affect quality of hardened bearing components. Dwell time selected depends on the shape and material of the component, its heating method and baseline microstructure. The dwell at the austenitizing temperature has a lower effect than the temperature value.

The outcomes of hardening depend also on the speed of heat dissipation. For bearing steel, the cooling rate must be very high for a temperature range of approximately 650 °C and below. The cooling efficiency of different environments depends mainly on thermal conductivity, specific heat, evaporating heat, viscosity of the hardening environment and amount of dissolved gases. The cooling process in water is very fast and is used to harden bearing balls. Different ingredients are added into quenching water; some of which increase the cooling capacity and some of them slow it down.

Due to the lower cooling rate, thus a smaller temperature gradient between the surface and core of the component being hardened, it is more convenient to cool bearing components in mineral oils rather than in water. The most suitable medium for a common hardening environment in terms of cooling rate is J4 bearing oil that can achieve the maximum cooling rate of 65 °C/s at surface temperature of 550 °C. Cooling in an AS140 salt solution (a mixture of KNO_3 $NaNO_3$) is used to equalise the temperature at 150 °C between the surface and core of the component. After this cooling, the component continues to cool down in oil or is finally cooled down in water. Increased cooling rate causes higher susceptibility of the hardened component to develop cracks, resulting from the higher temperature difference between the surface and the core of the hardened component, creating internal tensions.

Hardening and tempering of rings is one of the most important operations in production of roller bearings and it should ensure dimensional stability in addition to the required hardness of 60 to 63 HRC. Dimensional stability is necessary for subsequent technological operations needed to achieve the correct geometry of a finished bearing and stability of these dimensions in long-term operation. When bearing steels are hardened, martensite or a structure with a specific volume different than the original martensite is formed.

Internal tensions occur because various structures develop in various volumes and at different stages in terms of time and temperature due to the temperature gradient. Tensions arising from differences in temperature between the component surface and its core are referred to as thermal tensions. Structural tensions originate from the difference in specific volumes of the initial austenite and martensite formed or in other stages.

Temperature tensions can be affected by reducing or extending the process of heating, especially by preheating. Structural tensions depend on chemical composition of steel and course of the heat treatment.

Dimensional changes in chrome bearing steel increase with a higher hardening temperature. Tempering will reduce the increase in volume, and the reduction is higher with a higher hardening temperature.

In bearing rings, hardening and tempering influence their ovality. Ovality is related with volume changes only a little. It originates from the technology, resulting from an uneven distribution of internal tensions before hardening and uneven heating and cooling.

Cooling rate in the hardening process has an impact on volume change in components. The higher is the rate the higher is the deformation and the higher is also the difference in length between the states after hardening and tempering. The dimensional changes (Fig. 5) that occur after heat treatment are caused by the lack of stability of the microstructure of hardened and tempered bearing steel in the given operating conditions [Vasilko, 1998]. This is the result of permanent changes in instable structural stages of martensite and residual austenite. Therefore, stabilization of dimensions in hardened and tempered bearing components depends on the degree of super saturation of a solid solution - martensite and the residual austenite content, i.e. on the microstructure as well as on operating conditions, temperature, time and tensions.

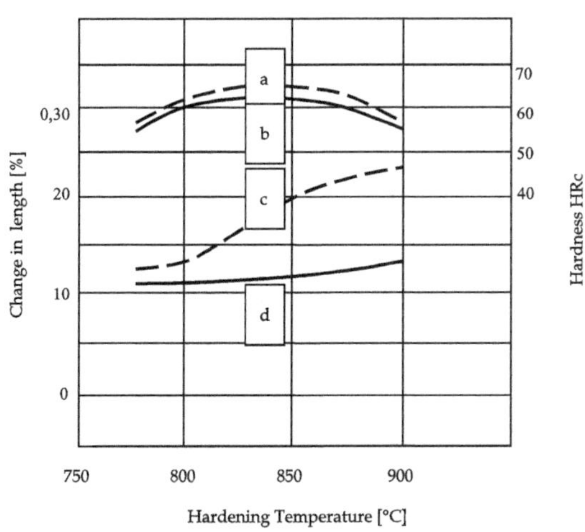

Figure 5. Effect of hardening temperature on change in length and hardness after hardening and tempering [8]: a – hardness after hardening; b – hardness after hardening and tempering 150°C; c – change in length after hardening; d - change in length after hardening and tempering 150°C

With this heat treatment, we try to get a fine martensitic structure of components, as shown in Fig. 6 - microstructure of 100Cr6 steel; properly tempered; martensite and fine, evenly distributed carbides.

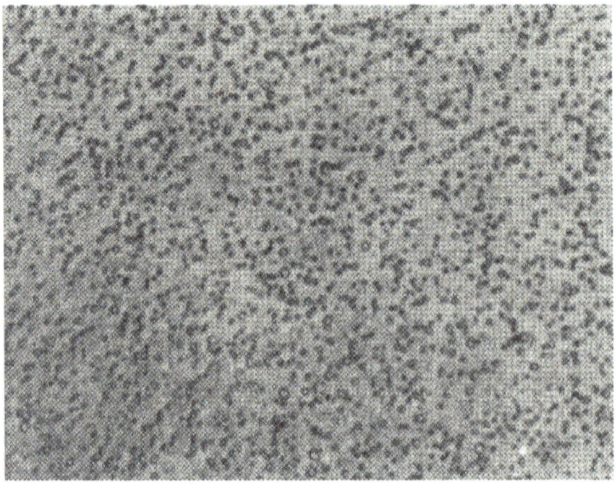

Figure 6. Microstructure of 100Cr6 bearing steel formed by martensite and fine carbides

To harden and temper bearing components, furnace equipment is used that can increase the level and quality of heat treatment and improve work productivity (Fig. 7) [Vasilko, 1998]. The company, where optimisation was implemented, uses a renovated hardening line, later fitted with computer control, and tempering is done on PP017/50 device (Fig. 8) [ZVL, 2008]. Furnace equipment can be operated also by the computer system, making easier the process of controlling and inspecting the heat treatment. Records on various parameters, such as temperature and time, are kept, making it possible to get back to them even after a longer period of time (Table 5) [ZVL, 2008].

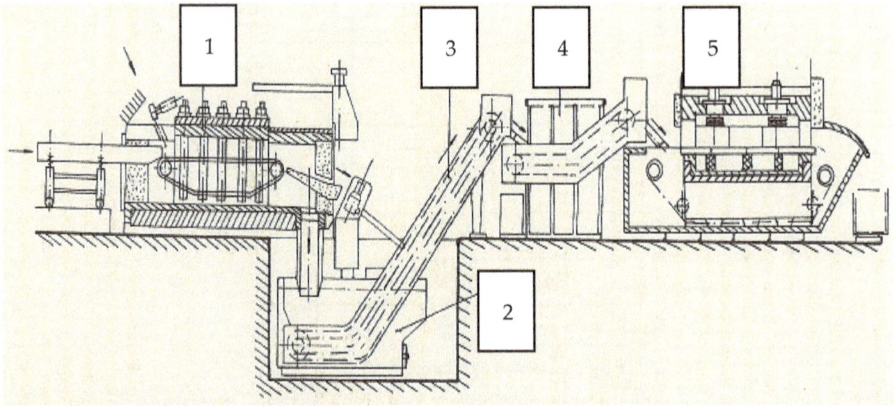

Figure 7. Diagram of furnace equipment for hardening and tempering bearing components (1 - hardening furnace, 2 - hardening tank, 3 - carrier, 4 - washing machine, 5 - tempering furnace).

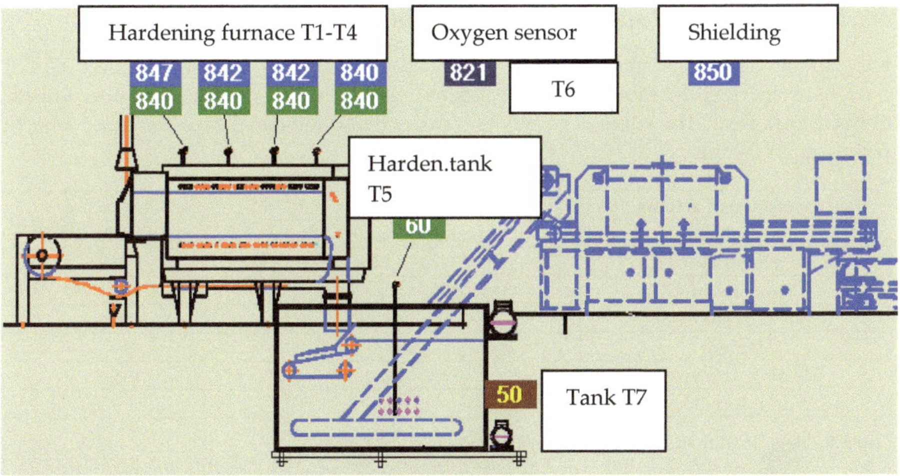

Figure 8. Diagram of the computer-controlled hardening line

Temperature – hardening furnace [°C]				Temperature - hardening tank [°C]	Temperature–exhaust gas oxygen sen. [°C]	Tempera-ture - tank [°C]	Time [hour:min]
T1	T2	T3	T4	T5	T6	T7	t
842	842	842	844	57	822	50	9:49
834	842	843	845	57	822	50	9:50
832	842	843	845	57	822	50	9:51
832	841	843	844	57	822	50	9:52
835	841	842	842	57	822	50	9:53
838	841	842	840	57	821	50	9:54
841	841	841	839	57	822	50	9:55
843	841	840	840	57	821	50	9:56
842	841	840	842	57	821	50	9:57
838	841	841	844	57	821	50	9:58
836	841	843	844	57	821	50	9:59
836	841	843	844	57	821	50	10:00
838	841	843	842	57	821	50	10:01
841	841	843	841	57	821	50	10:02
843	841	842	839	57	821	50	10:03
841	841	841	839	57	821	50	10:04
838	841	841	842	57	820	50	10:05
836	841	840	844	57	820	50	10:06
837	841	840	844	57	820	50	10:07
839	841	842	843	57	821	50	10:08
842	841	843	842	57	820	50	10:09

Table 5. Sample output from the computer-controlled hardening line with measuring data

4.2. Heat treatment of standard bearing rings

For standard bearings, the rings are hardened and tempered in accordance with the conditions specified in Table 6. Heat treatment is followed by grinding and super finishing of functional areas. The value of ovality of hardened and tempered rings achieved after this processing is 0.2 mm.

Description of actions	Hardening	Tempering
Device Name	hardening furnace Ø 100	furnace PP017 / 50
Zone temperature [°C]	840 ± 5	170 ± 5
Oil temperature [°C]	50 – 80	–
Method of placement	1 row	freely
Variator	3 – 5	–
Heating time [min]	60 – 80	155 ± 5
Output [pcs/hour]	1 080	1 080

Table 6. Heat Treatment Technological Procedure

Hardness tests are done on selected pieces after hardening and tempering. The required hardness after hardening is 63.5 to 65.5 HRc. The required hardness after tempering is 60 to 63 HRc. After tempering, the bearing rings are inspected for ovality and microstructure.

When bearing material is being heat treated, oxidation occurs at common heating. To prevent oxidation of the surface, the heat treatment is done in a controlled atmosphere consisting of nitrogen.

4.3. Optimised heat treatment and grinding mode

The heat treatment method (Table 6) followed by grinding as described in 4.2 is not sufficient to achieve the required ovality of 0.1 mm needed by the customer for the special bearings. It was necessary to develop, technologically master and verify a different, additional method of heat treatment and subsequent grinding of the rings, which would guarantee lower internal tensions, deformations and ovality values, whereas the required hardness of the rings should remain unchanged. Tempering is known to be used to eliminate internal tensions. This had been already done after hardening (see Section 4.2, Table 6), but resultant values of ovality were inadequate. It is also known that hardness of tempered parts decreases when tempering is done as described in Table 6, so any further tempering using this process was not possible.

Considering these facts and drawing from our experience, a procedure was suggested to ensure lower levels of ovality from 0.2 to 0.1 mm:

1. Heat treatment (hardening and tempering) as in 4.2, Table 6
2. Pre-grinding of functional surfaces in bearing rings, but not to reach the final value; only with a partial use of the total allowances for grinding – rough grinding
3. Re-tempering of the rings using the 155 °C/65′ process, i.e. under different technological conditions than in 4.2. The technological conditions – see Table 7.
4. Fine grinding of functional surfaces of the rings

Description of action	Tempering
Device Name	PP017 / 50
Zone temperature [°C]	155 ± 5
Method of placement	freely
Heating time [min]	65 ± 5
Output [pcs/hour]	1 080

Table 7. Technological procedure of additional tempering

Table 5 shows that two parameters have been modified for additional tempering. The first one is the tempering temperature, which is now lower, i.e. 155±5 °C. This is a substantial change that will ensure lower internal tensions and ovality reduced by 50% against the standard design. The second one is the heating time, which is 65±5 minutes. It is a sufficient time. However, if a longer time was preserved, the effect would be the same without any risk of lower hardness values. Tempering, referred to as "artificial ageing" is followed by hardness tests on selected pieces. The required hardness after tempering remains constant of 60 to 63 HRc.

Ovality on finished rings was measured using the Talyrond 73 device, see Figure 9. The measuring device ensures very high accuracy, better than 0.1 μm.

5. Result and discussion

Hardening and tempering has an impact on ovality in rings. The dimensional changes that occur after heat treatment are caused by the lack of stability of the microstructure of hardened and tempered bearing steel in the given operating conditions [Vasilko, 1998]. The aim of the heat treatment was to obtain a fine martensitic structure of components. A microstructure composed mainly from martensite is formed by hardening components made from bearing steel. The martensitic microstructure usually contains a low percentage of residual austenite in terms of volume Rate of heating adjusted to the hardening temperature, or inclusion of pre-heating, can affect the value and distribution of thermal tension in the bearing ring being heated. Structural tensions that arise during heat treatment by hardening and tempering are determined by the chemical concept of the steel used and heat treatment process parameters. Dimensional changes that occur during the tempering process may be a consequence of ε-carbide precipitation, decomposition of residual austenite, cementite precipitation, dislocation substructures welded together and re-distribution of residual tensions after mechanical processing [Perez et al., 2009].

To ensure production requirements on special bearings for this customer (see Table 4) it was necessary to optimise both the dimensional parameters and the heat treatment method and subsequent machining in order to reduce internal tensions. Modified heat treatment mode and subsequent fine grinding made it possible to fulfil the requirement of lower ovality from 0.006 to 0.003 mm. The useful value of the bearings improved, too, with these modifications in heat treatment and adjusted dimensional parameters:

- reduced friction and tear and wear
- lower noise of the equipment when these bearings are used
- lower operating temperature even when running the equipment in
- minimum run-in time
- possibility to set initial stretch more reliably
- running smooth even in the first hours of operation
- slight increase in dynamic load (due to optimised shape of ring and tapered roll raceways)

Figure 9. Measuring ovality using the Talyrond 73 measuring device

Extract from a chemical composition protocol - chemical analysis of 100Cr6 used is shown in Table 8. The material corresponds to the prescribed values.

Tested element	Prescribed values	Measured values
C	0,9 – 1,1	1,02
Mn	0,3 – 0,5	0,35
Si	0,15 – 0,35	0,28
Cr	1,3 – 1,65	1,51
P	max. 0,027	0,01
S	max. 0,03	0,014
Ni	max. 0,30	0,06
Cu	max. 0,25	0,09
Ni+Cu	max. 0,5	0,15

Table 8. Chemical composition of 100Cr6 bearing material in % by weight

Extract from a metallographic analysis protocol under DIN 17230, DIN 50602, SEP 1520, and corporate standard is shown in Table 9. The material corresponds to the prescribed values.

Tested property	Max. prescribed values	Measured values
HRc hardness	60 – 63	62
Post-grinding heating	none parts	none parts
Microstructure	3 – 6	5
Carbide mesh	5,3	5,2
Carbide streakiness - closed	6,2	6,0
Carbide streakiness - free	7,3	7,1
Sulphides – SS	1,3	1,2
Oxides – OA	3,3	3,1
Oxides – OS	6,2	6,0
Oxides – OG	8,3	8,1

Table 9. Metallographic analysis of 100Cr6 bearing material used

When a verification series was manufactured, the bearings were tested for endurance using a ZT1 testing station. There are 20 bearings tested simultaneously; up to the fifth discarded bearing; 90% of the bearings must withstand 1 million revolutions. Basic dynamic load rating prescribed in the catalogue is assessed. The attained value was 201% compared to the catalogue.

6. Conclusion

The herein described method of achieving lower internal tensions with effect on reduced deformations and ring ovality made it possible to attain ovality of finished rings of 0.003 mm compared to the standard requirements on ovality of 0.006 mm. This result was achieved with a satisfactory assessment of chemical, metrological and metallographic analysis with a further effect on improved endurance of bearings. This manufacturing method has been tested several times since then and the desired effect was confirmed. Subsequently, this procedure has been applied also to other types of the so-called thin-walled bearings for others customers, in particular from the automotive industry, who demanded stricter ovality values. The possibility to use this technology has been confirmed in all cases.

Note: Compliance with and check of prescribed processing condition are essential for heat treatment. An appropriate way how to improve checks during the heat treatment is to use a computer-controlled hardening line. Obviously, this incurs higher input costs. According to information from the company where this optimised method of heat treatment was implemented, the system started to be used for a typical hardening line and the benefit in terms of quality exceeded their expectations.

Author details

Anton Panda, Jozef Jurko and Iveta Pandová
Technical University of Košice, Slovak Republic

Acknowledgement

The authors would like to thank in words the grant agency for supporting research work and cofinancing the projects: VEGA #1/0047/2010.

7. References

Blecharz, P.; Štverková, H. Product Quality and Customer Benefit. International Symposium on Applied Economics, Business and Development, Dalian, China. Communication in Computer and Information Science. BERLIN, SPRINGER-VERLAG, 2011, Volume 208, pp. 382-388. ISSN 1865-0929, ISBN 978-3-642-23022-6.

Firm ZVL & ZKL Publ. No. TKS 1/96 S. (1996). Assembly, Disassembly and Failures of Roller Bearings, Management Art, 56 p., Žilina.

Firm ZVL AUTO spol. s r.o. Prešov (2008), Internal documentationfor tapered roller bearings production, Prešov

Jech, J. (1983). Thermal Processing of Steel. 384 p., SNTL Praha.

Panda, A. (2011). Automotiv production from development to serial production. Monograph, 212 p. FVT TU Košice, ISBN 978–80–553–0636–0, Prešov.

Panda, A.; Dupláк, J. & Jurko, J. (2011). Analytical expression of T-vc dependence in standard ISO 3685 for cutting ceramic. In: Key Engineering Materials, 317-322 p. Vol.480-481. Trans Tech Publications, ISSN 1013-9826, Zurich, Switzerland.

Panda, A.; Dupláк, J. & JURKO, J. (2011). Analytical expression of *T-vc* dependence in standard ISO 3685 for sintered carbide. *In: International Conference on Computer Science and Automation Engineering, CSAE 2011,*10-12.6.2011, Shanghai, art. no. 5952440, pp. 135-139, Publisher: Institute of Electrical and Electronics Engineers, ISBN 978-142448725-7, Shanghai, China.

Panda, A.; Dupláк, J.; Jurko, J. & Behún, M. (2012). Comprehensive Identification of Sintered Carbide Durability in Machining Process of Bearings Steel 100CrMn6. In: Advanced Materials Research, Trans Tech Publications, vol. 340, p.30-33, ISSN 1022-6680, Zurich, Switzerland.

Panda, A.; Jurko, J.; Džupon, M. & Pandová, I. (2011). Optimalization of Heat Treatment Bearings Rings With Goal to Eliminate Deformation of Material. Chemické listy, Material in Engineering Practice 2011, vol. 105, Herľany, p.459-461, Special issue, Asociation of Czech chemical society Praha, HF TU Košice, ISSN 0009-2770, Praha.

Panda, A.; Jurko, J. & Gajdoš, M. (2009). Accompanying phenomena in the cutting zone machinability during turning of stainless steels. International Journal Machining and Machinability of Materials, INDERSCENCE, ISSN 1748-5711, Switzerland.

Panda, A.; Jurko, J. & Gajdoš, M. (2011). Study of changes under the machined surface and accompanying phenomena in the cutting zone during drilling of stainless steels with low carbon content, 113-117 p. Metallurgy No.2, vol. 50, ISSN 0543-5846 Zagreb, Croatia.

Panda, A.; Vasilko, K. & Dupláк, J. (2011). Evaluation of *T-vc* dependence in standard ISO 3685 for selected cutting materials. *In: Asia-Pacific Conference on Wearable Computing Systems,* pp. 129-132, *APWCS2011, 19-20.3.2011,* vol.2, Changsha, China, IEEE, ISBN 978-1-4244-9870-3, Changsha, China.

Perez, M.; Sidoroff, Ch.; Vincent, A. & Esnouf, C. (2009). Microstructural evolution of martensitic 100Cr6 bearing steel during tempering: From thermoelectric power measurements to the prediction of dimensional changes. Acta Materialia 57, pp. 3170–3181.

Vasilko, K. (1988). Roller Bearings, 540 p., Alfa, Bratislava.

Chapter 5

Thermochemical Treatment of Metals

Frank Czerwinski

Additional information is available at the end of the chapter

1. Introduction

Surface engineering represents the technically attractive and economically viable method aimed at improving the superficial layer of materials. Since the material surface controls the service life in many applications, the objective is to develop a wide range of functional properties that are different from the base substrate including physical, chemical, electrical, electronic, magnetic or mechanical. Being a part of surface engineering, the thermochemical treatment employs thermal diffusion to incorporate non-metal or metal atoms into a material surface to modify its chemistry and microstructure (Fig. 1). The process is conducted in solid, liquid or gaseous media with one or several simultaneously active chemical elements. For majority of thermochemical treatments the mechanism includes a decomposition of solid, liquid or gaseous species, splitting of gaseous molecules to form nascent atoms, absorption of atoms, their diffusion into a metallic lattice and reactions within the substrate structure to modify existing or form new phases. Since in industrial scale processes the entire part is subjected to high temperatures, surface diffusion is superimposed on changes within the material volume that for some treatments may involve phase transformations and this adds to the complexity.

Historically, the thermochemical treatment was limited to machined parts, forgings and castings with an application in machinery, automotive, tooling, oil drilling, mining and defence [1]. The key processes covered nitriding, carburizing and their combinations. Similarly, steel was in practice the only material subjected to the modification. To enhance the process predictability and repeatability, the conventional gas nitriding was refined and the alternative technique of ion (plasma) nitriding was introduced. In quest for the perfect process, the plasma technology is still a subject of continuous improvement and developed techniques of post discharge nitriding or active screen plasma nitriding may serve as examples [2]. In the meantime, the thermochemical modifications included other processes such as boronizing, aluminizing, chromizing or thermo-reactive diffusion, exploring vanadium, molybdenum and other carbide-forming elements. Although they never achieved the application level of nitriding, they successfully serve many niche markets.

In recent decades, an application of the thermochemical treatment expanded to alloys with exotic chemistries [3], nonferrous metals like aluminum [4] and also refractory metals. Numerous hybrid processes were developed where thermochemical diffusion is a part of the multi-step treatment involving coating, cladding, laser processing etc. While the conventional applications still dominate, it is seen an expansion of the thermochemical treatment to novel manufacturing techniques such as micro-scale fabrication, fuel cells [5] or electronics [6].

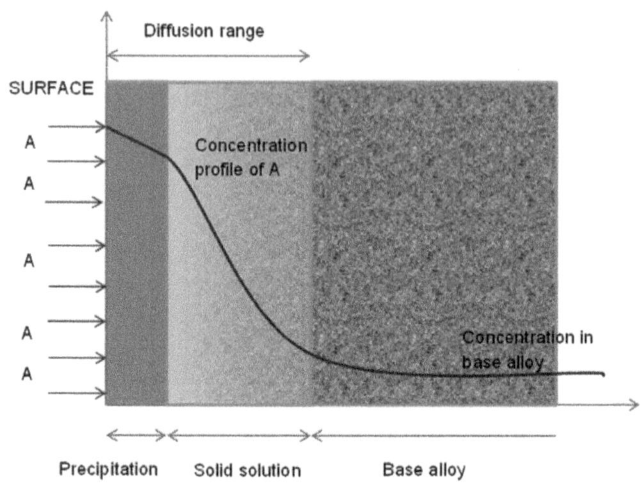

Figure 1. Principles of thermochemical treatment showing a distribution of the chemical element A inside an alloy along with typically modified sub-surface areas

This chapter covers major aspects of the thermochemical surface treatment of metals and alloys. A mixture of engineering fundamentals and recent global scientific developments should not only be useful for professionals from metallurgy and materials area but also for experts from other fields of engineering.

2. Nitriding

Nitriding has been and continues to be the major thermochemical treatment which along with ferritic nitrocarburizing represents the dominant volume of industrial surface modification technologies. The treatment leads to an incorporation of nitrogen into the surface of steel while it is in ferritic state. In commercial applications, the typical modified zone is up to 200-300 μm thick, rarely exceeding 600 μm. Its impact on surface hardness distribution, in terms of the maximum value and penetration depth, as compared with other heat and thermochemical treatments, is shown in Fig. 2. There is no additional heat treatment required following nitriding and the component surface experiences an increase in hardness, wear resistance, improved corrosion resistance and fatigue life.

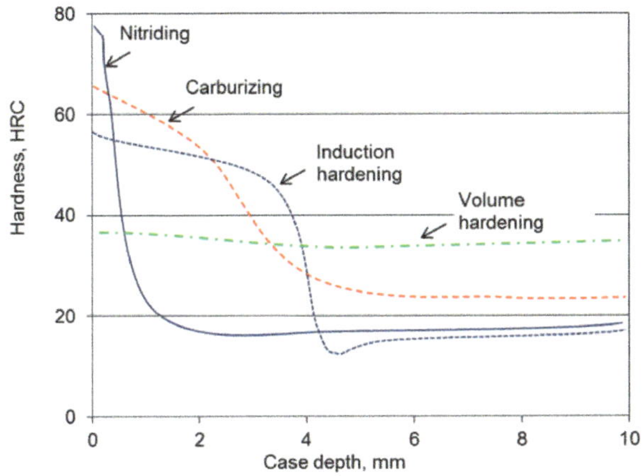

Figure 2. Hardness depth profiles for selected thermal and thermochemical treatments, emphasizing differences in the maximum hardness and penetration depth

2.1. Nitriding technologies available at present

To implement nitriding, several technologies, exploring different sources of nitrogen, were commercialized.

2.1.1. Gas nitriding

Gas nitriding was patented in 1913 and 1921, and is carried out usually at temperatures of 550-580 °C in a box furnace or fluidized bed in an atmosphere filled with partially dissociated ammonia [1]. The advantages of the fluidized bed are the near-ideal temperature uniformity through the entire gas-particle volume and fast heating rate [7]. For gas nitriding the fundamental reaction is the catalytic decomposition of ammonia to form the nascent (elemental) nitrogen:

$$NH_3 = \left[N\right] + 3/2H_2 \quad (1)$$

The control parameters include time, temperature and gas dissociation rate. In production environment, the latter is periodically measured and adjusted. The inherent feature of conventional gas nitriding is that the superficial concentration of nitrogen cannot be precisely monitored. As a result the structure of nitrided layer and the entire process are often missing predictability and repeatability.

The *controlled gas nitriding Nitreg®*, employs a mixed-gas atmosphere, composed of ammonia and an additive gas [8]. As opposed to conventional gas nitriding, the process is controlled not by the dissociation rate but by a different parameter, called the nitriding potential of the furnace atmosphere. The nitriding potential is expressed as the ratio of partial pressures of ammonia and hydrogen:

$$K_n = \frac{pNH_3}{\sqrt{(pH_2)^3}} \tag{2}$$

where: pNH_3 is the partial pressure of ammonia and pH_2 is the partial pressure of hydrogen

An advantage of effective control through the nitriding potential, expressing in more uniform nitrided case for complex geometries, is accompanied by general disadvantages of the gas process such as masking difficulties to prevent nitriding, requiring copper plating or painting with protective pastes and the special surface activation necessary for stainless steels or alloys generating a passive oxide films.

2.1.2. Liquid salt nitriding

Liquid nitriding, developed in 1940's, is conducted in the fused salt bath containing either cyanides or cyanates. A typical commercial bath is composed of a mixture of 60-70% sodium salts {96.5% NaCN, 2.5% Na_2CO_3, 0.5% NaCNO} and 30-40% potassium salts {96%KCN, 0.6%K_2CO_3, 0.75% KCNO, 0.5% KCl} [9]. The commercial equipment for salt nitriding, along with gas and plasma technologies is shown in Fig. 3. The major advantage is the short cycle time due to intense heating and the high reactivity of the medium. Several methods exist to accelerate further the nitriding rate, such as bath additions of sulphur or melt pressurizing. Typically, for low-alloy steel the cycle time lasting 1.5 h at the operating temperature of 565 °C produces a case of 0.3 mm thick. The salt-bath technology has also a number of negative features, such as the bath toxicity and poor quality of the nitrided surface.

Figure 3. Commercial technologies of liquid salt bath, gas and plasma nitriding (with permission from Rubig GmbH)

2.1.3. *Plasma (ion) nitriding*

Plasma nitriding, called also *ion nitriding,* was invented by Wehnheldt and Berghause in 1932 but became commercially viable as late as in 1970's. It uses the glow discharge phenomenon to introduce nascent nitrogen to the surface of an alloy and its subsequent diffusion into subsurface layers [10]. An example of modern installation is shown in Fig. 4. Plasma is formed in a vacuum using a high-voltage electrical energy to accelerate nitrogen ions which bombard the alloy surface [11] (Fig. 5a). The advantages of ion nitriding include the low temperature, short saturation time and simple mechanical masking. The unique advantage is surface-activation sputtering. Due to the sputtering effect of positive ions in the glow discharge, the protective oxide, inherent for surfaces of stainless steels, aluminum or titanium alloys, is removed. Thus, nitrogen atoms can be moved from the plasma to the material sub-surface. In the conventional direct-current system the nitrided component is subjected to the high cathode potential and plasma forms directly on the component surface. This may create disadvantages such as the temperature non-uniformity with a possibility of overheating, sensitivity to the part geometry, causing edge effect and a possibility of surface damage due to arcing.

Figure 4. Modern line of commercial plasma nitriding (with permission from Rubig GmbH)

To overcome the latter limitation, different approaches were investigated. The *post discharge nitriding,* where the nitrided part at an electrically floating potential is kept at the nitriding temperature by use of an external heater, was so far not adopted by industry [12]. Another technique, called *active screen plasma nitriding,* was invented in 1999 and has some commercial applications [2]. As shown in Fig. 5b, an essence of the new process is in applying the high cathodic potential to a screen surrounding the nitrided part which becomes the real cathode, replacing in this role the nitrided part. Therefore, plasma forms on the active screen, heating it up. Then, a radiation from the screen heats the nitrided part to the required temperature [13]. The plasma, forming on the screen, is composed of a mixture of ions, electrons and other active nitriding species which are forced to flow over the nitrided part by the designed gas circulation. Thus, complex geometries obtain the uniform nitrided layer and even blind holes are affected by diffusion and effectively nitrided.

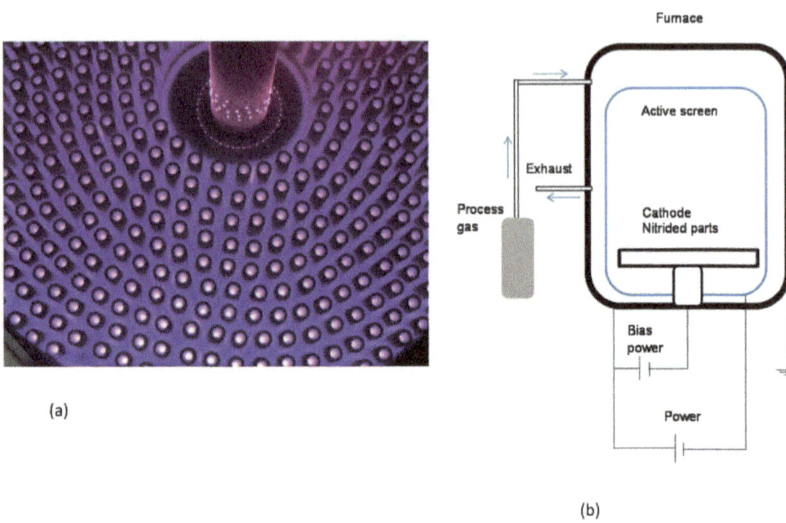

Figure 5. Plasma nitriding: (a) view of components during Ultraglow® process (with permission from Advanced Heat Treat Corp.) (b) schematics showing a concept of active screen plasma nitriding

2.1.4. Laser nitriding

In the last two decades, *laser nitriding* has been investigated as an alternative nitriding method [14]. As explained in Fig. 6, during a direct laser synthesis a material is placed in the reactive gas environment and irradiated with the laser light. Nitrogen is fed through a nozzle into the melt pool. On a time scale of hundreds of nanoseconds, the high intensity pulse-laser irradiation of $I \approx 10^8$ W/cm^2 in ambient nitrogen atmosphere is capable generating of 1- 1.5 μm thick thick nitrided layer.

2.1.5. Beam ion implantation

At limited scale, *beam ion implantation* can be used to incorporate nitrogen into a material surface. The conventional ion beam implantation, applied at room temperature, is capable to modify chemistry of relatively thin layers of materials. While using a beam of nitrogen ions with an energy of up to 1 MeV at room temperature, a continuous nitride layer of the order of 1 μm can be synthesized. There are, however, techniques exploring elevated temperatures of up to 600 °C or hybrids such as plasma immersion implantation or low voltage plasma immersion implantation, allowing generating thicker layers [15]. A comparative test with the beam ion implantation, plasma ion implantation, ion nitriding and gas nitriding of AISI 304 stainless steel created the same microstructure with nitrogen in iron solid solution [16]. After treatment, conducted at 400 °C for 0.5 h and 1 h, both the beam ion implantation and the plasma ion implantation produced over 1 μm thick layer, enriched in nitrogen to 20-30 at%, while ion nitriding and gas nitriding produced layers with a thickness below 1 μm and the lower nitrogen concentration.

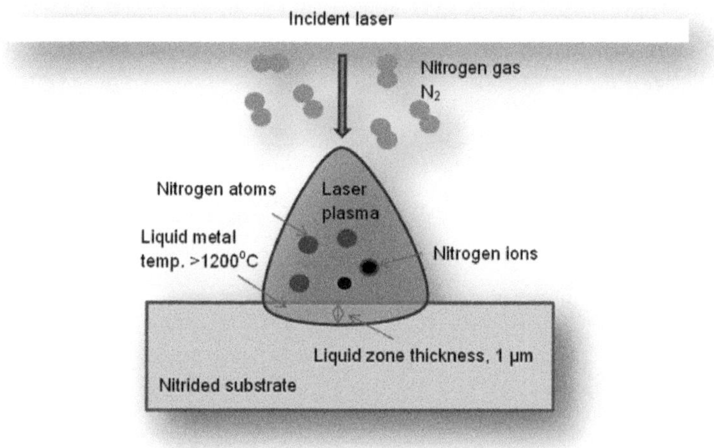

Figure 6. Principles of laser nitriding

2.2. Process theory

The Fe-N phase diagram provides essentials for nitriding of iron and low carbon steels. It consists of several solid solutions of N in α-Fe and γ-Fe, stable chemical compounds (γ'-Fe_4N_{1-z}, ζ -Fe_2N) and metastable phases (α' – martensite, α'' – $Fe_{16}N_2$) [17]. The bcc lattice of α-Fe can dissolve up to 0.4 at% of N without substantial straining with nitrogen atoms occupying octahedral interstices in a random matter. After the content of nitrogen dissolved in pure iron exceeds 2.4 at%, the γ' nitrogen martensite with a structure similar to the carbon martensite, is formed. At that nitrogen level, the bcc lattice experiences tetragonal straining and nitrogen atoms occupy 1/3 of the possible octahedral interstices [18] [19]. The nitrogen austenite phase can dissolve up to 10.3 at% of nitrogen and its atoms are randomly located in octahedral interstices of the fcc Fe lattice. It is considered that the Fe-N solid solution can be conceived as composed of two interpenetrating lattices: the sublattice for Fe-atoms and the sublattice for N atoms [20]. When sites in Fe sublattice can be considered as fully occupied, sites in N sublattice, which constitutes of octahedral interstices of the Fe sublattice, are partly occupied by N atoms and partly by vacancies. The γ'- Fe_4N_{1-z} nitride has a cubic elementary cell, formed by an fcc sub-lattice of Fe atoms with ordered arrangement of nitrogen atoms in central octahedral interstices. It has a narrow range of homogeneity within 19.3-20 at% at 590 °C. A schematic of the crystal structure of γ'-Fe_4N where nitrogen atoms occupy a quarter of the octahedral sites, surrounded by the shadowed octahedral, is shown in Fig. 7a [21]. On the other hand, the ε nitride has a variable stoichiometry of ε-Fe_2N_{1-z} with a structure based on fcc Fe, in which nitrogen atoms stay in octahedral sites and form a diamond type sub-lattice. The ε nitride has the largest range of homogeneity in Fe-N system, reaching from 15 to 33 at% of nitrogen. For some materials, not typical precipitation may occur; nitriding of Fe 2 at% Si alloy led to silicon nitride precipitates formed inside the ferrite grains and along grain boundaries [22]. The precipitates were amorphous with a stoichiometry of Si_3N_4. The amorphous nature is explained by thermodynamics due to the fact that the precipitation process occurred very slowly due to the very large volume misfit between the nitride and matrix.

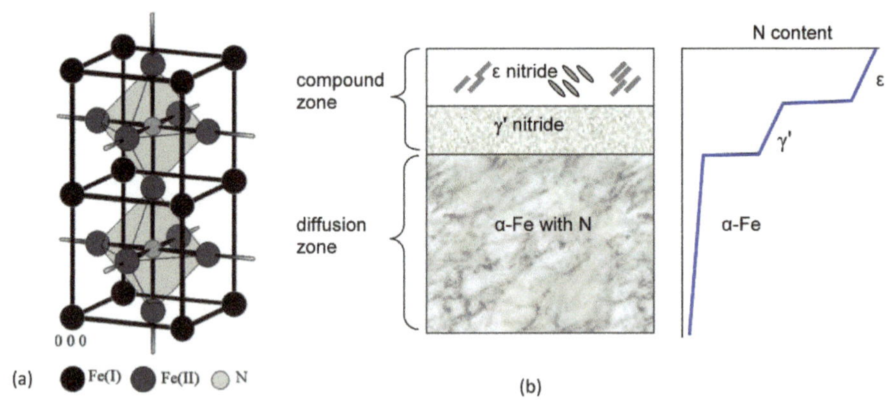

Figure 7. Schematics of: (a) crystal structure of γ'-Fe_4N showing two unit cells [21]; (b) phase distribution within a nitrided case on steel and accompanied nitrogen depth concentration.

During nitriding, a compound layer, composed of iron nitrides ε and/or γ', is formed at the steel surface. Beneath the compound layer a diffusion zone extends in the ferrite matrix in which nitrogen is dissolved interstitially. The heat effect of slow cooling after nitriding or the separate heating cycle lead to formation of the γ'- Fe_4N_{1-z} nitride which, in turn, increases the nitrogen content in the ε-Fe_2N_{1-z} nitride. Morphologically that process can change a ratio between sub-layer thicknesses within the compound layer at the expense of ε-Fe_2N_{1-z} or cause a precipitation of γ'- Fe_4N_{1-z} phase within the ε-Fe_2N_{1-z} layer, as shown in Fig. 7b. The particular depth structure of nitrided layer depends on the substrate chemistry and nitriding process, and examples for liquid, gas and plasma technologies are shown in Fig. 8.

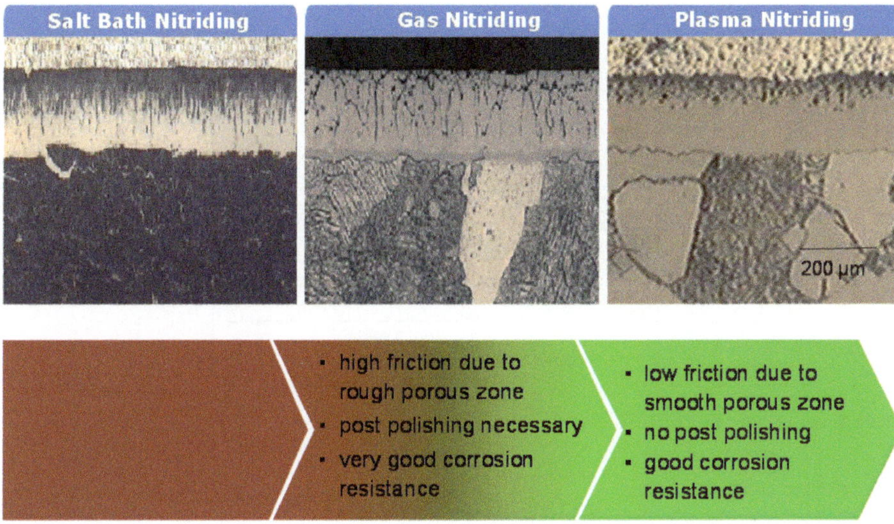

Figure 8. Microstructures of steel after liquid salt, gas and plasma nitriding along with some characteristic features aligned in a direction of improvement (with permission from Rubig GmbH)

2.3. Nitrided layer and its effect on substrate properties

Nitriding changes primarily the surface related properties. However, the presence of nitrided case affects also properties of the material volume beneath the nitrided case and the entire component.

2.3.1. Nitrided case depth

The quality control after nitriding is performed by (i) measuring the superficial hardness and its depth profile (ii) determining the nitrided case depth and (iii) an assessment of the cross-sectional microstructure. To provide unambiguous specifications on engineering drawings, two terms of nitrided case depths were introduced [23]. The *total case depth*, sometimes called simply as the *case depth*, is defined as the dark-etching sub-surface zone as determined metallographically on the component cross section. For alloys, which do not easily respond to etching or do not exhibit the sharp transition between the base material and diffusion zone, the total case depth is defined as a depth below the surface at which the microhardness is 10% higher than that of the base steel beneath it. The *effective case depth* is defined as the case depth where hardness exceeds certain values, defined either by the engineering drawing or standard. For typical nitriding steels, that hardness level is 50 HRC as converted from microhardness, which can be directly measured on the cross section [23]. As seen in Fig. 9, for the particular nitrided case there may be substantial differences between the effective and total case depths.

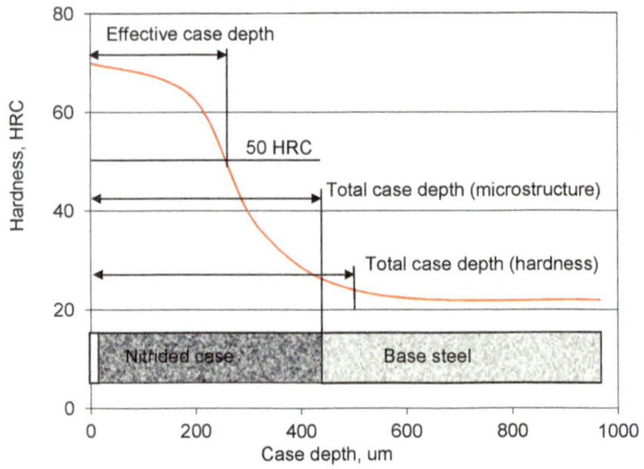

Figure 9. Schematics explaining measurements of total and effective case depths after nitriding

2.3.2. *Role of compound zone*

Within the nitrided case, the compound layer thickness and its integrity are of primary importance for component service performance. Due to its chemical stability the compound layer improves the corrosion resistance. In case of high nitrogen concentration the compound layer may be excessively porous and brittle so often it may also peel from the substrate increasing scuffing. After peeling, it may cause damage to tightly fit wear couples or contamination to processed products. Thus, it may be undesirable. Of specific applications, the compound layer is detrimental to gear life, particularly when gears experience misalignment during service [24]. Therefore, depending on nitriding class, certain thickness of compound layer is permitted [23]. For example, aerospace applications have strict restrictions where only trace amount of the compound zone, below 2 μm, is acceptable [25]. In some applications its presence is not permitted at all. Two examples of nitrided steels without the compound layer are shown in Fig. 10.

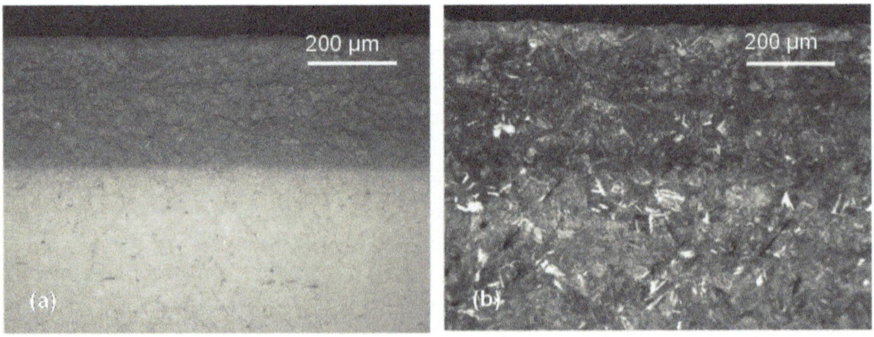

Figure 10. Microstructure of nitrided case on steel with absent or negligible compound layer: (a) fine-grained steel with a clear interface with the base steel; (b) coarse-grained steel with no clear interface with the base steel

Although a thickness of the compound zone depends on substrate chemistry, the essential control of compound layer formation is through altering the process parameters. In gas nitriding, the two-stage process, as developed by Carl Floe in 1953, is used to minimize the compound zone thickness [26]. The first stage ensures the rapid formation of the compound layer and the second stage arrests a formation of the compound layer without allowing the diffusion zone to be de-nitrided. The first stage runs with ammonia gas at a dissociation rate of about 30% at 495 °C, followed by increasing temperature to 563 °C and the dissociation rate to 75-85%. For plasma nitriding, the role of chamber atmosphere in the compound layer formation is critical as detailed in Table 1. The post nitriding removal of the compound layer is costly and requires lapping, honing, grinding or polishing. Since mechanical methods introduce stress, subsequent stress relieving may be necessary. Another alternative of the compound layer removal is by chemical etching in cyanide solutions.

It is known that small additions of oxidizing species to plasma or gas nitriding have a beneficial effect on nitrided layer formation since the presence of oxygen increases the layer growth rate and stabilizes the ε-compound layer [27]. When applying a post-oxidation step after nitriding the cohesive, homogeneous layer of iron oxide grows which further improves the corrosion resistance (Fig. 11) [28]. When the oxide layer is essential for plain carbon steels it is also important for Cr-containing stainless steels. An unalloyed steel with 1-2 μm thick oxide layer exhibits a wide range of passivation by the corrosion current and high breakdown potential. The effect of the oxide layer on the ε-compound layer is often compared to the effect of CrO_2 passivation film on a surface of stainless steel.

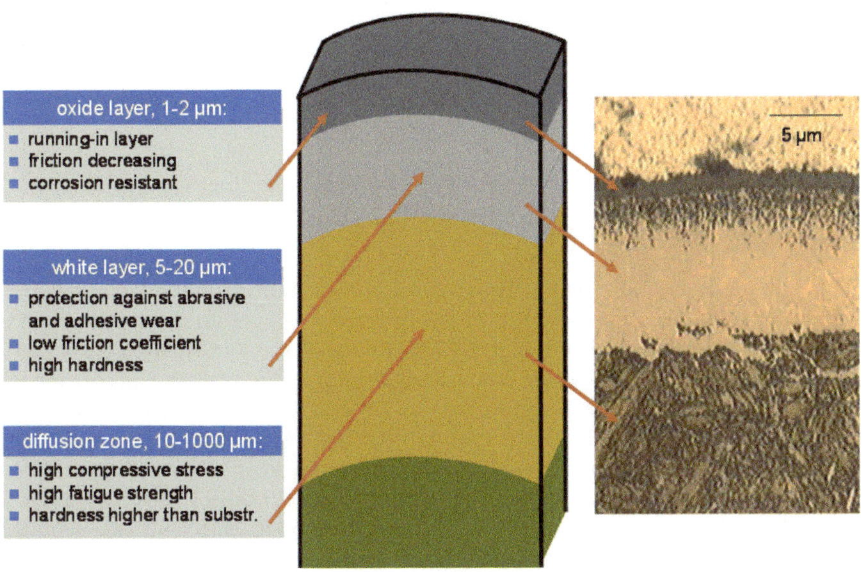

Figure 11. Microstructure after Plusox™ nitriding with major functional zones and their performance characteristics (with permission from Rubig GmbH)

Compound layer	N_2 (%)	H_2 (%)	CH_4 (%)
Negligible	8-15	85-92	-
Thin, continuous	20-35	65-80	-
Thick, continuous	40-60	40-60	-
Extra thick	75-79	11-25	1-2

Source: Rubig GmbH

Table 1. Compound layer control during plasma nitriding

2.3.3. Dimensional changes

Although dimensional changes during nitriding are relatively small, they are definitely measureable. Therefore for precision parts, the dimensional change must be considered during the manufacturing process to compensate the nitriding related increase (Fig. 12a).

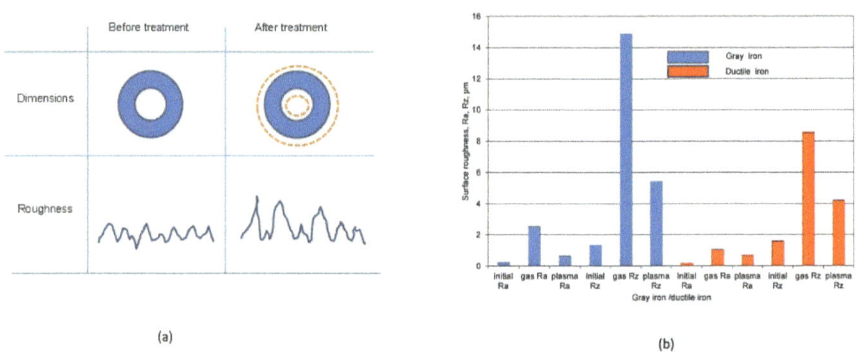

Figure 12. Schematics emphasizing changes of dimensions and surface roughness after nitriding (a) and influence of gas and plasma nitriding on roughness parameters of grey and ductile cast irons (b - based on data from [30] [29])

The volume increase depends on the quantity of the absorbed nitrogen. In some cases, changes related to incorporated nitrogen are superimposed on structural transformations taking place within steel at nitriding temperatures. For the hardened and inadequately tempered steel, a decomposition of still existing austenite can increase a proportion of the volume change. In case of martensitic age-hardened steels, some shrinkage takes place and reduces the overall dimensional changes. In addition, there are also changes in surface topography. As a result of nitrogen absorption an increase in surface roughness occurs. As shown for the case of nitriding-nitrocarburizing, surface roughening depends on the process type and substrate material. In the case of grey and ductile cast irons, the former was especially sensitive to surface roughening with 10 times increase of roughness parameters (Fig. 12b). Moreover, plasma process produced generally much smoother surfaces as compared to the gas process [29] [30]. Thus, to bring the roughness to its initial value, post-nitriding polishing may be required. Also, stress from previous mechanical or thermal treatment may lead to part distortion during nitriding. To provide the stress-free component, prior to nitriding the stress relief should be conducted at temperature at least 50 ^{o}C higher than the nitriding process.

2.3.4. Fatigue life and internal stress

In general, nitriding improves the characteristics of high cycle fatigue. The increase in high cycle fatigue strength is caused by a formation of the strong nitrided layer and a generation of compressive macro-stress. During failure, the initiation of cracks occurs at the interface between the nitrided layer and the matrix. For low cycle fatigue the decrease in durability was recorded due to cracking of the stronger nitrided layer at high stress loadings of repeated nature [31]. Two different behaviours during fatigue tests with smooth and notched samples were distinguished. For nitrided samples with a notch, acting as stress riser, faster crack initiation was recorded. No direct correlation was identified between the hardness and internal macro-stress. Moreover, an increase in fatigue life is not always correlated with the nitrided case thickness. Other factors, such as the nitrided layer structure and steel composition exert effect as well. According to Ericsson [32], the contributing factors of stresses generation during nitriding are different thermal expansion coefficients for the phases present, growth and precipitation stresses for nitrides as well as stresses due to the nitrogen composition gradient. In addition, the thermal stress may also be involved. There are results obtained by the conventional tilt and grazing incidence X-ray diffraction methods, that the residual stress in the compound layer is of tensile nature and changes with depth within the first 2 μm of the 10 μm thick compound layer [33]. The constant nitrogen concentration within the first 2 μm thickness of the surface layer does not support the concentration gradient in stress generation cause as claimed above.

2.4. Steels applicable to nitriding

In general, nitriding is applicable to a wide variety of carbon steels, low alloy steels, tool steels, stainless steels and cast irons. For optimum properties after nitriding, however, there are steels with chemistries, particularly designed for this purpose. They contain strong nitride-forming elements such as Al, Cr, Mn, Mo and V. There is a limitation on carbon content which should not exceed 0.5%, as most nitride-forming elements also form stable carbides which limit binding of nitrogen. When differences in hardness depth profiles for carbon and alloyed steels are essential (Fig. 13a), there are also substantial differences between individual grades, designed for nitriding (Fig. 13b). The especially high surface hardening is achieved with steels containing Al, forming AlN nitrides. However, additions of Al, typically in the range of 1%, cause steel brittleness. The nickel nitriding steels containing aluminum develop higher core strengths than do nickel-free nitriding grades. Nickel also increases the toughness of the nitrided case [34]. The base steel properties are of importance to provide the support for nitrided case, especially in applications where components carry high compressive and bending stresses. A selection of steels designed for nitriding is listed in Table 2.

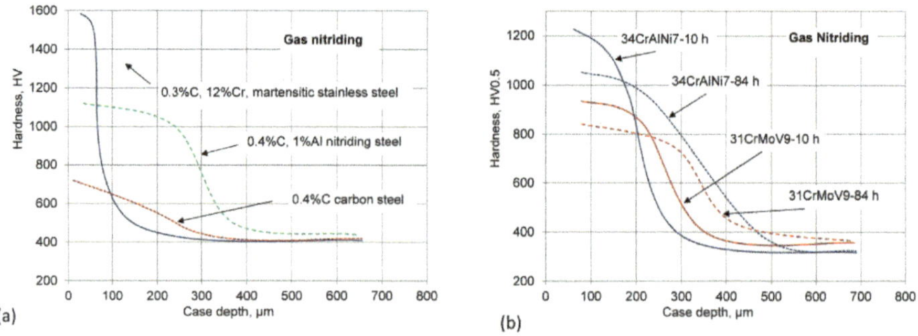

Figure 13. Effect of the steel composition on hardness depth profiles after nitriding: (a) comparison between carbon and alloyed grades; (b) comparison between two nitriding grades {(b) based on [41] with permission from Schmolz + Bickenbach }

The deep nitrided steel, 32CrMoV13, has an application for aerospace bearings of main shafts or jet engines, operating under high speed, high temperature and limited lubrication. This application requires the typical nitrided case of more than 600 µm, achieved after gas process in the temperature range of 525-550 °C for up to 100 h [35]. For that case depth, the surface hardness ranges from 730 to 830 HV30. The compound layer of approximately 30 µm thick is removed by grinding from rolling races and working surfaces. Within the diffusion layer, two zones may be distinguished. First is the 100 µm thick zone, adjacent to the compound layer, depleted of carbides, likely due to their dissolution and precipitation more stable nitrides or carbonitrides. The second zone, adjacent to the base steel, is precipitates free. It is considered that the good service performance of nitrided layer and its superior rolling-contact properties are achieved due to semi-coherent precipitates of nano-size nitrides and high compressive residual stresses. An example of Al-free steel with V used for camshafts is OvaX200 steel [36]. After 20 h of gas nitriding at 490-510 °C, The OvaX200 steel reaches the surface hardness of 850 HV1000.

The key concern during nitriding of stainless steels is to retain their corrosion resistance. The low temperature nitriding of austenitic grades leads leads to formation of nitrogen expanded austenite. This so-called *S-phase* exhibits an increased hardness and wear resistance combined with excellent corrosion resistance, inherent for stainless steel [37]. It is claimed that the active-screen plasma nitriding brings some benefits to nitriding of stainless steel. According to that mechanism, the material sputtered from the active screen and re-deposited on the nitrided surface plays an important role in the nitriding process. Namely, the sputtered atoms react with nitrogen containing species in the plasma and deposit on the steel surface. Then, the deposited iron nitrides decompose, releasing nitrogen which, in turn, diffuses into the sub-surface layer [38].

Nitriding is also applicable to maraging steels which have very low carbon content, typically below 0.03%, and are strengthened by the precipitation of intermetallic compounds, taking place at a temperature of approximately 480 °C. The purpose of nitriding the maraging steels is to increase their wear resistance. Since the temperature of conventional gas nitriding exceeds the aging temperatures of maraging steels, the plasma nitriding is more suitable. The plasma nitriding at a temperature as low as 450 °C for 10 h, in case of the Fe-18Ni-8.8Co-5Mo-0.4Ti-0.1Al steel, generates the nitrided layer with a thickness of 120 μm and surface hardness of 800 HV [39]. At the same time, it does not cause structural changes in the substrate, thus preserving the original microstructure.

The plasma nitriding with its surface sputtering effect, removing the surface oxide layer, is suitable for nitriding of high entropy alloys, containing large volumes of oxide-forming elements, such as Al, Cr, Si and Ti [3]. For the particular composition of $Al_{0.3}CrFe_{1.5}MnNi_{0.5}$ a dual-phase structure of fcc and bcc develops after homogenisation at 1100 °C [40]. After plasma nitriding at 525 °C for 45 h, the 80 μm thick layer, with a peak hardness of 1300 HV, is generated. Both the bcc and fcc phases experience uniform nitriding and main nitrides are AlN, CrN and $(Mn, Fe)_4N$. The alloy reaches the wear resistance 49-80 times higher than the nitrided conventional steels.

Steel	C	Mn	Si	Cr	Mo	Ni	V	Al
Nitralloy 135M AMS 6470	0.38-0.45	0.5-0.8	0.2-0.4	1.4-1.8	0.35-0.45	-	-	0.85-1.2
Nitralloy G	0.35	0.55	0.3	1.2	0.2	-		1.0
Nitralloy N AMS 6475	0.2-0.26	0.5-0.7	0.2-0.4	1.0-1.25	0.2-.3	3.25-3.75	-	1.1-1.4
Nitralloy EZ	0.35	0.8	0.3	1.25	0.2	0.2Sc	-	1.0
34CrAlNi7 DIN 1.8550	0.30-0.37	0.4-0.7	≤0.4	1.0-1.3	0.15-0.25	-	-	0.8-1.2
41CrAlMo7 DIN 1.8509	0.38-0.45	0.4-0.7	≤0.4	1.5.1.8	0.2-0.35	-	-	0.8-1.2
34CrAlMo5 DIN 1.8507	0.30-0.37	0.4-0.7	≤0.4	1.5-1.8	0.15-0.25	0.85-1.15	-	0.8-1.2
15CrMoV5-9 DIN 1.8521	0.13-0.18	0.8-1.1	≤0.4	1.2-1.5	0.8-1.1	-	0.2-0.3	-
31CrMoV9 DIN 1.8519	0.27-0.34	0.4-0.7	≤0.4	2.3-2.7	0.15-0.25	-	0.1-0.2	-
31CrMo12 DIN 1.8515	0.28-0.35	0.4-0.7	≤0.4	2.8-3.3	0.3-0.5	-	-	-
32CrMoV13 AMS6481	.29-0.36	0.4-0.7	0.1-0.4	2.8-3.3	0.7-1.2	-	0.15-0.35	-
OvaX200	0.14-0.17	1.2-1.4	0.15	2.1-2.3	0.45-0.55	0.45-0.55	0.15-0.25	-
5Ni – 2Al	0.20-0.25	0.25-0.45		0.4-0.6	0.2-0.3	4.75-5.25	0.08-0.15	1.8-2.2

Table 2. Selection of steels, designed for nitriding (weight %) [41] [36] [34] [26]

2.5. Nitriding of titanium alloys

The competitiveness of titanium alloys is due to their high strength to weight ratio, heat and corrosion resistance. At the same time, the low surface hardness and wear resistance along with poor high-temperature oxidation resistance are seen as their major disadvantages. Nitriding is one of many treatments aimed at improving their tribological characteristics.

2.5.1. Nitriding techniques applicable to titanium

In principle, all major nitriding techniques are applicable to titanium. A disadvantage of gas nitriding is the high temperature of 650-1000 °C required, long time of up to 100 h and reported fatigue life reduction. For Ti-6Al-4V alloy a typical compound layer of 2-15 μm forms with a surface hardness between 500 and 1800 HV [42]. The plasma nitriding of titanium alloys is conducted at temperatures of 400-950 °C and substantially shorter time from 0.5 to 32 h, generating a compound layer with a thickness of approximately 50 μm. A reduction in fatigue strength may be eliminated by lowering the nitriding temperature. The ion beam nitriding, using nitrogen at temperatures of 500-900 °C for up to 20 h, produces 5-8 μm thick compound layer with microhardness of 800-1200 HV on Ti-6Al-4V alloy. Also laser nitriding is applicable to titanium but surface case has a tendency to cracking. An attempt was made to apply the diode laser gas nitriding technique to Ti6Al4V alloy, commonly used for rotors and blades of engines in power generation [43]. The laser surface melting of the substrate surface in a mixture of nitrogen and argon leads to an increase in surface hardness up to 1300 HV0.2 although the outcome depends on process parameters.

2.5.2. Microstructure development during nitriding

The formation of the nitrided layer on titanium involves several reactions taking place at the gas/metal interface and within the metal. At the nitriding temperature, below the Ti polymorphic transformation, the α-Ti phase exists. First, the nitrogen absorbed at the surface diffuses inward titanium, forming the interstitial solution of nitrogen in the hcp titanium phase α-Ti(N) and building the nitrogen concentration gradient. After exceeding the solubility limit, the Ti_2N phase is formed. During further increase in the nitrogen concentration at the gas/metal interface, TiN is formed as specified below:

$$\alpha-\mathrm{Ti} \rightarrow \alpha-\mathrm{Ti}(\mathrm{N}) \rightarrow \mathrm{Ti_2N} \rightarrow \mathrm{TiN} \tag{3}$$

After slow cooling, the precipitation in the diffusion zone is possible. The simplified morphological schematic, emphasizing the growth sequence, is shown in Fig. 14.

2.5.3. Nitriding of other refractory metals (Zr, Nb, Mo, W, Ta)

Many refractory metal nitrides offer an attractive combination of high electrical conductivity and good corrosion resistance. For Mo-0.5%Ti and pure Mo alloys the inward diffusion of nitrogen is the rate controlling step. After gas nitriding at 1100 °C they reach a hardness of 1800 HV and the surface layer consists of two regions with the outer layer composed of γ-

Mo_2N and the inner layer of β-Mo_2N [44]. In Ti containing alloys an internal nitrided layer is additionally formed with a hardness of 800 HV which contains the fine 0.4 nm thick plate-like, coherent particles of TiN.

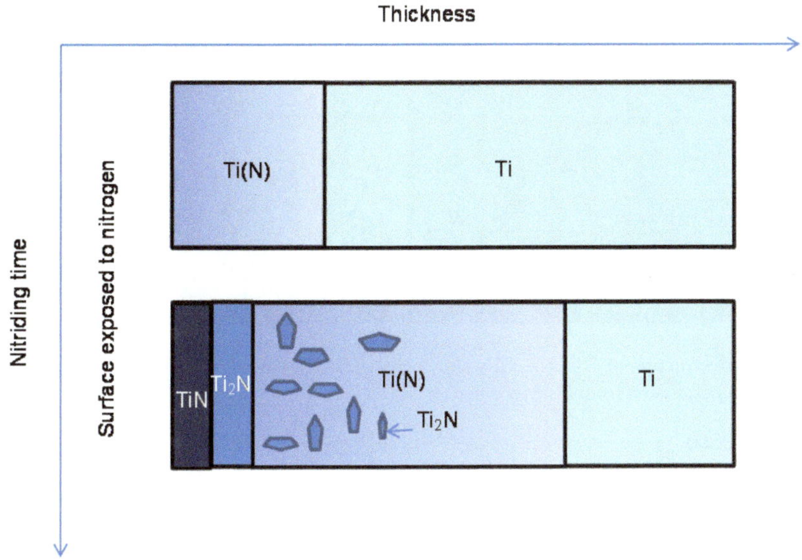

Figure 14. Schematics of the morphology development during nitriding of titanium

The nitrided niobium has a potential application in particle accelerators [6]. The niobium-made cavity of the superconducting radio-frequency accelerator operates at very low temperature of 1.9K to achieve the sufficient performance. After nitriding of niobium, its surface is transformed to δ-NbN with the critical temperature of 17K as opposed to 9K for pure Nb. Such a change shows a promise to raise the accelerator operating temperature. The laser nitriding was found to be effective for niobium.

The refractory alloy composed of Nb, 30%Ti and 20%W, called Tribocor 532N, and produced by conventional melting techniques, benefits from nitriding as well [45]. Since titanium has the highest affinity to nitrogen of all elements in the alloy, TiN forms preferentially on the nitrided surface. During the next step, as the alloy becomes depleted in Ti, the Nb nitride starts growing. As shown in Fig. 15, its surface microstructure changes completely. The nitrided surface achieves high hardness with superior corrosion resistance (Fig. 16). Among many applications, including an environment of molten magnesium and aluminum alloys [46], the nitrided Nb-30Ti10W alloy was considered in proton exchange membrane fuel cells for the bipolar plate. The plate serves to electrically connect the individual cells in a stack and to separate and distribute the reactant and product stream [5].

The nitrided zirconium has application potentials for cathodes in arc heaters, mainly due to its high erosion resistance under those service conditions [47]. After nitriding in a microwave plasma generator, the golden color zirconium nitride forms on its surface. In addition to metals, also ceramics of yttria-stabilized zirconia benefit from nitriding [48]. As a result of high temperature plasma nitriding, the complex layer, composed of tetragonal cubic zirconia, zirconium nitride and oxynitride, is created.

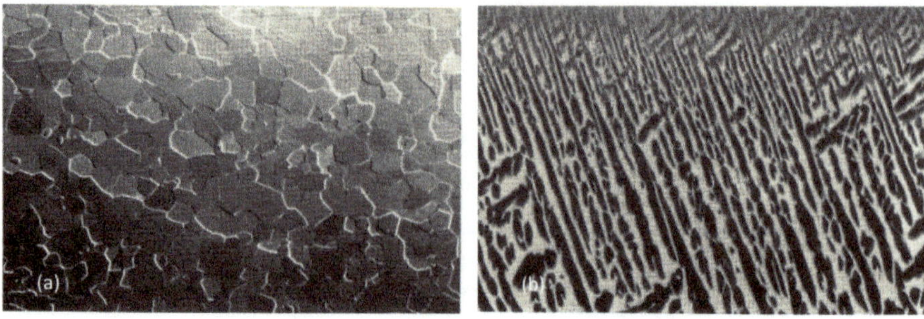

Figure 15. Microstructure of Tribocor 532N alloy before and after nitriding. Approximate magnification 500x [45] (with permission from Elsevier Science)

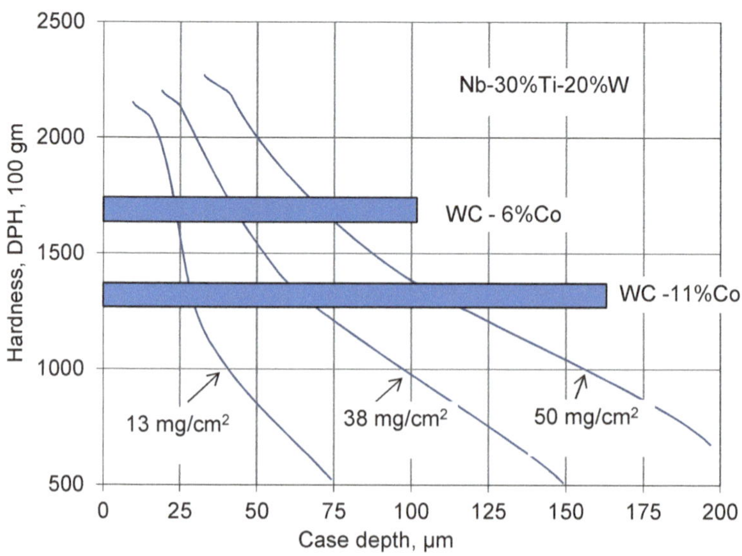

Figure 16. Hardness versus depth profile for three levels of nitrogen absorption in Tribocor 532N alloy [45] (with permission from Elsevier Science)

2.6. Nitriding of aluminum

One of the key limitations in the application of aluminum alloys is their low wear resistance. Therefore, the purpose of aluminum nitriding is similar as in the case of titanium, i.e. to increase the surface wear resistance. At present, Al-Si alloys are used for wear applications or a hard surface layer must be created. The latter is most often achieved by Al_2O_3 alumina coatings.

The conventional plasma nitriding of aluminum alloys is conducted at temperature of 500 °C for 20 h to form 1-2 μm thick layers. Since nitrogen is virtually insoluble in aluminum, during nitriding only the compound layer of AlN is formed. AlN is known for its high hardness of 1400 HV, high thermal conductivity and high electrical resistivity. In addition to nitriding, AlN can be formed on Al surface by several other techniques including evaporation or ball milling [49] [50]. The process of Al nitriding is slow since the diffusion of N in Al is the rate controlling step. To increase the nitriding rate, the grain size of Al should be refined. It suggests that the growth of AlN is controlled by grain boundary diffusion, hence increasing grain boundary density increases a number of fast diffusion paths. Another way of speeding up the AlN growth is by alloying additions and a presence of 1wt% of Ti is effective. Under the same nitriding conditions of temperature 500°C and time of 20 h the 3 μm thick layer with co-precipitates of TiN grows as compared to 1-2 μm thick AlN in case of Ti absence [49].

An alternative technique used for Al nitriding is the electron beam excited plasma (EBEP) [51]. The technology is sustained by the electron impact ionization with an energetic electron beam being a source of low pressure plasma. When employing this method it is possible to create on AA5052 alloy after 45 min at 570 °C the AlN layer with a thickness of 5 μm and pillar shape grains. At the interface with the substrate a spinel $MgAl_2O_4$ grows, affecting the adhesion of major AlN layer. Also inductively coupled RF plasma reactor is effective for nitriding of Al-Cu alloy 2011 [52]. At temperature of 400 °C the time of 36 h is needed to create a protective layer. During plasma nitriding of Al-Cu alloys, Al_2Cu precipitates increase the nucleation rate and growth of the nitrided case [53]. The crystallographic coherence between AlN and Al_2Cu enhances the formation of AlN nodules and islands. The layer formation is also accelerated by the solid-state interaction between Al_2Cu and penetrating nitrogen to form interfacial boundaries, acting as nitrogen diffusion paths. The plasma nitriding is relatively effective way to increase the corrosion resistance of aluminum. There are data pointing out that a treatment at 500 °C for 20 h leads to improvement during both the immersion test in 3.5% NaCl and the polarization test [54].

In addition to an improvement in surface characteristics, the presence of AlN layer affects the bulk properties of aluminum. According to [55], the plasma nitriding of Al-Si-Mg alloy causes decrease in yield, ultimate tensile strength, elongation and stress relaxation rate. It was explained that stress created at the interface between the AlN and Al substrate contributed to the premature failure. Other failure contributors are argued to be defects created by diffusion of nitrogen into the lattice.

A deficiency of AlN films, formed by plasma nitriding, is their tendency to cracking and delamination due to large compressive stresses and the property difference between the AlN layer and the Al substrate. As a possible solution of increasing the AlN adhesion, a combination of barrel nitriding and plasma nitriding is proposed [4]. The barrel nitriding is performed as a pre-treatment before the plasma nitriding. In addition to Al_2O_3 and Al-50wt%Mg powders, nitrogen gas is introduced and the content is heated to 630 °C.

3. Nitrocarburizing (ferritic nitrocarburizing)

During nitrocarburizing, nitrogen and carbon are supplied to the surface of steel in ferritic state at temperatures usually between 500 and 580 °C. The general classification of thermochemical treatments involving nitrogen and/or carbon is shown in Fig.17. According to some terminology, the high temperature equivalent of ferritic nitrocarburizing is called as austenitic nitrocarburizing. There is also a term of ferritic carburizing, describing the carburizing process at temperatures of the ferritic state.

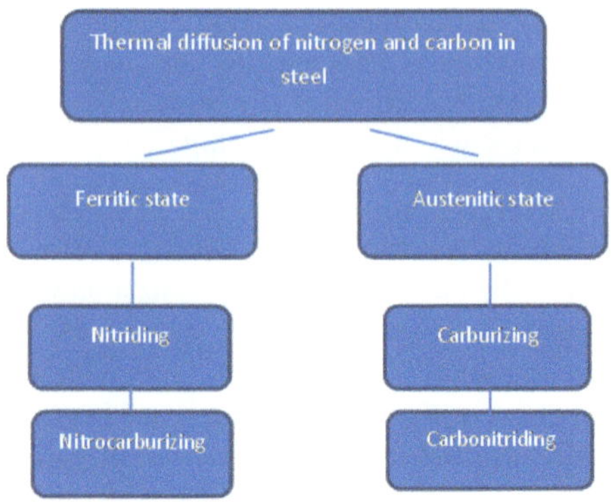

Figure 17. Classification of basic thermochemical treatments involving nitrogen and carbon

3.1. Process technology

The salt bath nitrocarburizing by Tufftride® is performed in a mixture of alkali cyanate and alkali carbonate in the temperature range of 480-630 °C [56]. The gas nitrocarburizing has been developed as a cleaner alternative to the salt bath technology. Besides ammonia, required to supply the nascent nitrogen, the nitrocarburizing atmospheres contain carbon-bearing additives like exothermic and endothermic gases CO_2, CO and H_2 as products of the dissociation of methanol. There is also a hybrid treatment, integrating the low temperature plasma nitriding and additions of carburizing species to the plasma media to cause the simultaneous incorporation of nitrogen and carbon. However, using the glow discharge of plasma containing nitrogen and carbon species it is difficult to produce a single ε-$Fe_{2\text{-}3}$(N,C) phase compound layer on engineering steels [57]. Instead, the plasma nitrocarburizing generates a compound layer with mixed phases of ε-$Fe_{2\text{-}3}$(N,C) and γ'-Fe_4(N,C), known to be detrimental in tribological applications, especially under impact loads. According to [58], excluding for low-carbon steel or to some extent for medium carbon unalloyed steel, gas nitrocarburizing does not produce compound or diffusion layers faster than gas nitriding. Moreover, the properties of nitrocarburized parts are not always superior to those obtained by nitriding and the nitrocarburizing process is more difficult to control.

There is a substantial difference between gas and plasma nitrocarburizing when considering an environmental aspect. As shown in Table 3, in addition to drastic reduction of CO, CO_2 and NO_x emission, reaching 500-5000 times by plasma technology, the total gas consumption is at least 10 times lower than in the gas process [57].

Emissions	Unit	Plasma	Gas
Amount of gas used	m^3/h	0.6	6.0
Total carbon emission via CO/CO_2	mg/m^3	504	137253
Total amount of NO_x gas	mg/m^3	1.2	664
Output of residual carbon bearing gas	mg/h	302	823518
Output of residual NO_x gas	mg/h	54	3984

Table 3. Emission data for plasma and gas nitrocarburizing [57]

Among different nitrocarburizing techniques, a special attention is paid to low pressure processes, performed in a mixture of NH_3 and CO_2. Since CO_2 containing gas has a high oxygen potential, especially under low pressure conditions, oxygen atoms accelerate the formation of $Fe_3(N,C)$ and contributes to the growth of adherent nitrocarburized layers [59]. The process named Nitreg-ONC® is based on Nitreg® technology but generates a modified complex compound layer which contains an increased concentration of carbon, oxygen and sulphur [60]. As a result, surface retains its high wear resistance, anti-scuffing and anti-seizing properties. Another advantage is the substantially increased corrosion resistance which for carbon steels reaches a level comparable with stainless grades. The increased corrosion resistance is associated with the superficial oxide structure which is not penetrable by corrosive fluids. In general, the treatment is considered a superior alternative to chrome plating.

3.2. Surface layer structure

During nitrocarburizing of iron, the microstructural evolution of the compound layer starts with the formation of carbon-rich cementite and develops into the direction of nitrogen-richer and carbon-poorer phases of ε and γ' [61]. Both steps are a consequence of higher solubility of nitrogen in α-Fe than carbon and lower rate of nitrogen transfer from the gas into the solid phase. The compound layer is typically composed of carbonitrides of iron ε-$Fe_3(N,C)_{1+x}$ and γ'-$Fe_4(N,C)_{1-z}$ along with θ-Fe_3C cementite (Table 4) [62] [63]. As during nitriding, beneath the compound layer the diffusion zone forms with carbon and nitrogen being dissolved in the ferritic matrix. It is well documented that the best properties are achieved when the compound layer contains predominantly the single ε phase (Fig. 18a). The compound layer, typically in the range of 20 μm, leads to significant improvements in hardness, wear and corrosion resistance. A presence of ammonia in gas nitrocarburizing atmosphere affects the compound layer structure and a presence of cementite Fe_3C. During ferritic carburizing of iron at a temperature of 550 °C in gas atmospheres containing a certain amount of NH_3, massive layers of cementite Fe_3C can be grown [64]. In order to generate thicker layers, the nitrocarburizing process is conducted at temperatures exceeding the Fe-N eutectoid point of 590 °C. After austenitic plasma nitrocarburizing at 700 °C for 3 h of 0.45% C steel the layer contains mainly the ε-$Fe_{2\text{-}3}$(N,C) phase but unlike after ferritic nitrocarburizing process, the austenite layer forms between the ε phase and diffusion zone (Fig. 18b) [57].

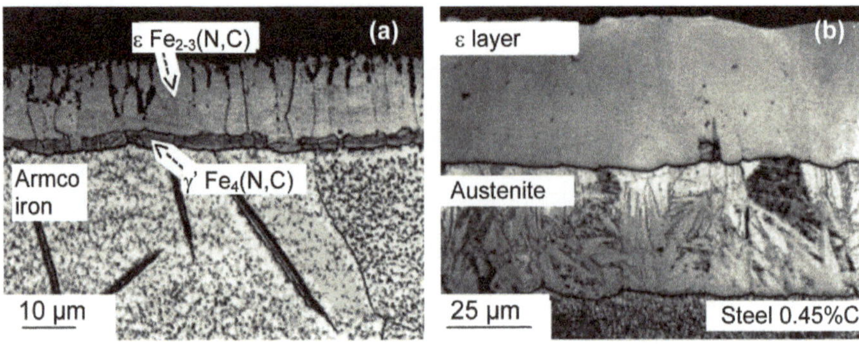

Figure 18. Microstructure differences after low temperature and high temperature plasma nitrocarburizing in atmosphere of 87% N_2 + 8% H_2 + 5% CO_2: (a) Armco iron, 570°C for 3h; (b) 0.45% C steel, 700°C for 3 h [57] (with permission from Elsevier Science)

There are benefits to surface corrosion resistance after combining the plasma nitrocarburizing and oxidizing [65]. The carbonitrided SKD61 steel with a 10 μm thick compound layer (predominantly ε-$Fe_{2\text{-}3}$(N,C) and small proportions of γ′- Fe_4(N,C)) and 200 μm thick diffusion layer subjected to plasma oxidation at 500 °C for 1 h creates 1-2 μm thick magnetite Fe_3O_4 layer on top of the compound layer [66]. According to the anodic polarization test, a significant improvement in the steel corrosion resistance is achieved.

Phase	N (at. %)	C (at. %)	Crystallography	Atom arrangement	Reference
α-Fe	0-37	0-0.02	Fe bcc,	N, C in octahedral sites	[17]
θ-Fe_3C	0	25	Fe complicated orthorhombic	C in bicapped trigonal prisms	[64] [67]
γ′-Fe_4N_{1-z}	19.4-20	<0.7	Fe fcc,	N ordered in central octahedral sites	[61] [68] [69]
ε-$Fe_3(N,C)_{1+x}$	15-33	0-8	Fe hcp,	N ordered in octahedral sites	[61] [68] [69]

Table 4. Characteristics of phases in Fe-N-C system at 580-590 °C

3.3. Applications

In addition to application of nitrocarburizing to carbon and nitriding steels to increase their surface hardness and tribological performance it is also used to stainless steels and special alloys. After low-temperature plasma nitrocarburizing at 450 °C of austenitic stainless steel AISI 304, the dual layer structure grows with a nitrogen-enriched layer on top of a carbon enriched layer. Both layers are free of nitride and carbide precipitates [70]. In addition to increased surface hardness up to 1500 HV and improved wear resistance, the corrosion resistance is also positively altered. There is a difference in corrosion resistance between processes conducted at various temperatures below 450 °C. When treatments conducted at 380°C and 415°C lead to similar properties, increasing temperature to 430 °C causes slightly higher corrosion resistance. The latter is attributed to the formation of a small amount of chromium nitride in the nitrogen-enriched surface layer. The overall improvement in corrosion resistance after nitrocarburizing of AISI 304 stainless steel is thought to be due to the extremely large supersaturation of an upper part of the nitrogen-enriched layer with both nitrogen and carbon. Also sintered Astaloy CrM® + 0.3% C, nitrocarburized in a salt bath at 580 °C for at least 2 h, experiences an increased corrosion resistance [71]. Its surface layer after the treatment is dominated by the ε-iron carbonitride $Fe_{2\text{-}3}(CN)$.

4. Carburizing

The objective of carburizing is to enrich surface layers of steel or other alloys with carbon. To achieve the sufficient carbon solubility and penetration depth the treatment is carried out at relatively high temperatures of 900-950 °C. As a result, steels, which do not have the sufficient carbon content within their volume, obtain the hard surface. The reduced carbon content is deliberately selected to retain the core toughness.

The endothermic carburizing atmospheres consist of a mixture of carburizing ingredients such as CO and CH_4 and decarburizing ones such as CO_2 and H_2O. To control the process, the carburizing potential of the furnace atmosphere requires the measurement of all the gas constituents CO, CO_2, CH_4 and H_2O. The driving force for carburizing is determined by the gradient between potentials of carbon in the furnace atmosphere and carbon at the steel surface. The key reactions of carburizing involve [72]:

$$2CO \rightarrow C_{(g-Fe)} + CO_2 \tag{3}$$

$$CH_4 \rightarrow C_{(g-Fe)} + 2H_2 \tag{4}$$

$$CO + H_2 \leftrightarrow C_{(g-Fe)} + H_2O \tag{5}$$

A variety of applications of steel carburizing were explored for decades with typical examples of automotive gears. This includes also stainless steels, in particular the ferritic and austenitic stainless grades. Recently, the carburizing process creates a growing attention in area of martensitic stainless steels. A comparison of hardness depth profiles for all three families of stainless steels is shown in Fig. 19. A relatively novel, low temperature gas carburizing at 470°C increases the surface hardness of AISI 316 austenitic stainless steel from 200 HV to 1000 HV through extreme supersaturation of up to 12 at.% carbon in the solid solution [73] [74]. After treatment, two types of carbides M_5C_2 and M_7C_3 form with long needles or laths morphology, exhibiting the special orientation relationship with the austenitic matrix (Fig. 20). It is claimed that the carburizing technique which combines the superplastic deformation and the carbon diffusion generates a thicker layer and substantially higher hardness [75]. For duplex stainless steel JIS US329J1 the surface hardness of 1648 HV is achieved as compared to 1300 HV for conventional carburizing. The plasma carburizing of AISI 410 stainless steel in a gas mixture of 80% H_2 + 20% Ar with 0.5-1% of CH_4 by volume, leads to surface hardness of 600-800 HV with no evidence of reduced corrosion resistance [76].

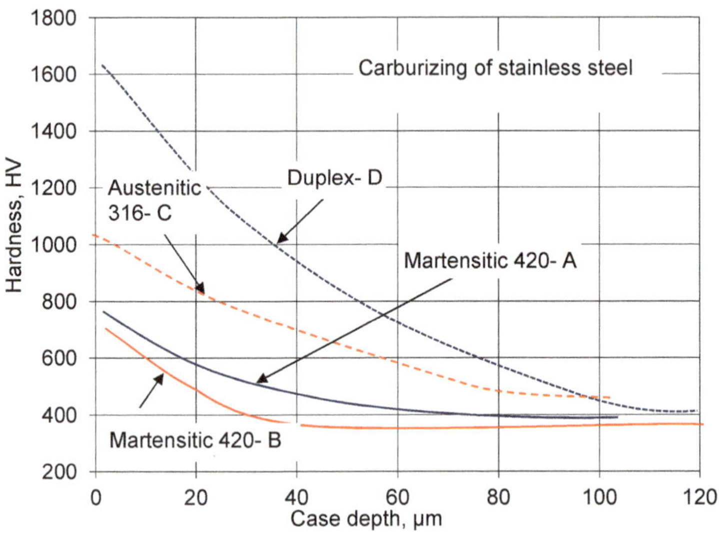

Figure 19. Hardness depth profiles for carburized stainless steel of different grades: A – AISI 420 martensitic stainless steel, carburized in low temperature plasma at 450 °C for 4 h in 1% CH_4 [76]; B – as A but CH_4 concentration of 0.5%; C – AISI 316 austenitic stainless steel, gas carburized at 470 °C for 246 h [74]; D –JIS SUS329J1 duplex stainless steel, superplastically deformed and carburized in powder at 950 °C for 8 h [75] (with permission from Elsevier Science)

In the area of non-ferrous alloys, carburizing is used to increase the wear resistance of some titanium alloys. As a result of double-glow plasma carburizing of the Ti2AlNb orthorhombic alloy, the layer of 40 μm with a hardness of 1051 HV and decreasing carbon content develops [77]. Also plasma carburizing of pure titanium in hydrogen free atmosphere is capable of creating the superficial carburized layer with special characteristics [78]. Of novel applications, carburizing of silicon is portrayed as an inexpensive *in situ* method of forming graphene on silicon wafer [79]. The process is seen as an alternative to the silicon technology.

During carburizing of silicon with carbon, pre-deposited from a carbon source a 3C-SiC(111) film forms because it is well lattice-matched with Si(110). The buffer layer of 3C-SiC(111) consists of hexagonal arrays that act as templates for graphene nucleation and growth.

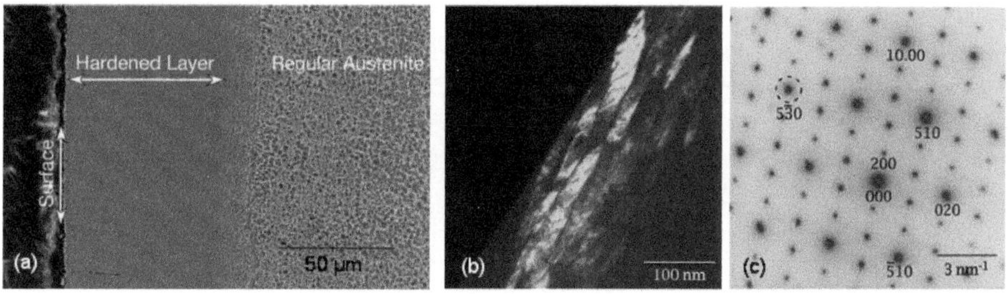

Figure 20. Optical microstructure of AISI 316 austenitic stainless steel after carburizing at 470 °C for 246 h (a) [74] and TEM image with selected area electron diffraction pattern showing carbide morphology (b, c) [73] (with permission from Elsevier Science)

5. Carbonitriding

Carbonitriding is a process similar to carburizing whereby a source of nitrogen is added to the carburizing atmosphere which results in simultaneous incorporation of carbon and nitrogen into alloy surface. Sometimes carbonitriding is confused with nitrocarburizing. It is usually a two-step treatment, conducted at temperatures of 800-940 °C in an environment containing both carbon and nitrogen and is followed by quenching. At carbonitriding temperatures, which are substantially higher than that used during nitriding or nitrocarburizing, steel is in the austenitic state, having high solubility of carbon. To improve toughness, quenching is followed by the second step of low-temperature tempering or stress relieving. At the processing stage, nitrogen inhibits diffusion of carbon, resulting in thinner case, improves hardenability and forms nitrides. After treatment, a presence of nitrogen in carburized steel increases hardness, wear resistance and delays tempering. The latter is of importance for elevated temperature applications. Carbonitriding is widely accepted for surface improvement of plain carbon steels, having low hardenability. According to the comparative study of both processes, carbonitriding and nitrocarburizing develop the compressive stress and are associated with the size and shape distortion [80]. However, nitrocarburizing causes lower compressive stress and size/shape distortion, as is the case for SAE 1010 steel.

Since carbon and nitrogen form with titanium the hard carbides and nitrides, carbonitriding is applicable to titanium and its alloys. In case of laser gas assisted carbonitriding of Ti-6Al-4V alloy, the 55 μm thick layer composed of TiC_xN_{1-x}, TiN and TiC phases grow [81]. In case of pure titanium, carbonitriding at 850 °C for 5 h forms the near-surface layer of carbonitrides and thick layer of α-stabilized solid solution of titanium with nitrogen and oxygen [82]. As the partial nitrogen pressure changes from 105 Pa to 100 Pa and to 10 Pa the surface hardness decreases and composition alters to $TiC_{0.25}N_{0.75}$ to $TiC_{0.50}N_{0.50}$ and $TiC_{0.52}N_{0.48}$, respectively.

6. Boronizing

During boronizing, called also boriding, the surface layer of material is saturated with boron. The process is performed in solid, liquid or gaseous medium and is applicable to any ferrous material as well as to alloys of Ni, Co or Ti. In case of steel it is carried out at temperatures between 840 and 1050 °C for up to 10 h creating borides FeB and Fe_2B, which have a needle-like structure and hardness reaching 2000 HV. In addition to improving wear resistance, boronizing enhances also the corrosion resistance and oxidation resistance at temperatures of up to 850 °C. The main disadvantage of boronizing is the brittleness of the compound layer, especially the FeB phase.

6.1. Application range

For high-carbon-bearing steel AISI 5100, boronizing in solid medium of B_4C, SiC and KBF_4 at temperatures 800-950 °C for up to 8 h creates the single phase layer of Fe_2B with a saw tooth morphology and hardness reaching 1800 HV [83]. The growth rate of boride layer is controlled by boron diffusion in the Fe_2B layer with the boronizing activation energy of 106 kJ/mol. For M2 high speed cutting steel, boronizing at 850-950 °C for up to 8 h, produces the smooth and compact layer with a thickness of up to 130 μm and hardness of 1600-1900 HV [84]. For tool steels the high hardness associated with a presence of borides causes a substantial reduction in toughness [85]. When applied to AISI 304L stainless steel by laser technology, boronizing develops the FeB, Fe_2B, Cr_2B, $Cr_{23}C_6$, Fe_3C and B_4C phases with surface hardness reaching 1490-1900 HV [86].

Boronizing is applicable to titanium alloys and a pack process at 950°C creates a compact, uniform layer composed of TiB_2 and TiB compounds [87]. Also, boronizing of pure nickel in powder-pack at 850-950 °C for up to 8 h creates the 237 μm thick surface layer composed of Ni_2B, Ni_5Si_2 and N_2Si phases with a hardness exceeding 980 HV [88]. The laser boronizing of nodular iron increases hardness five times and produces the fine-crystalline, homogeneous structure of iron borides [89]. The commercial boronizing Titancote™B generates a diffusion layer of complex borides with a thickness of 10-200 μm and hardness of 1600-1800 HV with applications in tooling, oil, gas or general components [90]. In addition to titanium, also other refractory metals such as tantalum, niobium, tungsten and also cobalt-chromium alloys benefit from boronizing. One of many advantages is increasing the surface strength without negatively affecting a biocompatibility.

6.2. Treatments with a boronizing step

The two stage treatment called *borochromizing* consists of chromium plating followed by diffusion boronizing and heat treatment. After powder boronizing of the 20 μm thick chromium coating on C45 carbon steel at 950 °C for 4h, the microstructure, thickness and microhardness are similar to the boride layer [91]. An additional treatment with laser, creates a solid solution or boride eutectics with martensite, reducing maximum hardness to 850 HV. An example of the boride layer grown on pure chromium after boriding in a solid medium at 940 °C for 8h, is shown in Fig. 21a [92]. The process of borochromizing can also be conducted, exploring exclusively thermodiffusion and the duplex salt bath immersion. During such a treatment, chromizing at 1050 °C is followed by boronizing at 950-1050 °C [93]. For DIN 1.2714 steel the treatment leads to a variety of phases such as CrB, Cr_2B, FeB and Fe_2B with the boron diffusion in the pre-chromized layer being the rate controlling step. The single-stage *boroaluminizing* is practiced in the gas phase at temperatures of 850-900 °C with controlled ratios of BF_3 and AlF_3 [94].

Borocarburizing is another two-step process where carburizing is followed by boronizing to generate boronitrides. It was proven that carburizing preceding boronizig reduces brittleness of boronized layers since the hardness gradient between iron borides and the carburized substrate becomes shallower. For 17CrNi6-6 steel, heat treated with laser after borocarburizing, three zones are distinguished, iron borides FeB+Fe_2B of the modified morphology the hardened carburized zone (heat affected zone) and the carburized layer without heat effect [95]. The laser heat-treated borocarburized layer is characterized by higher hardness than the carburized layer, which is attributed to the presence of FeB and Fe_2B phases. For low carbon steels containing Cr and Ni, the borocarburized layer of FeB and Fe_2B with a microstructure shown in Fig. 21b, reached a hardness of 1500-1800 HV with a sub-layer zone being in the range of 700-950 HV [96]. An advantage of the borocarburized layer is in the higher frictional resistance as compared with the single treatment of either boronizing or carburizing. As an extension of borocarburizing, carbonitrided surfaces may be subjected to boronizing hence creating complex (B+C+N) diffusion layers [97]. Although *borocarbonitriding* shows a tendency to reduce the depth of iron borides zone and the microhardness gradient across the surface the resultant wear resistance is higher than that after individual processes. Another benefit of borocarbonitriding is borocarbonitriding is the lower lower temperature and shorter time in comparison with borocarburizing.

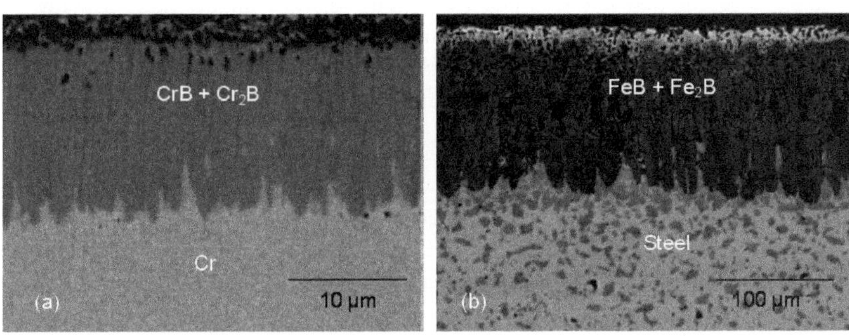

Figure 21. Cross-sectional microstructure after boronizing: (a) pure chromium, solid medium, 940 °C, 8 h [92]; steel 0.15%C, 1.69%Cr and 1.53%Ni, 930 °C, 20 h [96] (with permission from Elsevier Science)

7. Chromizing

The purpose of diffusion chromizing is to enrich surface layers of an alloy with chromium. As other diffusion processes it may be carried out by powder pack, salt bath or fluidized bed. The compound surface layer is formed by a reaction between the carbide former, such as Cr deposited on the surface and carbon in the substrate. The outcome shows similar properties to coatings produced by CVD and PVD. In some sources, the process is divided into soft chromizing, when carbon content in a substrate is below 0.1% and hard chromizing for the carbon content in a substrate exceeding 0.3%. As negative features of chromizing, the shallow penetration depth and the distinct interface with the substrate are often quoted. Both features are caused by the diffusion kinetics of chromium in steel.

7.1. Process and applications

The typical chemistry of chromizing powder consists of 30% of ferrochromium (71%Cr, 0.03%C and Fe as a balance), 2.5% ammonium chloride activator and 67.5% of alumina powder filler [98]. The diffusion depth depends on the temperature and substrate chemistry. It obeys the parabolic rate law and increases with chromizing time and carbon content in the matrix. For temperature of 950 °C and reaction time of 9 h, the diffusion layer thickness reaches 13.2, 22.5 and 27.0 for AISI 1020, 1045 and 1095 steels, respectively. The growth mechanism of chromium diffusion coatings on ferrous alloys was intensively studied in 1980s [99] along with the role of pack geometry, substrate composition, type of halide activator, inner filler, time temperature and chromium source.

Chromizing kinetics can be improved by a combination of conventional thermochemical process with recently developed surface mechanical attrition treatment. The latter aims at refining grains of surface layers into a nanometer range by the repeated plastic deformation such as high velocity ball impacting or mechanical grinding [100]. When a 20 μm thick surface layer with grain size of 10 nm was formed on AISI H13 tool steel, it provided a substantial enhancement of chromium diffusion. The two-step thermochemical treatment of chromizing, with the first step conducted within the stability limit of nano-structures at 600 °C for 2 h, followed by the second-step treatment at 1050 °C for 4 h, created the 30 μm thick layer with a gradient of chromium concentration. The layer contained $(Cr,Fe)_{23}C_6$ and $(Cr,Fe)_2N_{1-x}$ particles with a size below 200 nm.

7.2. Treatments with a chromizing step

Chromizing is often combined into a two-step treatment with nitriding, nitrocarburizing or boronizing. For AISI 1010 steel, nitrocarburized at 572 °C for 2 h, and subsequently chromized by the pack method in a powder of ferrochromium, ammonium chloride and alumina at 1000°C for up to 4h, the layer thickness reaches up to 13 μm with a hardness of 1800 HV [101]. The layer consists of Cr_2N and $(Cr,Fe)_2N_{(1-x)}$ phases. In another example, AISI 1045 steel was first nitrided with 2 μm thick compound layer and hardness of 740 HV and then chromized in powder mixtures consisting of ferrochromium, ammonium chloride and alumina at 1000°C for 2 h [102]. Chromizing of nitrided layer resulted in formation of Cr_2N chromium nitride and Fe_3N iron nitrides. Although an increase in hardness was observed, it did not lead to an improvement in wear resistance. When combining chromizing with boronizing, pack chromium treatment of previously boronized bearing steel provides high wear resistance, particularly in sliding applications [103].

8. Thermo-reactive diffusion

The *thermal diffusion* (TD), *thermo-reactive deposition/diffusion* (TRD) or *TD-Toyota diffusion process* is a high temperature treatment which generates a surface layer of carbides on steel as well as other carbon-containing materials such as nickel or cobalt alloys. In the treatment, carbon in the steel substrate diffuses into the deposited layer with a carbide-forming element such as vanadium, niobium, tantalum, chromium, molybdenum or tungsten. Then, the diffused carbon reacts, forming a compact, metallurgically-bonded coating with a thickness of up to 20 μm. The process is carried out at temperatures from 800 to 1250 °C for up to several hours. Due to the high temperature, steel requires bulk hardening either directly from the TD temperature or after the separate re-heating cycle. The typical hardness of vanadium carbide coatings, obtained using the salt bath TD process, reaches 3200-3800 V [104]. Also, niobium carbide NbC coatings exhibit the high hardness, wear resistance and low friction coefficient along with the high melting point. Coatings are produced by the steel immersion in the molten bath consisting of borax ($Na_2B_4O_7$), boric acid (B_2O_3) and ferro-niobium at 900-1100 °C for up to 10 h [105]. The diffusion of elements from niobium carbide coating to the steel and from the substrate to the coating was found to control the process kinetics for a bath containing more than 10% of ferro-niobium.

9. Hybrid thermochemical treatments

There are a number of surface modification technologies where thermochemical process is a single step in the multi-step treatment. An example of such hybrid is a concept of creating functionally graded materials, exploring a combination of coating and thermochemical treatment (Fig. 22). The single step process of deposition of thick coating with the high hardness is often difficult since they develop microcracks due to a generation of high internal stress [106]. Functionally graded materials offer new strategies for the implementation of high-performance structures in engineering components. They are comprised of continuous or discontinuous varying composition and/or microstructure over definable geometric orientations or distances. As a result they exhibit some unique properties which are beneficial for specific engineering applications. For example, the use of functionally graded systems in high-temperature components can enhance the adhesion and thermo-mechanical response of ceramic coatings deposited on metallic substrates [107].

The difference in phase transformation temperatures between the steel substrate and the Fe-10%Ni electrolytic deposit is an important factor of the thermal treatment proposed [108] [109]. At temperatures below 727 °C, the steel containing 0.9% C is composed essentially of pearlite, i.e. α + Fe_3C. At the same time, the temperature of the α - γ transformation of the coating is approximately 680 °C. By selecting the temperature between 680 and 727 °C, the thermal diffusion treatment can be conducted at the coexistence of α (substrate) - γ (coating) diffusion couple. By contrast, during annealing at a temperature above 727 °C, both the steel substrate and the coating are composed exclusively of austenite (γ). The co-existence of the α - γ or the γ - γ diffusion couples leads to the essentially different redistribution of carbon across the coating thickness and the surface region of the steel substrate.

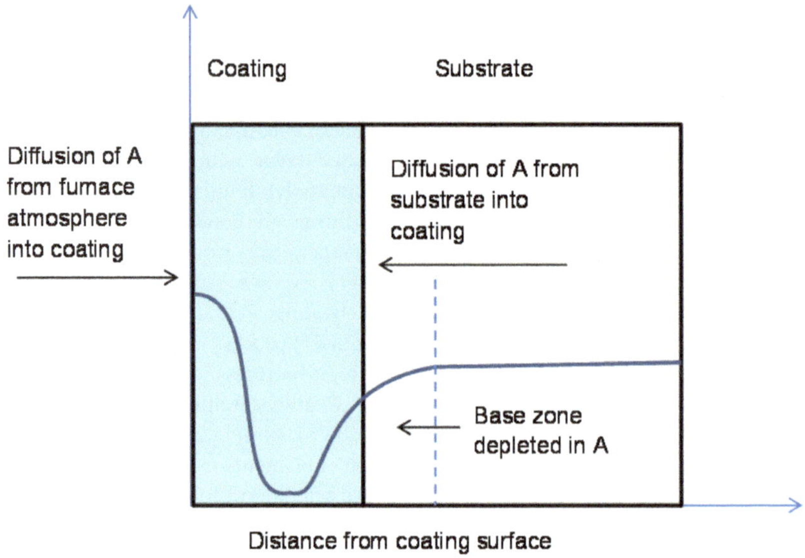

Figure 22. Concept of thermochemical treatment of a coating, exploring simultaneous diffusion from an environment and from the substrate

9.1. Carburizing and diffusion annealing at a coexistence of the α - γ diffusion couple

During annealing at temperatures below α-γ transformation of the steel substrate, the coating already contains a significant amount of the γ phase, which is the solid solution of Ni in γFe (fcc). At the same time, the substrate during annealing remained fully pearlitic. Thus, at the annealing temperature, cementite coagulated, dissolved and acted as a source of carbon, diffusing towards coating. Due to a relatively low temperature, the transport of large amounts of carbon for long distances within coating was difficult. After 30 min of annealing at 710 °C, the mean square root displacement of the carbon in austenite is as low as 32 μm. However, the carbon concentration gradient within the coating caused substantial modifications of the microstructure formed after cooling from annealing temperatures.

An example of cross-sectional image of the microstructure formed after cooling from the two-phase (α-γ) range of coating, is shown in Fig. 23a. The coating cooled from the one-phase region γ comprises continuously graded microstructures caused primarily by differences in carbon content at the coating-gas and coating substrate interface. Since the substrate was not transformed, the decarburization is seen as a thin ferritic layer, adjacent to fully pearlitic microstructure. The hardness depth profile for the treatment performed at 710 °C, is represented by the lower curve in Fig. 24. An increase in hardness is seen in the region adjacent to the substrate and to the outer surface, due to diffusion of carbon from these two directions.

9.2. Carburizing and diffusion annealing at a coexistence of the γ - γ diffusion couple

Significantly different changes in coating microstructure are observed after annealing at temperatures higher than the α-γ transformation of the steel substrate [109]. For example, at 1000 °C the diffusion coefficient of carbon in austenite D_C^{γ} is equal to 2.5×10^{-11} m^2s^{-1} which corresponds to the mean root square displacement of almost 270 μm after 30 min. This means that carbon is capable penetrating the entire coating thickness.

At temperatures above 727 °C, diffusion of carbon within the substrate, towards the substrate-coating interface, takes place in the austenite. As a result, the distribution of carbon in the substrate after cooling has a significantly different character than that described for α-γ diffusion couple. In general, the substrate does not show a ferritic layer but a continuously graded microstructure composed of ferrite and pearlite with an increasing contribution of pearlite, while moving inward from the substrate-coating interface. After 30 min annealing at 1000 °C, the ferritic and pearlitic region is approximately 400 μm thick.

Carburizing at 920 °C allows a higher enrichment of the coating in carbon and the higher hardness after cooling as showed by two upper curves in Fig. 24. The lower hardness in the regions close to the substrate and the outer surface can be explained on the basis of microstructural observations (Fig. 23b). While the coating carburized at 710 °C has a microstructure of acicular ferrite and bainite, the coating carburized at 920 °C is composed of martensite and retained austenite [109]. The high volume fraction of retained austenite in the regions close to the substrate and the outer surface caused the lower hardness.

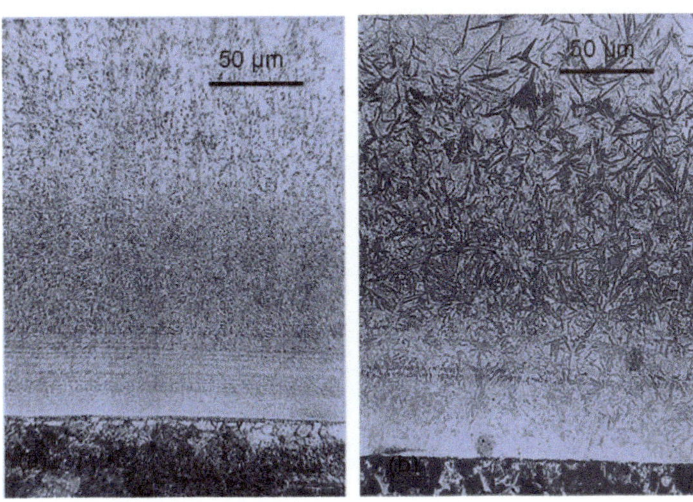

Figure 23. Microstructure of Fe-10%Ni coating on steel substrate after carburizing at temperatures of 670 °C (a) and 920 °C (b) [109] (with permission from Springer Verlag)

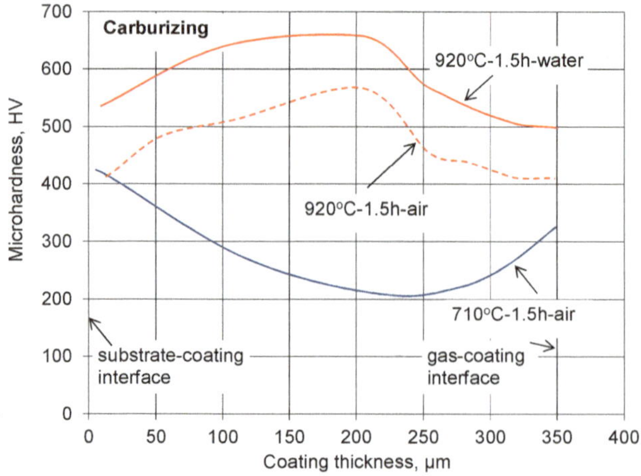

Figure 24. Hardness depth profile within Fe-10%Ni coating on steel substrate after carburizing [109] (with permission from Springer Verlag)

9.3. Carbonitriding and diffusion annealing at a coexistence of the α-γ diffusion couple

This hybrid treatment explores simultaneous carbonitriding at the coating gas interface and carburizing at the coating-substrate interface. At carbonitriding temperatures the substrate also acts as a source of carbon and, in fact, during these processes the flux of the element causing hardening (C,N) is moving from two interfaces the substrate/coating and gas/coating [109]. The resultant microstructure is shown in Fig. 25.

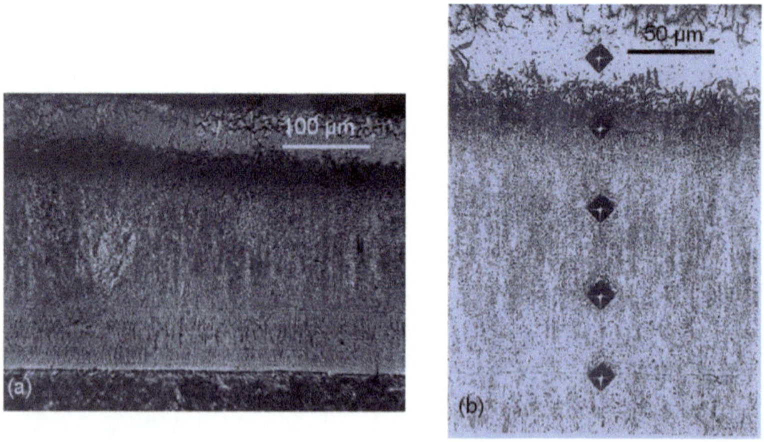

Figure 25. Microstructure of Fe-10%Ni coating after nitrocarburizing at 670 °C for 1.5 h in solid medium [109] (with permission from Springer Verlag)

The microhardness profile across the coating exhibits the maximum located in the sub-surface region (Fig. 26). A comparison with the corresponding microstructure indicates that the hardness peak is caused by a layer of carbonitrides, typically situated in the near-surface region. It should be emphasized that during carbonitriding, the microstructural changes in the coating are accompanied by the changes in the substrate. The extent of those changes is essentially the same as that described previously for diffusion annealing.

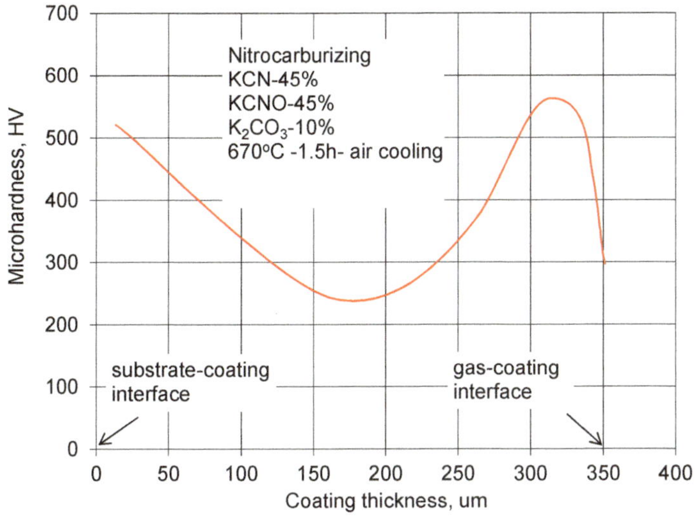

Figure 26. Hardness depth profile within Fe-10%Ni coating on steel substrate after nitrocarburizing [109] (with permission from Springer Verlag)

10. Summary

This chapter shows a variety of surface modification technologies, exploring the phenomenon of thermochemical diffusion. Although an idea of the thermochemical treatment originated at the beginning of the 20th century, it is still a subject of scientific research. At the commercial level, there is a continuous improvement of existing technologies, expansion to novel treatments and a search for unique applications. Of particular interests are hybrids which explore a combination of conventional thermochemical processes with new techniques of surface engineering, including surface deformations, cladding, coatings or laser modifications. In practice, a selection of the optimum technique depends on the component size, geometry, material chemistry, service requirements and the process economy. In recent years, also an environmental aspect is getting a growing attention. The key to benefit from opportunities created by thermochemical treatments is knowledge of capabilities of each technology for a particular substrate material under specific service conditions and its implementation at the stage of a component design.

Author details

Frank Czerwinski*

CanmetMATERIALS, Natural Resources Canada, Hamilton, Ontario, Canada

11. References

[1] D. Pye, Practical Nitriding and Ferritic Nitrocarburizing, Materials Park, Ohio: ASM International, 2003.

[2] J. Georges, "US Patent, Nitriding process and nitriding furnace therefore, Plasma Metal SA". Patent 5,989,363, 1999.

[3] J. Yeh, S. Chen, J. Lin, J. Gan, T. Chin, T. Shun, C. Tsau and S. Chang, "Nanostructured high-entropy alloys with multi principal elements:Novel alloy design and outcomes," *Advanced Engineering Materials,* vol. 6, pp. 299-303, 2004.

[4] M. Yoshida, M. Okumiya, R. Ichiki, C. Tekmen, W. Khalifa, T. Y. and T. Hara, "A novel method for the production of AlN film with high adhesion on Al substrate," *Journal of Plasma Fusion Research,* vol. 8, pp. 1447-1450, 2009.

[5] M. Brady, K. Weisbrod, C. Zawodzinski, I. Paulauskas and R. Buchanan, "Assessment of thermal nitriding to protect metal bipolar plates in polymer electrolyte membrane fuel cells," *Electrochemical and Solid State Letters,* vol. 5, no. 11, pp. A245-247, 2001.

[6] S. Singaravelu, J. Klopf, G. Krafft and M. Kelley, "Laser nitriding of niobium for application to superconducting radio-frequency accelerator cavities," *Journal of Vacuum Science and Technology B,* vol. 29, no. 6, p. 6, 2011.

[7] A. Turk, O. Ok and C. Bindal, "Structural characterization of fluidized bed nitrided steels," *Vacuum,* vol. 80, no. 4, pp. 332-342, 2005.

[8] M. Korvin, W. K. Liliental, C. D. Morawski and G. J. Tymowski, "Design of Nitrided and Nitrocarburized Materials," in *Handbook of Metallurgical Process Design Edited by G.E. Totten, K. Funatani and L. Xie,* New York, Marcel Dekker Inc., 2005, pp. 545-590.

[9] J. Dossett and H. Boyer, Practical Heat Treating, Materials Park, Ohio: ASM International, 2006.

[10] Aerospace Materials Specification 2759-8A, Ion Nitriding, 2007.

[11] Advanced Heat Treat Corp, 2012. [Online]. Available: www.ahtweb.com.

[12] A. Ricard, J. Deschamps, J. Godard, L. Falk, H. Michel and A. Eng, "Nitrogen atoms in ArN2 flowing microwave discharge for steel surface nitriding," *Materials Science and Engineering A,* vol. 139, pp. 9-14, 1991.

[13] C. Li, T. Bell and H. Dong, "A study of active screen plasma nitriding," *Surface Engineering,* vol. 18, no. 3, pp. 174-181, 2002.

[14] P. Schauf, "Iron nitrides and laser nitriding of steel," *Hyperfine Interactions,* vol. 111, pp. 113-119, 1998.

* Corresponding Author

[15] D. Manova, D. Hirsch, J. Gerlach, T. Hoche, S. Mandl and H. Neumann, "Nitriding of Fe-Cr-Ni steel by low energy ion implantation," *Surface and Coating Technology,* vol. 11, pp. 2443-2447, 2008.
[16] R. Wei, J. Vajo, J. Matossian, P. Wilbur, J. Davis, D. Williamson and G. Collins, "A comparative study of beam ion implantation, plasma ion implantation and nitriding of AISI 304 stainless steel," *Surface and Coatings Technology,* vol. 83, no. 1-3, pp. 235-242, 1996.
[17] H. Wriedt, N. Gokcen and R. Nafziger, *Bulletin of Alloy Phase Diagrams,* vol. 8, p. 355, 1997.
[18] N. Kardonina, A. Yurovskikh and A. Kolpakov, "Transformations in the Fe-N system," *Metal Science and Heat Treatment,* vol. 52, no. 9, pp. 457-467, 2010.
[19] T. Woehrle, A. Leineweber and E. Mittemeijer, "The shape of Nitrogen concentration-depth profiles in gamma prime layers growing on alfa iron substrates; The thermodynamics," *Metallurgical and Materials Transactions,* vol. 43A, pp. 610-618, 2012.
[20] B. Kooi, M. Sommers and E. Mittemeijer, "An evaluation of the Fe-N phase diagram considering long range order of N atoms in gamma prime Fe4N and epsilon Fe2N," *Metallurgical and Materials Transactions A,* vol. 27A, pp. 1996-1061, 1996.
[21] T. Gressman, M. Wohlschlogel, S. Shan, U. Welzel, A. Leineweber, E. Mittemeijer and Z. Liu, "Elastic anisotropy of gamma prime Fe4N and elastic grain interaction in gamma prime Fe4n(1-n) layers on alfa iron: first-principles calculations and diffraction stress measurements," *Acta Materialia,* vol. 55, pp. 5833-5843, 2007.
[22] S. Meka, K. Jung, E. Bischoff and E. Mittemeijer, "Unusual precipitation of amorphous silicon nitride upon nitriding Fe-2at%Si alloy," *Philosophical Magazine,* vol. 92, no. 11, pp. 1435-1455, 2012.
[23] Aerospace Materials Specification 2759-10A, Automated Gaseous Nitriding Controlled by Nitriding Potential, 2006.
[24] J. Davis, Gear Materials, Properties and Manufacture, Materials Park, Ohio: ASM International, 2005.
[25] S. E. D. Engineering Directorate, "Process specification for ion nitriding, NASA," PRC 2004 REV A, 2006.
[26] V. Homberg and C. Floe, Nitralloy and nitriding including the new Floe process, Nitralloy Corporation, 1954.
[27] Z. Kolozsvary, "Influence of oxygen in plasma nitriding," *International Heat Treatment and Surface Engineering,* vol. 3, no. 4, pp. 153-158, 2009.
[28] P. Extended, "www.rubig.com," Rubig Engineering, 2012. [Online].
[29] E. Rolinski, A. Konieczny and G. Sharp, "Influence of nitriding mechanisms on surface roughness of plasma and gas nitrided/nitrocarburized gray cast iron," *Heat Treating Progress,* vol. 3, pp. 39-46, 2007.
[30] E. Rolinski, A. Konieczny and G. Sharp, "Nature of surface changes in stamping tools of gray and ductile cast iron during gas and plasma nitrocarburizing," *Journal of Materials Engineering and Performance,* vol. 1059, pp. 1-8, 2009.

[31] V. Terentiev, M. Michugina, A. Kolmakov, V. Kvedaras and V. Ciuplys, "The effect of nitriding on fatigue strength of structural alloys," *Mechanika,* vol. 2, no. 64, pp. 12-22, 2007.

[32] T. Ericsson, "Residual stress caused by thermal and thermochemical surface treatments," *Advances in Surface Treatments,* vol. 87128, pp. 87-113.

[33] T. E. R. Watkins, C. Klepser and N. Jayarman, "Measurement and analysis of residual stress in epsilon iron nitride layers as a function of depth," *Advances in X-ray analysis,* vol. 43, pp. 31-38, 2000.

[34] "Nickel Alloy Nitriding Steels, Nickel Alloy Steels Data Book, Section 4, Bulletin A," *The International Nickel Company,* 1968.

[35] D. Girodin, "Deep nitrided 32CrMoV13 steel for aerospace bearings applications," *NTN Technical Review,* vol. 76, pp. 24-31, 2008.

[36] Ovako Ltd, 2012. [Online]. Available: www.ovako.com.

[37] S. Sun, X. Li and T. Bell, "X-ray diffraction characterization of low temperature plasma nitrided austenitic stainless steel," *Journal of Materials Science,* vol. 34, pp. 4793-4802, 1999.

[38] S. Corujeira Gallo and H. Dong, "New insight into the mechanism of low-temperaure active-screen plasma nitriding of austenitic stainless steel," *Acta Materialia,* vol. 67, pp. 89-91, 2012.

[39] K. Shetty, S. Kumar and P. Rao, "Ion nitriding of maraging steel (250 grade) for aeronautical applications," *Journal of Physics: Conference Series,* vol. 100, pp. 1-6, 2008.

[40] W. Tang, M. Chuang, S. Lin and J. Yeh, "Microstructure and mechanical properties of plasma nitrided Al0.3CrFe1.5MnNi0.5 high entropy alloy," *Metallurgical and Materials Transactions A,* vol. 43A, pp. 2390-2400, 2012.

[41] "Edelstahl Witten-Krefeld GMBH," Nitrodur nitriding steels, [Online]. Available: www.schmolz-bickenbach.co.za.

[42] A. Zhecheva, W. Sha, S. Malinov and A. Long, "Enhancing the microstructure and properties of titanium alloys through nitriding and other surface engineering methods," *Surface & Coating Technology,* vol. 200, pp. 2192-2207, 2005.

[43] A. Lisiecki and A. Klimpel, "Diode laser gas nitriding of Ti6Al4V alloy," *Archives of Materials Science and Engineering,* vol. 31, no. 1, pp. 53-56, 2008.

[44] M. Nagae, S. Okada, M. Nakanishi, J. Takada, Y. Hiraoka, Y. Takemoto, M. Hida, H. Kuwahara and M. Ki Yoo, "Nitriding of dilute Mo-Ti alloys at a low temperature of 1373 K," *International Journal of Refractory Metals and Hard Materials,* vol. 16, no. 2, pp. 127-132, 1998.

[45] Editorial, "A new hard material for corrosive environments," *Sealing Technology,* vol. 1998, no. 50, pp. 10-12, 1998.

[46] F. Czerwinski, Magnesium Injection Molding, New York: Springer Verlag, 2008.

[47] A. Momozawa, S. Taubert, S. Nomura, K. Komurasaki and Y. Arakawa, "Nitriding of zirconium cathode for arc-heater testing in air," *Vacuum,* vol. 85, no. 5, pp. 591-595, 2010.

[48] R. Milani, R. Cardoso, T. Belmonte, C. Figueroa, C. Perottoni, J. Zorzi, G. Soares and I. Baumvol, "Nitriding of yttria-stabilized zirconia in atmospheric pressure microwave plasma," *Journal of Materials Research,* vol. 24, no. 6, pp. 2021-2028, 2009.

[49] P. Visuttipitukul, T. Aizawa and H. Kuwahara, "Advanced plasma nitriding for aluminum and aluminum alloys," *Materials Transactions,* vol. 44, no. 12, pp. 2695-2700, 2003.
[50] M. Lee, S. Endoch and H. Iwata, "A basic study on the solid-state nitriding of aluminum by mechanical alloying using a planetary ball milling," *Advanced Powder Technologies,* vol. 8, pp. 291-297, 1997.
[51] L. Liu, A. Yamamoto, T. Hishida, H. Shoyama, T. Hara and H. Tsubakino, "Microstructure of nitrided AA5052 aluminum alloy formed by electron beam excited plasma technique," *Materials Transactions,* vol. 46, no. 3, pp. 687-690, 2005.
[52] J. Pongsopa, P. Visuttpitukul and B. Paosawatyanyong, "Surface hardening of aluminum-copper 2011 byRF plasma nitriding process," in *8th International Conference on Fracture and Strength of Solids 2010,* Kuala-Lumpur, Thailand, 2011.
[53] P. Visuttipitukul and T. Aizawa, "Plasma nitriding design for aluminum and aluminum alloys," *Surface Engineering,* vol. 22, no. 3, pp. 187-195, 2006.
[54] M. Moradshahi, M. Mahamoudi, M. Amiri and A. Dabizadeh, "The effect of time and temperature on increasing the corrosion resistance of pure aluminum by DC plasma nitriding," *Metallurgical Analysis,* vol. 31, no. 1, pp. 7-13, 2011.
[55] I. Ghauri, R. Ahmad, F. Mubaric, N. Afzal, S. Ahmed and R. Ayub, "Effects of plasma nitriding on the tensile properties of Al-Mg-Si," *Journal of Materials Engineering and Performance,* pp. 1-4, 2011.
[56] J. Bosslet and M. Kreutz, "www.durferrite.de," [Online].
[57] T. Bell, Y. Sun and A. Suhadi, "Environmental and technical aspects of plasma nitrocarburizing," *Vacuum,* vol. 59, pp. 14-23, 2000.
[58] W. Liliental, L. Maldzinski, T. Tarfa and G. Tymowski, "Nitrocarburizing vs nitriding in industrial applications," *Nitrex Metal,* vol. www.Nitrex.com, pp. 1-6, 1999.
[59] M. Nikolova, P. Danev, I. Dermendjiev and D. Gospodinov, "Vacuum Oxy-nitrocarburization of ultra-fine electrolytic iron," *Procedia Engineering,* 2011.
[60] N. M. T. Inc., "www.surfacefinishing.com," Nitrex Metal Technologies Inc., 2012. [Online].
[61] T. Woehrle, A. Leineweber and E. Mittemeijer, "Microstructural and phase evolution of compound layers growing on alpha iron during gaseous nitrocarburizing," *Metallurgical and Materials Transactions A,* vol. 43, pp. 2401-2413, 2012.
[62] E. Mittemeijer and M. Sommers, "Thermodynamics, kinetics and process control of nitriding," *Surface Engineering,* vol. 13, no. 6, pp. 483-487, 1997.
[63] M. Sommers and E. Mittemeijer, *Surface Engineering,* vol. 3, pp. 123-137, 1987.
[64] T. Gressmann, M. Nikolussi, A. Leineweber and E. Mittemeijer, "Formation of massive cementite layers on iron by ferritic carburizing in the additional presence of ammonia," *Scripta Materialia,* vol. 55, pp. 723-726, 2006.
[65] S. Hoppe, "Fundamentals and applications of teh combination of plasma nitrocarburizing and oxidizing," *Surface and Coatings Technology,* vol. 98, no. 1-3, pp. 1199-1204, 1998.
[66] I. Lee and K.-H. Jeong, "Plasma post oxidation of plasma nitrocarburized SKD 61 steel," *Journal of Materials Science and Technology,* vol. 24, no. 1, pp. 136-138, 2008.

[67] M. Nikolussi, A. Leineweber and E. Mittemeijer, "Microstructure and crystallography of massive cementite layers on ferrite substrates," *Acta Materialia,* vol. 56, pp. 5837-5844, 2008.

[68] H. Du, *Journal of Phase Equilibria,* vol. 14, pp. 682-693, 1993.

[69] J. Kuntze, Nitrogen and Carbon in Iron and Steel, Berlin: Academic Verlag, 1990.

[70] Y. Sun, "Enhancement in corrosion resistance of austenitic stainless steels by surface alloying with nitrogen and carbon," *Materials Letters,* vol. 59, pp. 3410-3413, 2005.

[71] M. Teimouri, M. Ahmadi, N. Pirayesh, M. Khoee, H. Khorsand and S. Mirzamohammadi, "Study of corrosion behaviour of nitrocarburized sintered Astaloy CrM+C," *Journal of Alloys and Compounds,* vol. 477, no. 1-2, pp. 591-595, 2009.

[72] S. Sahay and K. Mitra, "Cost model based optimization of carburizing operation," *Surface Engineering,* vol. 20, no. 5, pp. 379-384, 2004.

[73] F. Ernst, Y. Cao and G. Michal, "Carbides in low-temperature carburized stainless steel," *Acta Materialia,* vol. 52, pp. 1469-1477, 2004.

[74] Y. Cao, F. Ernst and G. Michal, "Colossal carbon supersaturation in austenitic stainless steels carburized at low temperatures," *Acta Materialia,* vol. 51, pp. 4171-4181, 2003.

[75] I. Jauhari, S. Rozali, N. Masdek and O. Hiroyuki, "Surface properties and activation energy analysis for superplastic carburizing of duplex stainless steel," *Materials Science and Engineering A,* vol. 466, pp. 230-234, 2007.

[76] C. Scheuer, R. Cardoso, R. Pereira, M. Mafra and S. Brunatto, "Low temperature plasma carburizing of martensitic stainless steel," *Materials Science and Engineering,* vol. A539, pp. 369-372, 2012.

[77] D. Zhu, W. Liang, Q. Miao, K. Guo and L. Li, "Tribological behaviour of Ti2AlNb by surface plasma carburizing," *Advanced Materials Research,* Vols. 482-484, pp. 914-918, 2012.

[78] Z. Qin, S. Rong and X. Rong, "GDA and ToF-SIMS of plasma carburized layer on pure titanium with hydrogen free," *Advanced Materials Research,* Vols. 403-408, pp. 24-27, 2012.

[79] S. Cho, H. Kim, M. Lee, D. Lee and B. Kim, "Direct formation of graphene layers on top of SiC during the carburization of Si substrate," *Current Applied Physics,* vol. 12, no. 4, pp. 1088-1091, 2012.

[80] V. Campagna, V. Bowers, D. Northwood, X. Sun and P. Bauerle, "Comparison of carbonitriding and nitrocarburizing on size and shape distortion of plain carbon SAE1010 steel," *Surface Engineering,* vol. 27, no. 2, pp. 86-91, 2011.

[81] B. Yilbas, S. Akhtar, A. Matthews, C. Karatas and A. Leyland, "Microstructure and thermal stress distributions in laser carbonitriding treatment of Ti-6Al-4V alloy," *Journal of Manufacturing Science and Engineering, Transactions of the ASME,* vol. 133, no. 2, p. 021013, 2011.

[82] O. Yashiv, "Surface hardening of titanium by noncontact thermodiffusion carbonitriding," *Materials Science,* vol. 44, no. 5, pp. 659-664, 2008.

[83] M. Ipek, G. Celebi Efe, S. Ozbek, S. Zeytin and C. Bindal, "Investigation of boronizing kinetics of AISI 51100 steel," *Journal of Materials Engineering and Performance,* vol. 21, no. 5, pp. 733-739, 2012.

[84] I. Ozbek and C. Bindal, "Kinetics of borided AISI M2 high speed steel," *Vacuum,* vol. 86, no. 4, pp. 391-397, 2011.
[85] P. Jurci and M. Hudakova, "Diffusion boronizing of H11 hot work tool steel," *Journal of Materials Engineering and Performance,* vol. 20, no. 7, pp. 1180-1187, 2011.
[86] J. Deebasree, R. Thirumurungesan, P. Shankar and R. Suba Rao, "Laser boriding of austenitic type 304L stainless steel," in *International Symposium fo Research Students on Materials Science and Engineering,* Chennai, India, 2004.
[87] L. Lei, F. Li, X. Yi, Z. Xi and Z. Fan, "Pack RE-boronizing of TC4 titanium alloy at lower temperature," *Heat Treatment of Metals,* vol. 37, no. 2, pp. 101-105, 2012.
[88] D. Mu, B. Shen, C. Yang and X. Zhao, "Microstructure analysis of boronized pure nickel using boronizing powders with SiC as dilutant," *Vacuum,* vol. 83, no. 12, pp. 1481-1484, 2009.
[89] M. Paczkowska, "Laser boronizing and its potential application," *Welding International,* vol. 26, no. 4, pp. 251-255, 2012.
[90] "Boronizing -Titancote B," Richter Precision Inc, [Online]. Available: www.richterprecision.com.
[91] A. Bartkowska, A. Pertek, M. Jankowiak and K. Jozwiak, "Laser surface borochromizing of C45 steel," *Archives of Metallurgy and Materials,* vol. 57, no. 1, pp. 211-214, 2012.
[92] M. Usta, I. Ozbek, C. Bindal, A. Ucisik, S. Ingole and H. Liang, "A comparative study of borided pure niobium, tungsten and chromium," *Vacuum,* vol. 80, pp. 1321-1325, 2006.
[93] M. Aghaie-Khafri and M. Mohamadpour, "A study of chromo-boronizing on DIN 1.2714 steel by duplex surface treatment," *Journal of Metals,* vol. 64, no. 6, pp. 694-701, 2012.
[94] Z. Zakhariev and R. Petrova, "Gas phase reactions during simultaneous boronizing and aluminizing of steels," *Journal of Alloys and Compounds,* vol. 196, pp. 59-62, 1993.
[95] M. Kulka, N. Makuch, A. Pertek and A. Piasecki, "Microstructure and properties of borocarburized and laser-modified 17CrNi6-6 steel," *Optics and Laser Technology,* vol. 44, no. 4, pp. 872-881, 2011.
[96] A. Pertek and M. Kulka, "Characterization of complex (B+C) diffusion layers formed on chromium and nickel-based low carbon steel," *Applied Surface Science,* vol. 202, pp. 252-260, 2002.
[97] M. Kulka and A. Pertek, "Characterization of complex (B+C+N) diffusion layers formed on chromium and nickel-based low carbon steels," *Applied Surface Science,* vol. 218, pp. 113-122, 2003.
[98] W. Lee and J. Duh, "Evaluation of microstructures and mechanical properties of chromized steels with different carbon content," *Surface and Coatings Technology,* Vols. 177-178, pp. 525-531, 2004.
[99] G. Meier, C. Cheng, R. Perkins and W. Bakker, "Diffusion chromizing of ferrous alloys," *Surface and Coatings Technology,* Vols. 39-40, no. C, pp. 53-64, 1989.
[100] S. Lu, Z. Wang and K. Lu, "Enhanced chromising kinetics of tool steel by means of surface mechanical attrition treatment," *Materials Science and Engineering A,* vol. 527, pp. 995-1002, 2010.

[101] O. Ozdemir, S. Sen and U. Sen, "Formation of chromium nitride layers on AISI 1010 stel by nitro-chromizing process," *Vacuum,* vol. 81, no. 5, pp. 567-570, 2007.

[102] F. Hakami, M. Heydarzadeh, J. Rasizadeh and M. Ebrahimi, "Chromizing of plasma nitrided AISI 1045 steel," *Thin Solid Films,* vol. 519, no. 20, pp. 6783-6786, 2011.

[103] S. Sen and U. Sen, "The effect of boronizing and borochromizing on tribological performance of AISI 52100 bearing steels," *Industrial Lubrication and Tribology,* vol. 61, no. 3, pp. 146-153, 2009.

[104] R. Arter, "Japanese technology finds a home in Indiana," *Tooling and Production,* vol. 56, no. 7, pp. 72-74, 1990.

[105] M. Azizi and M. Soltanieh, "Kinetic study of niobium carbide coating formation on AISI L2 steel using thermo-reactive deposition technique," *IJE Transactions B,* vol. 23, no. 1, pp. 77-85, 2010.

[106] F. Czerwinski, "The microstructure and internal stress of Fe-Ni nanocrystalline alloys electrodeposited without a stress reliever," *Electrochimica Acta,* vol. 44, pp. 667-675, 1998.

[107] V. Lee, D. Stinton, C. Berndt, F. Erdogan, Y. Lee and Z. Mutasim, "Concept of functionally graded materials for advanced thermal barrier coating application," *Journal of American Ceramic Society,* vol. 79, p. 3003, 1996.

[108] F. Czerwinski, "Creating functionally graded microstructures by electrodeposition and diffusion annealing," *Plating and Surface Finishing,* vol. 87, no. 6, pp. 68-71, 2000.

[109] F. Czerwinski, "Diffusion annealing of Fe-Ni alloy coatings on steel substrate," *Journal of Materials Science,* vol. 33, pp. 3831-3837, 1998.

Chapter 6

Methodology of Thermal Research in Materials Engineering

Anna Biedunkiewicz

Additional information is available at the end of the chapter

1. Introduction

The mechanism of nanostructural carbides synthesis, and chemical activity of nanoparticles in oxidizing environments, even occurring in trace amounts in high-purity gases, require application of precise methods and performing investigations for a wide range of parameters. The methodology of investigation and the way of determination of selected synthesis conditions have been described in this work. Appropriate selection of investigation methodology enables understanding of process mechanisms, performing quantitative analysis and then correct determination of their conditions.

Good basis for analysis of the processes proceeding with participation of solids are kinetic studies. Kinetic studies can be carried out under isothermal or non-isothermal conditions. The transitions with participation of solid reagents usually proceed in many stages. Each step should be treated as an independent process. The measurements indispensable for identification of process stages and reagents are usually carried out by methods of thermal analysis. There are elaborated different methods of kinetic studies. They are the subject of ongoing discussion [1].

Methods of process kinetics are of great significance for materials engineering. In work [2], for example, kinetics of carbothermal synthesis of β-SiC was investigated; in work [3,4] kinetics and mechanism of carbothermal reduction of MoO_3 to Mo_2C; in work [5] kinetics of thermal decomposition of NH_4VO_3; in work [6] kinetics of nanometric ceramic materials synthesis in argon and their oxidation in dry air were investigated.

Kinetic studies of manufacturing process of carbides of the metals, e.g. titanium, vanadium, niobium, tantalum or silicon, are of particular significance. These metal carbides belong to the group of ceramic materials known as conventionally hard materials, high wear and oxidation resistance. This results from the character of chemical bound and crystallographic structure. The fabrication of ultrafine-grain ceramics by powder- metallurgical processes involves a number of serious difficulties. In particular, it is necessary to use ultrafine powders and to optimize sintering conditions so as to prevent grain growth. Due to their high reactivity of such powders, the process must be run in an inert atmosphere. The composites containing nanostructural carbides show higher strengthening in comparison to their microstructural equivalents [7].

Synthesis of these materials is most often carried out by carbothermal reduction of oxides or precursors containing metal-oxygen-carbon conjugation [1,2,8,9,10]. Other attractive synthesis routes are processes of decomposition of organometallic precursors or synthesis carried out with participation of hydrocarbons, eg. CH_4 or C_6H_6, and salts of transition metals [11,12]. Carbothermal synthesis of the carbides proceeds through stages of oxides formation and than oxycarbides formation being the intermediate products of the syntheses. In the latter case the intermediate products are low-stoichiometry carbides of high vacancies concentration resulting from stability range of these phases in equilibrium metal-carbide system.

Presence of oxygen atoms in MC_xO_{1-x} lattice or high concentration of carbon vacancies in MC_{1-x} lattice cause reduction in hardness and wear and oxidation resistance [12,13].

The synthesis stages leading to elimination of above mentioned lattice defects and obtainment of high-stoichiometry, oxygen stripped carbides, characterized by best properties, proceeds in temperature range above 1000ºC.

At the example of decomposition processes investigations process kinetics of the decomposition of NH_4VO_3 has been analysed. The issues concerning kinetics of those processes have been presented at the example of thermal decomposition of NH_4VO_3 to V_2O_5 in dry air. Vanadium carbides, carbonitrides and borides are known ceramic materials [14-16]. NH_4VO_3 could be used in these processes as a precursor. The results of investigations on thermal decomposition of NH_4VO_3 in dry air have been presented. The base of kinetic description of these processes were termoanalytical TG-DSC measurements. They allowed identifying of intermediate and final products, distinguishing stages of the process, determination of their temperature ranges and acquisition the quantitative description. Process kinetics of the stages were described by Kissinger's method, isoconversional method and applying Coats- Redfern equation.

Obtaining the carbides of high carbonization degree, remaining the right grains size and properties, is essential. Selection of parameters meeting these requirements is difficult. Kinetic studies have significance during investigations of conversions proceeding with use of nanomaterials [17]. In the process of TiC synthesis by sol-gel method, described in work [6], the intermediate product is low stoichiometric MC_{1-x} ($x \leq 0.3$). Carbides nucleate and grow in the carbon matrix. The process is carried out in argon. Describing kinetics, Coats-Redfern's equation was applied. Kinetic models of stages were identified basing on statistical evaluation and compliance to a large extent of conversion degrees of stages calculated and determined from termoanalytical measurements. While building the kinetic models of the processes, the results of measurements were treated as statistic values. A system of a complex analysis of measurements results was developed with the use of artificial neuron networks. Kinetics and by the same, the mechanism of the process of oxidation of nanocrystalline carbides in form of powder were tested, and they were subjected to evaluation based on the comparison of the oxidation rate values.

The possibility to remove the carbon matrix in reaction with oxygen was considered during analysis of the kinetics of the process of TiC_x/C nanocomposite oxidation in dry air. It was proved that it was not possible to purify the obtained nc-TiC by burning in the air the carbon matrix, contained in the system.

2. Thermal decomposition of NH_4VO_3

2.1. Methods kinetic analysis

It is assumed that the rate of non-catalytic, heterophase reactions depends on temperature, conversion degree and pressure

$$r = \frac{d\alpha}{dt} = \phi(T,\alpha,P) \tag{1}$$

After separation of variables we obtain

$$\frac{d\alpha}{dt} = k(T)f(\alpha)h(P) \tag{2}$$

Under thermogravimetric measurements conditions $h(P) \approx 1$ [18]. Then

$$\frac{d\alpha}{dt} = k(T)f(\alpha) \tag{3}$$

Separation of variables means that k(T) function should not depend on conversion degree, and function f(α) should not depend on temperature. Function k(T) is described by Arrhenius equation

$$k(T) = A\exp\left(-\frac{E}{RT}\right) \tag{4}$$

k(T) function maintains its exponential character in relatively narrow range of temperature. Assuming T=0 as integration limit this range is exceeded since k (T) tends asymptotically to the limit values (at $T \to o$ and $T \to \infty$).

By combining (3) and (4) we obtain

$$\frac{d\alpha}{dt} = A\exp\left(-\frac{E}{RT}\right)f(\alpha) \tag{5}$$

Thus, for isothermal conditions we have

$$g(\alpha)=\int_{0}^{\alpha}\frac{d\alpha}{f(\alpha)}=k(T)\mathrm{t} \tag{6}$$

For non-isothermal conditions, at a linear heating rate of sample $\frac{dT}{dt}=\beta$ we obtain

$$g(\alpha)=\frac{A}{\beta}\int_{T\alpha=0}^{T\alpha=1}\exp\left(-\frac{E}{RT}\right)dT \tag{7}$$

The integral has no analytical solution. An important approximate solution is the Coats-Redfern equation [12]

$$\ln\left[\frac{g(\alpha)}{T^2}\right]=F(T)=\ln\left[\frac{AR}{\beta E}\left(1-\frac{2RT_m}{E}\right)\right]-\frac{E}{RT} \tag{8}$$

Based on experimental data for each stage the form of the f(α) or g(α) function most consistent with the experimental data and the parameters of Arrhenius equation A and E should be determined.

During standard thermogravimetric measurements the temperature of the sample, the TG, DTG and HF functions and the mass spectra of evolved gaseous products are recorded. Solid products are identified by XRD method. On this basis the division of the process into stages is made and α(T) dependencies are determined for the stages. The methods of measurement results elaboration and kinetic models recommended by ICTAC Kinetics Comittee are given in work [1]. These include in particular Kissinger's method and isoconversional method.

2.1.1. Kissinger's method

The basis of Kissinger's method are the parameters describing the process rate $\dot{}$ (T_m, α_m), determined at different heating rates of samples [14]. For maximum $\frac{d^2\alpha}{dt^2}=0$. By differentiating equation (5) we obtain

$$\frac{d^2\alpha}{dt^2}=\left\{\frac{E\beta}{RT_m^2}+Af'(\alpha_m)\exp\left[-\frac{E}{RT_m}\right]\right\}\left(\frac{d\alpha}{dt}\right)_m=0 \tag{9}$$

Since $\left(\frac{d\alpha}{dt}\right)_m\neq 0$, therefore

$$\frac{E\beta}{RT_m^2}=-Af'(\alpha_m)\exp\left[-\frac{E}{RT_m}\right] \tag{10}$$

There should be added that the formula (10) requires that $f'(\alpha_m)$<0. In this method therefore only kinetic models fulfilling this condition can be used.

By transforming (10) and adapting the result to the measurements performed at different heating rates of samples the Kissinger equation (11) is obtained

$$\ln\left(\frac{\beta_i}{T_{m,i}^2}\right) = \ln\left[-\frac{AR}{E} f^{'}(\alpha_m)\right] - \frac{E}{RT_{m,i}} \quad (11)$$

While performing calculations, $\ln\left(\frac{\beta_i}{T_{m,i}^2}\right) \div \frac{1000}{T_{m,i}}$ charts are prepared. Then the activation energy E and the value of expression $\ln\left[-\frac{AR}{E} f^{'}(\alpha_m)\right] = B$ are calculated by linear regression method.

Knowing B, $-Af^{'}(\alpha_m)$ is calculated from equation (12)

$$-Af^{'}(\alpha) = \frac{E}{R}\exp(B) \quad (12)$$

To calculate A one needs to know the kinetic model for a given stage (form of the function f(α)). The kinetic model most consistent with the experimental data, of the tested models, was established by analyzing the trajectories of Y(T) function depending on α(T). The conversion degrees α(T) for the stages were estimated on the basis of experimental data, whereas the Y(T) function was calculated from the formula

$$Y(T) = \frac{Af'(\alpha_m)}{\beta_i} \int_{T_{\alpha=0}}^{T_{\alpha=1}} \exp[-\frac{E}{RT}]dT \quad (13)$$

For the stage A $f^{'}(\alpha_m) \approx const.$ The integral was calculated numerically. There should be added that at constant activation energy, trajectories of $\frac{1}{\beta i}\int_{T\alpha=0}^{T\alpha=1} \exp\left[-\frac{E}{RT}\right]dT$ functions depending on the conversion degree, for different βi, should be the same. The ranges $T_{\alpha=0} \div T_{\alpha=1}$ for given βi must be determined experimentally. At higher heating rates of the samples there should be wider temperature range $T_{\alpha=0} \div T_{\alpha=1}$. This is consistent with DTG and HF charts. After determining the form of the function f(α) the derivative $f^{'}(\alpha_m)$ is calculated. Knowing $Af^{'}(\alpha_m)$ and $f^{'}(\alpha_m)$ Arrhenius coefficient A is calculated.

2.1.2. *Isoconversional method*

In the the isoconversional method equation (5), obtained after separation of variables, is also used. Differentiating (5) with respect to ∂T^{-1} we obtain

$$\left[\frac{\partial \ln\left(\frac{d\alpha}{dt}\right)}{\partial T^{-1}}\right]_{\alpha} = \left[\frac{\partial \ln k(T)}{\partial T^{-1}}\right]_{\alpha} + \left[\frac{\partial f(\alpha)}{\partial T^{-1}}\right]_{\alpha} \tag{14}$$

For α = const we have

$$\left[\frac{\partial \ln\left(\frac{d\alpha}{dt}\right)}{\partial T^{-1}}\right]_{\alpha} = -\frac{E_{\alpha}}{R} \tag{15}$$

Integrating (15), taking into account that $\frac{dT}{dt} = \beta$, and generalizing the result for different β_i we obtain

$$\ln\left[\beta_i \left(\frac{d\alpha}{dt}\right)_{\alpha,i}\right] = const - \frac{E_{\alpha}}{RT_{\alpha.i}} \tag{16}$$

Integration constant, according to equation (5), is equal $\ln\left[f(\alpha)A_{\alpha}\right]$. Equation (16) therefore takes the form

$$\ln\left[\beta_i \left(\frac{d\alpha}{dt}\right)_{\alpha,i}\right] = \ln\left[f(\alpha)A_{\alpha}\right] - \frac{E_{\alpha}}{RT_{\alpha,i}} \tag{17}$$

It should be added that the introduction of the coefficients A_{α} and E_{α} depending on conversion degree does not comply with the principle of separation of variables. Equation (17) is used in the differential isoconversional method. The basis of the reaction rate calculations are reaction rates $\left(\frac{d\alpha}{dt}\right)_{\alpha,i}$ determined for selected conversion degrees and different β_i and the corresponding to them temperature values $T_{\alpha,i}$.

In the case of the integral isoconversional method the basis for calculations, for isothermal conditions, is equation

$$g(\alpha) = A\exp\left(-\frac{E}{RT}\right)t \tag{18}$$

Activation energy, in this case, is calculated from the formula

$$\ln t_{\alpha,i} = \ln\left[\frac{g(\alpha)}{A_{\alpha}}\right] + \frac{E_{\alpha}}{RT_{\alpha}} \tag{19}$$

The time of obtaining the assumed conversion degree at different temperatures was designated by $t_{\alpha i}$. The charts of $\ln t_{\alpha,i} \div \frac{1000}{T_i}$ are constructed. The parameters of equation (19) are calculated by linear regression method. For non-isothermal conditions there is no analytical solution. During the calculations of activation energy there are used approximate equations of general form

$$\ln\left(\frac{\beta_i}{T_{\alpha,i}^B}\right) = const - C\frac{E_\alpha}{RT_\alpha} \tag{20}$$

In this work the Kissinger's-Akoshira-Sunose equation was used

$$\ln\left(\frac{\beta_i}{T_{\alpha,i}^2}\right) = const - \frac{E_\alpha}{RT_\alpha} \tag{21}$$

In this case the charts of $\ln\left(\frac{\beta_i}{T_{\alpha,i}^2}\right) \div \frac{1000}{T\alpha}$ are constructed. The parameters of this equation are calculated by linear regression method.

2.2. Experimental

NH_4VO_3 from Fluka was used as a substrate. Decomposition process was carried out in dry air (Messer, Germany) containing 20,5% vol. O_2 rest N_2. Impurities occurred in amounts: H_2O < 10 vpm, CO_2 < 0,5 vpm, NO_x < 0,1 vpm, hydrocarbons < 0,1 vpm. Thermogravimetric measurements were carried out on TG–DSC Q600 (TA Instruments) apparatus. Gaseous products of proceeding transitions were identified by mass spectrometry method. Pfeifer Vacuum ThermoStar GDS 301 apparatus was used. Solid products were identified by XRD method. X'Pert Pro apparatus from PANalytical with a copper X-ray tube with current voltage 35 kV and intensity 40 mA was used. Spectra processing and analysis was performed using X'Pert HighScore 1.0 software with incorporated ICDD spectra library.

2.3. Results

During the TG-DSC measurements weighed amounts of the sample in the order of 20 mg were used. The temperature of the sample, TG, DTG and HF functions, and mass spectra of gaseous products were registered in time. In all series the temperature of samples changed linearly in time. It was found that thermal decomposition of NH_4VO_3 in dry air proceeds in the three endothermic stages according to

$$6\,NH_4VO_3 \rightarrow (NH_4)_3V_6O_{16} \rightarrow (NH_4)_2V_6O_{16} \rightarrow V_2O_5 \tag{22}$$

The theoretical, total mass loss of the sample equals 22,29 wt%. The mass losses in the stages, referred to the initial mass of the sample, equal: 11,48 wt% in stage I, 4,404 wt% in stage II, and 6,66 wt% in stage III. The intermediate products (for control also the final product) were obtained under isothermal conditions, in the temperature ranges of their occurrence. As the final product V_2O_5 (ICDD card 85-0601), and as the second intermediate product $(NH_4)_2V_6O_{16}$ (ICDD card 79-205) were obtained. The intermediate product formed in step I was identified on the basis of the mass balance (there is no pattern of this compound in ICDD directory). In all the stages and at different heating rates of the samples evolved: NH_3, H_2O, NO and N_2O resulting from oxidation of NH_3. In the gas phase NO_2 did not occur. The results of this step of research are given in [19].

2.3.1. *Analysis of influence of sample heating rate on the course of the process*

The influence of sample heating rate on the course of the process was examined on the basis of the TG_u , DTG and HF functions. The measurements were carried out at sample heating rates of: 1; 1,5; 2; 2,5; 3; 3,5; 4; 4,5; 5; 6; 7 ; 8 and 10 K min^{-1}. The isothermal measurements were carried out at the sample heating rate equal to 2 K min^{-1}. Basing on the TG-DSC measurements the division of the process into stages was made and the conversion degrees for the stages were determined. There should be added that the data sets concerning conversion degrees, determined for the stages, are the basis for the description of process kinetics as in [1,17,20-23].

2.3.2. *Analysis of TG_u, DTG and HF charts*

The trajectories of TG_u, DTG and HF plots are presented in separate figures. In Figure 1 TG_u functions in temperature registered during the measurements are presented.

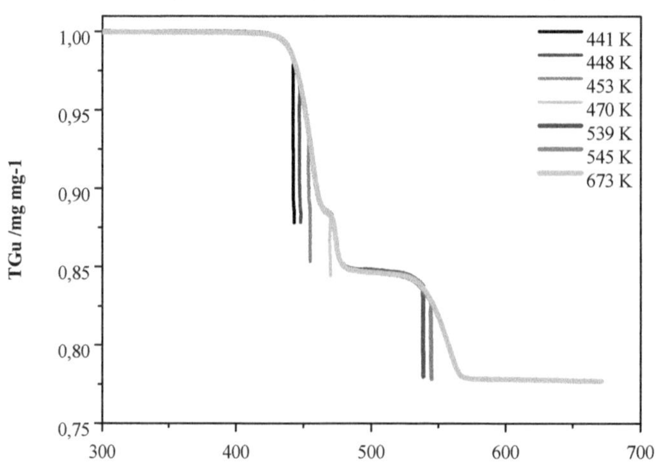

Figure 1. TG_u functions in temperature. Decomposition of NH_4VO_3 in dry air [19].

The linear segments correspond to the isothermal conditions. The samples were heated isothermally until the stable mass has been reached (about 30 min).

In case of the investigated process, under isothermal conditions, the results for the stage could by obtained only at a few temperatures, while at higher temperatures the results were obtained at high conversion degrees. For these reasons, the course of investigated process was not further examined under isothermal conditions.

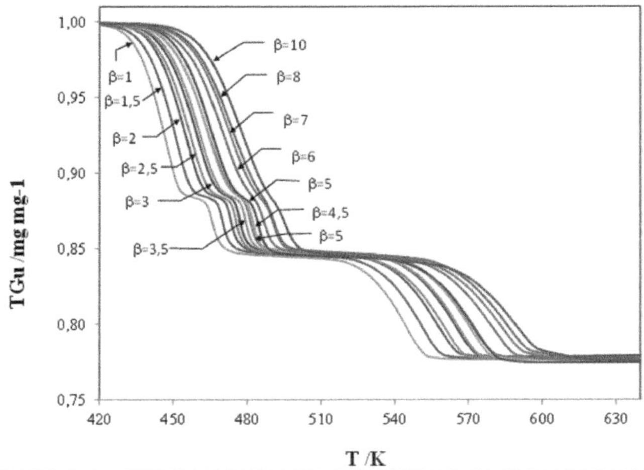

Figure 2. Plot of TG_u functions in temperature. Decomposition of NH_4VO_3 in dry air [19].

In Figure 2 the plots of TG_u function in temperature obtained under non-isothermal conditions are presented. The single curve is formed of about 20 thousand points. The determined TG_u values should depend only on temperature and heating rate of the samples. This was confirmed by neural networks method in reference [24]. All the series were considered simultaneously. The computer software Statistica Neural Network was used. Figure 3 shows an example of DTG plots for selected heating rates of the samples.

It follows that at higher β_i the temperature range $T_{\alpha=0} \div T_{\alpha=1.}$ increases (Fig.3). The temperature range of the process in stage is an important parameter.

Along with the increase of sample heating rates the DTG function plots are shifted into the higher temperature range. It is also visible that at the higher sample heating rates stages I and II are overlapped to a larger extent, and the final segments of plots, corresponding to the stage III, are not monotonic. As mentioned before, this was attributed to the oxidation of formed earlier V_2O_{5-x} to V_2O_5. The T_m temperatures corresponding to peaks of DTG function plots (necessary for Kissinger's method) have been determined. The results are listed in Table 1.

The values of α_m, for T_m temperatures given in Table1 were retrieved from data sets of $\alpha(T)$. There are also given the temperature ranges, determined basing on DTG plots, for the stages.

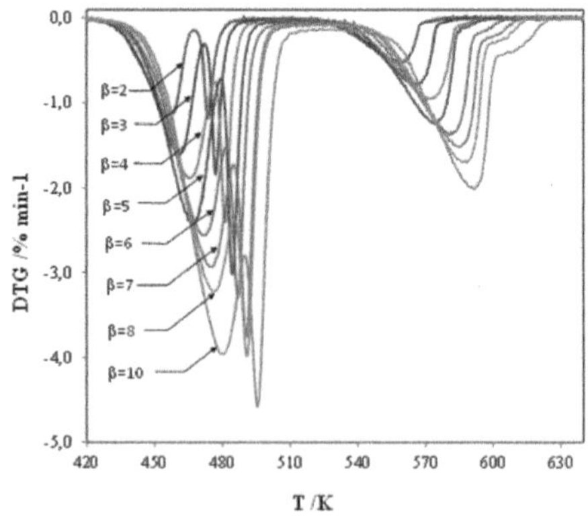

Figure 3. Plots of DTG function in temperature. Decomposition of NH_4VO_3 in dry air [19].

β Kmin^{-1}	Stage I			Stage II			Stage III		
	T_m /K	α_m	ΔT /K	T_m /K	α_m	ΔT /K	T_m /K	α_m	ΔT /K
1	447,83	0,721	410,75-457,95	465,65	0,384	457,95-490,65	543,93	0,617	509,25-566,15
1,5	452,62	0,742	405,55-462,85	469,87	0,415	462,85-489,45	550,38	0,653	513,25-572,15
2	457,48	0,761	413,15-467,55	473,58	0,476	467,55-491,35	558,2	0,705	516,25-580,45
2,5	458,06	0,698	412,85-469,65	475,6	0,443	469,65-494,65	560,65	0,701	518,15-583,45
3	460,18	0,683	415,65-472,15	477,16	0,495	472,15-496,15	565,44	0,729	522,95-590,75
3,5	462,25	0,703	415,65-474,15	479,51	0,474	474,15-497,15	567,21	0,737	523,75-590,75
4	465,12	0,689	413,25-476,15	481,41	0,517	476,15-503,15	572,81	0,744	526,55-596,75
4,5	465,32	0,689	417,25-476,95	481,9	0,38	476,95-500,35	572,17	0,743	526,15-595,15
5	467,29	0,714	418,45-479,35				574,3	0,736	530,45-597,95
6	471,44	0,698	416,45-481,85				580,7	0,719	533,45-609,25
7	475,68	0,741	417,15-485,25				588,2	0,83	537,85-614,15
8	476,22	0,71	421,65-485,25				587,06	0,723	536,25-620,75
10	480,09	0,716	423,15-490,15				591,89	0,73	538,85-625,35

Table 1. List of data for Kissinger's method. Thermal decomposition of NH_4VO_3 in dry air.

In Figure 4 the HF function in temperature has been presented for selected example values of β.

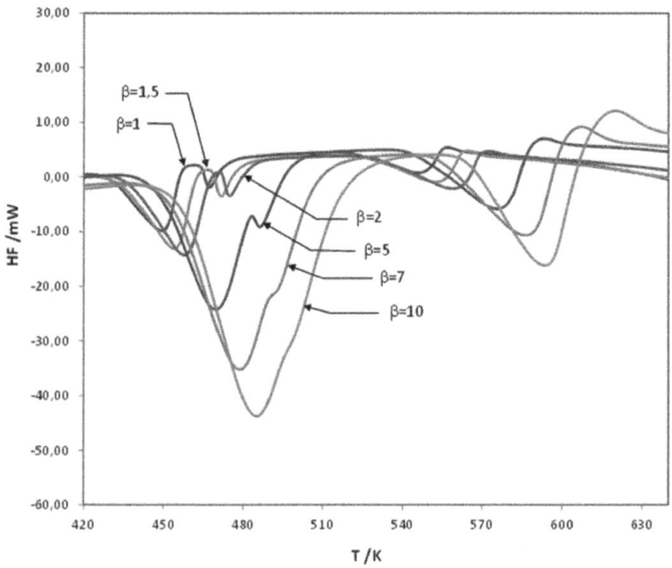

Figure 4. Plots of HF functions in temperature. Decomposition of NH_4VO_3 in dry air.

It is visible from the course of HF plots that in the case there are three endothermic stages referred to the NH_4VO_3 decomposition. The HF plots end with the exothermic transition attributed to the oxidation of small amount of V_2O_{5-x} to V_2O_5. Along with the increase of sample heating rate stages I and II overlap. Peaks of HF function are shifted, with respect to the DTG peaks, by a few degrees into the range of higher temperature. There should be added that NH_3, H_2O, NO and N_2O evolved in all the stages, also at higher simple heating rates.

2.3.3. *Analysis of α(T) plots for determined stages*

The conversion degrees for stages were calculated from the following formula

$$\alpha(T) = \frac{m_0 - m}{m_0 - m_k} \tag{23}$$

The data concerning the TG_u function were the basis for calculations. During the division of the process into stages also DTG and HF plots were taken into account [22-24].

In Figure 5 the results obtained for the stages are presented. One curve was formed of a few thousand points.

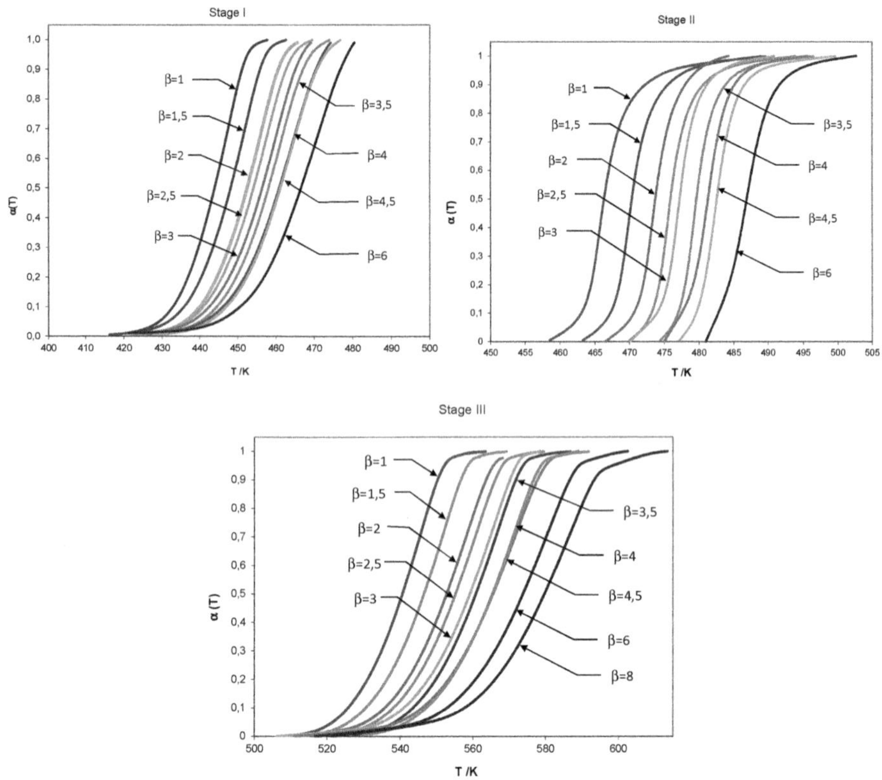

Figure 5. Plots of α(T) depending on temperature for the stages. Thermal decomposition of NH_4VO_3 in dry air.

2.4. Result analysis

2.4.1. Kissinger's method

In Figure 6 the plots of $\ln\left(\frac{\beta_i}{T_{m,i}^2}\right) \div \frac{1000}{T_{m,i}}$ for the stages, obtained on the basis of the data from Table 1, are presented.

The parameters for Kissinger's equation (11) were calculated by linear regression method. The computer software Statistica 6.0 was used. The results are listed in Table 2.

Stage	E kJ mol^{-1}	B*	Af'(α_m) 1 min^{-1}	r_p	model	-f '(α_m)	A 1 min^{-1}
I	117,66	19,496	4,133 E12	0,979	A2	2,476	1,669 E12
II	164,48	30,178	2,37 E17	0,989	A4	6,762	3,504 E16
III	113,75	12,623	4,137 E9	0,985	A2	2,596	1,594 E9

Table 2. List of the results calculated by Kissinger's method. Thermal decomposition of NH_4VO_3 in dry air. * constant in Kissinger's equation (11)

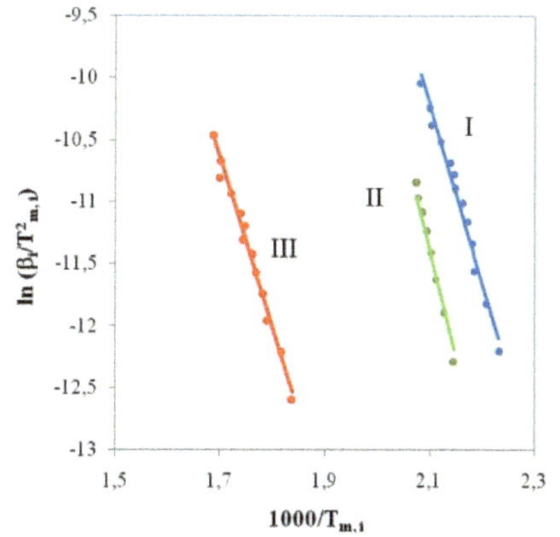

Figure 6. Plots of $\ln\left(\frac{\beta_i}{T_{m,i}^2}\right) \div \frac{1000}{T_{m,i}}$ for the stages. Thermal decomposition of NH_4VO_3 in dry air.

The kinetic models for the stages were determined analyzing the courses of the function

$$Y(T) = \frac{Af'(\alpha_m)}{\beta_i} \int_{T_{\alpha=0,01}}^{T_{\alpha=0.99}} \exp\left(-\frac{E}{RT}\right) dT.$$

There should be added, that the values of α_m determined for the stages changed marginally for different β. Therefore, there could be assumed that $f'(\alpha_m) \approx const$. In figure 7 plots of Y(T) function for the stages depending on α(T), for β = 3K min^{-1}, are presented as an example.

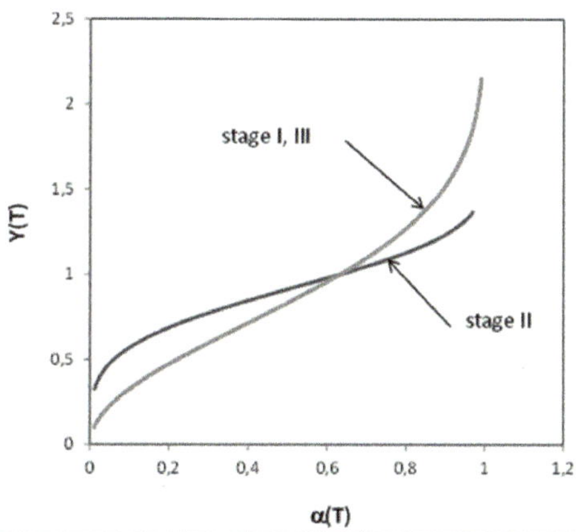

Figure 7. Plots of Y(T) function for the stages depending on α(T) for β = 3K min^{-1}. Thermal decomposition of NH_4VO_3 in dry air.

For the stages I, III the most consistent with experimental data, of the tested models, was the A2 model.

$$g(\alpha)=\left[-\ln(1-\alpha)\right]^{\frac{1}{2}} \tag{24}$$

$$f(\alpha)=2(1-\alpha)\left[-\ln(1-\alpha)\right]^{\frac{1}{2}} \tag{25}$$

$$f^{'}(\alpha_m)=-2\left[-\ln(1-\alpha_m)\right]^{\frac{1}{2}}-\frac{1-\alpha_m}{\alpha_m\left[-\ln(1-\alpha_m)\right]^{\frac{1}{2}}} \tag{26}$$

Whereas for stage II this was the A4 model

$$g(\alpha)=\left[-\ln(1-\alpha)\right]^{\frac{1}{4}} \tag{27}$$

$$f(\alpha)=4(1-\alpha)\left[-\ln(1-\alpha)\right]^{\frac{3}{4}} \tag{28}$$

$$f^{'}(\alpha_m)=-4\left[-\ln(1-\alpha_m)\right]^{\frac{3}{4}}-\frac{3(1-\alpha_m)}{\alpha_m\left[-\ln(1-\alpha_m)\right]^{\frac{1}{4}}} \tag{29}$$

The selection of the model was confirmed making calculations by Coats – Redfern method. The selected models and calculated values of $f^{'}(\alpha_m)$ are given in Table 2.

2.4.2. *Isoconversional method*

The basis for calculations by this method were dependencies α(T) determined for the stages at different heating rates of the samples (Fig. 5). First, the activation energies were calculated for $\alpha_{m,i}$ corresponding to the inflection points of α(T) curves (average values of α_m given in Table 1). In this case, formula (21) should take the form of equation (11). In Figure 8 the obtained plots of $\ln\left(\frac{\beta_i}{T^2_{\alpha_m^{'},i}}\right)\div\frac{1000}{T_{\alpha_m^{'},i}}$ are presented.

The results of calculations are given in Table 3. This method is less accurate than the Kissinger's method

Stage	α_m	E /kJmol^{-1}	B*	r_p
I	0,713	117,3	19,401	0,989
II	0,460	148,2	26,05	0,987
III	0,721	122,3	14,426	0,988

Table 3. List of the results of activation energies calculations by isoconversional method for the inflection points of α(T) curves. *) constant in equation (21).

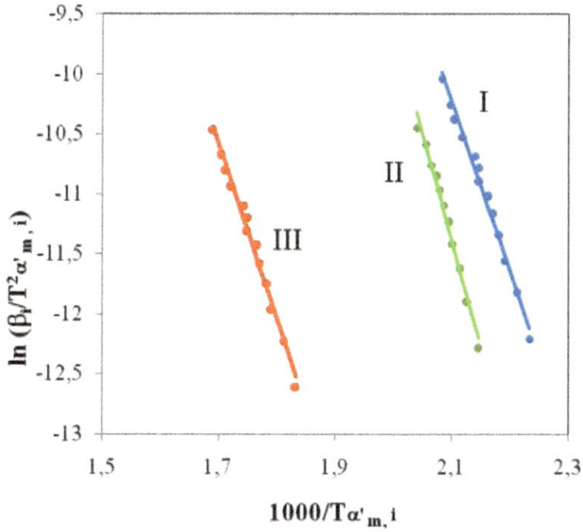

Figure 8. Plots of $\ln\left(\frac{\beta_i}{T^2_{\alpha_m^{'},i}}\right) \div \frac{1000}{T_{\alpha_m^{'},i}}$ for the stages. Isoconversional method. Conversion degrees equal to $\alpha_{m,i}$.

The calculations for other values of conversion degree were also performed. Should be noted that α(T) tend asymptotically to the limit values. In these ranges $T_{\alpha,i}$ could not be determined with sufficient accuracy. Therefore, the calculation was performed 0,1<α < 0,90.

The plots of $\ln\left(\frac{\beta_i}{T^2_{\alpha,i}}\right) \div \frac{1000}{T_{\alpha,i}}$ for the stages are shown in Figure 9.

The values of activation energies calculated by the isokinetic method are given in Table 4.

α_i	Stage I			Stage II			Stage III		
	E kJ mol⁻¹	B*	r_p	E kJ mol⁻¹	B*	r_p	E kJ mol⁻¹	B*	r_p
0,1	140,45	27,00	0,986	163,02	30,083	0,996	147,58	21,299	0,991
0,2	131,69	24,303	0,988	161,22	29,493	0,998	137,71	18,709	0,990
0,3	128,45	23,06	0,991	158,71	28,788	0,992	126,45	16,012	0,991
0,4	125,65	22,099	0,995	155,72	27,987	0,994	130,05	16,577	0,989
0,5	123,13	21,256	0,999	156,50	28,187	0,994	127,46	15,849	0,986
0,6	120,29	20,351	0,989	154,54	27,58	0,989	125,48	15,276	0,987
0,7	117,56	19,489	0,988	156,70	28,065	0,985	122,94	14,597	0,985
0,8	114,75	18,61	0,988	158,2	28,365	0,982	121,83	14,227	0,987
0,9	112,38	17,826	0,989				119,92	13,679	0,983

Table 4. The activation energies calculated for the stages by the isoconversional method. Thermal decomposition of NH_4VO_3 in dry air.* constant in equation (21)

In the case of investigated process, the E for the stages I and III (asymmetric plots of DTG and HF), changed continuously along with the change of conversion degree. In the case of stage II (symmetric plots of DTG and HF) E was practically constant, similar to that determined by Kissinger's method. That is the isokinetic method compensates the impact of β_i on the course of the process by changing the activation energy. Theoretically [25] much more interesting is the possibility to compensate the impact of β_i, at a constant activation energy, with use of temperature ranges $T_{\alpha=0} \div T_{\alpha=1}$, the values determined experimentally.

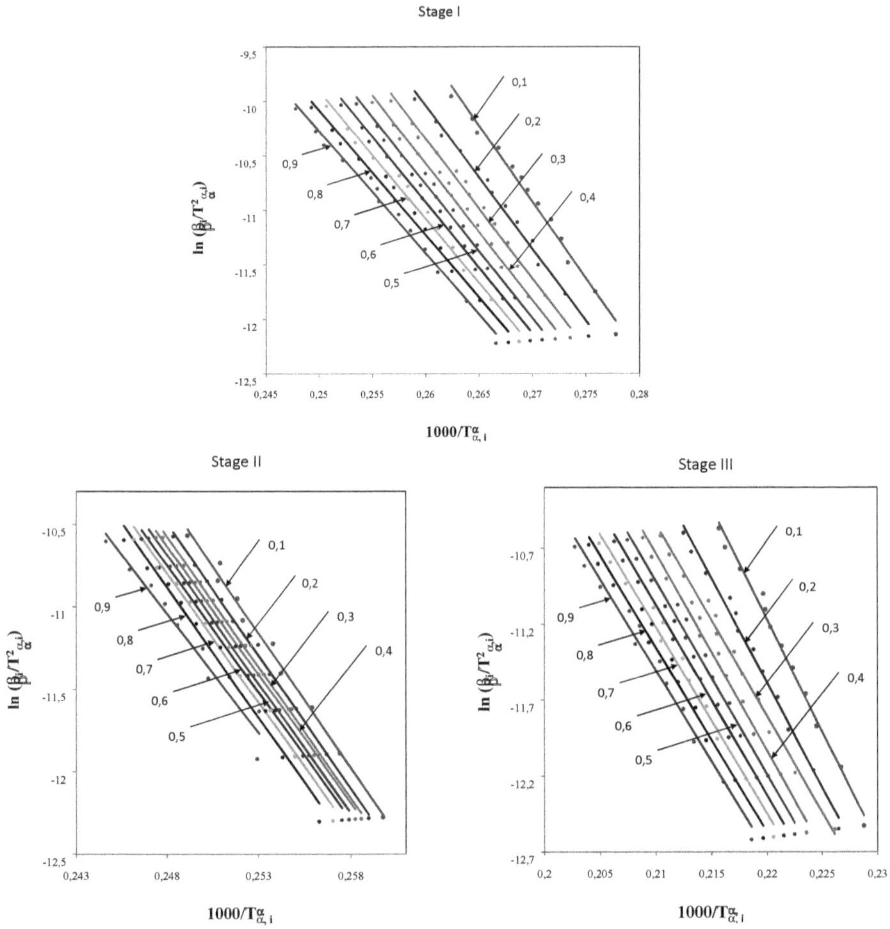

Figure 9. Plots of $\ln\left(\frac{\beta_i}{T^2_{\alpha,i}}\right) \div \frac{1000}{T_{\alpha,i}}$ for the stages. Thermal decomposition of NH_4VO_3 in dry air.

2.4.3. *Coats – Redfern method*

During the calculations by this method the sets of α(T) presented in Figure 5 were also used. The activation energies E, determined for the stages by Kissinger's method (for α_m values belonging to the sets of α(T)), were taken as the base values. The plots of $\ln\left[\frac{g(\alpha)}{T^2}\right] \div \frac{E}{RT}$ were constructed. Among the known kinetic models the ones the most consistent with the experimental data were selected; for stages I and III model A2, and for stage II model A4. For the selected model and different β_i the values of the coefficient A were calculated for the stage in the following manner. First Z(T) was calculated from the formula

$$Z(T) = \ln\left[\frac{AR}{\beta E}\left(1 - \frac{2RT_m}{E}\right)\right] = \ln\left[\frac{g(\alpha)}{T^2}\right] + \frac{E}{RT} \tag{30}$$

Then lnA was calculated according to

$$\ln A = Z(T) + \ln\left[\frac{\beta E}{R\left(1 - \frac{2RT_m}{E}\right)}\right] \tag{31}$$

The calculations of Z(T) were performed for the range $0{,}01 \le \alpha \le 0{,}99$. The values of T_m and the temperature ranges for the stages are given in Table 1. However the calculated values of A are given in Table 5. For all the stages the obtained values of coefficient A were higher than the values determined by Kissinger's method. There should be added that the proportion between the values was remained.

Using average values of A the activation energies were tested. The criterion was the best conformity of trajectories of α(T) plots, calculated and determined experimentally, in the whole range. The conversion degrees were calculated as follows. First g(α) was calculated from the formula

$$g(\alpha) = T^2 \exp\left[F(T)\right] \tag{32}$$

Formula (32) is suitable for different models. Whereas the method of calculating conversion degrees depends on the form of kinetic model. For example, for model A2

$$g(\alpha) = \left[-\ln(1-\alpha)\right]^{\frac{1}{2}} \tag{33}$$

thus

$$\alpha = 1 - \exp\left\{-\left[g(\alpha)\right]^2\right\} \tag{34}$$

The values of activation energy calculated in this way are given in Table 5.

β K min^{-1}	Stage I; A2; E=117,66			Stage II; A4; E=164,48			Stage III; A2; E=113,75		
	ΔZ(T)	A E12 min^{-1}	E* kJ mol^{-1}	ΔZ(T)	A E17 min^{-1}	E* kJ mol^{-1}	ΔZ(T)	A E9 min^{-1}	E* kJ mol^{-1}
1	19,81	5,91	114,10	29,98	1,468	157,48	12,72	4,97	110,2
1,5	19,58	7,23	114,36	29,34	1,721	157,85	12,46	5,73	110,5
2	19,22	6,71	114,56	29,31	2,227	-	-	-	-
2,5	19,12	7,61	114,60	28,85	1,436	157,00	12,06	6,52	110,6
3	19,11	9,02	115,06	28,87	2,150	158,20	12,05	7,54	111,1
3,5	18,82	7,85	114,56	28,57	1,860	157,90	11,76	6,70	110,5
4	18,83	9,15	115,26	28,51	1,908	158,50	11,76	7,59	111,5
4,5	18,61	8,27	114,56	28,31	1,845	157,88	11,57	7,10	110,5
5	18,57	8,84	114,76	28,34	2,113	158,80	11,50	7,31	110,6
6	18,41	8,99	115,16	27,98	1,770	158,10	11,33	7,48	111,1
7	18,24	8,85	115,36	27,81	1,742	158,90	11,19	7,56	111,3

Table 5. The kinetic parameters obtained by Coats–Redfern method. Thermal decomposition of NH_4VO_3 in dry air.

The parameters A and E determined by this method for different β_i remain nearly constant.

The corrected activation energies are slightly lower than the ones determined by Kissinger's method; in the case of stage II they are practically equal to the values calculated by isokinetic method. The result is interesting because it shows that the activation energy determined by Kissinger's method can be regarded as representative for the whole set of α(T) related to the stage. It should also be emphasized that determining the triad of g(α), A and E directly from the Coats-Redfern equation, usually the good results are not obtained [1,21,24]. That is it is a matter of calculation methods, and not of the Coats-Redfern equation.

Using the kinetic parameters given in table 5 the verifying calculations were performed. For the selected β_i the trajectories of α(T) and r(α,T) functions were determined. The process rate was calculated from the formula

$$r(\alpha,T) = A\exp\left(-\frac{E}{RT}\right)f(\alpha) \tag{35}$$

While calculating f(α) the values of α(T) obtained from the Coats-Redfern equation (formulas (32)–(34)) were used. In Figure 10 the results obtained for stage I and II (β= 2, 4 and 6 K min^{-1}) are presented as an example.

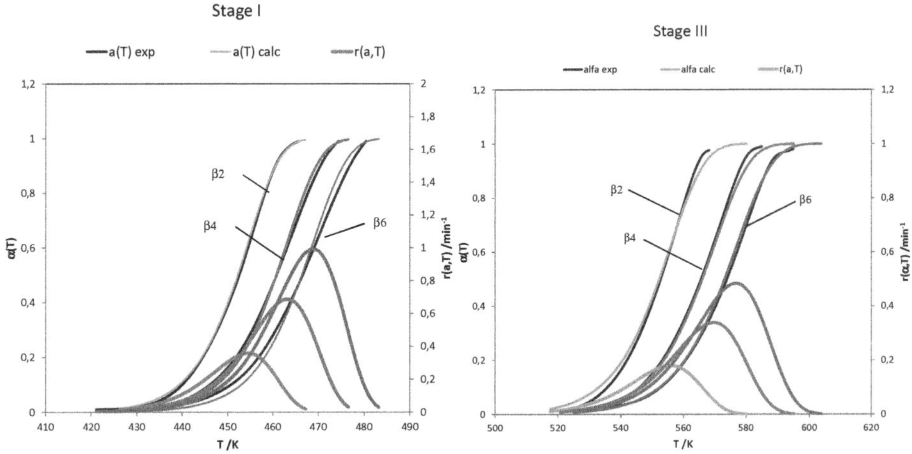

Figure 10. Plots of α(T) and r(T, α) for the stage I and III. Thermal decomposition of NH_4VO_3 in dry air.

The results are fairly well consistent with the experimental data; at the experimentally determined temperature ranges the process rates change from zero for α(T) = 0, reaching the maximum values at the points of inflection of α(T) curves, and then decrease to zero at $\alpha(T) \to 0$. The obtained results show that in our study Coats-Redfern equation was of great importance. Due to the analytical form it is also easy to use in the calculations of the kinetics of heterophase non-catalytic processes.

3. Heat treatment of TiCx/C. Carbonisation of nc-TiCx

Manufacturing, storage and use of nc-TiC in form of powder for sintering or co-deposition processes involves the possibility of occurrence of free carbon in the system. Carbon can be a by-product of the synthesis process (pyrolysis of hydrocarbons) and remain in equilibrium in the TiC-C system or be the product of oxidation of TiC in the air according to the mechanism proposed by Schimada [26], whereby the oxygen dissolves in the carbide and the layer containing oxycarbides is formed.

$$TiC + (3/2)\,x\,O_2 \rightarrow TiC_{1-x}O_x + CO_2 \tag{36}$$

In second stage, in the layer amorphous TiO_2 is formed and elemental carbon is produced [27-31]

$$TiC_{1-x}O_x + (1-x/2)\,O_2 \rightarrow TiO_2 + (1-x)C \tag{37}$$

The produced carbon changes the state of the surface of TiC particles. A low oxygen content is the most important prerequisite for high sintering activity of the nanocrystalline TiC powders, especially in the sintering processes, synthesis of the nanocomposites for example nc-TiC in metallic matrices.

In the manufacturing of nanomaterials by sol-gel method the second, high temperature, i.e. at temperatures above 1400 K, stage is essential. The first stage of this method is the sol-gel technique. This stage is carried out at lower temperature. The intermediate product of pyrolytic decomposition of PAN-DMF-$TiCl_3$ is formed - the powder containing nanocrystalline TiCx in carbon matrix [6]. Nanocrystallites of titanium carbide are characterized by high fraction of lattice defects, i.e., vacant carbon sites, presence of oxygen and/or nitrogen. In the second stage carbonisation and purification of TiC_x/C composite in argon takes place. It is essential to obtain the materials with high values of the C/Ti ratio, while maintaining the proper, nanometric grain size, in order to obtain the most favourable properties such as hardness and oxidation resistance. The selection of parameters meeting these requirements is difficult.

There was assumed that the good basis are kinetic studies. They allow to determine the intermediate and final products, distinguish the stages of the process, determine the temperature ranges of their courses and obtain a quantitative description.

The kinetic measurements were carried out using TG-DSC-MS technique. Nanoparticles size, lattice parameter, chemical and phase composition before and after heat-treatment were determined with the following techniques: XRD (Philips PW3040/x0 X'Pert Pro), HRTEM (JEM 3010), SEM (JEOL JSM 6100), EDX (Oxford Instruments, ISIS 300), XPS (SIA 100 Cameca), total carbon measurement (MULTI EA2000, AnalyticJena), and the presence of free carbon was estimated by Raman Spectroscopy. The measurements were carried out under non-isothermal and isothermal conditions. The advantage of this method is the possibility of continuous registration of measured values and use of small weighed amounts of samples during measurements. The procedure is illustrated at the example of heat treatment in argon of the nanocomposite powder containing nc-TiC_x ($x \leq 0.7$) in carbon matrix obtained by sol-gel method. The method ensures maintaining the small size of crystallites, by physically limiting the volume available for their growth, and the matrix prevents agglomeration of particles and oxidation during storage and transport.

At a certain temperature range, and under certain conditions, the reaction of carbonisation is possible

$$TiC_x + yC \rightarrow TiC_{x+y} \tag{38}$$

Heating of the samples in an argon atmosphere can lead to the growth of crystallites. The aim of this study was to develop conditions for annealing of the composite, under which, as a result of carbonisation, TiC_x reaches a high stoichiometric composition, i.e. $x > 0.8$, at the minimal growth of crystallites. Obtaining such optimal material properties can be controlled by selecting the appropriate temperature and rate of the process.

The thermogravimetric measurements were carried out at sample heating rates of 10, 20, and 50 $Kmin^{-1}$. The mass of the samples were ca. 40 mg. During the measurements the samples were heated up to 1473, 1573, 1673 and 1773 K. These temperature values correspond to the isothermal conditions. The samples were heated under isothermal conditions for 6 h. High-purity argon 'Alphagaz 2 Ar' from Air Liquide (H_2O\ppm/mol, O_2\0.1 ppm/mol, C_nH_m\0.1 ppm/ mol, CO\0.1 ppm/mol, CO_2\0.1 ppm/mol, H_2\0.1 ppm/mol) was used during experiments. With regard to reactivity of O_2 the special attention during measurements was given to possibility of oxidation of components of the system during the process run. Argon purge flow rate during measurements was set at 100 cm^3/min. The part of the results in works [17,32] were presented. Below the complete analysis description were introduced.

3.1. Kinetic analysis

The kinetic description of the process was based of thermogravimetric measurements. The results of the measurements are presented on the plots of sample temperature, TG, DTG and HF function dependencies on time (Fig. 11).

Initially the measurements were carried out under non-isothermal conditions at a linear change in sample temperature, then at the transitional regime, and finally under isothermal conditions (Fig.11a). It should be added that these results were the basis of the description of the process. The theory of kinetics under non-isothermal conditions require that this function depends only on the sample heating rate and temperature. The results were evaluated by neural networks method. Neural networks, used to analyze the non-isothermal measurements, were previously described in [20-23]. The TG_u function was the described (dependent) variable, and the sample heating rate and temperature were the describing (indepedent) variables. All the measurement series were examined simultaneously. The received network was GRNN 2/11310. Statistical analysis of this network is given in Table 6.

Parameter	Tr	Ve	Te
S.D. Ratio	0.01716	0.01926	0.01734
Correlation	0.999874	0.999845	0.99987

Table 6. Statistical estimation of GRNN 2/11310 network

In columns 2, 3, 4 the statistical evaluation of training (Tr), verification (Ve) and testing (Te) subset is listed.

A high accuracy was obtained. The TG dependencies determined experimentally could have been used in kinetic calculations.

The essential operation is the division of the process into stages. Basing on the presented measurement results four stages of the process were identified (Fig. 12). In the endothermic stage I, proceeding with a weight loss, the release and desorption of volatile products, contained in the samples after the first stage, occurred. In the exothermic, second stage, marked with the symbol II, proceeding with a samples weight gain, the oxidation of non-carbonised nc-TiC_x/C by oxygen present in trace amounts in argon occurred. In the endothermic third stage, labeled by III, simultaneously proceeded the pyrolysis of organic admixtures and carbonisation of nc-TiC_x/C. The pyrolysis and carbonisation were treated as concurrent reactions. Stage IV (exothermic) proceeded with sample weight gain. It concerns the oxidation of carbonised nc-TiC with oxygen contained in argon. The process proceeded at temperature above 1573 K.

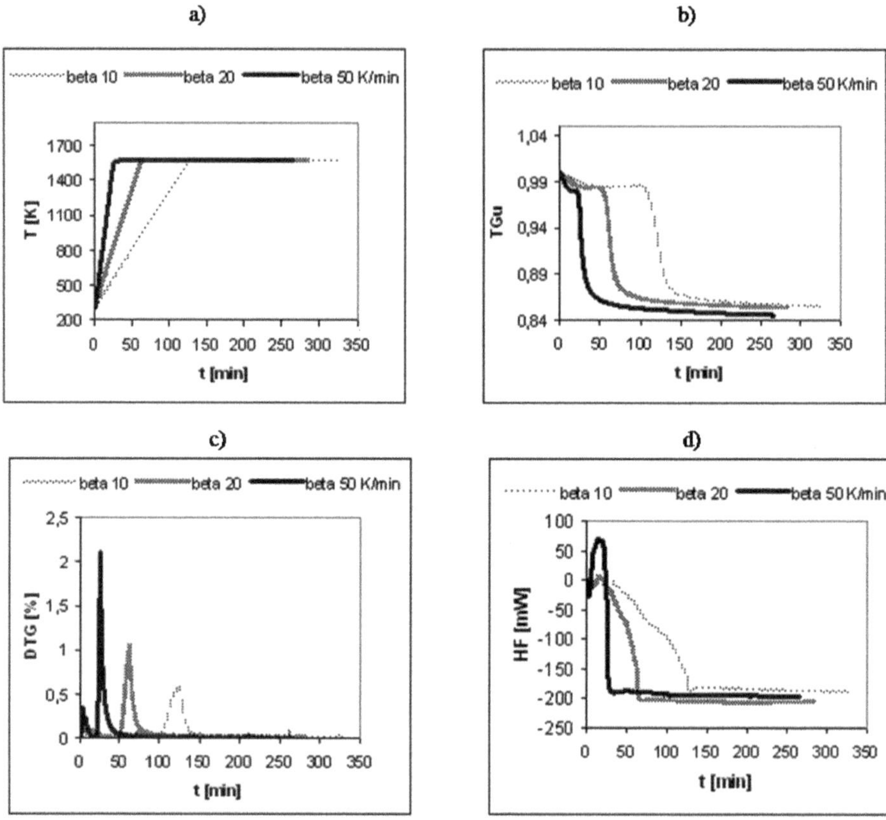

Figure 11. Plots of T, normalized TG function and DTG and HF in time. Heat treatment of nc-TiC/C in argon. a) temperature, b) normalized TG, c) DTG, d) HF

During the thermogravimetric measurements the evolved gaseous products were identified by mass spectrometry method. The compounds produced in the reaction with the oxygen present in trace amounts in argon, are described in detail, because these processes could affect the quality of the obtained, carbonised nc-TiC. CO_2-m/e44, CO-m/e28 NO-m/e30 and NH_3-m/e17 have been identified. NH_3 formed as a result of pyrolytic decomposition of carbon compounds was the precursor of nitric oxide. CO_2-m/e44 mass spectrum, is shown in Figure 13 as an example. To facilitate the analysis of the results the normalized TG_u function is also plotted.

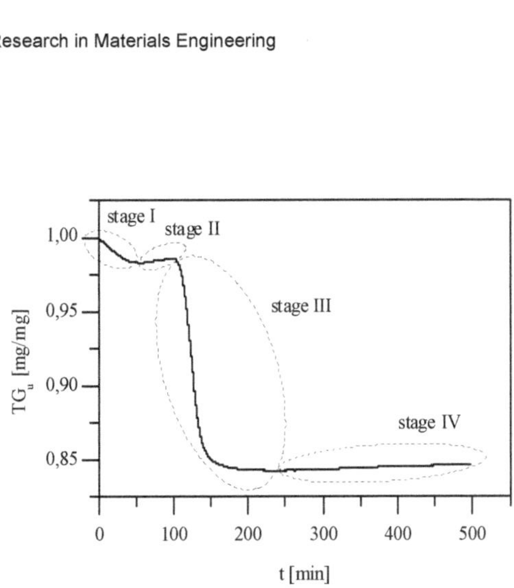

Figure 12. Plots of normalized TG functions in time for β=10 K/min and 6h at 1673 K recorded during the heating of nc- TiC_x/C in argon.

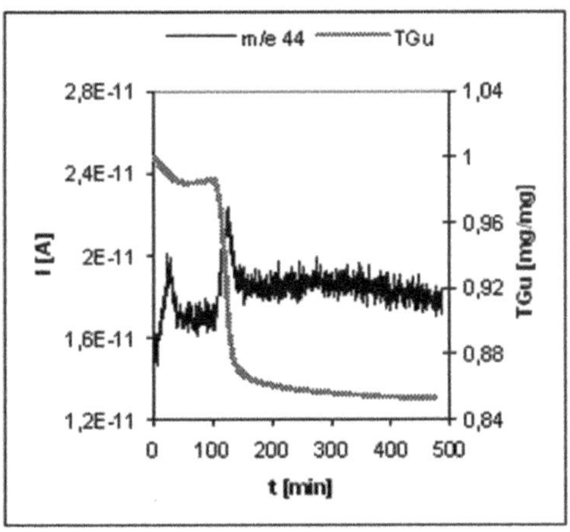

Figure 13. Mass spectra of CO_2 and normalized TG function plots. Heat treatment of nc-TiC/C in argon at the heating rate 10 $Kmin^{-1}$.

The analysis of mass spectra allowed the more accurate characterisation of distinguished stages. In stage I volatile products were evolved. CO_2, CO and NO were also formed. In this stage H_2, NH_3, HCN and CH_3–CH_3, CH_2=CH_2 and CH≡CH also evolved. In stage II oxidation of non-carbonised nc-TiC_x/C occurred. During the course of stage III CO and CO_2 were emitted. They were attributed to the oxidation of released hydrocarbons. HCN, CH_3–CH_3, CH_2=CH_2, and CH≡CH were also formed.

The correct kinetic description was hindered by evolution of secondary products. Therefore the values of $\alpha(T)$ determined for each stage required an independent evaluation.

3.2. The results of calculations

For distinguished stages, basing on the TG function, the conversion degree was calculated. The formula (23) was used.

During the measurements weight changes of the samples were as follows: in stage I: 10 $Kmin^{-1}$ (1.6%), 20 $Kmin^{-1}$ (1.7%) and 50 $Kmin^{-1}$ (2%). In stage II, the sample weight changes equaled: 10 $Kmin^{-1}$ (0.089%), 20 $Kmin^{-1}$ (0.047%). Whereas in whole the range of the course of stage III weight changes were: 10 $Kmin^{-1}$ (13.47%), 20 $Kmin^{-1}$ (13.23%) and 50 $Kmin^{-1}$ (13.66%). The obtained relations of $\alpha(T)$ for stages I and II are shown in Figure 14.

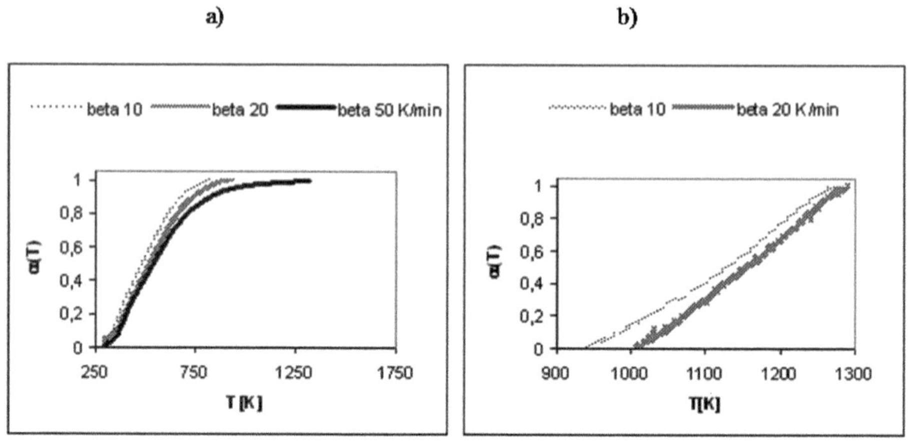

Figure 14. Temperature dependencies of $\alpha(T)$ function; a) stage I, b) stage II.

In accordance with the theory of kinetics of heterogeneous processes the plots of $\alpha(T)$ are shifted into higher temperature range along with the increase in samples heating rate.

Due to the measurements in different regimes the dependencies of $\alpha(T)$ determined for stage III required a more detailed discussion. The conversion degree in stage III was changing regularly in time (Fig. 15a). Irregular changes were observed in the trajectories of the curves of conversion degree dependencies on temperature (Fig. 15b). It was also found that in the transient area a significant increase in conversion degree took place.

The results presented in Figure 16a were elaborated according to the rules of non-isothermal processes theory and in Figure 16b according to the isothermal processes theory.

The plots of $\alpha(T)$ determined for non-isothermal conditions concern the pyrolysis process.

At low temperatures, there is no carbonisation of nc-TiC, and at temperature above 1573 K pyrolysis proceeded much faster than carbonisation. Under the non-isothermal conditions, the transition from one temperature range to the other was short. As a result the influence of carbonisation on the recorded samples weight loss was not revealed.

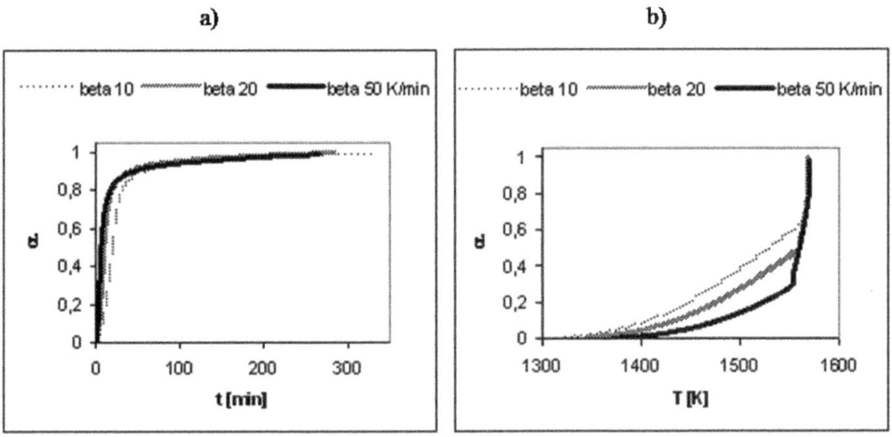

Figure 15. Total conversion degree. Stage III; a) time dependency b) temperature dependency.

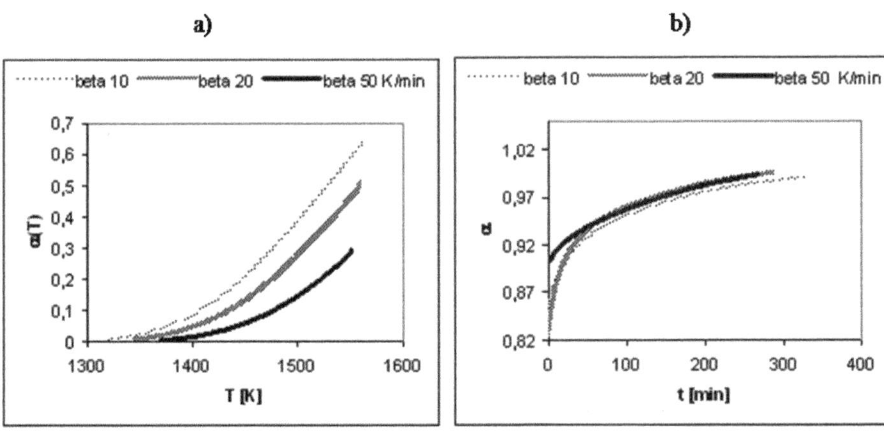

Figure 16. Conversion degree dependencies on temperature and time. Purification and carbonization of nc-TiC/C in argon. Stage II. a) non-isothermal conditions b) isothermal conditions.

The dependencies of $\alpha(T)$ determined for stages I, II and III under non-isothermal conditions, at a linear heating rate of the samples, were evaluated by neural networks method. The $\alpha(T)$ was the described variable and the sample heating rate and temperature were the describing variables. For each stage all the measurements series were analyzed simultaneously. The results are listed in Table 7

According to the theory of non-isothermal processes kinetics, the $\alpha(T)$ dependencies, determined for the stages, describe with high accuracy two parameters: sample heating rate and sample temperature. These results could therefore be used in further calculations, i.e., during the identification of kinetic models (determination of the form of g(α) function)) and during determination of Arrhenius parameters A and E.

Parameter	Stage I , MLP 2/2			Stage II			Stage III, MLP 2/15		
	Tr	We	Te	Tr	We	Te	Tr	We	Te
S.D. Ratio	0.0361	0.0355	0.0364	0.0269	0.0274	0.0284	0.0728	0.0820	0.0784
Correlation	0.999	0.999	0.999	0.999	0.999	0.999	0.997	0.997	0.997

Table 7. Estimation of α(T) dependencies determined for stage I, II and III with use of artificial neural networks method.

First, the values of these parameters were estimated by linear regression method. Each series were analysed separately. The obtained results for stage III are given, as an example, in Table 8.

β [K/min]	r*	F	E [kJ/ mol]	A [1/min]	T_m [K]	Δα	ΔT
10	-0,988	50029,7	315,09	8,48 E09	1530	0.005-0.639	1321-1564
20	-0,988	47245,7	327,77	3 E10	1559	0.005-0.508	1345-1560
50	-0,989	20219,0	375,62	1,95 E12	1590	0.004-0.29	1370-1551

Table 8. List of kinetic parameters. Stage III, model F1.
r* - correlation coefficient, F – Snedecor statistic

Using the determined values of A and E parameters, the values of α(T) were calculated from the Coats-Redfern equation. They were compared with the data determined from the measurements. The systematic error in the order of 4.5% has been noted. The accuracy was improved by correcting the value of E parameter. There was required that the error in the series, i.e. the mean square error between the values determined from the measurements and calculated $\ln\left[\frac{g(\alpha)}{T^2}\right]$, was close to zero. The calculations for the remaining stages were performed in the same way. The results are given in Table 9.

The results have been verified. Using the kinetic parameters given in Table 9 the conversion degrees were calculated from the Coats-Redfern equation for the e stages and compared to the ones determined from measurements. As an example, in Figure 17 the results for stage I are shown. A good consistency was obtained.

The kinetic parameters determined for the stages have been used for simulation calculations. The α(T) and r(α,T) dependencies on temperature and sample heating rate were investigated. The results are presented in Figure 18.

The determined dependencies are in accordance with the theory. With the increase in sample heating rates the α(T) curves are shifted into the higher temperature range. The process rate increases from zero for $\alpha(T)$ equal zero, reaches the maximum in temperature Tm, and then usually decreases to zero at $\alpha(T)\to 1$. The temperature ranges for stages runs are consistent with the ones determined experimentally. The $\alpha(T)$, and $r(\alpha,T)$ plots for a stage do not come to an end, because at high conversion degrees process ran according to the different kinetic models.

stage	Model	$g(\alpha)$	β [K/min]	A [1/min]	E [kJ/mol]	T_m [K]	$\Delta\alpha$	ΔT [K]
I	D3	$\frac{3}{2}\left[1-(1-\alpha)^{\frac{1}{3}}\right]^2$	10	2,2 E-6	15,8	487,4	0,03-0.9989	299-826
			20	1,2E-6	16,5	524,9	0,035-0,9969	299-937
			50	3,5E-7	16,3	539,5	0.005-0.9995	298-1319
II	F2	$(1-\alpha)^{-1}-1$	10	6,81 E4	125,25	1128	0.002-0.999	935-1271
			20	7,74 E6	160,8	1156	0.002-0.999	1007-1291
III	F1	$[-\ln(1-\alpha)]$	10	8,48 E09	306,8	1530	0.005-0.639	1321-1564
			20	3 E10	318,8	1559	0.005-0.508	1345-1560
			50	1,95 E12	364,5	1590	0.004-0.29	1370-1551

Table 9. List of kinetic parameters for the stages.

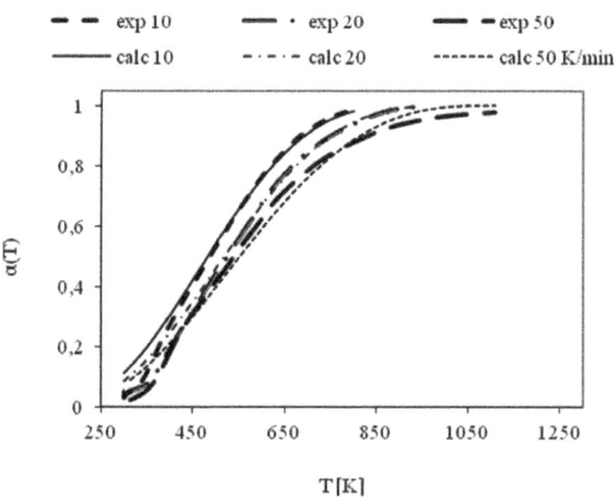

Figure 17. Comparison of $\alpha(T)$ calculated and determined from experiments. Stage I, model D3.

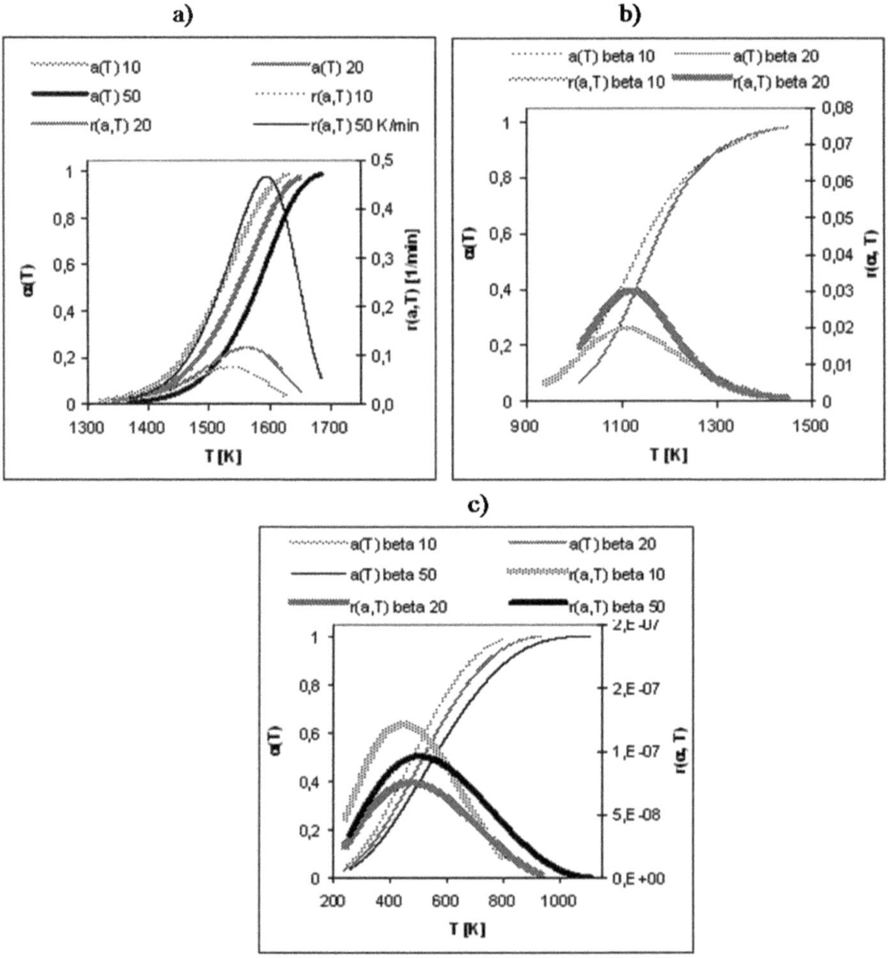

Figure 18. Plots of α(T) and r(α,T) functions; a) stage I, model D3; b) stage II, model F2; c) stage III, model F1.

Under the isothermal conditions the last phase of stage III and whole the stage IV proceeded. The results presented earlier, obtained in the series up to 1573, 1673 and 1773 K, were not sufficient for description of the course of stage III under isothermal conditions. They have been complemented by additional measurements. During the investigations samples were heated up to 1343, 1503, 1543 and 1623 K using the following heating rates: 10, 20 $Kmin^{-1}$.

While elaborating these results the calculations for stages I, II and III were also performed. There were obtained similar results as before (Table 9). These data are not provided. At the high conversion degrees in stage III the change of kinetic model (form of g(α) function) took place. Under the isothermal conditions, at lower conversion degrees, the process first proceeded according to the F1 model (similarly as under the non-isothermal conditions), then at high conversion degrees (higher than 0.98) according to the D3 (three-dimensional diffusion Jander's model). This concerns the removal of remaining products of pyrolysis.

The charts of α (T) were obtained from about 25,000 measurement points for each case. The value of k(T) was calculated by linear regression method for the subsequent sets, each containing 1100 values of the g(α) function. The results were evaluated using several measures. The values of R^2 measure were determined. The calculation results for all measurement series are given in Table 5. The temperature, the mean values of k(T) determined for the entire sets, the time of obtaining the isothermal conditions t_1, counted from the beginning of the measurement, and the conversion degree α_1 obtained for this time are given.

series	β [K/min]	T [K]	k*10³ [1/min]	t_1	α_1	R^2
1443 K	10	1440,9	13,61	158,4	0,7092	99,72
	20	1440,9	14,41	99,98	0,69	99,92
1503 K	10	1501,1	14,75	203,92	0,9553	99,68
	20	1501	35,18	92,85	0,872	99,59
1543 K	10	1541,4	9,39	201,72	0,9331	99,72
	20	1541,3	12,48	142,69	0,9539	99,87
1573 K	10	1568,5	6,94	199,19	0,9303	99,36
	20	1568,2	8,3	131,15	0,938	99,85
1623 K	10	1620,2	40,63	162,6	0,9653	99,91
	20	1619,7	61,84	91,8	0,9615	99,95
1673 K	10	1669,4	54,28	160,2	0,9699	99,69
	20	1669,1	104,31	86	0,9883	99,39
1773 K	10	1770,5	219,28	154,2	0,9831	99,72
	20	1770	242,24	83,6	0,9831	99,52

Table 10. The results of kinetic calculations for isothermal conditions. F1 model.
R^2 – statistical measure

In the considered range of temperature two processes preceded simultaneously; pyrolysis of organic compounds, contained in the raw samples and proceeding with their participation carbonisation of nc-TiC_x/C. The bounded carbon remains in the system. As a result the lesser sample mass losses were observed. In lower temperature proceeds also pyrolysis, as shown by the values of k(T) given in Table 10. Carbonisation starts at temperature of about 1541 K and becomes a dominating process at temperature of about 1570 K. At 1610 K pyrolysis becomes a dominating process again. With regard to carbonisation effectiveness, the process should be carried out at temperature of about 1570 K. The minimum on the curve k(T) has been observed in Figure 19.

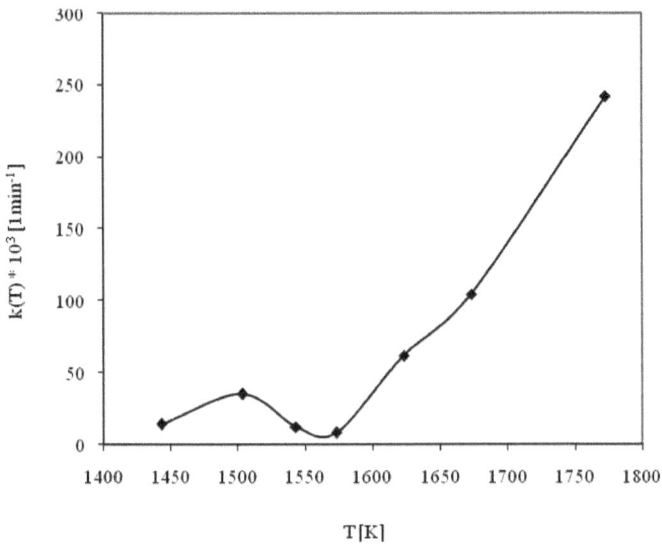

Figure 19. Dependency of reaction rate constant on temperature for β = 20 K/min. F1 model ($g(\alpha) = [-\ln(1 - \alpha)]$); ♦– mean value of k(T)

In the same manner were obtained the results presented in Table 11, concerning the desorption process of volatile products after the completion of carbonisation process.

series	β [K/min]	T [K]	k*10³ [1/min]	t_2	α_2	R^2
1543 K	10	1541,4	0,255	402,3	0,989	97,82
	20	1541,3	0,354	326,2	0,9933	95,03
1573 K	10	1568,5	1,47	221,76	0,9412	99,59
	20	1568,2	2,34	150,98	0,9486	99,65
1673 K	10	1669,4	9,5	160,2	0,9699	99,69
	20	1669,1	16,07	87	0,9892	99,45
1773 K	10	1770,5	41,08	154,18	0,9831	99,66
	20	1770	49,36	84,1	0,9854	99,57

Table 11. The results of kinetic calculations for isothermal conditions. D3 model

In the third column the values of k(T) are given, and in the subsequent columns time t_2 from which desorption becomes the dominant factor, and the corresponding conversion degree.

Under the measurement conditions the oxidation of carbonised nc-TiC, by the oxygen contained in trace amounts in argon, occurred in series up to 1673 and 1773 K. This process proceeded according to the R2 model (reaction at the interface, cylindrical symmetry) or R3 model (reaction at the interface, spherical symmetry). The weight gain of the sample was less than 1%. Inhibition of oxidation process of carbonised nc-TiC can be explained on the basis of the mechanism of Shimada [26], the formation on the surface of nc-TiC particles of amorphous TiO_2 layer, blocking access of oxygen to the reaction zone. The given values of k(T) indicate that the samples obtained at heating rates of 10 K/min oxidized slower than at the heating rates of 20 and 50 K/min. It is also visible that the final temperature of the process affects the properties of carbonised nc-TiC. Basing on the series up to 1673 and 1773 K there was found that after the completion of carbonisation and purification process of nc-TiC, at higher temperatures oxidation of the carbonised nc-TiC by oxygen present in argon in trace amounts takes place.

The removal of carbon from the matrix and carbonisation of nc-TiC_x proceeded most preferably in a series up to 1573 K, at the heating rate of 20 K/min. The results of investigations concerning this series are therefore given.

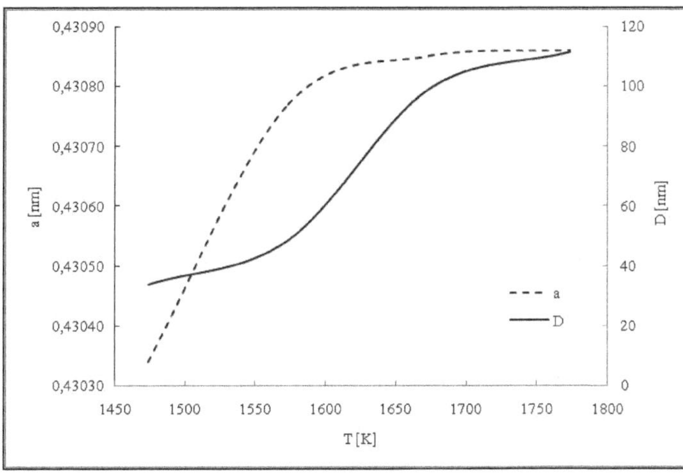

Figure 20. Dependency of nc-TiC mean lattice parameters (a) and mean particles diameters (D) on temperature [32].

Carbonisation resulted in an increase of lattice parameter of titanium carbide. The largest increase in lattice parameter was observed for the series up to 1573 K. Under these conditions, the average particle size was in the order of 40 nm. Mean values of lattice parameters of nc-TiC and the average particle size determined after the carbonisation processes are shown in Figure 20. The measurement of crystallites size by Scherrer method and on the basis of TEM images showed that the mean size of TiC crystallites after carbonisation was approximately 30% higher in relation to the size before the heat treatment process at temperature of 1573 K. The results of microscopic examination were confirmed by the results of the X-ray diffraction. The analysis of chemical composition and phase composition showed an increase in the fraction of carbon in titanium carbide from $TiC_{\sim .68}$ to $TiC_{\sim .8 \div .85}$ and removal of carbon from the matrix. The results of this step of research are given in [6,32].

4. Oxidation of the nc-TiC_x/C and nc-TiC_x

The possibility of implementing purification of TiC_x/C composites by burning out the elementary carbon, composing matrix, was considered. The results of oxidation of nc-TiC_x/C system have been presented. Oxidation of the TiC_x/C powders, being an intermediate product of sol-gel synthesis, and of the TiC_x powders obtained by reduction with hydrogen was investigated. The reduction of TiC_x/C powders with hydrogen aimed at removing from the system the carbon from the matrix. Purification with hydrogen according to the reaction $C+H_2 \rightarrow CH_4$ was carried out at temperature of 1173K, under pressure of 16MPa, for 4.5 h [6,32]. The measurements were carried out using thermogravimetric method, under non-isothermal conditions. The samples unreduced with hydrogen were studied at the following heating rates: 5Kmin^{-1} (sample weight of 18.749 mg), 10 Kmin^{-1} (sample weight of 16.049 mg), 15 Kmin^{-1} (sample weight of 13.322 mg), 20 Kmin^{-1} (sample weight of 15.908 mg). The powders after reduction with hydrogen instead were studied at 5 Kmin^{-1} (sample weight of 17.768 mg), 10 Kmin^{-1} (sample weight of 17.174 mg), 15 Kmin^{-1} (sample weight of 17.544 mg), 20 Kmin^{-1} (sample weight 17.544 mg). During the measurements temperature of samples, TG, DTG, and HF were recorded. In one series several dozen thousands of each variable values were recorded. During the measurements the linear change of sample temperature over time was maintained, as required by the theory of non-isothermal kinetics. The normalized TG curves of the samples unreduced and reduced with hydrogen are shown in Figure 21.

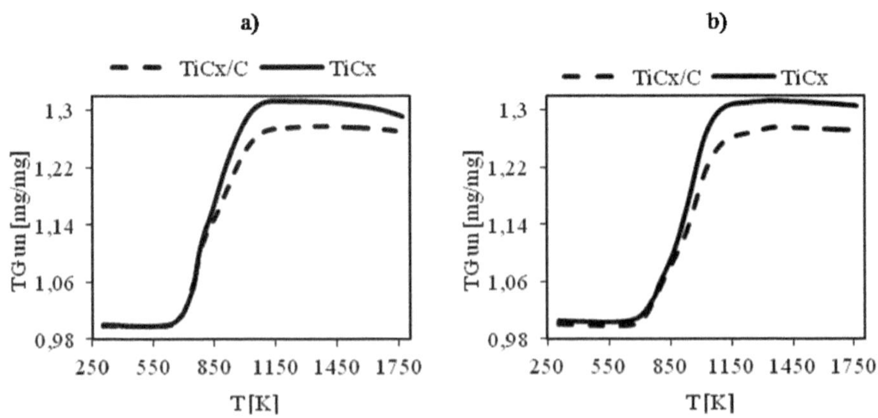

Figure 21. Plots of TG_u curves for TiC_x/C samples and TiC_x (after reduction with H_2) samples, a) 5, b) 20 Kmin^{-1}

The normalized TG_u curves of the samples reduced with hydrogen are shifted upward to the same degree for different heating rates (ΔTG in the order of 0.03 mg), because in the TiC_x powders (after reduction with hydrogen), the relative content of titanium increased as a result of removing elemental carbon by acting with hydrogen. This resulted in greater weight gain of TiC_x during the oxidation in comparison with TiC_x/C. In both cases the weight gain of the sample started at temperature in the order of 600 K. This means that under these conditions the oxidation of nc-TiC_x started.

In the DTG curves, there are three peaks (Fig. 22). The first peak is associated with start of the oxidation of nc-TiC_x. The subsequent weight loss is associated with the burning out of the elemental carbon, produced during the oxidation of the nc-TiC_x. This process proceeds simultaneously with the further oxidation of nc-TiC_x. The next peak concerns the oxidation of unreacted nc-TiC_x. The elemental carbon, contained in nc-TiC_x/C samples unreduced with hydrogen, burns out in the final stage of the process.

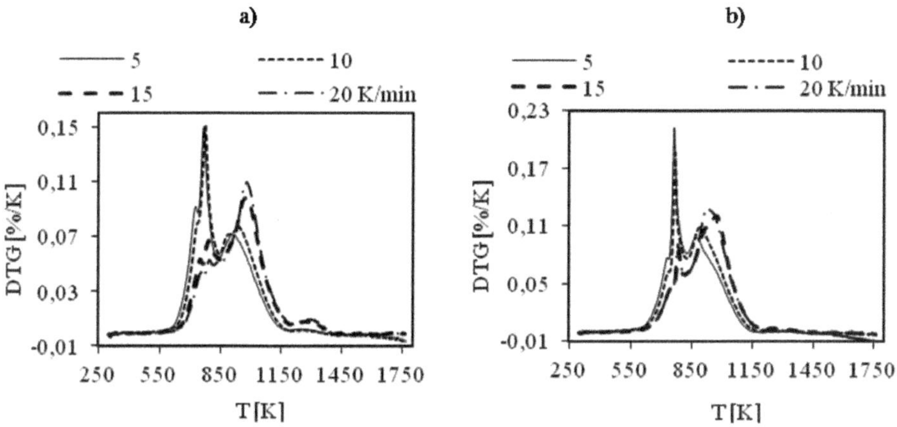

Figure 22. Dependency of DTG curves on temperature, a) for TiC_x/C, b) for TiC_x (after reduction with H_2)

Two distinct maxima were observed in the HF plots (Fig. 23). The first one is associated with the beginning of oxidation process of nc-TiC_x and the second one with burning out of the elemental carbon, produced during the oxidation of the nc-TiC_x, and further course of the nc-TiC_x oxidation. The apparent increase in the value of HF function, in the final stage of the process, is associated with the terminating oxidation of nc-TiC_x.

The results have been confirmed by the identification of CO_2, formed in the system, by mass spectrometry. There was also found that while increasing sample heating rates the plots of mass spectra of CO_2, originating from the nc-TiC oxidation process and from burning out of the formed carbon, overlapped. The carried out experiments have shown that the nc-TiC, obtained by sol-gel process, cannot be purified by burning out in the air the carbon admixtures contained in the system.

The performed studies indicated also the possibility of occurring during the description of oxidation of ceramic nc-TiC/C powders in the air, some difficult issue related to the simultaneous proceeding, in a certain range of temperature, of metal carbide oxidation and burning out of the carbon. The following manner of kinetics description of the both concurrent reactions, based on thermogravimetric studies, has been proposed.

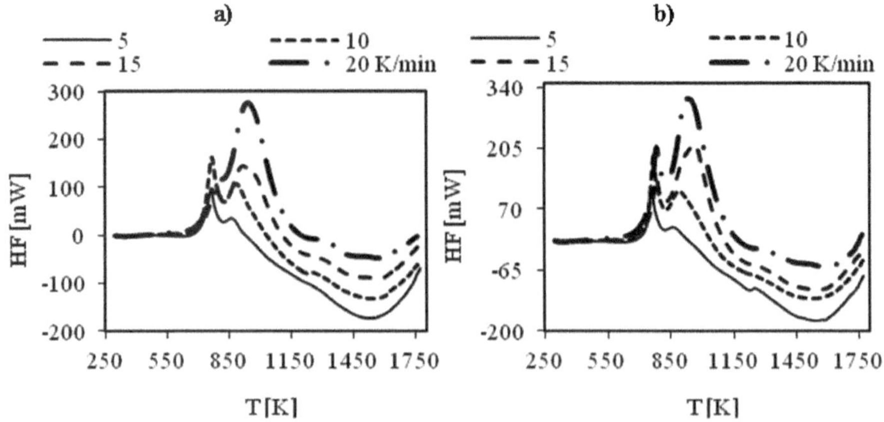

Figure 23. Dependency of HF on temperature, a) for TiC_x/C, b) for TiC_x (after reduction with H_2)

To obtain the normalized TG_u curve, corresponding to the oxidation process of TiC_x in the whole range of temperature the neural networks method was applied. Basing on the results registered before burning out of the formed carbon and after the end of this process the network was fitted. Then the fitted network was used to generate the segment of normalized TG_u curve for the temperature range in which both transformations proceeded simultaneously. The multi-layer MLP networks were used. The described variable was TG_u function, and the describing variable was temperature. Each measurement series was analysed separately. Using these models sections of TG curves corresponding to the oxidation process of nc-TiC, in the temperature range in which this process proceeded simultaneously with the burning out of the carbon, were generated. The TG_u plots generated by the network and determined experimentally are shown in Figure 24.

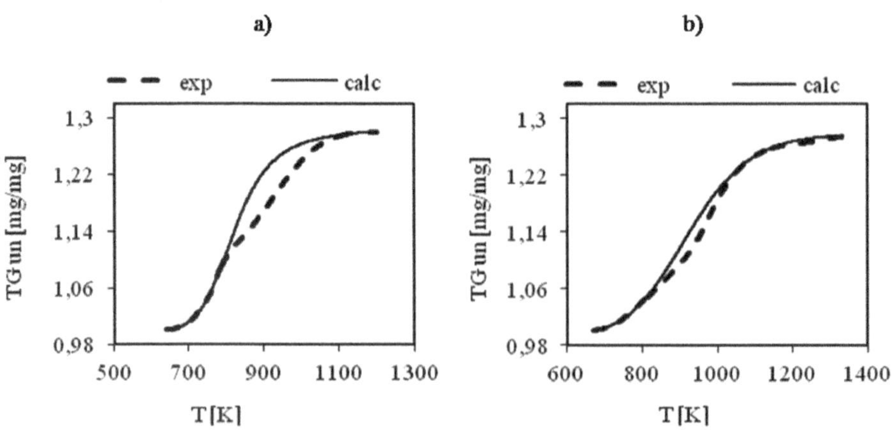

Figure 24. Plots of TG_u function, calculated and experimental. Oxidation of nc-TiC_x/C samples, unreduced by hydrogen, in air, a) 5 $Kmin^{-1}$, b) 20 $Kmin^{-1}$

On the basis of TG_u curves, complemented by the results of calculations, the $\alpha(T)$ dependencies for the oxidation process of nc-TiCx in the whole temperature range were determined.

The conversion degree for the process of burning out the elemental carbon was determined as follows. By subtracting the experimental values from the calculated TG_u values, ΔTG was determined, and then, integrating numerically, TG_u curves for the process of burning out the carbon were determined. The $\alpha(T)$ dependencies obtained for both processes, are presented in the form of graphs in Figure 25.

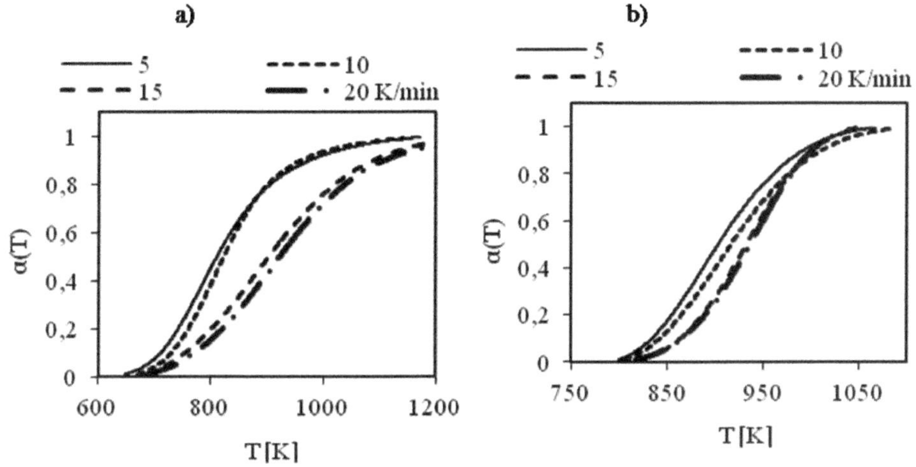

Figure 25. The $\alpha(T)$ dependencies. Oxidation of unreduced nc-TiC_x/C in air, a) oxidation of nc-TiC_x, b) burning out of elemental carbon

According to the theory, along with the increase in sample heating rates, the plots of $\alpha(T)$ are shifted into the higher temperature range.

The conversion degree for the process of burning out the carbon in the matrix (the second temperature range) was calculated in the same way. In the case of the oxidation process of nc-TiC_x (after reduction with hydrogen) two stages occurred: nc-TiC_x oxidation and burning out of the elemental carbon formed at the beginning of the oxidation process of nc-TiC. The $\alpha(T)$ dependence was determined for all the stages in the same way. The determined $\alpha(T)$ dependencies were the basis of kinetic studies. The Coats-Redfern equation was used. For all the stages kinetic models and Arrhenius parameters were determined (Table 12).

These data contain the full information about the kinetics of analysed processes. There should be noted that the kinetic parameters (the forms of $g(\alpha)$ functions and the values of A and E) determined on the basis of experimental data, should correspond to their physicochemical meaning. In the analysed case, the F2 model [$g(\alpha) = (1-\alpha)^{-1}-1$], having a theoretical justification, was used, and the determined activation energy values are similar to those found in many chemical reactions.

Sample	Conversion	Model	B [K/min]	A [1/min]	E [kJ/mol]	T_m [K]	$\Delta\alpha$	ΔT [K]
TiC_x/C	TiC	F2	5	26343,35	88,31	810,45	0,01-0,99	648-1147
			10	291522,00	99,20	822,77	0,01-0,99	671-1141
			15	16689,12	86,70	903,79	0,02-0,99	691-1245
			20	22863,39	89,26	923,22	0,01-0,99	690-1295
	$C_{elemental}$	F2	5	1,76E+10	185,04	902,35	0,02-0,99	803-1058
			10	8,27E+09	177,71	916,13	0,02-0,99	809-1085
			15	4,74E+12	222,13	932,37	0,02-0,98	827-1046
			20	7,1E+12	223,18	936,23	0,01-0,99	817-1042
	C_{matrix}	F2	15	5,24E+16	391,91	1243,06	0,02-0,98	1142-1366
			20	7,22E+16	392,43	1246,63	0,02-0,99	1138-1373
TiC_x after reduction	TiC	F2	5	1,01E+06	106,33	809,11	0,01-0,99	645-983
			10	1,37E+05	95,18	838,82	0,01-0,99	649-1134
			15	1,70E+05	98,51	892,95	0,01-0,99	673-1153
			20	3,19E+05	100,63	886,78	0,01-0,99	668-1144
	$C_{elemental}$	F2	5	6,98E+09	178,41	902,90	0,02-0,99	803-1063
			10	6,44E+09	174,85	908,97	0,02-0,99	810-1078
			15	1,55E+12	214,49	933,09	0,02-0,99	832-1063
			20	5,19E+11	204,20	930,08	0,02-0,99	830-1071

Table 12. List of kinetic data for the transformations in oxidation processes of titanium carbide samples before and after the reduction with H_2.

Basing on the obtained results an analysis of the process has been performed. The r(α,T) dependencies on sample heating rates and sample temperature for the stages were studied. The plots of r(α,T) obtained for the nc-TiC_x/C not-reduced with hydrogen are shown in Figure 26.

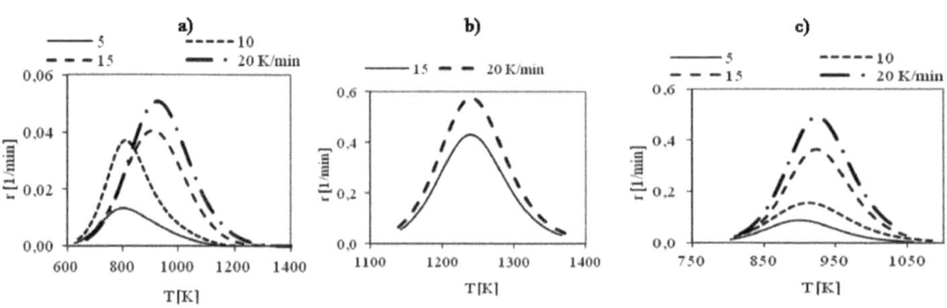

Figure 26. Plots of r(α,T). Oxidation of nc-TiC_x/C unreduced with hydrogen; a) oxidation of nc-TiC_x, b) burning out of the carbon formed during the oxidation process of nc-TiC, c) burning out of the carbon contained in the samples

According to the theory of kinetics of non-isothermal processes the reaction rate should increase along with the increase in sample heating rates. This condition is not well fulfilled for the oxidation process of nc-TiC_x unreduced with hydrogen (Fig.26a). This means that the samples used in measurement series differed. For the both processes of burning out the carbon the results consistent with theory were obtained. There should be noted that according to the theory, in each stage the process rate increases from zero for $\alpha(T)=0$, reaches the maximum in temperature Tm, and then decreases to zero at $\alpha(T)\rightarrow 1$. The temperature ranges determined for the stages runs on the basis of calculations are consistent with the ones determined experimentally.

The analogous plots obtained for the oxidation process in air of nc-TiC_x reduced with hydrogen are shown in Figure 27.

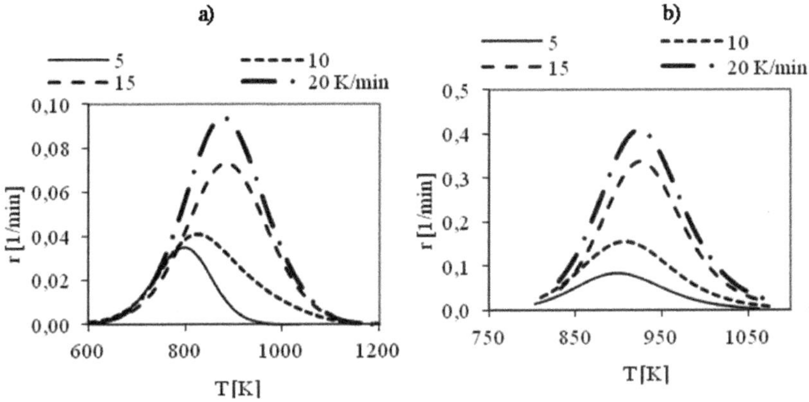

Figure 27. Plots of r(α,T). Oxidation in air of nc-TiC reduced with hydrogen; a) oxidation of nc-TiC_x, b) burning out of elemental carbon formed during the oxidation of nc-TiC_x

In this case, full consistency with the theory was obtained. The plots in Figure 27a show that nc-TiCx after reduction with hydrogen was uniform, the process rate increased along with the increase in sample heating rate, and the maximum was shifted into the higher temperature range. In both cases, the process rate, according to the theory, increases from zero for $\alpha(T)=0$, reaches its maximum at the temperature T_m, and then decreases to zero at $\alpha(T)\rightarrow 0$. The temperature ranges determined for the stages runs in measurement series on the basis of calculations are consistent with the ones determined experimentally.

The obtained results show that the proposed description method of the kinetics of two reactions proceeding simultaneously in a certain range of temperature, allows obtaining a satisfactory accuracy. This method was developed for the needs of processes of oxidation in air of nanocrystalline TiC and nanocomposites of TiC/C with varying carbon content in the matrix, for evaluating the protective qualities of carbon matrix, and also to evaluate and compare the resistance to oxidation of carbide ceramics [17]. In Figure 28 the use of the kinetics knowledge for comparative evaluation of the rate of TiC/C nanocomposite oxidation, depending on the carbon content in the matrix is shown as an example.

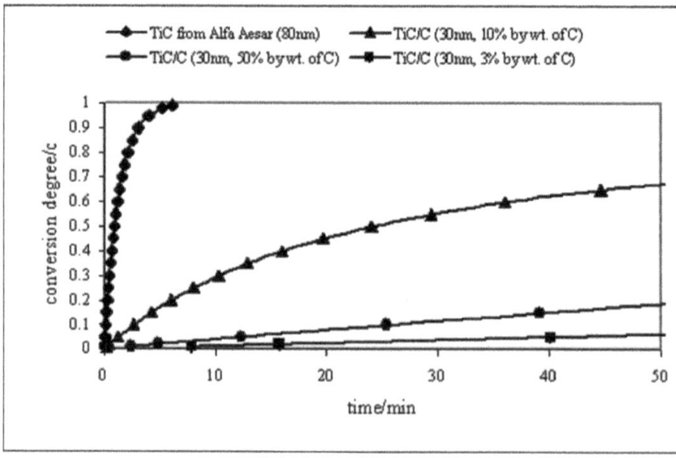

Figure 28. Dependence of conversion degree on time - $\alpha(t)$. T= 823 K. Oxidation of TiC commercial and TiC/C nano-composites (30 nm) in air. 50, 10, 3 % by wt. of the carbon contents in composites respectively [17].

5. Conclusions

The results of thermal decomposition of NH_4VO_3 in dry air have been presented. The measurements were carried out by TG – DSC method. The gaseous products were determined by MS method. Solid products were identified by XRD method. On the basis of measurement results the division of the process into stages has been made and the temperature ranges for stage courses and changes of sample masses in stages were determined. There was demonstrated that decomposition of NH_4VO_3 proceeds according to the following equation

$$6NH_4VO_3 \rightarrow (NH_4)_3 V_6O_{16} \rightarrow (NH_4)_2 V_6O_{16} \rightarrow 3V_2O_5$$

In all the stages at different sample heating rates NH_3, H_2O, NO and N_2O were evolved, which were formed as a result of NH_3 oxidation. NO_2 did not occur among the evolved gases. There should be added that N_2O was formed mainly during the stage II and III.

While performing the measurements the emphasis was placed on the possibility of obtaining experimental data for description of kinetics of investigated process, in accordance with ICTAC Kinetics Committee recommendations.

In the case of the investigated process, the necessary results for isothermal conditions were not obtained because the measurements for the stage could be performed only in a few temperatures, while at higher temperatures the results were obtained at high conversion degrees.

For non-isothermal conditions the needed data have been obtained. Kinetic calculations were performed using Kissinger's method, isoconversional method and Coats-Redfern method. Applying Kissinger's method the activation energies were determined and the kinetic models were assigned for the stages. The stages I and III are well-described by model A2, and the stage II by model A4. It has been also shown that the influence of heating rate of the sample on the course of the process can be compensated, at constant activation energy, by temperature range $T_{\alpha=0} \div T_{\alpha=1}$, which is the value determined experimentally. The temperature ranges were given for the stages.

In case of the investigated process applying the isoconversional method the following results were obtained. The E values for the stages I and III (assimetric plots of DTG and HF) changed constantly along with the change of conversion degree. However, in the case of stage II (symmetric plots of DTG and HF), E was practically constant. It seems probable that the isoconversional method compensates the influence of β_i on the course of the process by changing the activation energy. Theoretically much more interesting is the possibility to compensate the β_i, at constant activation energy, with temperature range $T_{\alpha=0} \div T_{\alpha=1}$.

While carrying out the calculations using the Coats-Redfern method the activation energies determined by Kissinger's method were used as the base values. Almost constant values of the A and E parameters were obtained for the stages for different heating rate of the samples. The verifying calculations were performed. The $\alpha(T)$ and $r(\alpha,T)$ dependencies were determined. The good consistency with experimental data was obtained. The obtained results show that the Coats-Redfern equation is of great importance for the studies of the kinetics of heterogeneous non-catalytic processes

Describing kinetics of the TiC_x carbonisation and their oxidation, Coats-Redfern's equation was applied kinetic models of stages were identified based on statistical evaluation and compliance to a large extent, of degrees of transformation for stages calculated and determined from measurements. Building the kinetic models of processes, the results of measurements were treated as statistic values. A system of a complex analysis of measurements results was developed with the use of artificial neurone networks. Based on the TG curves four stages have been distinguished. The first, endothermic stage proceeding with mass loss, corresponded to desorption of volatile products, D3 model. The second, exothermic stage proceeding with mass growth, was assigned to oxidation of uncarbonized nc-TiC_x/C by the oxygen present in argon at trace level, F2 model. The third endothermic stage, proceeding with mass loss, referred to carbonization of nc- TiC_x/C and pyrolysis of organic compounds, contained in the raw samples. The pyrolysis of admixtures and carbonization of nc-TiC_x/C proceeded simultaneously. After completing the carbonization process at the temperature above 1573 K, oxidation of carbonized TiC_x/y by oxygen present in argon at trace level was observed, R2 model.

The third, basic stage preceded in non-isothermal and isothermal conditions; at lower conversion degrees F1 model (first-order reaction) and at the higher conversion degrees (above 0.98) D3 model (three-dimensional diffusion, spherical symmetry, Jander equation) was applied.

Adapting to the description of the processes which took place with the participation of nc-TiC_x/C the parameters for the process of purification were determined together with the simultaneous carbonization of nc-TiC_x in argon, in conditions which make impossible their coalescence and growth to micron sizes. Kinetics and by the same, the mechanism of the processes of oxidation of nanocrystalline TiC_x in form of powder were tested, and they were subjected to evaluation based on the comparison of the rate of oxidation.

6. Nomenclature

GRNN	- Generalized-Regresion Neural Network
MLP	- Multilayer Perceptron
MS	- Mass Spectrometry
SEM	- Skanning Elektron Microscopy
TEM	- Transmission Electron Microscopy
TG – DSG	- Thermogravimtry and Differential Scanning Calorymetry
XRD	- X-Ray Diffraction
A	- pre-exponential Arrhenius factor ($1 min^{-1}$)
B	- const
C	- const
E	- apparent activation energy ($J\ mol^{-1}$, $kJ\ mol^{-1}$)
$f(\alpha)$	- conversion function dependent on mechanism of reaction
$f'(\alpha_m)$	- derivative of $f(\alpha)$ function for maximal reaction rate
ϕ (T, α,, P))	- temperature, conversion degree and pressure function
$g(\alpha)$	- integral form of kinetic model
h(P)	- pressure function
k(T)	- reaction rate constant ($1\ min^{-1}$)
m_0	- initial sample mass for the stage (mg)
m	- current sample mass for the stage (mg)
m_k	- final sample mass after for the stage (mg)
r	- reaction rate ($1\ min^{-1}$)
R	- gas constant ($J\ mol^{-1} K$)
r_p	- correlation coefficient
t	- time (min)
T	- temperature (K)
$T_{\alpha=0}$, $T_{\alpha=1}$	- initial and final temperature of the stage (K)
T_m	- maximum conversion rate temperature for a stage (K)
$T_{\alpha'_m}$	- temperature referring to α'_m
$\Delta T''$	- temperature range (K)
TG_u, TG_{un}	- normalized TG
α	- conversion degree
α_m	- conversion degree for maximum rate of a stage
α'_m	- average conversion degree for rate maxima of a stage for different heating rate
$\Delta\ \alpha$	- conversion degree range
β	- heating rate ($K\ min^{-1}$)
Subscripts	
m	- maximum rate of a stage
i, j	- order

Author details

Anna Biedunkiewicz
Institute of Materials Science and Engineering West Pomeranian University of Technology, Szczecin, Poland

Acknowledgements

This work it concerns were partially generated in the context of the MULTIPROTECT project, funded by the European Community as contract No. NMP3-CT-2005-011783 under the 6th Framework Programme for Research and Technological Development. Financial support of part of the work by the Ministry of Science and Higher Education within the Projects No N N507 444334, 2008–2011 and No NR15-0067-10/2010-2013, is gratefully acknowledged.

The scientific support of my teacher professor Jerzy Straszko throughout this work is gratefully acknowledged.

7. References

[1] Vyazovkin S, Burnham AK, Criado JM, Perez-Maqueda LA, Popescu C, Sbirrazzuoli N (2011) ICTAC Kinetics Committee Recommendation for Performing Kinetic Computations on Thermal Analysis Data. Thermochimica acta, 520: 1-19
[2] Weimer AW, Nilsen KJ, Cochran GA, Roach RP (1993) Kinetics of carbothermal reduction synthesis beta silicon carbide. AIChE j. 39:493-503
[3] Chaudhurry S, Mukerjee SK, Vadya NV, Venugopal (1997) Kinetics and Mechanism of Carbothermal Reduction of MoO_3 to Mo_2C, J alloy comp. 261(1-2): 105-113
[4] Patel M, Subrahmanyam J (2008) Synthesis of Nanocrystalline Molybdenum Carbide (Mo_2C) by Solution Route. Mater research bulletin. 43: 2036-2041
[5] Wanjun T, Yuwen L, Xi Y, Cunxin W (2004) Kinetic Studies of the Calcination of Ammonium Metavanadate by Thermal Methods. Ing eng chem res. 43 (9): 2054-2059
[6] Biedunkiewicz A (2009) Aspects of Manufacturing of Ceramic Nanomaterials of TiC/C, TiC, TiC-SiC-C and Ti(C,N)-Si(C,N)-Si_3N_4 Type by Sol-Gel Method; Szczecin: West Pomeranian University of Technology, 177p.
[7] Liu CH, Li WZ, Li HD (1996) TiC/Metal Nacreous Structures and Their Fracture Toughness Increase, J. mater res. 11(9): 2231-2235
[8] Thorne K, Ting S, Chi CJ (1992) Reduction of Titanium Dioxide to Form Titanium Carbide. J. mater sci. 27: 4406-4414
[9] Koc R, Folmer JS (1997) Carbothermal Synthesis of Titanium Carbide Using Ultrafine Titania Powders. J. mater sci. 32(12): 3101-3111
[10] Lu Q, Hu J, Tang K, Deng B, Qian Y, Zhou G, Liu X (1999) The co-Reduction Route to TiC Nanocrystallites at Low Temperatures. Chem phys lett. 314(1-2): 37-39
[11] Burakowski T, Wierzchoń T (1999) Surface Engineering of Metals: Principles, Equipment, Technologies, New York: CRC Press, 592p.
[12] Ohring M (1992) The Materials Science of Thin Films, New York: Academic Press, Inc. Harcourt Brace Jovanovich Publishers, 704p.

[13] Voitovich VB, Laverenko VA (1991) Oxidation of Titanium Carbide of Differency Purity, Soviet metallurgy met ceram. 30 (11): 927-932
[14] Jin Y, Liu Y, Wang Y, Ye J (2010) Synthesis of (Ti, W, Mo, V)(C, N) Nonocomposite Powder From Novel Precursors. Int refract j hard mater. 28: 541-543
[15] Yeh CL, Wang HJ (2011) Combustion Synthesis of Vanadium Borides. J Alloys Compounds, 509: 3257-3261
[16] Mutin PH, Popa AF, Vioux A, Delahay G, Coq B (2006) Nonhydrolytic Vanadian-Titania Xerogels: Synthesis, Characterization, and Behavior in the Selective Catalytic Reduction of NO by NH_3. Applied catal b: environ; 69 (1-2): 49-57
[17] Biedunkiewicz A, Gordon N, Straszko J, Tamir S (2007) Kinetics of Thermal Oxidation of Titanium Carbide and Its Carbon Nano-composites in Dry Air Atmosphere. J therm anal cal. 88 (3): 717-722
[18] Burnham AK, Weese RK, Wernhoff AP, Maienschein JL (2007) A Historical and Current Perspective on Predicting Thermal Cookoff behavior. J therm anal calorim. 89: 407-415
[19] Biedunkiewicz A, Gabriel U, Figiel P, Sabara M (2012) Investigations on NH_4VO_3 Thermal J therm anal cal., 108:965–970
[20] Coats AW, Redfern JP (1964) Kinetic Parameters From Thermogravimetric Data. Nature, 201: 68-69
[21] Vyazovkin S, Wight CA (1999) Model-Free and Model-Fitting Approaches to Kinetic Analysis of Isothermal and Nonisothermal Data. Thermochim acta, 340-341: 53-68
[22] Biedunkiewicz A, Strzelczak A, Chrościechowska J (2005) Non-isothermal Oxidation of TiCx Powder in Dry Air. Pol j chem technol; 7 (4): 1-10
[23] Biedunkiewicz A, Strzelczak A, Możdżeń G, Lelątko J (2008) Non-isothermal Oxidation of Ceramic Nanocomposites Using the Example of Ti-Si-C-N Powder: Kinetic Analysis Method, Acta mater. 56: 3132-3145
[24] Straszko J., Biedunkiewicz A, Strzelczak A (2008) Application of Artificial Neural Networks in Oxidation Kinetic Analysis of Nanocomposites, Pol j chem technol, 10(3):21-27
[25] Galway AK (2003) What is Mean by the Term "Variable Activation Energy" when Applied in the Kinetic Analyses of Solid State Decompositions (Cryptolysis Reaction)? Thermochim acta, 397:249-268
[26] Shimada S, Kozeki M (1992) Oxidation of TiC at Low Temperatures, J mater sci. 27: 1869-1875
[27] Shimada S (2002) A Thermoanalytical Study on the Oxidation of ZrC and HfC Powders with Formation of Carbon. Solid state ionics, 149 (3-4) : 319–326
[28] Shimada S (1996) A Thermoanalytical Study on Oxidation of TiC by Simultaneous TG-DTA-MS Analysis, J meter sci. 30:673-677
[29] Shimada S (2001) Interfacial Reaction on Oxidation of Carbides with Formation of Carbon, Solid state ionics, 99: 141-142
[30] Shimada S (2001) Formation and Mechanism of Carbon-containing Oxide Scales by Oxidation of Carbides (ZrC, HfC, TiC), Mat sci forum 369-372: 377-384
[31] Gozzi D, Montozzi M, Cignini PL (1999) Apparent Oxygen Solubility in Refractory Carbides, Solid state ionics 123: 1-10.
[32] Biedunkiewicz A (2011) Manufacturing of Ceramic Nanomaterials in Ti–Si–C–N System by Sol–Gel Method, J sol-gel sci technol. 59: 448-455

Chapter 7

Carbonitriding of Materials in Low Temperature Plasma

Angel Zumbilev

Additional information is available at the end of the chapter

1. Introduction

Nitriding and carbonitriding are used as basic processes for details and tools surface strengthening, during which a layer is formed on their surface, containing nitrogen or a combination of nitrogen and carbon.

When these two methods are used in conventional gas furnaces or in salt baths, the thickness of the layer or the composition of the resultant layers could not be reliably regulated, which necessitates varying with the potentials of nitrogen and carbon in the gas mixture or the liquid medium. The percentage of nitrogen and active carbon is defined by a few parameters, namely, temperature and composition of the gas mixture, and the possibilities for variation are limited.

During the process of carbonitriding and nitriding in glow discharge plasma these difficulties are resolved, which is an essential advantage of this method. The usage of glow electric discharge is a perspective method for nitriding and carbonitriding of materials in modern machine building.

The works [1,2,3,4,5] consider mainly the mechanism of building, the structure and the properties of the nitrided layers, obtained in low-temperature plasma, and present the peculiarities of carbonitriding in glow electric discharge. There is lack of data concerning carbonitriding in glow discharge plasma in an actuating medium, consisting of ammonia and gas corgon (82% Ar and 18% CO2), and there also lacks sufficient information of any comparative investigations between the two processes – nitriding and carbonitriding.

It is established in the works [3,5] that, when in the process of carbonitriding propane-butane is used as a carbon-carrier, the phase composition of the combined zone in the carbonitrided layer could not be precisely regulated. Better results could be obtained when using a mixture of methane and argon [15]. In metal welding the role of the protective gas is often taken by carbon, which contains both argon and carbon dioxide in a particular ratio.

One of the aims of the present paper is to investigate the possibility to use gas corgon not only in welding but also as an indirect carbon-carrier in the process of simultaneous saturation of a metal surface with nitrogen and carbon (carbonitriding) at low temperatures. The small percentage (18%) of carbon dioxide in the gas corgon makes it possible to regulate the amount of carbon, introduced into the vacuum camera.

Despite the numerous investigations, conducted with the use of ammonia, nitrogen or a mixture of nitrogen and hydrogen as saturating media, there is no an integrated model yet, representing the mechanism of nitriding and carbonitriding in glow discharge. There are two principle methods concerning the question of forming diffusion capable nitrogen and carbon atoms on the surface of the treated articles.

According to the first method, iron nitrides are formed initially, which then dissociate into lower substances and release nitrogen, in its turn diffusing into the treated material. According to the model, developed by Kölbel, it is assumed that, as a result from the pulverization of iron in the glow discharge, iron nitrides rich of nitrogen are formed and they deposit on the surface of the treated articles. The deposited nitrides decompose and release nitrogen, which diffuses into the interior of the material. The availability of ions could increase the number of the centers of chemisorption. In nitrogen and hydrogen containing atmospheres the process goes with the participation of NH-radicals, which, after taking hydrogen, turn into the very active radical NH_2. In result from the interaction between the iron and the neutral nitrogen atoms or radicals on the surface of the substrate iron nitrides are formed, which release diffusion capable nitrogen.

According to the second method, diffusion capable nitrogen is directly formed on the surface of the treated articles, i.e., without the preliminary forming of iron nitrides. Materials, having cathode, anode or floating potential, have been treated by the comparatively new method of nitriding in two-step vacuum-arc discharge under low pressure. The fact, that the samples with a positive potential can be nitrided renounces the idea of forming iron nitrides in the gas medium, as a surface, subject to electron impact does not pulverize [15]. The high activity of the saturating medium in this case is due to the neutral nitrogen atoms. The process depends only on the concentration of atomic nitrogen and the temperature of the article, while the electron/ion impact plays the role of a convenient tool for ensuring the temperature needed for the process. By investigating the area of the dark cathode space of a direct current glow discharge in the presence of hydrogen the work for electron detachment from the iron is reduced and thus the absorption of the nitrogen atoms is facilitated . Ultimately, during the process of glow discharge nitriding, atomic nitrogen is formed on the surface of the articles, which, depending on their temperature, diffuses into their interior.

Since the beginning of the 1970s glow discharge sources have been used predominantly in the field of investigating alloys. The scientific literature suggests a great number of applications based on glow discharge spectroscopes, not supposed so far, including polymeric mass-spectroscopy, sensitive assessment of nano-materials, as well as analysis of very thin (<0.1 μm) layers [6, 7, 8, 9]. The glow discharge optical emission spectroscopy (GDOES) is an atomic emission process for carrying out deep profile analysis. It combines pulverization and atomic emissions in order to enable an extremely fast and sensitive analysis. The plasma is generated in the chamber by applying voltage between the anode and the cathode with the availability of argon under low pressure. The ionized argon atoms cause pulverization in the area of the sample. The deposited atoms are excited in the plasma and radiate photons with characteristic wave lengths.

The GDOES is usually used for defining surface coatings, hidden connections, and deep profiles. The technique suggests quick, reliable and economically effective decisions. It suggests additional information for the rest of the surface analysis methods.

Ensuring high quality of manufactured products is directly related to increasing their reliability and durability, which, in turn, are determined to a large extent by the internal stresses in the details.

One of the basic methods of increasing the wear resistance of details is the purposeful improvement of their surface layer properties by means of mechanical, thermal, chemical-thermal and other types of hardening treatment.

Since the values of the internal stresses are often below the limit of flow of the corresponding material, their measuring is highly demanding to the measuring equipment. There are plenty of methods for defining the internal stresses and they can be divided into the following two groups: destructive methods – the methods of disassembling, of hanging down (the slack method), of drilling, boring and trimming; non-destructive methods – the Roentgen method, the magnetic method, the ultrasound method and the neutron rays method.

The Roentgenographic method allows registering submicroscopic changes in the distances between the atoms corresponding to the measured planes in the crystal lattice of the grains for a mono-crystal material. It is a completely non-destructive method. Because of the limited depth of penetration of the X-rays, which, for steel is $l \leq 20\mu m$, only the tense state of the closest to the surface layer is registered. The calculation principle used here allows determining of only two-axial internal stresses, parallel to the surface.

The distance between the atoms in the crystal lattice is normally about several nanometers. The wave length λ of the X-rays is also several dozens of nanometers, i.e., these quantities are of the same order. Therefore the Roentgen rays are considered to be among the most reliable for investigating the crystal structure.

The aim of the present work is to investigate the influence of the carbonitriding in a low-temperature plasma in an actuating medium consisting of ammonia and corgon (82% Ar and 18% CO_2) over the surface hardness, the total thickness of the carbonitrided layer, as well as its influence on the type and size of the formed compressive residual stresses on the surface and also to study the distribution of the nitrogen and carbon in depth of the formed layer of Armco-Fe and 25CrMnSiNiMo steel.

2. Methodology of investigation

2.1. Investigated materials and modes of thermal treatment

The investigated materials (Armco-Fe, 25CrMnSiNiMo steel) differ significantly by the availability and amount of alloying elements in them. The chemical composition of the materials mentioned above is checked by the equipment for automatic analysis "Spectrotest" [Table 1]. The low percentage of sulphur /< 0.015%/ in them guarantees a high level of hardness and toughness of the investigated materials.

Material	Chemical elements, weight percentage							
	C	Cr	Mo	Ni	P	Si	Mn	S
Armco-Fe	0.02	0.02	0.02	0.03	0.002	0.01	0.07	0.002
25CrMnSiNiMo	0.24	0.87	0.12	1.36	0.002	1.45	1.28	0.002

Table 1. Chemical composition of the materials

The requirement for a preliminary thermal treatment is imposed mainly by the following consideration: for achieving the desired mechanical parameters and structure, enabling a favorable process of nitrogen diffusion in depth. The investigated steel is thermally treated in a chamber furnace under modes [Table 2].

Material	t_{hard} (°C)	Cooling medium	$t_{temp.}$ (K)	$t_{temp.}$ (°C)	Cooling medium
25CrMnSiNiMo	900	Oil	873	600	Air

Table 2. Modes of preliminary thermal treatment

Treated this way, the samples are then subjected to ion carbonitriding in the installation "Ion – 20", according to the modes [Table 3]. Ammonia (NH_3) and corgon (82 % Ar и 18% CO_2) in different percentages are used as saturating gases. The temperature of treatment for the process of carbonitriding is 823K (550°C).

№ of the mode	Treatment	τ h	P_1 NH_3 Pa	P_2 corgon Pa	P total Pa	U V
1	nitriding	2	400	-	400	530
2	carbonitriding	2	360	40	400	470
3	carbonitriding	6	360	40	400	470
4	nitriding	6	400	-	400	530
5.	carbonitriding	6	200	200	400	415
6	carbonitriding	2	200	200	400	415
7	carbonitriding	4	280	120	400	435
8	nitriding	4	400	-	400	530
9	carbonitriding	2	350	350	700	380
10	carbonitriding	6	350	350	700	380

Table 3. Modes of nitriding and carbonitriding

2.2. Metallographic investigations

In order to clear out the morphological peculiarities of the nitrided and carbonitrided layers, metallographic analysis has been done.

When defining the structure and the thickness of the obtained layers metallographic pictures taken by means of a microscope – Axioscop – have been used.

The thickness of the nitrided and carbonitrided layer has been defined by the depth, to which hardness, equal to the core, has been achieved. Measuring the micro hardness has been done by means of a micro hardness-meter "Shimadzu" at a load of 0.98 N (100 g), following the Vikers' method.

2.3. Glow discharge optical emission spectroscopy

The process of defining the distribution of nitrogen and carbon in the nitrided and carbonitrided samples has been realised by means of the GDOES device GDA – 750, vom Spectruma-Analytik GmbH. The parameters of the glow discharge plasma are as follows: current - 20 mA, voltage - 800 V, plasma density - 10^{10}-10^{11} cm^{-3}, electron temperature 0.1 - 0.5 eV and plasma volume 15 cm^3. The turbomolecular pump (56 $\ell \cdot s^{-1}$) works constantly. The basic pressure is 10^{-6} Pa. The surfaces are polished beforehand in order to achieve congestion in the plasma sector of the GDOES device. The standardizing of the device has been carried out with a sample containing 7 weight per cents of nitrogen and 1 weight per cent of carbon.

2.4. Internal stresses

The investigation of the internal stresses in the carbonitrided samples is performed by means of a Roentgen diffraction-meter SET-X ENSAM, following the „$\sin^2\Psi$" method.

The direction of measuring is characterized by the angles Ψ: 0°, 14.96°, 21.42°, 26.57°, 31.09°, 35.26°, 37.23°, -10.52°, -18.43°, -24.09°, -28.88°, -33.21°, -37.27° and two angels ϕ=0 and ϕ=90 [Fig 1].

A powdered sample is used for standardizing the Roentgen diffraction-meter. Since the powdered sample is free of residual stresses, it allows checking and easily adjusting the device. In this particular case chromium Roentgen radiation Cr – Kα with a wave-length of λ=2.29Å was used. Information about the formed stresses is obtained at a distance of 7μm from the surface of the sample in a plane α - Fe {2 1 1}, i.e. the stresses are measured in α - Fe_n - phase just under compound zone.

By means of the Roentgen diffraction-meter the diffraction angles in the carbonitrided layers are measured. The data are introduced into the program „MATHLAB-2008". Through graphical representation of a straight line, built in the coordinates „2θ - $\sin^2 \Psi$", the value of the diffraction angle 2θ at $\sin^2 90^\circ$ is defined.

The residual stresses are defined by the following dependence:

$$\sigma_{\phi} = \frac{E}{2(1+\mu)}.\cot\theta\left(2\theta_{\psi=0} - 2\theta_{\psi=90}\right)\frac{\pi}{180}$$

The values of the elasticity constants in the given formula are chosen for non - carbonitrided steel: Poisson's ratio $\mu = 0.29$, elasticity modulus E = 210 GPA. The master diffraction angle is $2\theta = 156^{\circ}30'$ and $\theta = 78^{\circ}15'$. The miscount at defining stresses depends on the relative mistake $\Delta\theta/\theta$ at defining the angle θ. It is within 2 - 3%.

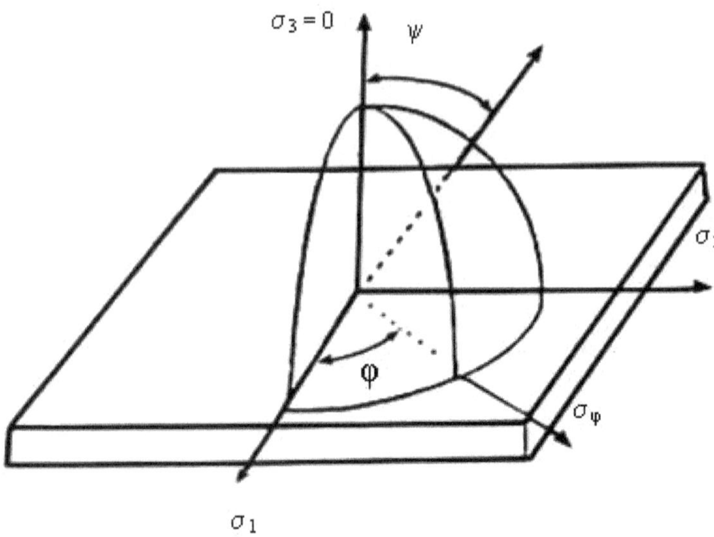

Figure 1. Characteristics of the direction of measuring by the angles Ψ and ϕ

3. Experimental results and analysis

3.1. Metallurgical analysis of carbonitriding fixtures

By measuring the micro hardness of thermally treated and ion-nitrided samples in depth, the maximum surface hardness - $HV_{0.1}$ - and the total thickness of the nitrided layer - δ_{tot} - have been defined; by means of a metallographic microscope the thickness of the combined zone - δ_{cz} - has been determined. The results are given in [Table 4].

It can be seen [Table 4] that after nitriding 25GrMnSiNiMo steel and Armco-Fe, under the modes [1, 8 and 4], a nitrided layer with different surface micro hardness, total thickness and combined zone thickness is obtained. Under these three modes of nitriding 25GrMnSiNiMo steel has a higher micro hardness but a lower total thickness of the layer and a thicker combined zone than Armco-Fe. This fact could be explained by the presence of alloying constituents in the steel, which actively participate in forming nitrides and strengthening the surface layer. They impede the diffusion of the nitrogen in depth, and in consequence, thinner layers with a thicker combined zone are obtained.

№ of the mode	25GrMnSiNiMo			Armco-Fe			
	$HV_{0.1}$ MPa	δ_{tot} μm	δ_{cz} μm	$HV_{0.1}$ MPa	δ_{tot} μm	δ_{cz} μm	U V
1	8500	170	6	3800	220	5	530
2	9400	160	5	4300	260	6	470
3	8600	290	8	4200	340	7	470
4	9800	300	11	4200	350	10	530
5	9200	240	8	4400	330	7	415
6	8900	150	5	3700	210	6	415
7	9300	210	6	4800	280	6	435
8	9500	230	7	4150	290	6	530
9	7400	140	4	5400	210	4	380
10	7500	230	5	5600	320	8	380

Table 4. Results from carbonitriding and nitriding of 25GrMnSiNiMo steel and Armco-Fe samples

During the process of carbonitriding of 25GrMnSiNiMo steel in a medium, consisting of 90% NH_3 + 8.2% Ar + 1.8 % CO_2 at the pressure of 400Pa, a layer with a lower micro hardness ($HV_{0.1}$= 9400 - 8600MPa), total thickness (160 - 290 μm) and combined zone thickness (5 - 8μm) is obtained, than after the process of nitriding without addition of a carbon-containing gas. This is most likely due to the small percentage of argon (8.2%) in the gas medium, since argon, because of its bigger atomic mass, has a strong pulverizing action. At the high coefficient of pulverizing the length of the free run of the pulverized atoms is bigger and the possibility for a backward diffusion of carbon and nitrogen is lower. A carbonitrided layer with a lower concentration of nitrogen and carbon is obtained. The more active pulverization does not allow the combined zone to grow and, as a result, a more deficient in nitrogen and carbon combined zone is obtained. In the diffusion zone of the carbonitrided layer of the steel detectable nitrided (carbonitrided) precipitations are not observed – [Fig.2.a].

After Armco-Fe carbonitriding under the same mode there are no similar dependences established during the process of forming the layer as the ones, described for 25GrMnSiNiMo steel. The obtained carbonitrided layer has a higher surface micro hardness ($HV_{0.1}$= 4200 – 4300 MPa), total thickness (260-340 μm) and combined zone thickness (6-7 μm). In the diffusion zone of a carbonitrided layer considerable amounts of nitrided (carbonitrided) precipitations are observed, mainly in the area of the grains. The precipitations originate at the boundary of the grains and propagate inward the volume. A bigger amount of precipitations is observed under the longer mode (6h) of carbonitriding – [Fig.2.c].

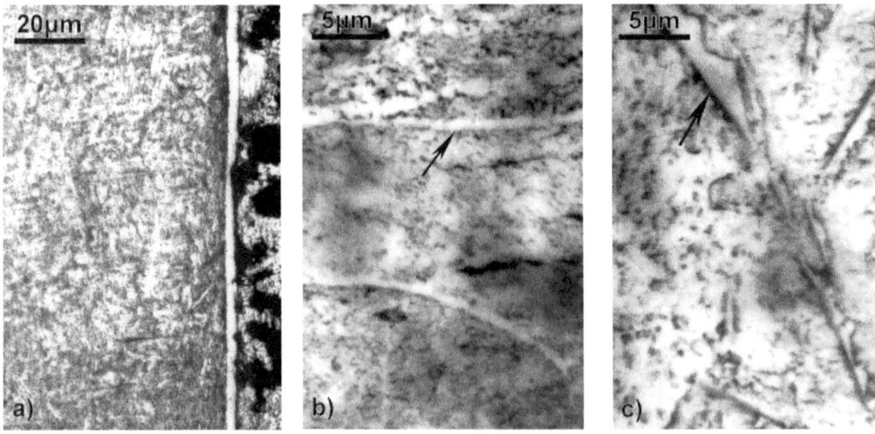

Figure 2. Microstructure of 25GrMnSiNiMo steel [a] and Armco-Fe [b, c] after carbonitriding at: t= 550°C, P_{NH_3} = 360Pa, $P_{82\%\ Ar+18\%\ CO_2}$ = 40Pa; a, c - τ = 6h, b - τ = 2 h

Together with the increase in the pressure of carbon ($P_{82\%\ Ar+18\%\ CO_2}$ = 200Pa, [Table 4] in a gas medium, in case of 25GrMnSiNiMo steel treatment carbonitrided layers are formed, having lower micro hardness ($HV_{0.1}$= 8900- 9200 MPa), total thickness (150-240 μm) and combined zone thickness (5-8 μm), than under the modes of treatment, considered so far. It is due to the increased activity of pulverizing, since the amount of argon in the gas medium is bigger - 41%. The higher rate of pulverizing leads to decreasing the probability for collisions between the atoms and ions, in consequence of which smaller amount of nitrogen and carbon is delivered to the surface. The microstructure analysis of the layer does not show detectable differences in the precipitations in the diffusion zone – [Fig.3 a-b].

The process of Armco-Fe carbonitriding under the same conditions ($P_{82\%\ Ar+18\%\ CO_2}$ = 200Pa, [Table 4], [modes 5-6] leads to obtaining a layer with higher micro hardness ($HV_{0.1}$= 4400 MPa) than in the process of nitriding - $HV_{0.1}$= 4200 MPa. It can be seen from [Fig.2.2c] that in the obtained carbonitrided layer there are carbonitrided (nitrided) precipitations with smaller sizes but in greater amount than in the layer, obtained after treatment with a bigger amount of ammonia (90% NH_3 + 8.2%Ar + 1.8 % CO_2).

After 25GrMnSiNiMo steel carbonitriding under the mode 7 [Table 4] in a gas medium of 70% NH_3 + 24.6%Ar + 5.4 % CO_2 at the pressure of 400Pa, a 210 μm layer with maximum micro hardness $HV_{0.1}$= 9300 MPa and combined zone thickness of 6 μm is obtained.

During the process of Armco-Fe carbonitriding under the same mode [mode 7], [Table 1] a layer with highest surface micro hardness of $HV_{0.1}$= 4800 MPa is obtained, in comparison to all the modes of treatment, considered so far. In the diffusion zone of the formed carbonitrided layer carbonitrided (nitrided) precipitations are observed [Fig. 2.3], which are of smaller sizes and in greater amount than the ones, obtained at using 90% NH_3 and 10 % corgon [Table 4,modes 2- 3,Fig.2.]; and bigger in size but in a smaller amount at carbonitriding with 50% NH_3 and 50 % carbon [Table 1, modes 5-6, Fig.3]. This is probably due to the diffusion of carbon in the diffusion zone of the carbonitrided layer as well. Together with the increase of the pressure of the gas medium to 700Pa (NH_3 - 350 Pa and corgon -350 Pa – [modes 9-10, Table 1] in the process of carbonitriding of the two materials under investigation, layers with essential differences are formed. After 25GrMnSiNiMo steel carbonitriding layers with a lowest surface micro hardness ($HV_{0.1}$= 7400 - 7500 MPa) and total thickness (140 -230 μm) in comparison to all the other modes of nitriding and carbonitriding are obtained. The white zone is non-uniform and broken – [Fig.4]. This is probably due to the increased amount of argon in the gas medium and the high density of the current at the higher pressure of the gas medium, as well as to the low voltage of the discharge [380V, Table 4.].

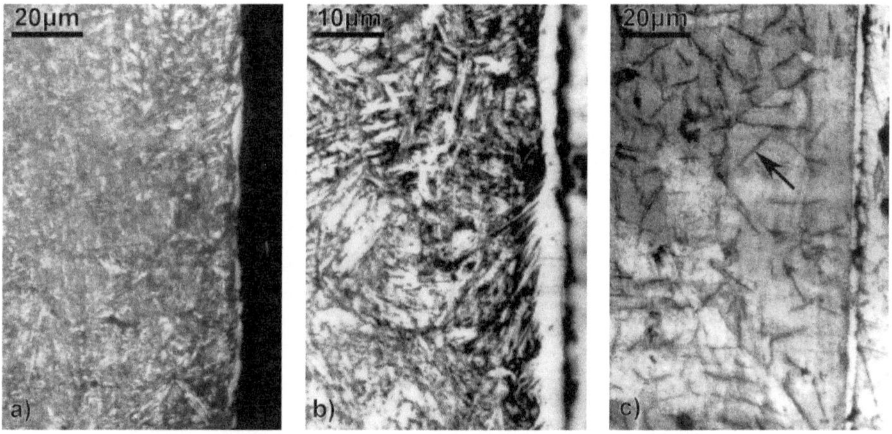

Figure 3. Microstructure of 25GrMnSiNiMo steel [a, b] and Armco-Fe[c] after carbonitriding at: t= 550°C, P_{NH3} = 200Pa, $P_{82\% Ar+18\% CO2}$ = 200Pa; a, c - τ = 2h, b - τ = 6h

After Armco-Fe carbonitriding under the same modes of treatment a layer with highest micro hardness ($HV_{0.1}$= 5400-5600 MPa) and lowest combined zone thickness (4 μm) in comparison to all the other modes of nitriding and carboniding is obtained. In the diffusion zone of the carbonitrided layer the smallest amount of precipitations with the smallest sizes in comparison to all the other modes of carbonitriding is observed.

Under the same mode of treatment, 25GrMnSiNiMo steel possesses higher micro hardness and smaller total thickness of the nitrided and carbonitrided layer, than Armco-Fe. After certain treatment of 25GrMnSiNiMo steel with an additionally introduced carbon-containing gas (carbon) in the ammonia medium at different percentage ratios layers with lower depth, surface hardness and combined zone thickness than in the process of nitriding are obtained.

After conducting the process of carbonitriding of the investigated materials with carbon (82 % Ar + 18 % CO_2) and ammonia, layers of small combined zone thickness are obtained. For

Armco-Fe the thickness is within 4 to 8 μm. During the process of nitriding of technical iron, a nitrided layer is formed, having the biggest combined zone [10 μm, mode 4, Table 4], which is not valid for the other modes of nitriding. The increase in the duration of the process and the decrease in carbon pressure lead to combined zone growth. During the process of 25GrMnSiNiMo steel carbonitriding this dependence remains the same. After 25GrMnSiNiMo steel nitriding a layer with a bigger combined zone is observed, than during the process of carbonitriding (6-11 μm against 4-8 μm). This is explained by the stronger pulverizing action of plasma as a consequence of the increase in current density and decrease in voltage after introducing carbon into the ammonia medium.

It could be noted that the use of carbon together with ammonia in the process of carbonitriding leads to decrease in the pressure of plasma discharge, which depends on the ratio between the two gases. When introducing carbon into the camera, the pressure of the discharge falls down. In order to reach the required temperature of carbonitriding in this case it is necessary to increase the current of the discharge. Thus the power of plasma remains the same. The increase of the current density leads to an increase in the pulverizing action of plasma, despite of the low voltage level. This is explained by the bigger amount of ions, bombing the surface of the detail. The bigger current density does not lead to an increase in the kinetic energy of the ions. The decrease in the pressure of the gases in the camera causes an increase in the discharge voltage and the kinetic energy of the ions increases at the same time as a result, though their amount remains unchanged. The coefficient of pulverizing could also be increased this way, which would lead to a decrease in the combined zone thickness.

3.2. Roentgenographic determination of internal stresses

By means of the Roentgen diffraction-meter the diffraction angles at different angles of rotation of the sample - Ψ и ϕ – are measured in the carbonitrided layers. The data are introduced into the program „MATHLAB-2008", by means of which graphs are built and the values of the angle 2θ for $\sin^2\Psi$, at Ψ=90° are calculated.

After defining the angle $2\theta^s$ at Ψ= 90^0 for all carbonitrided samples, residual compressive stresses in the carbonitrided layers have been calculated. The results are given in Table 5. Three ways of defining the diffraction angles have been used: the maximum intensity method - $\sigma^{\phi s}$, the chord method - $\sigma^{\phi c}$, and the body centre method - $\sigma^{\phi b}$.

Material	τ [h]	P_1 NH_3 [Pa]	P_2 corgon [Pa]	$HV_{0.1}$	δ_{tot} [μm]	δ_{cz} [μm]	$\sigma^{\phi s}$ [MPa]	$\sigma^{\phi c}$ [MPa]	$\sigma^{\phi b}$ [MPa]
Armco-Fe	4	280	120	480	280	6.5	-54	-28	-14
25CrMnSiNiMo	4	280	120	930	210	6.3	-584	-621	-521
25CrMnSiNMo	2	200	200	890	150	4.5	-829	-713	-655
25CrMnSiNiMo	2	350	350	740	140	4.1	-90	-63	-41

Table 5. Results from the obtained residual stresses

It can be seen [Table 5] that after ion carbonitriding of Armco-iron at $t_{nitr.}$ = 823K (550°C), P_{NH3} = 280Pa, τ = 4h, a carbonitrided layer with total thickness δ_{tot}=280μm, compound zone thickness δ_{cz} = 6,5 μm and maximum micro-hardness of 480HV0.1 is obtained. In the so formed layer compressive residual stresses occur. From the three methods of defining the diffraction angles the method of maximum intensity is the one from which the highest value of residual stresses ($\sigma^{\phi s}$ = - 54MPa) results. The compressive stresses, resulting after the ion carbonitriding in the surface layer of Armco-Fe are much lower than those of 25CrMnSiNiMo. During the process of Armco-Fe carbonitriding, a diffusion layer with bigger specific volume is formed, than it occurs with alloyed steel. This can be explained by the bigger total amount of nitrogen and carbon in depth the carbonitrided layer for 25CrMnSiNiMo steel [Fig.4].

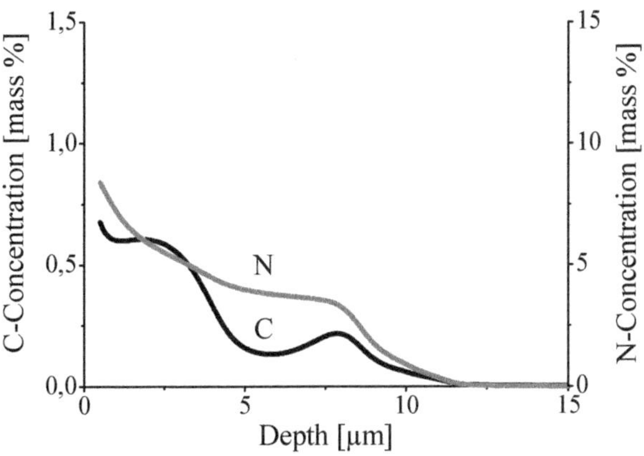

Figure 4. Nitrogen and carbon profile analysis according to GDOES in depth of a layer, carbonitrided at: t = 823K (550°C), $P_{1\ ammonia}$=280Pa, P_{carbon}=120Pa, τ = 4h, 25CrMnSiNiMo

Because of the lack of alloying elements in the Armco-Fe, no special carbonitrides are formed in it, and the difference in the specific volumes of the carbonitrided layer and the core material is therefore smaller. This leads to reducing the value of the residual stresses formed in the carbonitrided layer in Armco-Fe.

After ion carbonitriding of 25CrMnSiNiMo steel at $t_{nitr.}$ = 823K (550°C), P_{1NH3} = 200Pa, $P_{2carbon}$ = 200Pa, τ = 2h, a carbonitrided layer with total thickness δ_{tot} =150μm, compound zone thickness δ_{cz} = 4.5μm and maximum micro-hardness of 890HV0.1 is obtained. In thus obtained carbonitrided layer residual compressive stresses with the highest value of $\sigma^{\phi s}$ = - 829 MPa are formed.

With prolongation of the time of 25CrMnSiNiMo-steel carbonitriding from 2 to 4 hours and reducing the carbon pressure from 200 to 120 Pa while increasing ammonia pressure from 200 to 280 Pa, a carbonitrided layer with higher total thickness δ_{tot} = 210μm, compound zone thickness δ_{cz} = 6.3 μm, and higher maximum micro-hardness - 930HV0.1 - is formed. The resultant compressive stresses on the carbonitrided surface are lowered to $\sigma^{\phi s}$ = 584

MPa. The explanation can be found in the phase composition and the thickness of the compound zone (δ_{cz}=6.3 μm).

Together with increasing ammonia and carbon pressure from 200 to 350Pa for 2 hours' time of treatment a carbonitrided layer with total thickness δ_{tot} =140μm, compound zone thickness δ_{cz} = 4.1μm and maximum micro-hardness 740HV0.1 is obtained. In the carbonitrided layer, obtained this way, the smallest residual compressive stresses of $\sigma^{\phi s}$ = -90MPa are formed. The high pressure of the two saturating gases P_{tot} = 700 Pa does not activate the process of pulverizing and therefore a bigger amount of nitrogen and carbon is delivered to the surface. This probably leads to forming a bigger amount of micro-pores in the white zone [Fig.5], as a consequence of which the specific volume of the carbonitrided surface formed is bigger than the one in the core material.

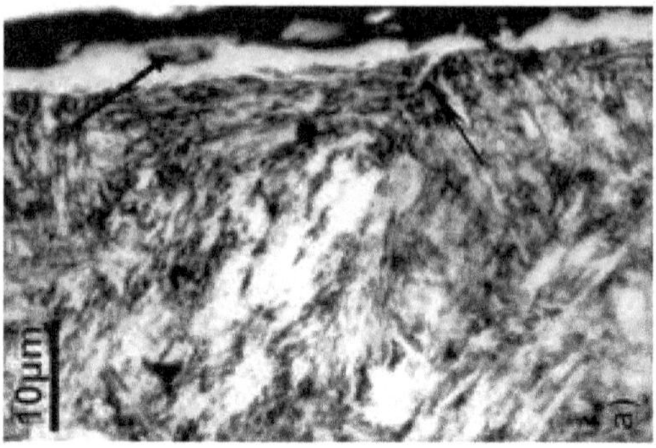

Figure 5. Microstructure of 25CrMnSiNiMo -steel after carbonitriding at: t= 823K (550°C), P_{NH3}=350Pa, $P_{82\%\ Ar+18\%\ CO2}$ = 350Pa, τ = 2h

In the process of 25CrMnSiNiMo-steel ion carbonitriding the increase of ammonia pressure (P_{1MH3} = 200-350 Pa) in the saturating medium forms a carbonitrided layer with approximately the same total thickness (δ_{tot} = 140-150μm) and compound zone thickness (δ_{cz}= 4.1 – 4.5μm), but with lower micro-hardness (890 – 740 HV0,1). This leads to a considerable decrease of the residual compressive stresses on the carbonitrided surface ($\sigma^{\phi s}$ = 829 –90MPa).

It can be noted that under the same modes of ion carbonitriding [modes 1 and 2, Table 5] the two materials under investigation form different residual compressive stresses on their surfaces ($\sigma^{\phi s}$ = 54–584 MPa). This significant difference is explained by the availability of alloying elements in the 25CrMnSiNiMo-steel, which, after the process of carbonitriding, form disperse carbonitrides. This leads to certain increase in the micro-hardness and in the specific volume of the carbonitrided layer and thence, to increase in the residual compressive stresses on the surface as well.

It can be noted from the conducted investigations that the chosen modes of ion carbonitriding form on the surface of the materials carbonitride layers with a bigger specific volume than on the core.

Depending on the concentration of nitrogen and carbon in the carbonitrided layer, as well as on the alloying elements contained in the materials, the specific volume of the surface changes; this, in turn, leads to forming residual compressive stresses of different values.

3.3. Results from the analysis and the investigations of 25CrMnSiNiMo steel

After nitriding under the mode 6 [Table 2] a layer is formed in 25CrMnSiNiMo steel with micro-hardness of $1072HV_{0,1}$, total thickness of 250 μm and combined zone thickness of 10μm. The distribution of the diffused in depth nitrogen is given in [Fig. 6].

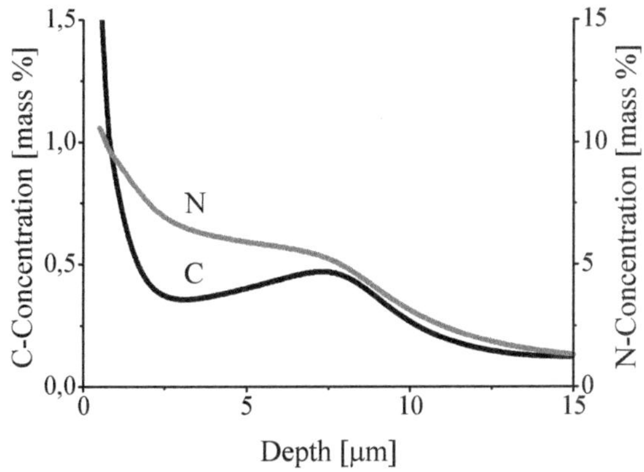

Figure 6. Carbon (C) and nitrogen (N) concentration in depth after carbonitriding of 25CrMnSiNiMo steel at: t = 550 ^{0}C, P_{NH3} = 360 Pa, $P_{82\%\ Ar + 18\%\ CO2}$ = 40 Pa, τ = 6 h

When, except for ammonia, 10% corgon is introduced into the chamber in addition, a carbonitrided layer with lower total thickness (290 μm) and combined zone thickness (8μm) than they are in the nitrided layer is obtained.

It can be noted that after carbonitriding in the media of 90% NH_3 + 8.2% Ar + 1.8 % CO_2 at 400Pa pressure, a layer with lower micro-hardness, total thickness and combined zone thickness, than after the process of nitriding, is obtained. This is probably due to the availability of carbon in the saturating medium, which, owing to its bigger atomic mass, has strong pulverizing action. With the high coefficient of pulverization the length of the free run of the pulverized atoms is bigger and the possibility for backward diffusion of the nitrogen and carbon is lower. Lower concentration of nitrogen (10%) and higher content of carbon by nearly 50% is obtained in the combined zone of the carbonitrided layer, compared to the nitrided one – [Fig.6].

It can be seen from [Fig.6] that the concentration of carbon (0.48%) has increased at the border between the basic material and the combined zone, while under the carbonitride zone gradual change of the carbon content has been observed. This can be explained by the simultaneous saturation of the surface both with nitrogen and carbon, where part of the nitrogen atoms is replaced by the carbon ones.

With the increase in the carbon pressure [P $_{82\%\ Ar\ +\ 18\%\ CO2}$ = 200 Pa, Table2, modes 3 and 4] in the gas medium, after carbonitriding of 25CrMnSiNiMo steel, carbonitrided layers are formed, having lower micro-hardness (890 – 920 $HV_{0,1}$), total thickness (150 – 240 μm) and combined zone thickness than under the rest of the modes of treatment. This is due to the increased activity of pulverization, as the amount of argon in the gas medium is higher – 41%. The higher degree of pulverization leads to decreasing the probability for collisions between the atoms and ions, as a result of which lower amount of nitrogen and carbon is delivered to the surface. Lower concentration of nitrogen (20%) and increased content of carbon with nearly 20% is obtained in the combined zone of the carbonitrided layer in comparison with the nitrided one – [Fig.7].

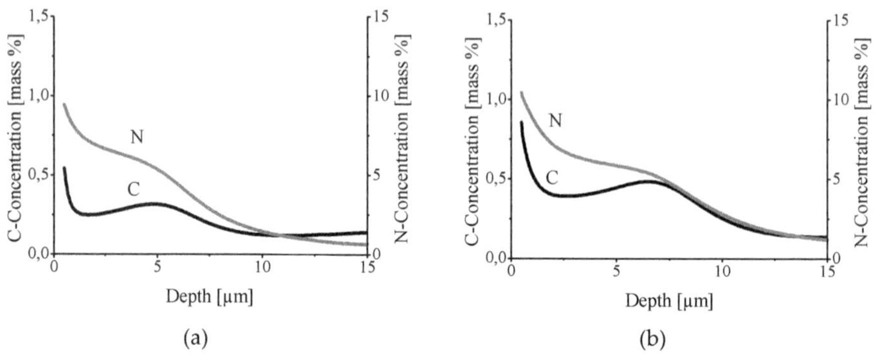

Figure 7. Distribution of carbon (C) and nitrogen (N) in depth after carbonitriding of 25CrMnSiNiMo steel at: a) t = 550 ^{0}C, P $_{NH3}$ = 200 Pa, P $_{82\%\ Ar\ +\ 18\%\ CO2}$ = 200 Pa, τ = 2h, b) t = 550 ^{0}C: P_{NH3} = 200 Pa, $P_{82\%\ Ar\ +\ 18\%\ CO2}$ = 200 Pa, τ = 6 h

It can be seen from [Fig.7.a].that at the end of the combined zone of the layer, at 5 μm depth, slight increase of the carbon up to 0.3 % is observed, while at the beginning of the combined zone the carbon is over 0.5 %. The concentration of the nitrogen in the carbonitrided zone decreases sharply, reaching at the end of the combined zone the level of 4.9 %, while the nitrogen on the surface is 9.1%.

With prolongation of the time of carbonitriding from 2 to 6h the micro-hardness and the combined zone thickness increase. Significant increase in the concentration of carbon in the combined zone can be seen from [Fig.8], where it achieves the level of 0.81% on the surface and slightly decreases at the end of the zone – down to about 0.5%. The distribution of the nitrogen in the carbonitrided zone decreases gradually. At the border between the diffusion zone and the combined zone the nitrogen concentration is 5.8 %, while on the surface it is 10.5%.

After carbonitriding of 25CrMnSiNiMo steel under the 5th mode of treatment [Table 2] in the gas medium of 70% NH_3 + 24.6 % Ar + 5.4 % CO_2 at 400 Pa pressure, a layer is obtained with total thickness 210 μm, maximum micro-hardness 930$HV_{0.1}$ and combined zone thickness 9 μm. The distribution of the nitrogen and carbon in depth of the carbonitrided zone changes gradually, staying almost the same up to 7.5 μm – [Fig.8].

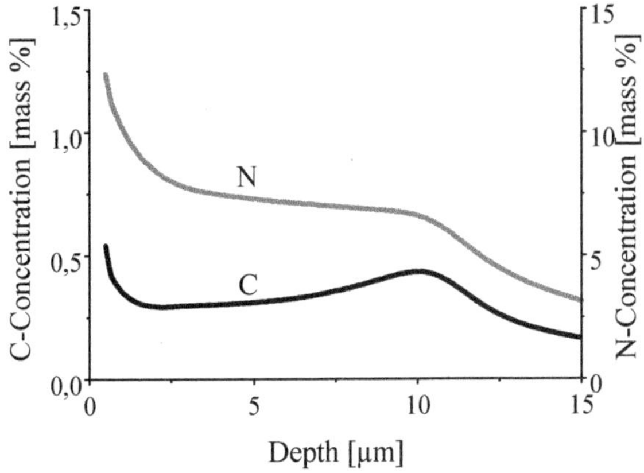

Figure 8. Distribution of carbon (C) and nitrogen (N)in depth after carbonitriding of 25CrMnSiNiMo steel at: t = 550 ^{0}C, P_{NH3} = 280 Pa, $P_{82\% Ar + 18\% CO2}$ = 120 Pa, τ = 4 h

Significant increase in the concentration of carbon at the beginning of the combined zone can be seen from [Fig. 8], where it reaches the level of 0.52% and slightly decreases at the end of the carbonitrided zone, going to about 0.48 %. The distribution of the nitrogen in the carbonitrided zone changes gradually and its concentration at the end of the combined zone reaches the level of 7%, while at the beginning of the combined zone it is about 12.2 %. Under this mode of treatment the highest level of nitrogen concentration 12.2 % is achieved in the combined zone and the most gradual change of the content of nitrogen and carbon in the formed layer occurs in comparison to all the other modes of ion carbonitriding of 25CrMnSiNiMo steel.

3.4. Armco-iron

In Armco-iron carbonitriding under the 4th mode of treatment (50%HN_3 + 41% Ar + 9 %CO_2) from Table 2 a layer with surface micro-hardness of 370$HV_{0.1}$, total thickness of 210 μm and combined zone thickness of 6 μm is obtained and the distribution of nitrogen and carbon in depth of the carbonitrided zone is given in [Fig. 9].

From [Fig.9] it can be seen that at the end of the combined zone at 6μm depth slight increase of the carbon content to 0.24 % is observed, while at the beginning of the combined zone the carbon content reaches 1 %. The figure shows that the distribution of carbon in depth of the combined zone is sharp to 3μm depth with concentration of 0.20 %. At the end of the carbonitrided zone the carbon concentration increases to 0.24 %. The nitrogen distribution change in the carbonitrided layer goes gradually. At the end of the combined zone (6μm) its concentration reaches 3.4 %, while at its beginning the concentration is 8.3%.

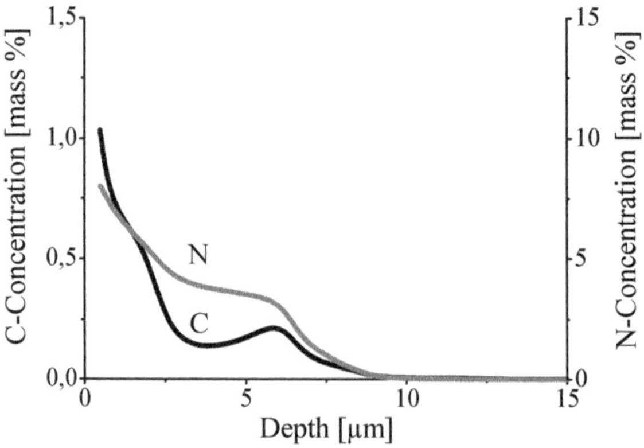

Figure 9. Distribution of carbon (C) and nitrogen (N)in Armco-iron after carbonitriding t = 550 ^{0}C, P_{NH3} = 200 Pa, $P_{82\% Ar + 18\% CO2}$ = 200 Pa, τ = 2

After ion carbonitriding of Armco-iron under the 5th mode (70%HN_3 + 24.6% Ar + 5.4 %CO_2), a layer with the highest surface micro-hardness $480HV_{0.1}$ is obtained, in comparison to all the other modes of treatment, which can be explained by the distribution of carbon and nitrogen in the carbonitrided zone – [Fig. 10a]. The figure illustrates that in depth of the combined zone increased carbon concentration of up to 0.68 % on the surface is observed, while it stays almost constant (0.65 %) at 3 μm depth. After that, at 6μm, sharp decrease of the carbon content is observed, together with its increase (0.25%) at the end of the carbonitride zone of the layer. The distribution of nitrogen in the carbonitride zone goes up gradually reaching 8.5 % on the surface, while at 3μm depth it is 5.1 %. It can be noted that under this mode of carbonitriding the nitrogen and carbon concentration changes more gradually and this concentration is higher in depth of the carbonitrided zone of the layer.

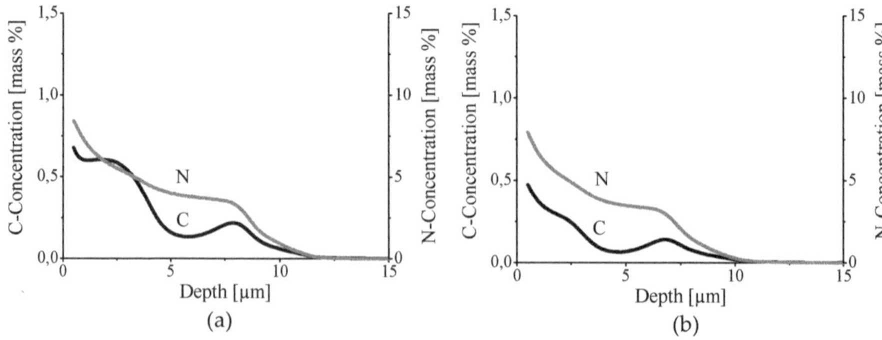

Figure 10. Distribution of carbon (C) and nitrogen (N) in Armco-iron, after carbonitriding at: a) t = 550 ^{0}C, P_{NH3} = 280 Pa, $P_{82\% Ar + 18\% CO2}$ = 120 Pa, τ = 4 h, b) t = 550 ^{0}C, P_{NH3} = 360 Pa, $P_{82\% Ar + 18\% CO2}$ = 40 Pa, τ = 2 h

[Fig.10.b] illustrates the distribution of carbon and nitrogen in Armco-iron after carbonitriding at: t = 550 ^{0}C, P_{NH3} = 360 Pa, τ = 2 h, $P_{82\% Ar + 18\% CO2}$= 40 Pa.

Under this mode of treatment a layer with surface micro-hardness of $430HV_{0.1}$, total thickness of 260 μm and combined zone thickness of 6μm is obtained. The concentration of carbon in depth of the layer is relatively low and reaches 0.48 % on the surface, while at 4.3 μm from the surface it has the lowest value (0.16 %). At the end of the combined zone the concentration slightly increases (0.21 %). The distribution of nitrogen in the carbonitrided zone goes up gradually, reaching on the surface the level of 8.1 % and decreasing to 3.9 % at 4.3μm from the surface. Under this mode of carbonitriding the carbon and nitrogen concentration changes gradually; however, their concentration is lower in depth of the combined zone than it is under the 5th mode of treatment [Table 2].

It can be noted that when in the process of carbonitriding the saturating medium contains bigger amount of CO_2 (9%), on the surface of Armco-iron combined zone with highest carbon concentration 1% is formed, while at (1.8 %) content of CO_2 the concentration is the lowest - 0.48%. From the modes of carbonitriding of Armco-iron under consideration the most uniform distribution of nitrogen and carbon in depth of the layer is observed under the 5th mode [Table 2]. The active role of argon for delivering carbon and nitrogen on the surface of the treated material is worth mentioning. As a result of its bigger atomic mass, argon has strong pulverizing action. With the high coefficient of pulverization the length of the free run of the pulverized atoms is bigger and the possibility for backward diffusion of carbon and nitrogen is lower. By changes in the pressure, as well as in the content of argon in the saturating medium, the backward diffusion of nitrogen and carbon can be regulated, thus making it possible to obtain layers with different features and properties.

On the basis of the conducted glow discharge optical emission spectral analysis of samples from Armco-iron and 25CrMnSiNiMo steel it is necessary to note that under all modes of ion carbonitriding carried out in ammonia and carbon medium layers are formed with concentration of carbon in the combined zone, which, for the 25CrMnSiNiMo steel is within 0.6 % - 1.4 %, while for Armco-iron it is between 0.45% and 1 %.

3.5. Dissociation and ionization of carbon dioxide and ammonia

It is necessary to note that two areas could be distinguished in the structure of the glow discharge: the zone of the discharge, where the processes of dissociation of the employed gases (CO_2, NH_3, Ar) occur, and the zone of the discharge, where reactions of recombination proceed.

Carbon dioxide dissociation has been investigated by a great number of authors both theoretically and experimentally by using various sources of plasma such as a microwave discharge, a plasma reactive burner or radio frequency arc discharge. Despite the numerous works the kinetic mechanism of CO_2 dissociation has not been studied very well yet [2.5]. Actually the mechanism of CO_2 dissociation is determined mainly by the parameters of the glow discharge (average energy of the electrons in the plasma) and the properties of the plasma gas (pressure, velocity of flow, energy) [10.11].

In the cathode space of the glow discharge [Fig.11] the ordered motion of electrons and the position of the positively charged ions is the predominant event, while in the anode section the chaotic motion of the electrically charged particles prevails. Electrons are detached from the cathode and they are accelerated in the direction towards the anode, acquiring energy, sufficient for dissociation and ionization of the atoms and molecules. The obtained positive ions are directed toward the cathode (C) and, colliding with its surface, they cause an emission of new electrons, while the secondary electrons, formed during the ionization, are accelerated by the field toward the anode (A).The cathode dark space in the structure of the glow discharge [Fig.11] includes the whole area of the cathode up to the next section of the negative glowing. This area is related to a big part of the voltage, called cathode fall of the potential. In this area the gas glowing is weak, as the energy of the electrons is higher than the maximum for excitation. This energy is sufficient for causing dissociation and ionization of the employed gases. The electrons, originated from the ionization of the atoms, are accelerated by the field and move toward the area of the negative glowing.

The gases in this area are in ionized state (plasma). The plasma results from the accelerated electrons, coming from the cathode dark space. At the moment the accelerated electrons impact the CO_2 molecule, it decomposes in result from the bond breakage, which leads to forming CO, C and O, expressed by the following reactions [12, 13]:

a. Dissociation at direct electron impact:

$$CO_2 + e^- \rightarrow CO_2^* + e^- \rightarrow CO + O - 532\ \text{kJ/ mol (energy of dissociation - 7.3 eV)} \quad (1)$$

$$CO \leftrightarrow C + O - 1076\ \text{kJ/ mol (energy of dissociation - 11.1 eV)} \quad (2)$$

$$CO_2 \leftrightarrow CO + 1/2O_2 - 281\ \text{kJ/ mol (energy of dissociation – 2.9 eV)} \quad (3)$$

$$O_2 \leftrightarrow 2O - 497\ \text{kJ/ mol (energy of dissociation – 5.12 eV)} \quad (4)$$

The symbol * corresponds to the state of high excitation. The addition of inert gases (He, Ar) into the CO_2, medium leads to increasing the average energy of the electron in the discharge. At sufficient concentration of CO_2, considerable reduction in the energy used for dissociation of one molecule is achieved.

It can be noted that with the increase in the vibration temperature the energy consumption for forming an atom reduces significantly since in these cases the decomposition of molecules is facilitated. However, the energy consumption for forming a carbon atom many times exceeds the energy for molecular dissociation. This is the case with CO. It can be explained first of all by the higher energy of dissociation of CO so that it considerably exceeds the average energy of the electron and the process of dissociation becomes a multiple-stage one.

b. dissociation bonding

$$\begin{aligned} CO_2 + e^- \rightarrow (CO_2^-)^* &\rightarrow CO + O \\ &\rightarrow C^- + 2O \\ &\rightarrow CO^- + O \end{aligned} \quad (5)$$

During the reaction the negative ions from CO_2 are minority [12]. The last reaction could lead to forming vibration excited CO_2 molecules by recombination of CO^- and O, causing dissociation in CO + O after that.

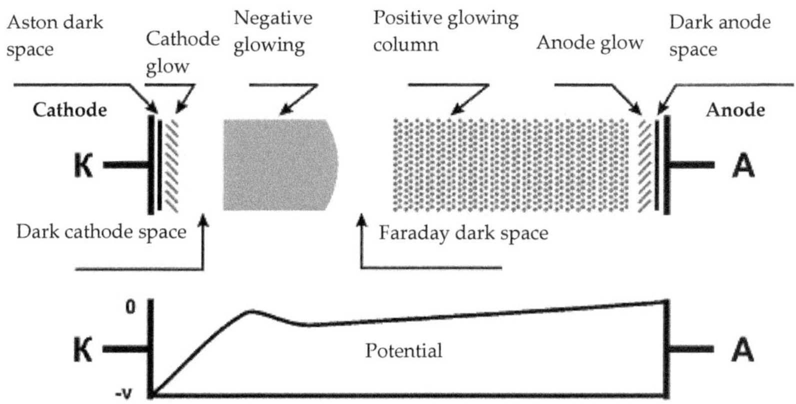

Figure 11. Glow discharge structure

c. ionization at electron collision

$$CO_2 + e^- \rightarrow CO_2^+ + e^- + e^- \text{ (energy of ionization – 13.8 eV)`} \tag{6}$$

For the initial process of ionization many different types of positive ions are obtained (CO^+, 0^+, 0_2^+), but the most important is CO_2^+, the others are usually neglected [11,12]. Recently the following reactions have been identified [1,5]:

$$\begin{aligned} CO_2^+ + e^- &\rightarrow CO_2, \\ CO_2^+ + e^- &\rightarrow C + O_2, \\ CO_2^+ + e^- &\rightarrow CO^- + O \end{aligned} \tag{7}$$

It can be noted that CO_2^+ molecules can be formed in two excited states and forming a stable CO_2^+ dominates. CO_2 dissociation is mainly due to vibration excitation caused by electron collisions. It is clear in this case that by heating the employed gases in imbalanced conditions higher efficiency of dissociation can be achieved, since then the introduced energy is not used in all degrees of freedom. As the molecular dissociation goes due to vibration excitation, the effective dissociation could occur at imbalanced conditions, in which the increase in the vibration excited states is higher than it is at balanced conditions. Therefore it is assumed that molecular dissociation occurs in imbalanced plasma with high vibration temperature. With the imbalanced gas the efficiency of the dissociation increases for two reasons. The first reason is the comparatively smaller energy, used for excitation of translational and rotational degrees of freedom. The second reason is the inharmoniousness of the molecules, leading to an increase of the relative number of vibration excited molecules. Due to it the same extent of dissociation of the molecules is obtained at lower temperature of vibration, than the temperature at the lack of inharmoniousness. It can be noted that in the zone of the negative glowing of the glow discharge both processes of dissociation and ammonia ionization occur, leading to obtaining nitrogen and hydrogen by the following probable reactions:

$$NH_3 \rightarrow HN_3^{+} + e^{-} \quad - 164.65kJ/mol \text{ (energy of ionization - 1.69 eV)} \tag{8}$$

$$NH_3^{+} \rightarrow N^{+} + 3H - 2298.29kJ/mol \text{ (energy of ionization and dissociation-23.67 eV)} \tag{9}$$

$$NH_3^{+} = N + 2H + H^{+} - 2216.96 \text{ kJ/mol (energy of ionization and dissociation - 22.83 eV)} \tag{10}$$

$$NH_3 = NH_2 + H^{+} + e^{-} - 1656.08 \text{ kJ/mol (energy of ionization and dissociation - 17.05eV)} \tag{11}$$

$$NH_2 = NH + H^{+} + e^{-} - 1582.16 \text{ kJ/ mol (energy of ionization and dissociation – 16.29 eV)} \tag{12}$$

$$NH = NH^{+} + e^{-} - 1242.47 \text{ kJ/ mol (energy of ionization and dissociation – 12.79eV)} \tag{13}$$

$$NH^{+} = N + H^{+} - 334.67 \text{ kJ/ mol (energy of ionization and dissociation – 3.44 eV)} \tag{14}$$

$$N = N^{+} + e^{-} - 1381.51 \text{ kJ/ mol (energy of ionization and dissociation – 14.22 eV)} \tag{15}$$

$$H_2 \leftrightarrow 2H - 434.95 \text{ kJ/ mol (energy of dissociation – 4.48 eV)} \tag{16}$$

$$N_2 \leftrightarrow 2N - 948.54 \text{ kJ/ mol (energy of dissociation – 9.77 eV)} \tag{17}$$

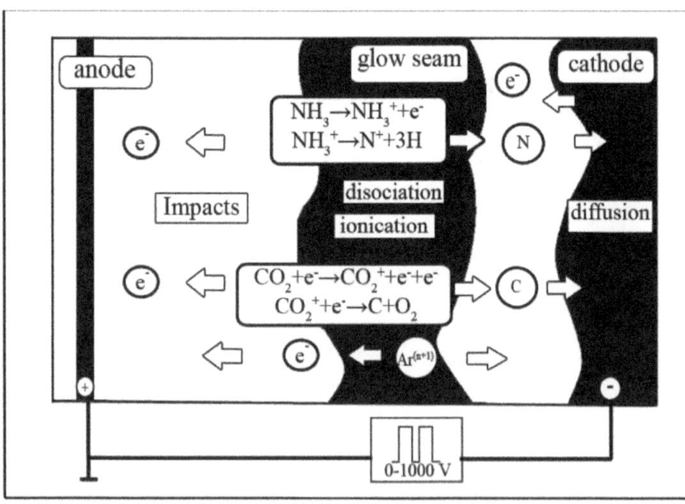

Figure 12. Reactions, going in close proximity of the cathode space in a medium of ammonia, carbon dioxide and argon

On the basis of the above exposed data the following probable mechanism of carbonitriding in a saturating medium of ammonia and corgon could be suggested:

As the energy of dissociation and ionization of ammonia and carbon dioxide is higher, in the area of the cathode fall of the glow discharge probably atomic carbon should form initially by the reaction $CO_2^{+} + e^{-} \rightarrow C + O_2$ and atomic nitrogen by the reaction $NH_3^{+} = N + 3H$. The dissociation of the ammonia molecule and the breakage of the carbon-oxygen bonds go close to the cathode in the zone of the negative glowing. Consequently, processes of ionization of the carbon (reaction $CO_2 + e^- \rightarrow CO_2^+ \rightarrow C^+ + 2O$) and nitrogen (reaction $NH_3^+ = N^+ + 3H$) atoms occur; these atoms impact the surface and diffuse at a certain distance into the material – [Fig.12]. The process of saturation of the metal surface depends only on the concentration of the atomic carbon and nitrogen in the plasma, as well as on the temperature of the article, while the electron or ion impact plays role for ensuring the necessary temperature of the details.

The availability of argon in the saturating medium in combination with CO_2, leads to increasing the average energy in the discharge, which, at sufficient concentration of CO_2, results in considerable reduction of the energy spent on the dissociation of a molecule.

In the process of carbonitriding part of the nitrogen atoms are replaced by the bigger carbon atoms, which causes forming the ε-phase in the carbonitrided layer. It is possible for two reasons: approximately the same ion radius of the nitrogen [r_{ion} = 13(5+1) pm, r_{ion} = 16(3+1) pm] and the carbon [r_{ion} = 16(4+1) pm]; the possibility for the carbon atoms to take up the vacant junctions.

4. Conclusion

1. It is established that, after ion carbonitriding of 25GrMnSiNiMo steel at the same temperature of treatment and time of saturation but different composition of the two gases in the vacuum camera (ammonia – 50, 70, 90, 100% and corgon - 50, 30 , 10%), the layers obtained have smaller thickness and micro hardness than after ion nitriding.
2. It is established that, the gases corgon and ammonia can be used in different percentage ratio in the role of saturating medium during the process of carbonitriding in glow discharge plasma.
3. It has been established that the glow discharge optical emission spectroscopy can be used for investigating carbonitrided layers formed in low-temperature plasma in ammonia and corgon medium.
4. It has been proved that after carbonitriding of the investigated materials at t=550 ^{0}C, P_{NH3} = 280 Pa, $P_{82\%\ Ar + 18\%\ CO2}$ = 120 Pa, τ = 4 h the most gradual change of the carbon and nitrogen content in the carbonitrided zone of the layer occurs.
5. Increased amount of carbon has been found both in the combined and in the diffusion zone of the carbonitrided layer.
6. After carbonitriding in low temperature plasma, difference between the values of the residual compressive stresses, obtained by the three ways of defining diffraction angles (the method of maximum intensity - $\sigma^{\phi s}$, the cord method - $\sigma^{\phi c}$, and the method of the body centre - $\sigma^{\phi b}$) has been established.
7. It has been established that, after the process of ion carbonitriding, the residual compressive stresses formed on the surface of the alloyed steel 25CrMnSiNiMo have a considerably higher value, than the value of these stresses in Armco-Fe.
8. It has been proved that in the process of ion carbonitriding of the investigated materials, the increase in the ammonia and carbon pressure in the vacuum chamber leads to a decrease in the value of the residual compressive stresses.
9. A probable mechanism of glow discharge carbonintriding in a medium of ammonia, carbon dioxide and argon is suggested. As a consequence of the dissociation and

ionization of the ammonia molecules and the carbon dioxide molecules, close to the cathode in the zone of the negative glowing diffusion capable nitrogen and carbon are obtained, which impact the surface and diffuse at a certain distance into the material. The process of saturation of the metal surface depends only on the concentration of the atomic carbon and nitrogen in the plasma and the temperature of the article, while the electron/ion impact plays role for ensuring the necessary temperature of the details.

Author details

Angel Zumbilev
Technical University of Sofia-Plovdiv Branch, Bulgaria

5. References

[1] Andreev S., V. Zakharov, V. Ochkin, S. Savinov, Plasma-Chemical CO_2 Decomposition in a Non-self-sustained Discharge With a Controlled Electronic Component of Plasma, Spectrochimica Acta Part A, 60 ,2004, pp. 3361-3369.
[2] Eletskii A., B.Smirnov, Dissociation of Molecules in Plasma and Gas: The Energy. Pure &Appl. Chem., Vol. 57, No. 9, pp.1235-1244, 1985.
[3] Fisher-Chatterjee P.,W.Eysell,u.a.,Nitrieren und Nitrocarburieren, Sindeifingen, Expert Verbag, 1994.
[4] Hoffman R.: Effects of Nitrogen in Metal Surfaces, Proceeding of an International Conference on Ion Nitriding, Cleveland, Ohio, USA 15-17, September 1989, pp. 23-30. Of an International Conference, Amtech 9-11, November, 2005, Rousse, pp 76-80.
[5] King Simon, J. Stephen, D. Price, Electron Ionization of CO_2, International Journal of Mass Spectrometry 272, 2008, pp. 154-164.
[6] Klages C., Glow Discharge Optical Emission Spectroscopy, GDOES, Analytik und Prufung in der Oberllachentechnik/Elementzusammensetzung, 1998. pp. 3-15.
[7] Kobayashi A., K. Osaki, C. Yamabe, Treatment of CO_2 Gas by High-energy Type Plasma, Vacuum 65, 2002, pp.475-479.
[8] Lampe Thomas, Plasmawärmebehandlung von Eisenwerkstoffen in stickstoff-und kohlenstoffhaltigen Gasgemischen, VDI-Verlag GmbH, Düsseldorf, 1985
[9] Paatsch W., D. Hodoroaba, GDOES – Ein Verfahren zur Messung von Wasserstoffprofilen an beschichteten Bauteilen, Federal Institute for Materials Research & Testing (BAM), Berlin, Germany, 2001, pp 1-5.
[10] Rainforth W., Glow Discharge Optical Emission Spectrometry (GDOES), the University of Sheffield, Issue 12, April, 2006.
[11] Rainforth W., The use of glow discharge optical emission spectrometry (GDOES) as a surface engineering tool, The University Sheffield, Issue 6, October, 2004.
[12] Rond C., A. Bultel, P. Boubert, B. G. Cheron, Spectroscopic Measurements of Nonequilibrium CO_2 Plasma in RF Torch, Chemical Physics, 354, 2008, pp.16-26.
[13] Toshkov B., Nitriding in low-temperature plasma, King, Sofia, 2004.
[14] Toshkov V., Theoretical and Practical Aspects of Nitriding of Iron and Iron-carbon Alloys in Low-temperature Plasma, (Ph.D. thesis, Sofia Technical University), 1997.
[15] Varhoshkov E., Steel carbonitriding in smouldering discharge, (Ph.D. thesis, Sofia Technical

Chapter 8

Effects of Heat Treatments on the Thermoluminescence and Optically Stimulated Luminescence of Nanostructured Aluminate Doped with Rare-Earth and Semi-Metal Chemical Element

Sonia Hatsue Tatumi, Alexandre Ventieri, José Francisco Sousa Bitencourt, Katia Alessandra Gonçalves, Juan Carlos Ramirez Mittani, René Rojas Rocca and Shiva do Valle Camargo

Additional information is available at the end of the chapter

1. Introduction

Ionizing radiation dosimetry plays a very important role in several fields, useful in the ordinary life, such as radiotherapy, nuclear medicine diagnosis, nuclear medicine, radioisotope power systems, earth science, geological and archaeological dating methods, etc.

The phenomenon of Thermoluminescence (TL) has been known since 1663, when Robert Boyle notified the "Royal Society" in London, which observed the emission of light by a diamond when it was heated in the dark [1]. Afterwards, a large number of scientists began to work with TL; as did Henri Becquerel, whose work described IR measurements spectra [2] and the effect of TL, too. Marie Curie, in 1904, noted that the TL properties of the crystals could be restored by exposing them to radiation from the radio element mentioned in her doctoral thesis. In the middle of the 1930's and 1940's, Urbach performed experimental and theoretical work with TL [1] and in 1945 Randall and Wilkins developed a first theoretical model [3] of thermoluminescent emission kinetics.

The use of thermoluminescence in dosimetry date from 1940, when the number of people working on places with radiation sources such as hospitals, nuclear reactors etc. exposed to ionizing radiations (γ-rays, X rays, α and β-particles, UVA and UVB) increased and efforts to develop new types of dosimeters began [4]. Among the pioneers of TLD we have Daniels,

1953, with LiF, Bjarngard, 1967 with $CaSO_4$ and Ginther and Kirk (1957) with CaF_2. After these works search in other materials such as natural fluorides or synthetic as, $LiBO_3$:Mn, CaF_2:Dy, $CaSO_4$ and $MgSiO_4$:Tb. They usually obtained as monocrystalline samples Czochralski, Bridgmann, etc.

However, Cameron, 1961, with their research on the application of LiF: Mg and Ti obtained the first thermoluminescence dosimeter (TLD-100) [5], which is still one of most popular TLD phosphor, due to tissue equivalent $Z_{eff} = 8.04$, which is an important characteristic for personal dosimetry. Akselrod et al., 1990 [6] carried out studies on TL properties of carbon doped Al_2O_3 (TLD-500), a very sensitive material to radiation exposure, showing few TL peaks, with dose interval of detection between 0.05 μ to 10 Gy and fading rate of 3% by year (when kept in the dark). This high sensitivity of the material is attributed to oxygen vacancies created during the crystal growth procedure; the electrons can be trapped at these vacancies creating F^- and F^+ centers, which act as recombination centers yielding a bright emission.

In order to increase the luminescence emission response with dose of some thermoluminescence dosimeters (TLD), heat treatments procedures were frequently performed. Halperin et al., 1959, [7] noted that the thermal treatment enhanced the intensity of various TL glow peaks of NaCl by factors of a few thousands; on prolonged heat treatment. They observed that the intensity of TL peaks locate above RT decreased, while those at lower temperatures continued to grow even after 80 hours of heat treatment at 550 °C. Mehendru, 1970 [8], studied the effects of heat treatment on the TL response of pure KCl; they associated the peaks at 95, 135, and 190 °C with the F centers, these last created due to the background divalent cation impurities, and with the first- and the second-stage F centers, respectively. Kitis et al., 1990 [9], studied the sensitization of LiF:Mg, Ti as a function of irradiation at elevated temperatures, pre-irradiation annealing, and post irradiation annealing between 150–400 °C; the results showed that the first and third conditions cause an enhancement of the sensitivity; after more two works about preheating and high temperature annealing on TL glow curves of LiF:Mg, Ti [10] and in LiF TLD100 [11] were published. Holgate, 1994, [12] investigated TL and radioluminescence (RL) spectra of calcium fluoride samples doped with neodymium and variations of spectra with Nd concentrations and thermal treatments were observed. Nowadays, it is possible to find oxides, sulfates, sulfides and alkali haloids doped with rare-earths and transition metals as commercial dosimeters.

The optically stimulated luminescence (OSL) was pioneered used to determine environmental radiation dose received by geological samples [13]. However, the idea of using OSL dosimetry was first suggested in 1956 [14]. The first experience was made using MgS, CaS, SrS and SrSe phosphors doped with different rare-earths [15,16]. Nanto et al., 1993 [17], investigated the OSL properties of single crystals of KCl:Eu; after, Akselrod et al. 1998 [18], proposed the use of Al_2O_3:C for OSL dosimetry, because the high sensitivity of the crystal to visible light. At the present time, the crystal is the principal OSL dosimeter [19-22]. Currently, there are various materials proposed for OSL dosimetry as KBr:Eu [23,24],

$LiAlO_2$:Tb, $Li_2Al_2O_4$:Tb, Mg_2SiO_4:Tb, Mg_2SiO_4:Tb, Co $CaSiO_3$:Tb [25]. Some morphological studies was introduced in the science of OSL materials, using high-resolution microscope (TEM) coupled to punctual electron diffraction analysis in nanoscale, showed that dopants formed nanocrystalline structure located at surface of the matrix elements as Al_2O_3:Mg, Yb:Er, Nd and $KAlSi_3O_8$:Mn [26-30].

Nowadays, luminescent dosimetry materials can be used in personal dosimetry, radiotherapy, nuclear medicine and diagnostic and environmental dosimetry they are widely used due to high sensitivity, linear response to the dose, the response is independent with radiation energy (within a certain range), their reusability, etc.

The aim of this chapter is to present a very comprehensive research about new materials consisting in aluminate crystals doped with rare-earths, for radiation dosimetry using TL and OSL.

Some features on fabrication of aluminates dosimeters will be shown relating the luminescence response according to the relative concentration of several rare-earths and transition metals. A study in nanoscale effects, size, shape and surface morphology using TEM, SEM, EDS and electron diffraction measurements will be shown too. The physicochemical properties of the doped materials are strongly related to the fabrication process as well the experimental parameters as temperature of thermal treatments, calcination time, heating rates, etc.

As materials science has developed down to nanoscale, the exceptional properties of nanoscaled rare-earth materials are only now being recognized and performed intentionally.

2. Experimental part

Polycrystalline powder samples of α-Al_2O_3:Er, Yb; Mg; Tb and Nd were obtained by sol-gel and Pechini process. In the sol-gel procedure stoichiometric amounts of tri-sec-butoxide of aluminum was dissolved in distilled water and hydrochloric acid. The dopants Er, Yb, Nd and Tb oxides were added during the sol stage, with different concentrations. Some portions of the resulting powder were calcinated at different temperature from 1200 to 1600 °C. Experimental parameters of the calcination process, as heating and cooling rates and set point values were varied, in order to verify the effect on the luminescence response.

Pechini is a chemical routine that produces, at the end of the stage, an organic polymer with metallic ions, which will be responsible for the formation of the desired material. The polymer is obtained after low temperature reaction among ethylene glycol, citric acid and aluminum nitrate. Once the polymer is ready, a number of heat treatments are carried out in order to (1) collapse the polymeric structure and allow the gradual oxidation reaction of the metallic ions with atmospheric oxygen, and (2) obtain the desired structure of the material. This technique is known to obtain uniform composition and controlled grain size distribution, due to the slow oxidation reaction and the viscosity of the polymer, which avoid precipitation.

The morphological characteristics of the samples were analyzed using a Philips CM200 TEM equipped with EDS operating at 160 keV, some Cu contamination from the sample holder can be observed in the all the EDS results, the samples were located at 400 mm from the source. The X-ray powder diffractions were recorded with the MiniFlex II model diffractometer of Rigaku Corporation.

TL and OSL measurements were performed in an oxygen-free nitrogen atmosphere using Daybreak Nuclear and Medical Systems Inc, model 1100-series TL/OSL reader and RISØ TL/OSL reader Model DA - 20.

TL was detected using the BG-39 (340-610 nm) optical filter and heating rate of 10 °C/s. OSL measurements were made using an array of blue (470 nm) LEDs for sample stimulation and detected in the UV with a Schott U-340 optical filter.

The irradiations were performed at RT in a ^{60}Co source with dose rate of 28.7 Gy/h and with a beta source (^{90}Sr/^{90}Y) coupled to the RISØ TL/OSL reader, with dose rate of 0.08 Gy/s.

3. Results and discussion

3.1. XRD

Figure 1a shows the XRD of pure alumina produced by sol –gel without any dopants. A very well agreement with standard α-Al_2O_3 pattern is observed (Figure1a). On the other hand, doped samples show additional peaks related with new structures formed by the dopants and the alumina matrix, for example, in the case of Tb doped samples it was verified the $Tb_3Al_5O_{12}$ crystalline structure (Figure 1b), and for Nd doped sample the AlNd structure (Figure 1c).

Sample doped with Er and Yb and calcinated at 1200 °C supplied a broad background, related to amorphous phase, and many other peaks associated to Yb_2O_3, Er_2O_3 and $Yb_3Al_5O_{12}$. For the sample calcinated at 1600°C it was not observed the background, but the peaks of Yb_2O_3, Er_2O_3 with predominance of the $Yb_3Al_5O_{12}$ (Figure 1d and e) were noted.

Figure 2 shows XRD patterns of alumina powder obtained by Pechini process. Applying crescent calcinations temperatures, we can obtain alumina in gamma and alpha phase (Figure 2a and 2b). The Mg addition promoted the formations of $MgAl_2O_4$ crystals (Figure 2c). When the doped sample is heated up to 1100 °C (Figure 2d), most of the material is converted to α-Al_2O_3, except for a few low intensity peaks related to the occurrence of magnesium spinel ($MgAl_2O_4$). This observation will be corroborated in the next section through TEM images.

3.2. Thermoluminescence

TL glow curves of samples obtained with different calcinations temperatures, and detected in UV and VIS regions are shown in Figure 3a and 3b, respectively. It can be seen that calcinations at 1600 °C favored the increase of 190 °C TL dosimetric peak and diminution on intensity of high temperature peak simultaneously. The TL intensity in VIS region is higher than those found in the UV one.

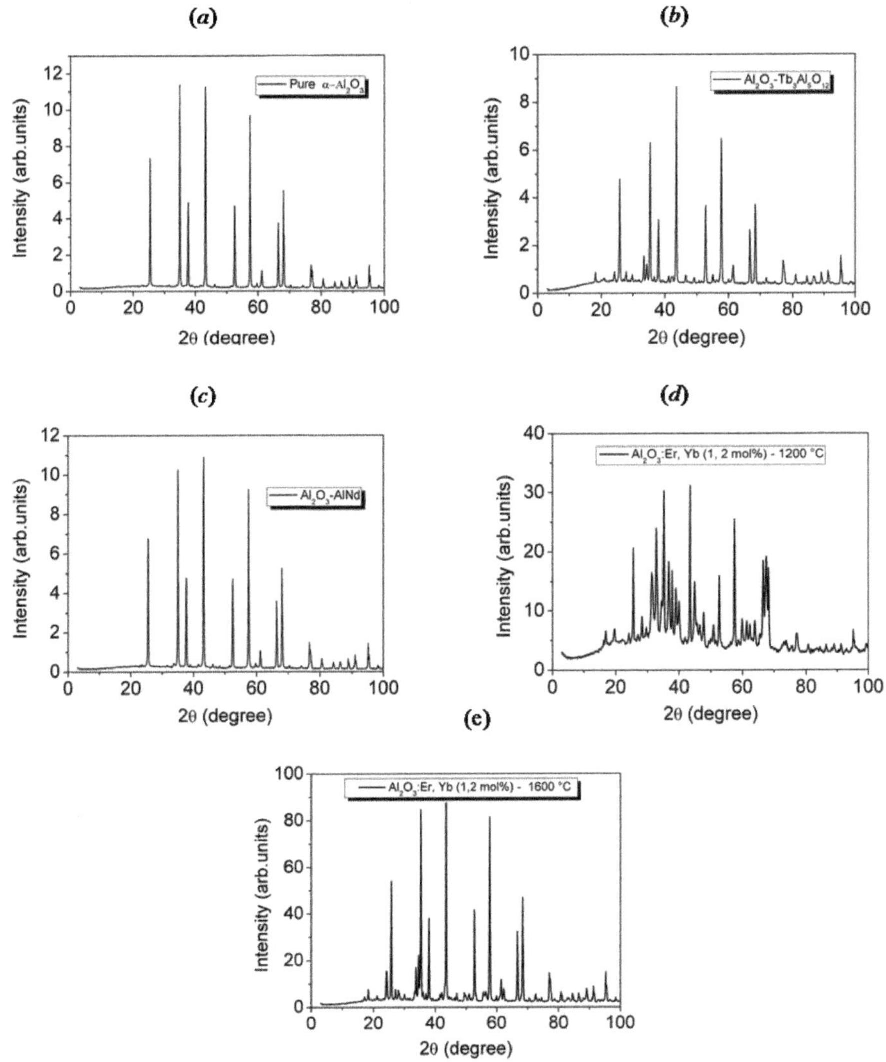

Figure 1. XRD pattern of the α-Al_2O_3 samples, a) undoped sample, b) doped with Tb, c) doped with Nd, d) Er and Yb (1 and 2 mol%) doped and calcinated at 1200 °C and (e) Er and Yb (1 and 2 mol%) and calcinated at 1600 °C.

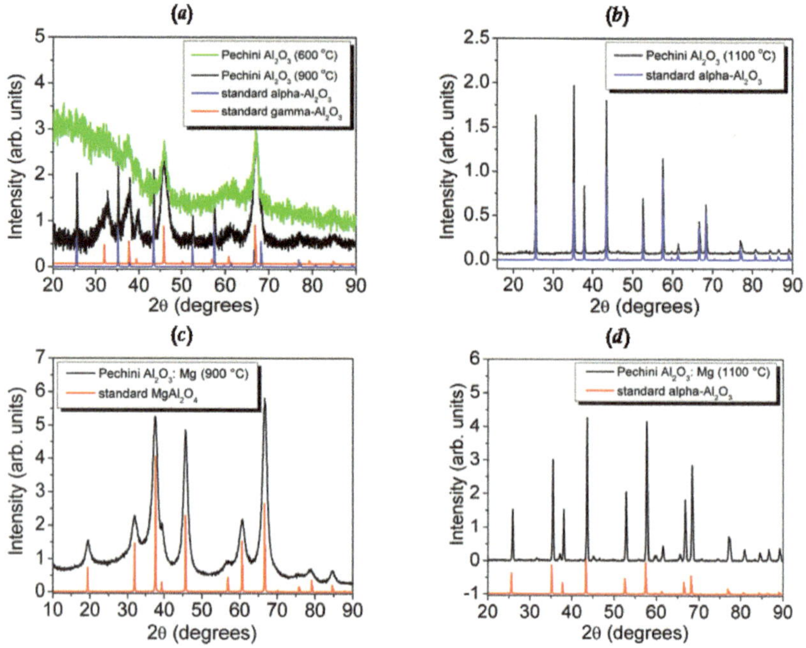

Figure 2. XRD patterns from Al_2O_3 samples obtained from Pechini routine. a) shows undoped samples annealed at 600 and 900 °C and standard patterns found for γ-Al_2O_3 and α-Al_2O_3. b) shows XRD from undoped sample annealed at 1100 °C and the standard α-Al_2O_3. Magnesium doped sample annealed at 900 °C is shown in (c) with the standard pattern of magnesium spinel ($MgAl_2O_4$). After annealing at 1100 °C, magnesium doped sample produced the XRD shown in (d).

Figure 3c and 3d show results of TL responses in UV and VIS regions with set point times varying of 30 min, 1, 2, 4 and 8 h. In the UV region, the thermal treatment of 4 h promoted a high increment of the 190 °C peak while in VIS region, the time was of 8 h. It is known from literature that the emission mechanism of these two luminescence regions is different. In the case of UV emission, the responsible is the F^+ center according the mechanism:

$$F + h^+ \rightarrow F^{+^*} \rightarrow F^+ + h\nu_{325nm} \text{ (stimulation process, TL/OSL)} \quad (1)$$

Where the recombination of the F center with a hole (h^+) generates an excited F^{+*}, which decays into the ground state (transition 1B → 1A) emitting a photon at 325 nm. Therefore, the calcinations at 1600 °C stimulated an increase of F centers concentration.

In the other case of VIS emission, it is believed that the luminescence occurs as follows [31, 32]:

$$F \rightarrow F^+ + e^- \text{ (irradiation process)} \quad (2)$$

$$F^+ + e^- \rightarrow F^* \rightarrow F + h\nu_{410-420nm} \text{ (heating process, TL)} \quad (3)$$

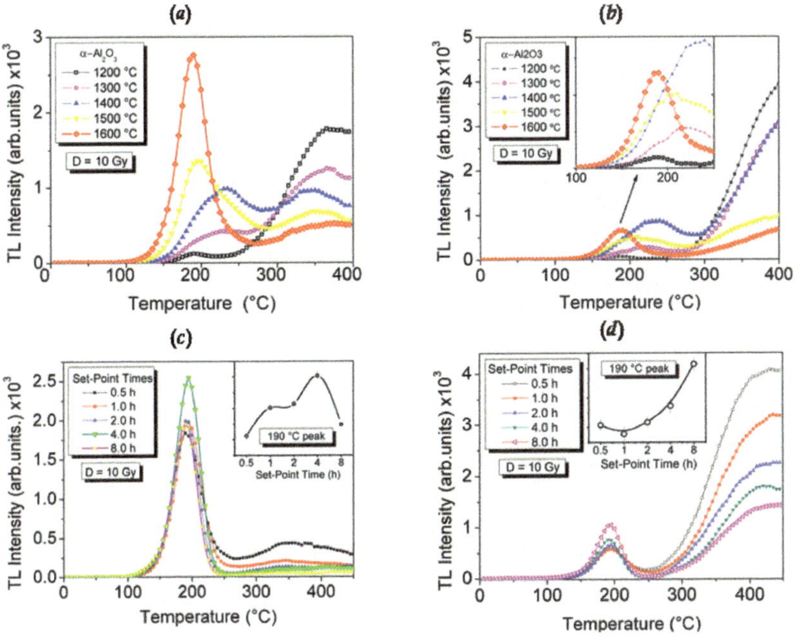

Figure 3. TL glow curves of pure alumina samples obtained by sol-gel process, study of effects of calcination temperatures and time on the TL response. a) UV emission with different calcinations temperatures; b) VIS emission with different calcinations temperatures; c) UV emission with different set-point times and d) VIS emission with different set-point times.

Where F center loses an electron after absorption of high energy radiation and become F^+ center. On thermal stimulation, the recombination of the electron and the F^+ center produces an excited F center (F*), which decays into its ground state (3P transition state → 3S) with the emission of photons at 410-420 nm.

Therefore, following our results the long set point time (~ 8 h) favored the electron traps formations, and in the case of UV emission we have the great rate of formation of hole traps after 4 hours of calcinations.

Figure 4 shows TL glow curves of pure alumina obtained with different heating and cooling rates. In all cases the slow rate of 3 °C/min supplied the best result, confirming that longer calcinations time can promote a better diffusion of the defects and ions and also eliminates the internal tension forces in the crystalline lattice, which can homogenize the crystal.

TL glow curves of the samples doped with Er and Yb are shown in the Figure 5a and 5b. Samples calcinated at 1200 °C supplied one peak at high temperature (Figure 5a), which did not increase proportionally to the dose and another peak at low temperature region with very low intensity. After calcinations at 1600 °C, two prominent peaks at 224 °C and 442 °C were observed (Figure 5b). For the sample doped with Er (1 mol %) and Yb (2 mol %), the peak temperature changed to 203 °C and an increment about 1.4 time in the TL intensity was observed. In all the samples, the TL response of the high temperature peak is not proportional to the dose. For samples doped with Tb (2.5 mol%) and Nd (2.5 mol%) an increased in UV intensity of the 190 °C peak was also noted, the first one increased 3.5 times and for Nd was 2.5 times, when compared to undoped one (Figure 5c).

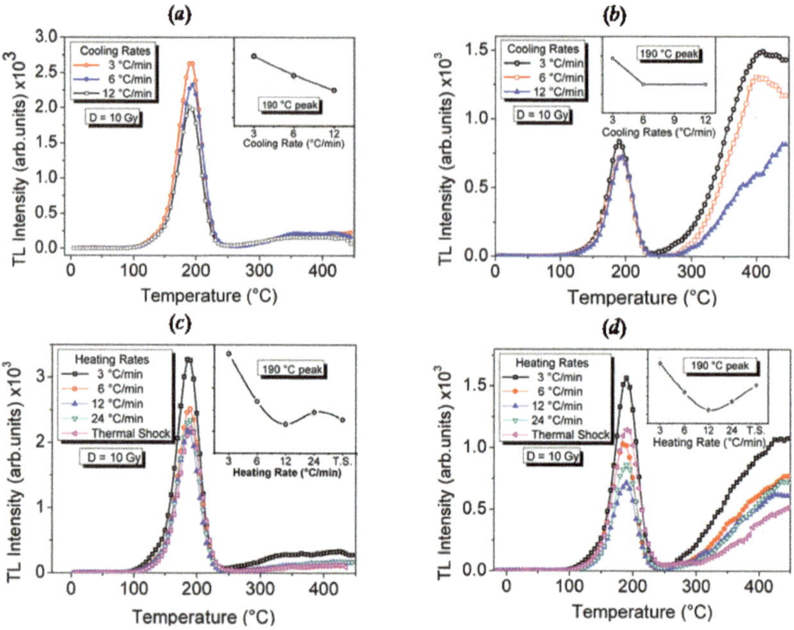

Figure 4. TL glow curve of pure alumina samples obtained by sol-gel process a) UV emission with different heating rates b) VIS emission with different heating rates; c) UV emission with different cooling rates d) VIS emission with

Figure 6 shows TL glow curves of pure and Mg doped samples obtained by Pechini process, the curves are slightly different from those obtained by sol-gel. High intensities for 190 °C TL peak from pure samples were obtained at 1100 °C for VIS region (Figure 6a) and at 1350 °C for UV one (Figure 6b). On sample doped with Mg high intensities were detected in both cases UV and VIS with calcinations at 1600 °C, the same value found in sol-gel samples.

Figure 6 shows TL glow curves for samples produced via Pechini method. As the calcination temperature increases, different observations can be made, depending on the composition and the measurement spectra. In Figure 6a, showing TL emission of undoped sample in the visible region, all the peaks intensities decrease for higher temperatures, but mainly low temperature (90 °C) and high temperature (410 °C) peaks. In this case, the best sample would be the one calcinated at 1350 °C, due to its high intensity and low competition among the trapping centers (more stable TL signal). High temperature treatments can destroy as well as create trapping and recombination centers, and that explains why some TL peaks may disappear, whilst others may rise or increase.

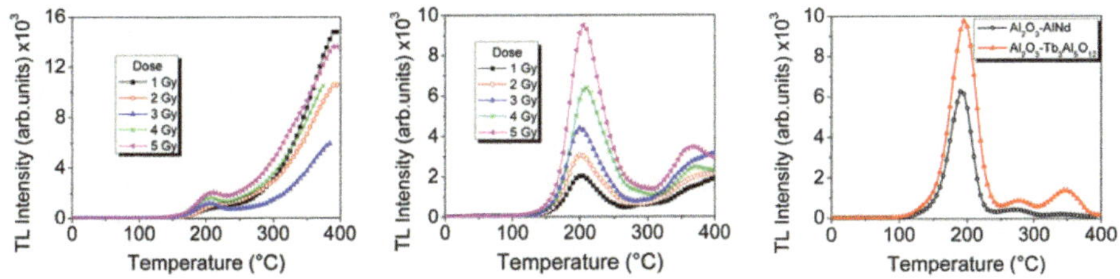

Figure 5. TL glow curve of α-Al_2O_3 obtained by sol gel: a) doped with Er, Yb (1, 2 mol%), calcinated at 1200 °C and irradiated with γ-rays, b) doped with Er, Yb (1, 2 mol%), calcinated at 1600 °C and irradiated with γ-rays and c) TL glow curve UV emission of sample doped with Nd (2.5 %) (black open circle) and Tb (2.5 %) (red triangle), both calcinated at 1600 °C and irradiated with 10 Gy.

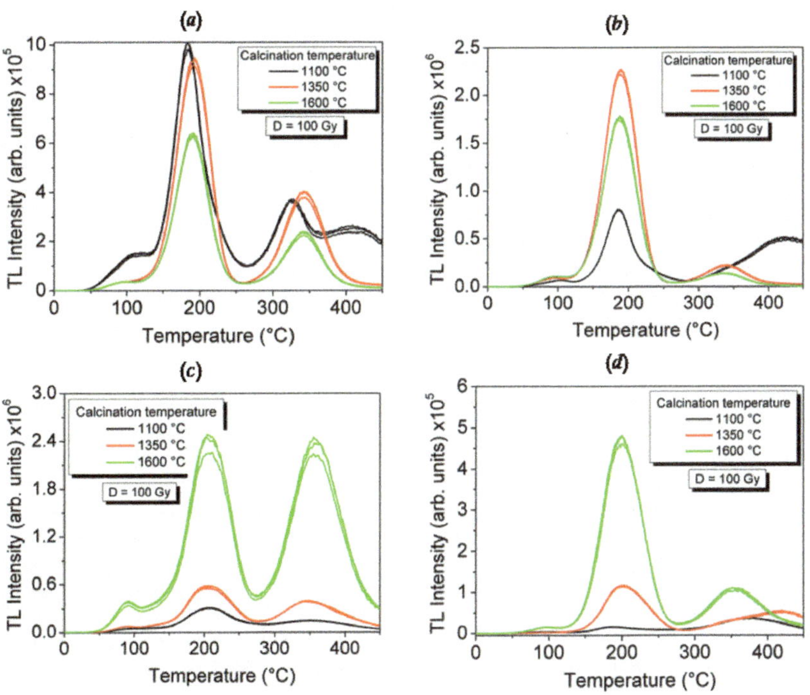

Figure 6. TL glow curves from Al_2O_3 (a and b) and Al_2O_3:Mg (c and d) samples obtained from Pechini routine and annealed at 1100, 1350 and 1600 °C. Measurements were taken in both visible (a and c) and UV (b and d) spectra.

Figure 6b is the TL emission of undoped sample in the UV region, which indicates that high temperature peaks tend to fade when the sample is calcinated at higher temperatures. Once again, the sample calcinated at 1350 °C showed the best glow curve. For the undoped sample, calcination at 1600 °C seems to increase the competition, which decreases the overall intensity. It is not likely that the high temperature is damaging the material, since the melting point is still too far away (around 2050 °C). Also, the high temperature may be causing the crystallites to grow, decreasing the surface area exposed to the incoming radiation, thus changing the trapping dynamics at some level.

The incorporation of magnesium atoms in the crystalline lattice made a great deal on the TL response of the samples. In the first place, both visible and UV emissions (Figures 6c and 6d, respectively) increased with the increasing of the calcination temperature, which was not observed for undoped samples. Secondly, the high temperature peak (355 °C) of the visible emission had its intensity increased by a factor of 3; a minor difference on the relation between the main dosimetric peak and the high temperature one was also observed for the UV emission.

It is important to observe that only samples calcinated above 1100 °C exhibited some appreciable TL emission; this means that α-Al_2O_3 acts as a better ionizing radiation sensor than other phases (δ and γ). For samples calcinated below that temperature (600 and 900 °C), most of the trapping and recombination centers may not yet be active.

One reason for the high luminescence of α-Al_2O_3:C comes from the theory of point defects. It is known that the synthesis of carbon doped alumina is done in a highly reductive atmosphere of carbon ions, resulting into a great production of oxygen vacancies in the crystalline lattice. If these vacancies are occupied by two electrons we have the formation of the neutral F center. Otherwise, if the vacancy is occupied by only one electron, we have the formation of F^+ center. In the latter situation, the presence of charge compensation is demanded for the F^+ center formation; in the case of the α-Al_2O_3:C it is C^{2+} impurity, which replaces Al^{3+} ion in the crystalline lattice. In our case we suppose that the Mg is acting as C impurity.

3.3. Thermoluminescence spectra

TL spectra of pure and doped with Tb, Er-Yb, and Nd alumina are shown in Figures 7a) to 7d) respectively. On visible region from 360 to 600 nm luminescence bands due to rare-earth elements are observed on doped samples. However all the samples pure and doped showed an intense luminescence band between 650 and 800 nm not identified yet in the literature. However, from fluorescence results, this band is usually associated to Cr^{3+} impurity incorporated in raw materials used for alumina production [33].

When the dopant are incorporated, other bands are detected related to the rare-earth elements (Figure 7b, 7c, and 7d). The Tb doped sample shows, in addition for the first at 694 nm, another broad band at 428 nm, with 187 nm of width, due to the transitions of Tb^{3+}. The results are in agreements with the transitions of $^5D_3$ and $^5D_4 \rightarrow {}^7F_j$ (j=1-6).

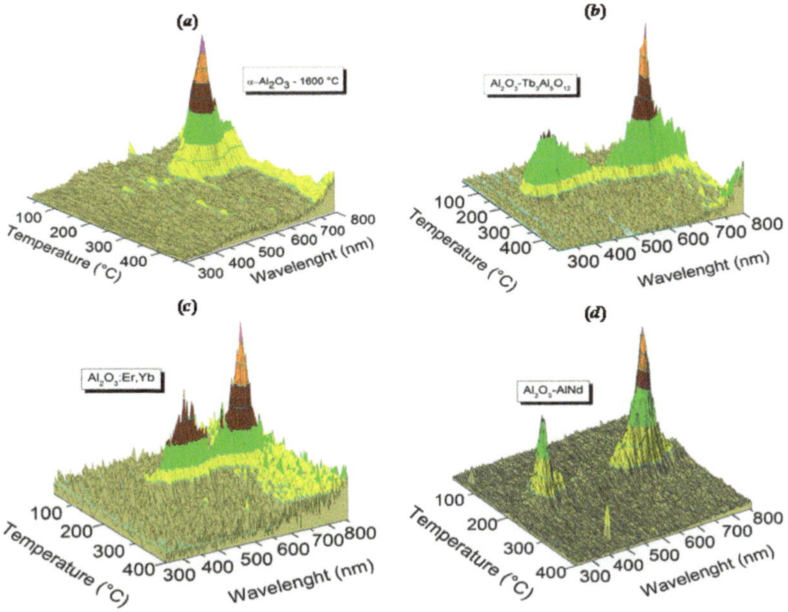

Figure 7. TL spectra of Alumina, a) undoped sample, b) Tb doped sample, c) Er and Yb doped sample, d) Nd doped sample.

In the case of Al_2O_3:Er:Yb sample, the band is centered at 528 nm (448-589 nm) and the emission mechanism can be related to these rare-earth elements. It is well known that Er^{3+} has high efficiency for the infrared to visible light conversion and cooperative sensitization properties. The $^4I_{11/2}$ (Er^{3+}) level and $^2F_{5/2}$ (Yb^{3+}) state are very closely matched in energy, thus the exposition from the 0.9 to 1.1 μm range will excite both Er^{3+} and Yb^{3+} ions. The visible emission can occurs because the Yb^{3+} transfers the excitation energy to Er^{3+} and the final state of this process is the population of the $^4F_{7/2}$ state and nonradiative relaxation to the $^4S_{3/2}$ level, from which green photon (547 nm) is emitted. The excitation routes for red emission at 660 nm are not clear yet; in this case the Yb^{3+} transfer energy to Er^{3+} and the red emission occurs in the $^4F_{9/2} \rightarrow {}^4I_{15/2}$ transition [34].

For Nd doped sample, it was noted emission at 396 nm (364-460 nm). It is known that Nd^{3+} can emit in UV region and the possible transitions related to these emissions are: $_4D^{5/2}$ $_4D^{3/2}$ $_4P^{3/2} \rightarrow$ $_4I^{9/2}$; $_4D^{5/2}$ $^4D_{3/2}$ $^4P_{3/2} \rightarrow {}^4I_{11/2}$ [35].

3.4. Optically stimulated luminescence

As seen in TL emission curves showed previously, the response of OSL signal also increased for samples calcinated at high temperatures, for both samples obtained by sol-gel (Figure 8a) and Pechini process (Figure 8b). Figure 8c and 8d show an example of OSL increment about 3 times, after calcination at 1600 °C.

OSL decays can be fitted by exponential functions [36], depending on the number of the traps involved in the process, for example for two traps we have:

$$I_{OSL} = I_1 \exp\left(-\frac{t}{\tau_1}\right) + I_2 \exp\left(-\frac{t}{\tau_2}\right) \tag{4}$$

Where I_{OSL} is the total OSL intensity, I_1 and I_2 are the initial intensities of exponentially decaying from faster and slower components of the shinedown curve; τ_1 and τ_2 the respective decay constants.

Sample – Pechini Process	Calcination temperature (°C)	τ_1 (s^{-1})	τ_2 (s^{-1})
Al_2O_3	1100	3.05±0.03	13.4±0.3
Al_2O_3	1350	2.415±0.005	11.11±0.04
Al_2O_3	1600	2.27±0.05	11.5 ± 0.2
Al_2O_3: Mg	1100	2.49±0.09	13.8±0.6
Al_2O_3: Mg	1350	2.6±0.1	12.6±0.9
Al_2O_3: Mg	1600	2.22±0.05	11.3±0.3
Sample Sol Gel Process			
Al_2O_3	1600	9.0±2.0	41±11
Al_2O_3 :Yb:Er	1600	22.5±1.0	-0-
Al_2O_3: Nd	1600	13.3±0.8	54±11
Al_2O_3:Tb	1600	8.7±2.2	45±6.0

Table 1. Values for decay constants obtained from theoretical fitting of the OSL curves of alumina (equation 4).

Table 1 shows the decay constants obtained for the studied samples. All the OSL decay curves can be fitted by second order exponential function, except for Er and Yb doped sample, which followed first order decay. For samples obtained by Pechini process the high temperatures calcinations diminished the constants values, the fast component value for both pure and Mg doped samples heated at 1600 °C supplied almost same constants decays. In the case of samples made by sol-gel, the Tb doped sample has constants similar to those determined by pure alumina. However, the incorporation of Nd, Er and Yb changed the constants values. The OSL curves of the samples obtained by sol-gel supplied greatest decay constants. Yang et al. [37] showed some OSL fitting results of α-Al_2O_3: C excited by green light and detected in the UV, and they found that τ_1 vary from 4.6 to 11.3 s^{-1} and τ_2 from 24.1 to 30.1 s^{-1}. They attributed the variations to different concentrations of F and F^+ centers in the sample. Nevertheless, our results indicated that τ values are constant and do not depend of dose values delivered to the samples, the results for τ_1 are similar to those found in the literature, however τ_2 are relatively higher.

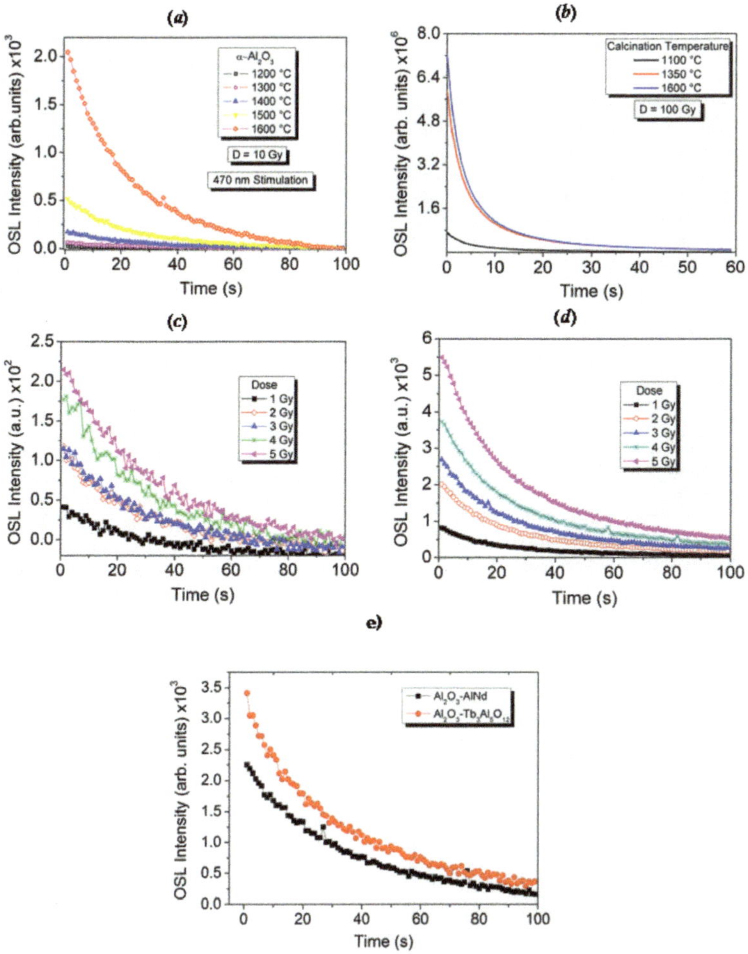

Figure 8. OSL curves of alumina, a) undoped samples obtained by sol-gel with different calcinations temperatures, b) undoped samples obtained by Pechini process with different calcinations temperatures, c) Er, Yb doped samples calcinated at 1200 °C and irradiated with gamma-radiations, d) Er, Yb doped samples calcinated at 1600 °C and irradiated with gamma-radiations and e) Nd and Tb doped samples calcinated at 1600°C.

3.5. TEM

According to TEM images, it was verified that all the doped samples present nanocrystals formations on the surface of alumina grains (Figure 9a). In Tb doped sample, it was verified the presence of $Tb_3Al_5O_{12}$, the chemical structure was determined by electron diffraction and EDS results (Figure 9b). Nanocrystals of AlNd are easily observed with its well developed faces (Figure 10a), in most of the case, the nanocrystals average size is about 200 nm, these aluminates composition was also verified by the EDS (Figure 10b).

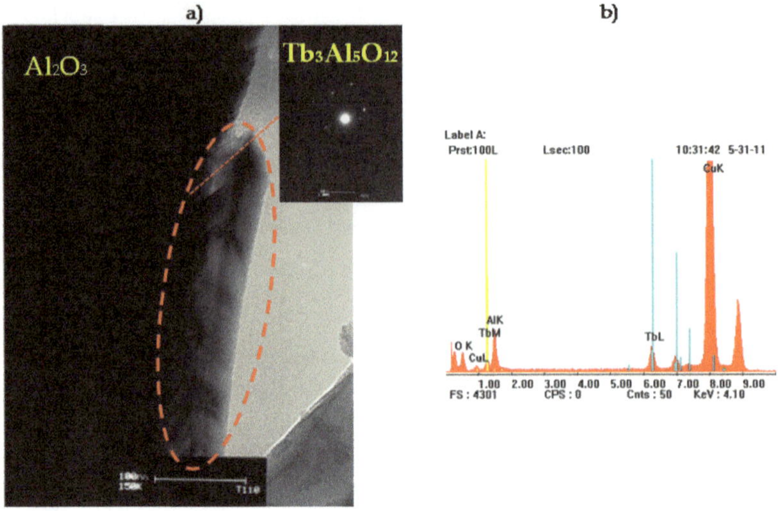

Figure 9. TEM image from Tb doped Al_2O_3 and electron diffraction showing the presence of nanocrystals of $Tb_3Al_5O_{12}$. b) EDS results with Tb peak.

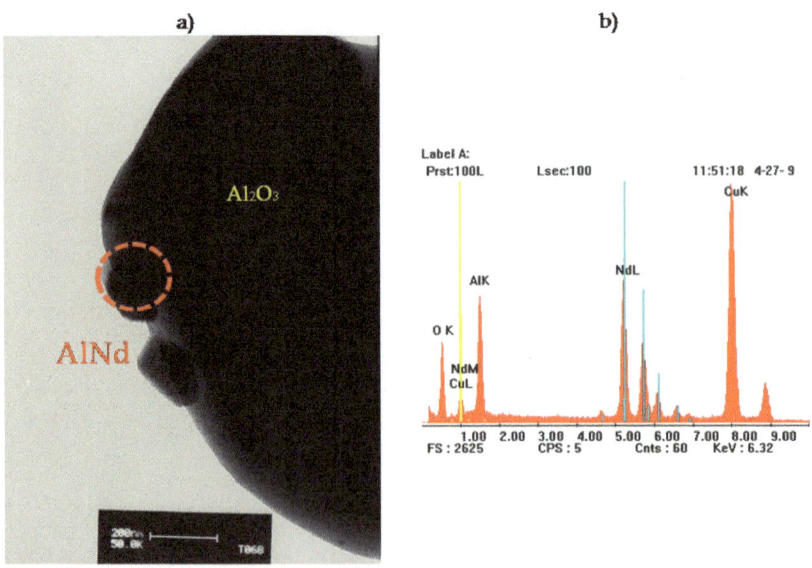

Figure 10. a) TEM image from Nd doped Al_2O_3, b) EDS results with Nd peak.

Figure 11a) shows an example of TEM images obtained for α-Al_2O_3 doped with 1mol% of Er and 2mol% of Yb, and calcinated at 1200 °C. EDS analysis show the presence of Er and Yb dopants in the α-Al_2O_3. Electron diffraction analysis identified the crystal as Yb_2O_3, however by XRD analysis showed plus two new composition in the samples doped and attributed to Er_2O_3 and $Yb_3Al_5O_{12}$ (Figure 1e). The results of TEM analysis give the following average nanocrystals diameters D = (36±2) nm for the sample calcinated at 1200 °C and D=(182±8) nm for sample calcinated at 1600 °C (Figures 9 d), these results and the homogeneity in the crystalline size suggests that the nanocrystal powder growth is depending on thermal treatment temperature. The growth and cluster formations of the nanocrystals are the consequence of the reduction of grain boundary area and therefore the total energy of the system.

In the case of Mg doped alumina there is a formation of Mg spinel (magnesium aluminate) nanocrystals dispersed on the surface of alumina grains, with size about 40 nm (Figure 12). It is considered that the high temperature calcination makes magnesium atoms to diffuse to the surface of the clusters, creating a thin layer of magnesium spinel, due to the high local concentration of the dopant.

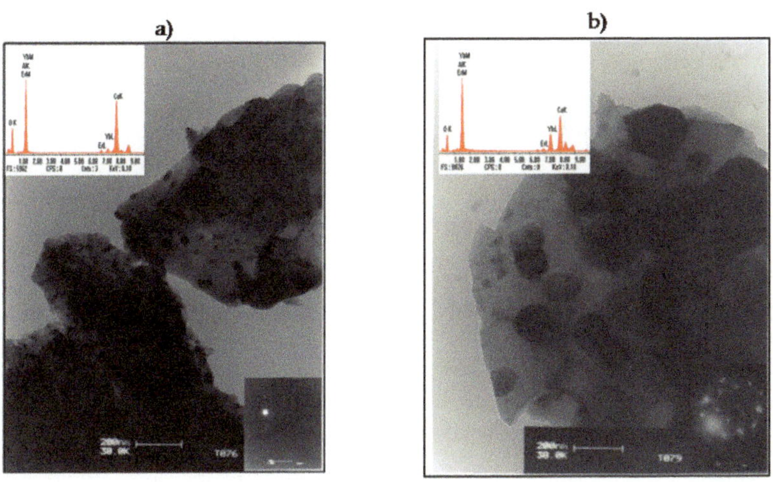

Figure 11. Results of TEM, EDS and Electron diffraction analysis obtained for for α-Al_2O_3 doped with 1 mol% of *Er* and 2 mol% of *Yb*, and calcinated at 1200°C and 1600°C, (*a*) TEM images sample calcinated at 1200°C, (*b*) TEM images sample calcinated at 1600°C.

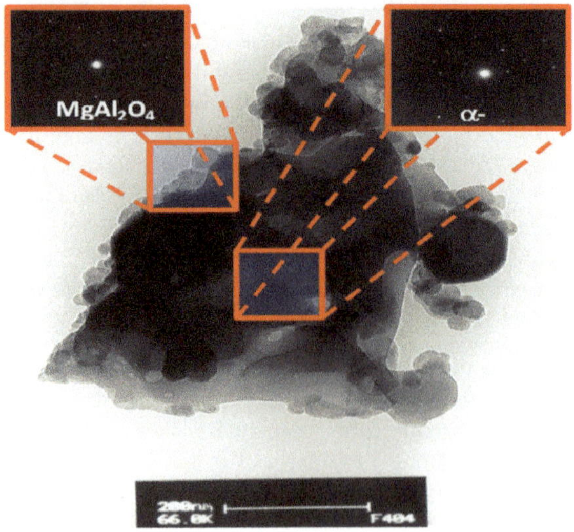

Figure 12. TEM image from magnesium doped Al_2O_3. It is considered that the high temperature annealing makes magnesium atoms to diffuse to the surface of the clusters, creating a thin layer of magnesium spinel.

4. Conclusions

Polycrystalline powder of α-Al_2O_3 was successfully obtained using sol-gel and Pechini process. A second phase was found in doped samples, forming nanocrystals of aluminates and lanthanides oxides on the surface of alumina grains. We did not observe the incorporations of the dopants inside the alumina structure. Strongly dependent of temperature calcinations, the nanocrystallinity of the sample was retained after calcinations at higher temperatures, and an increase in the crystallite size was already perceptible.

The average diameter of nanocrystals depended on the dopant specie: for Yb and Er doped samples, it was D = (36 ± 2) nm for the sample calcinated at 1200 °C and D= (182 ± 8) nm for one calcinated at 1600 °C, therefore increased by 5 times. The approximately size of the AlNd is 200 nm for sample calcinated at 1600 °C during 4 h and for $Tb_3Al_5O_{12}$ crystals the size is about 300 nm in the same calcination conditions.

The TL emission mechanism in the visible region can be related to F center and to the lanthanide (Ln) relaxation. During the irradiation the Ln^{3+} ion is reduced to Ln^{2+}. It is not completely known if the reduction is due to the transfer of an electron or a hole; however, the process of reduction of trivalent lanthanide ions by irradiation was previously verified in literature [18, 19]. During thermal stimulation an electron can recombine with the Ln^{2+} forming Ln^{3+*} in excited state, which emits a photons returning on this way to the ground state. In the case of Mg spinel, probably the Mg ions promoted the oxygen vacancies stabilization, improving the luminescence response in the visible spectra, causing the main peak to increase 5 times in comparison with the undoped sample. It is believed that the occurrence of the nanometric spinel layer created an interface between both materials ($Al_2O_3/MgAl_2O_4$) with high concentration of defects.

OSL shinedown curves, supplied by undoped samples calcinated to 1200 and 1600 °C, could be fitted by second for all the samples except to α-Al_2O_3:Yb, Er, which was fitted by first order exponential decay. TL intensity of 190 °C peak and OSL responses with the dose increased linear for low doses region, from 80 to 1000 mGy, and the minimum dose detected value was 5 mGy obtained for TL (UV) and 350 μGy for OSL α-Al_2O_3 + $Tb_3Al_5O_{12}$.

In summary, calcination conditions are of great importance for materials production that are being used as radiation sensors, once it greatly influences the stabilization of intrinsic defects, diffusion of dopants and the occurrence of new phases, due to the incorporation of dopants alongside the matrix, and others. These new phases also seem to play an important role in the luminescence emissions, due to the creation of new trapping and recombination centers, producing materials with unique properties that can be exploited to obtain better dosimeters.

Author details

Sonia Hatsue Tatumi *, Alexandre Ventieri, José Francisco Sousa Bitencourt, Katia Alessandra Gonçalves, Juan Carlos Ramirez Mittani, René Rojas Rocca and Shiva do Valle Camargo
Universidade Federal de São Paulo, Universidade de São Paulo, Centro Estadual de Educação Tecnológica Paula Souza, Brazil

Acknowledgement

The authors wish to thanks to FAPESP, CAPES and CNPq for financial support.

5. References

[1] Aitken MJ (1985) Thermoluminescence Dating. Academic Press. 3 p.
[2] Becquerel AH (1883) Maxima et Minima d'excitinction de al Phosphorescence sous l'Influence des Radiations Infra-rouge. Comptes Rendus 96: 1853-1856.
[3] Randall JT, Wilkins MHF (1945) Phosphorescence and Electron Traps. I. The study of Trap Distributions. Proc. Royal Soc. Lond. 184: 366-389.
[4] McKeever SWS (1985) Thermoluminescence of Solids. Cambridge University Press. 8 p.
[5] Cameron JR, Suntharalingam N, Kenney GN (1968) Thermoluminescent Dosimetry. University of Wisconsin Press.

* Corresponding Author

[6] Akselrod MS, Kortov VS, Kravetsky DJ, Gotlib VI (1990) High Sensitive Thermoluminescent Anion-Defect α-Al_2O_3: C Single Crystal Detectors. Radiat. Prot. Dosim. 33: 119-122.

[7] Halperin A, Kristianpoller N, Ben-Zvi A (1959) Thermoluminescence of X-Ray Colored NaCl Crystals. Phys. Rev. 116: 1081-1089.

[8] Mehendru PC (1970) Effects of Heat Treatment and Background Divalent Cation Impurity on the Growth of F Centers in Highly Pure KCl Crystals. Phys. Rev. B 1: 809-814.

[9] Kitis G, Charitidis C, Charalambous S (1990) Thermoluminescence Sensitization of LiF: Mg, Ti Under Heat Treatment Between 150 and 400 °C. Nuclear Instruments and Methods in Physics Research Section B: Beam Interactions with Materials and Atoms 51, 3: 263-268.

[10] Piters TM, Bos AJJ, Van der Burg B (1996) Effects of Annealing on Glow Peak Paramenters of LiF: Mg, Ti (TLD-100) Dosimetry Material. Radiat. Prot. Dosim. 84, 1-4: 201-206.

[11] Bhatt BC, Menon SN, Mitra R (1999) Effects of Pre- and Post-Irradiation Temperature Treatments on TL Characteristics and Radiation Induced Sensitisation of Various TL Peaks in LiF (TLD-100). Radiat. Prot. Dosim. 84, 1-4: 175-178.

[12] Holgate SA, Sloane TH, Townsend PD, White DR, Chadwick AV (1994) Thermoluminescence of Calcium Fluoride doped with Neodymium. Journal of Physics: Condensed Matter 6: 9255-9266.

[13] Huntley DJ, Godfrey-Smith DI, Thewalt MLW (1985) Optical Dating of Sediments: Nature 313: 105-107.

[14] Antonov-Romanovsky VV, Keirum-Markus IF, Poroshina MS, Trapeznikova ZA (1956) IR Stimulable Phosphors. Conference of Academy of Sciences of the USSR on the peaceful uses of Atomic energy. USAEC Report AEC-tr-2435 Pt.1: 239-250.

[15] Bräunlich P, Schaffer D, Scharmann A (1967) A Simple Model for Thermoluminescence and Thermally Stimulated Conductivity of Inorganic Photoconducting Phosphors and Experiments Pertaining to Infra-Red Stimulated Luminescence. Luminescence Dosimetry: 57- 63.

[16] Sanborn EN, Beard EL (1967) Sulfides of Strontium, Calcium and Magnesium in Infrared-Stimulated Luminescence Dosimetry. Luminescence Dosimetry: 183-193.

[17] Nanto H, Murayama K, Usuda T, Taniguchi S, Takeuchi N (1993) Optically Stimulated Luminescence in KCl: Eu Single Crystal. Radiat. Prot. Dosim. 47: 281-284.

[18] Akselrod MS, Lucas AC, Polf JC, McKeever SWS (1998) Optically Stimulated Luminescence of Al_2O_3. Radiat. Meas. 29: 391-399.

[19] McKeever SWS, Akselrod MS (1999) Radiation Dosimetry Using Pulsed Optically Stimulated Luminescence of Al_2O_3: C. Radiat. Prot. Dosim. 84: 317-320.

[20] McKeever SWS (2001) Optically Stimulated Luminescence Dosimetry. Nucl. Instr. Meth. Phys. Res. B 184: 29-54.

[21] Yukihara EG, McKeever SWS (2006) Spectroscopy and Optically Stimulated Luminescence of Al_2O_3: C using Time-Resolved Measurements. J. Appl. Phys. 100,083512 (9 pages).

[22] Itou M, Fujiwara A, Uchino T (2009) Reversible Photoinduced Interconversion of Color Centers in α-Al_2O_3 Prepared under Vacuum. J. Phys. Chem. 113: 20949-20957.

[23] Douguchi Y, Nanto H, Sato T, Imai A, Nasu S, Kusano E, Kinbara A (1999) Optically Stimulated Luminescence in Eu-doped KBr Phosphor Ceramics. Radiat. Prot. Dosim. 84: 143-148.

[24] Gaza R, McKeever SWS (2006) A Real-Time High-Resolution Optical Fibre Dosemeter Based on Optically Stimulated Luminescence (OSL) of KBr: Eu, for potential use during radiotherapy of cancer. Radiat. Prot. Dosim. 120: 14-19.

[25] Mittani JC, Proki´c M, Yukihara EG (2008) Optically Stimulated Luminescence and Thermoluminescence of Terbium-activated Silicates and Aluminates. Radiation Measurements 43: 323-326.

[26] Bitencourt JFS, Tatumi SH (2009) Synthesis and Thermoluminescence Properties of Mg^{2+} Doped Nanostructured Aluminium Oxide. Physics Procedia 2: 501-514.

[27] Bitencourt JFS, Ventieri A, Gonçalves KA, Pires EL, Mittani JC, Tatumi SH (2010) Comparison Between Neodymium Doped Alumina Samples Obtained by Pechini and Sol-gel Methods using Thermo-Stimulated Luminescence and SEM. Journal of Non-Crystalline Solids 356: 2956-2959.

[28] Gonçalves KA, Bitencourt JFS, Mittani JCR, Tatumi SH (2010) Study of Radiation Induced Effects in the Luminescence of Nanostructured Al_2O_3: Yb, Er crystals. IOP Conf. Series: Materials Science and Engineering 15: 012080.

[29] Pires EL, Tatumi SH, Mittani JCR, Caldas LVE (2011) TL and OSL Properties of $KAlSi_3O_8$: Mn Obtained by Sol-gel Process. Radiation Measurements 46: 473-476.

[30] Pires EL, Tatumi SH, Mittani JCR, Kinoshita A, Baffa O, Caldas LVE (2011) TL, OSL and ESR Properties of Nanostructured $KAlSi_3O_8$: Mn Glass. Radiation Measurements 46: 1492-1495.

[31] Summers GP (1984) Thermoluminescence in single crystal α-Al_2O_3: C. Radiat. Prot. Dosim. 8: 69-80.

[32] Lee KH, Crawford JH (1979) Luminescence of the F Center in Sapphire. Phys. Rev. 19: 3217-3221.

[33] Kristianpoller N, Rehavi A, Shmilevich A, Weiss D, Chen R (1998) Radiation Effects in Pure and Doped Al_2O_3 Crystals. Nuclear Instruments and Methods 141: 343-346.

[34] Ropp RC (1991) Luminescence and the Solid State. Elsevier. 401 p.

[35] Jaque D, Capmany J, Luo ZD, Garcia Sole J (1997) Optical Bands and Energy Levels of Nd^{3+} Ion in the $YAl_3(BO_3)_4$ Non-linear Laser Crystal. J. Phys: Condens. Matter 9: 9715-9729.

[36] McKeever SWS, Chen R (1997) Luminescence Models Radiation Measurements 27 5/6: 625-661.

[37] Yang X, Li H, Bi Q, Cheng Y, Tang Q, Su L, Xu J (2008) Growth of Highly Sensitive Thermoluminescent Crystal α-Al_2O_3: C by the Temperature Gradient Technique. Journal of Crystal Growth 310: 3800-3803.

Chapter 9

Post Deposition Heat Treatment Effects on Ceramic Superconducting Films Produced by Infrared Nd:YAG Pulsed Laser Deposition

Jeffrey C. De Vero, Rusty A. Lopez, Wilson O. Garcia and Roland V. Sarmago

Additional information is available at the end of the chapter

1. Introduction

Pulsed laser deposition (PLD) has become a potential method in fabricating highly quality superconducting thin films suitable for electronic applications such as in Josephson junction-based electronics and in second generation coated conductors [11, 12, 14, 20, 23]. PLD of high T_c superconductors generally utilize excimer lasers in the ultraviolet (UV) range [6, 11] . However, excimer lasers use toxic gases such as Cl and F for excitation. In contrast, flash lamp pumped Nd:YAG laser can provide stable power and better beam profile [14, 20]. Nd: YAG lasers are also easy to operate and have low maintenance costs [2, 14, 20]. To date, the third harmonic (355 nm) and fourth harmonic (266 nm) of the Nd:YAG has been used to grow high quality high-Tc superconducting films [12, 14, 20].

In UV PLD of Bi-Sr-Ca-Cu-O films, the substrates are usually heated to 800^0C followed by in-situ and ex-situ post heat treatments in gas atmospheres [1, 13, 22, 29]. Reports have shown that Bi- content of the film is greatly influenced by the substrate temperature and can be highly deficient at higher temperatures [1, 29]. In some cases, heat treatment during deposition results in the contamination of the film especially on Si substrates [6]. PLD of $YBa_2Cu_3O_{7-\delta}$ using UV lasers produce films with T_c of 90 K require substrate heating ranging from 700-800^0C in a background O_2 gas (pressure of ranging from 100 to 200 mTorr) [6, 7, 32]. This is usually done since Y-Ba-Cu-O is highly dependent on oxygen content. This process is either performed in-situ or ex-situ and part of the post heat treatment [4, 18]. These results implies that post heat treatment is necessary to homogenize the composition of the film and improve the critical temperarure T_c [1, 4, 19]

Recently, we reported the fabrication of micron thick $Bi_2Sr_2CaCu_2O_{8+\delta}$ (Bi-2212) and Yttrium doped Bi-22Y2 through PLD with a 1064 nm Nd:YAG laser [9, 10]. The films underwent heat treatment outside the PLD growth chamber to produce flat and highly c-axis oriented films with stoichiometries identical to the targets. In the case of Bi-2212, the measured T_c is only about 58 K and for Bi-22Y2, the highest T_c is 90.5 K at 25% Y concentration [9, 10]. Y substituted Bi-2212 films grown by IR PLD show drop in magnetoresistance and improved critical current density J_c [3].

The primary motivation of these previous works is to use the existing Nd:YAG laser in PLD experiments to avoid the complicated optics and gas systems of an excimer laser based PLD. Also, when fundamental wavelength of the Nd-YAG laser is used for deposition, films with the same chemical composition as the starting material can be fabricated, and therefore it is versatile in the deposition of multicomponent films. The film properties can be adjusted through heat treatment steps after deposition.

In this chapter, we examine effect of post heat treatment on the the morphology, composition, crystallinty of the $Bi_2Sr_2CaCu_2O_{8+\delta}$ and $YBa_2Cu_3O_{7-\delta}$ prepared by IR Nd:YAG PLD.

2. Infrared Nd:YAG laser ablation and post heat treatment

The laser system used in this study is a Q-switched Nd:YAG (Spectra Physics GCR 230) operating at 1064 nm at 10 Hz repetition rate with 8 ns pulsed duration. The deposition was performed in a stainless steel vacuum chamber continuously evacuated to maintain a pressure of 10^{-2} mbar. The solid state sintered target is placed 30 mm from the substrate and was rotated for uniform laser ablation. The Bi-Sr-Ca-Cu-O films were grown on (100) MgO substrate with laser fluence of 5.5 J/cm^2 , while Y-123 films on (100) $SrTiO_3$ single crystal substrates with laser fluence of 2.0 J/cm^2.These values of fluences are typical for PLD of Bi-Sr-Ca-Cu-O and Y-Ba-Cu-O [5, 8, 23, 33].The deposition was performed without substrate heating and no other gasses from external sources were introduced in the chamber during deposition.

For Bi-Sr-Ca-Cu-O films, two heat profiles in ambient air was performed. Some were partial melted at 880 °C and some were heated at 940 °C for 15 minutes and rapidly quenched to room temperature. Both of the heat profiles are seconded with annealing at 850 °C for 2 hour. In the case of Y-123, three heat treatment steps were performed, first the films were re-sintered at 900 °C for 12 hrs followed by heating at 1000 °C for 15 minutes and rapid thermal quenched to room temperature. The last heat treatment involves, heating on a tube furnace with oxygen at 930 °C for 12 hours, and annealed at 450 °C for 2 hours in ambient air.

Scanning electron microscopy (SEM) were used to examine the surface morphological features and composition of the films. Film thickness was determined by SEM cross- sectional imaging. X-ray Diffraction (XRD) were used to investigate the composition and crystal properties of the film. To verify superconducting property of the films, linear four point probe resistance measurement were performed.

3. Post deposition heat treatment effects

3.1. Bi-Sr-Ca-Cu-O films

SEM surface micrographs of partial melted and rapidly quenched Bi-2212 films grown with 5.2 J/cm^2 energy fluence at 10^{-2} mbar chamber pressure for 180 minutes is shown in Figure 1. Relatively smooth and flat films were obtained using partial melting and annealing treatment as can be seen in figure 1a. At higher magnification (film 1c), layering inherent to the

Bi-Sr-Ca-Cu-O is observed. The average thicknesses of the films are 9 μm. In figure 1b, films heated at 940 °C for 15 min and the rapidly quenched films show pronounced layering and terraces of Bi-Sr-Ca-Cu-O. This makes the film rougher compared to partially melted Bi-2212 films. The rapidly quenched films is thinner compared to partial melted films with an average thickness of about 2 μm. Both films show small amount of spheroidal particulates after heat treatment [10].

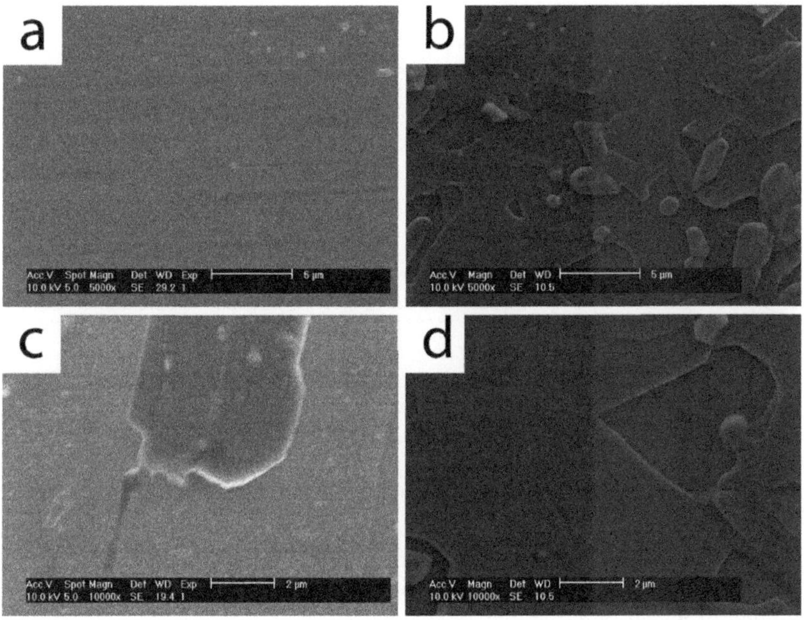

Figure 1. SEM surface micrographs of Bi-2212 films subjected to (a, c) partial melting and (b,d) rapid quenching. Both films were subquently annealed.

XRD measurements on the partial melted and rapidly quenched Bi-2212 films is shown in figure 2. The peaks are indexed using the card file no. 41-0317. Both films are highly c-axis oriented with minimal Bi-2201 impurity. However, sharper XRD peaks are observed for films subjected to partial melting indicating higher crystalline quality.

The rough morphology of films subjected to rapid quenching at very high temperature is due to very fast cooling of the Bi-2212 material. It has been observed that IR PLD Bi-2212 films require partial melting and annealing to allow uniform diffusion and migration of Bi-2212 materials on MgO substrate forming smoother film [10]. This attributed to the micron- size spheroidal grains trasferred on the substrate by the IR laser during deposition [10]. Hence, heat treatment is required to facilitate growth, flatten and densify the material producing much thinner films.

Figure 3 shows the resistance vs temperature measurement on the partial melted Bi-2212 films with transition temperature, T_C of about 79 K. In our previos report, partial melting with subsequent annealing for 10 hours results into T_c of only about 58 K. The shorter annealing

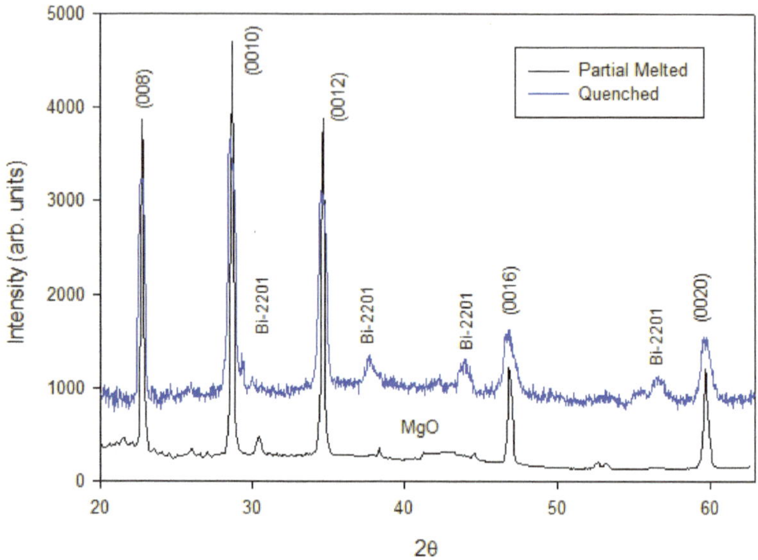

Figure 2. XRD pattern of Bi-2212 film grown by IR- PLD Nd:YAG subjected to post heat treatment. Both films are highly c-axis oriented films are achieved with post heat treatment. Sharper peaks were observed for partially melted films.

time improved the transition temperature. It has been reported that longer annealing time reduces oxygen content in the films [23]. In the case of rapidly quenched films, the measured T_c show almost similar value. Although rapid quenching produce rougher films it still posses T_c comparable with the partial melted films.

3.2. Y-Ba-Cu-O

Figure 4 shows the SEM micrographs of (a) a representative as- deposited films grown using $2.0J/cm^2$ laser energy fluence at 10^{-2} mbar deposition pressure for 180 minutes, and rapidly quenching films in ambient air at 1000 °C for 15 minutes, (b)without, and (c,d) with oxygen annealing. The as-deposited films show spheroidal grain with an average size of about 0.6 μm. The surface of the film also contain micro-cracks extending to about 10 μm in length. The surface of both rapidly quenched Y-123 films were covered with rectangular grains of different sizes. In fig. 4b, the rectangular grains have an average size of about 1.4 μm $\pm$0.3 x 1.3 μ $\pm$0.3, while film subjected to oxygen annealing (fig. 4c) have an average size of 1.7 μm $\pm$0.4 x 1.8 μm $\pm$0.2. The rectangular particles also grow on top of another particle (indicated by circle). Some of the rectangular particles also have different orientation including growth perpendicular to the basal plane (arrow) and parallel to the substrate surface (dashed arrow). In most UV PLD of Y-123, these rectangular shaped grains are presumed to be a- axis oriented [16, 17, 21, 28]. This is observed especially when a solid- state sintered target is used for the

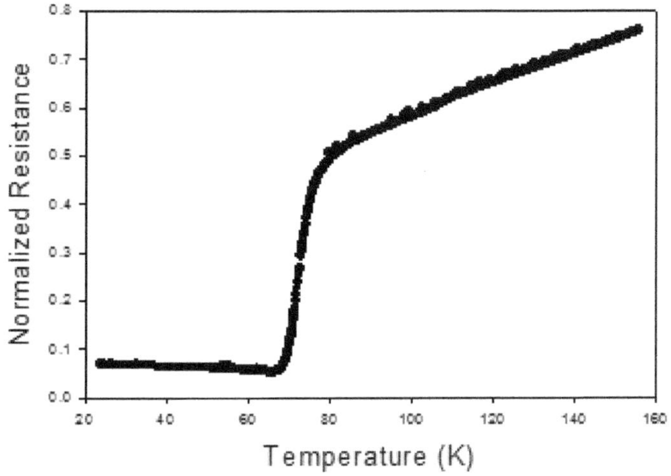

Figure 3. Resistance as a function of temperature of Bi-2212 film prepared by 1064 nm Nd:YAG pulsed laser deposition with post heat treatment.

deposition.However, the sizes which are comparably smaller compared to IR Nd:YAG Y-123 films [16].

At a higher magnification (fig. 4d), the surface of the oxygen annealed Y-123 film show that the grains interconnect laterally forming larger particulates with extended grain boundaries (indicated by squares). This is a common feature of Y-123 films especially when solid state sintered targets are used as the material source for the deposition [15, 21]. While the heat treatment helps in densifying the grains, it also aids in coalescence of grains and forming an alignment of the material relative to the substrate orientation [10]. The appearance of rectangular grains could be a result of incongruent melting of Y-123 at rapid thermal heating at 1000 °C and the thickness of the films. The surface of the film also indicates that the film is not fully melted at that temperature. This is partly due to micron-sized grains transferred by IR laser pulses that needs higher temperature to completely melt. This also attributed to the high melting temperature of Y-123 of about 1200 °C. Hence, we can infer from the SEM image that film is at the initial stage of growth.

The observed morphology of the Y-123 film is due to the micron sized particulates generated by IR PLD. It has been reported that large YBCO particulate are difficult to merged into YBCO films completely especially in the surface of a thick film where the mobility and heat exchange rate are lower than that in the substrate surface, hence they will hinder the c-axis growth of surrounding film [34]. Larger YBCO grains in the as-deposited films could provide sites for the nucleation of a-axis grains [25, 34]. Since large chunks of the target are able to arrive on the substrate surface forming inherently thick films, preferential a-axis formation is observed when you subject to heat treatment. In contrast, UV PLD generates ultrafine particles that lands on heated substrate forming thinner film with c-axis orientation. We are also unable

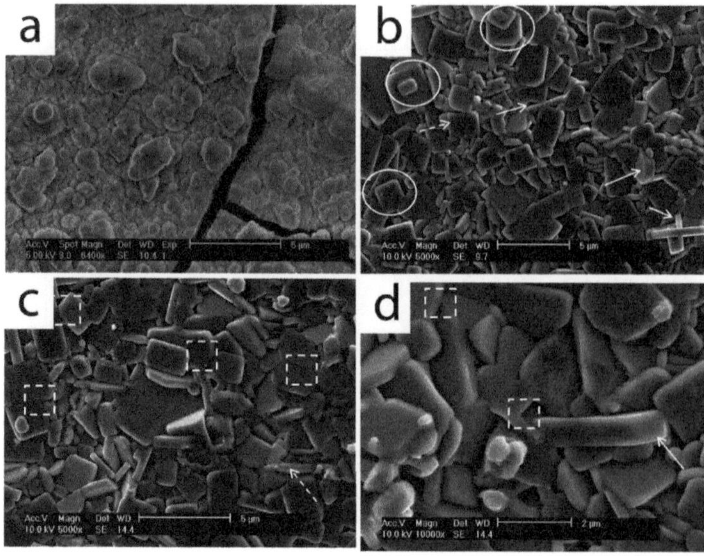

Figure 4. SEM micrograph of Y-123 film deposited by IR Nd: YAG PLD (a) as- deposited, and rapidly quenched at 1000 °C for 15 minutes to room temperature in ambient air (b) without and (c) with oxygen annealing. The window in (c), magnified by a factor of 2 at 2 μm scale, is shown in (d).

to see micro-cracks initially observed in the as-deposited films. This is due to annealing that result to removal of pores and enlargement of grains [9, 10]. The growth of a-axis grains also contribute to dispersion of micro-cracks [15].

Figure 5 shows the XRD spectra of Y-123 films without and with oxygen annealing. The XRD pattern for both films are composed of c-axis and a-axis oriented Y-123 grains. It contains no other diffraction peaks from the precursors of the Y-123 material. This indicates high phase purity of the film. The reflection corresponding to (006) and STO (200) is hard to distinguish due to the small lattice mismatch of Y-123 with STO (about 2%). The value of c-lattice parameters for film not subjected to oxygen annealing is about c= 11.77 while for oxygen annealed film, c= 11.68 . The value of the c-lattice for oxygen annealed film is similar to the oxygen rich Y-123 (c= 11.68) [30] . The variation in c-lattice can be attributed to the decrease in the fraction of oxygen gas in YBCO material [25]. Hence, oxygen annealing is necessary for the as-deposited films.

The mixture of a-axis and c-axis is due to the thickness of the film and partly due to incongruent melting of Y-123 at 1000 °C . It has been reported that Y-123 films possess mixture of a-axis and c-axis growth as a result of increasing film thickness [27]. In contrast to previous reports that the critical thicknesses to obtain crack free Y-123 is about 2.2 μm [31], we do not observed micro cracks on heated films [26]. This is due to granularity of the heated films suggesting that a higher temperature is needed to fully melt the films. Future work must be done by using higher melting temperature in oxygen atmosphere in order to improve the surface morphology and T_c of the Y-123 films.

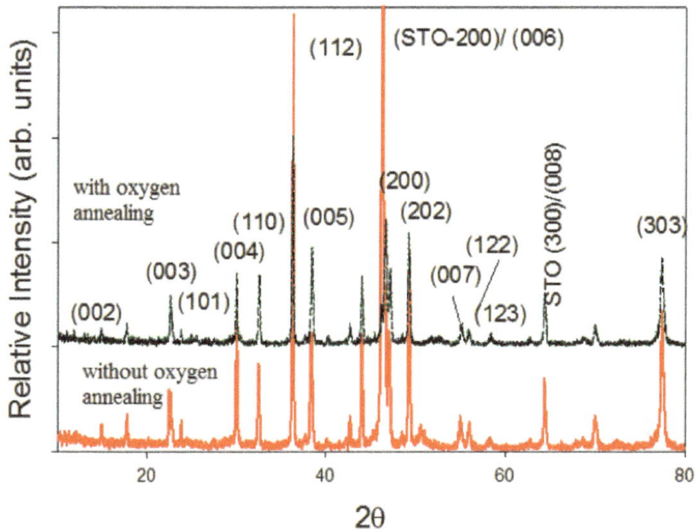

Figure 5. XRD spectra of post heated Y-123 films. Both films were rapidly quenched in ambient air, while (b) includes oxygen annealing

Figure 6 shows the resistance vs. temperature measurement on Y-123 films. Without oxygen annealing the film shows semiconducting behavior (fig. 6a) [24]. This attributed to oxygen deficiency in the film. In contrast, the oxygen annealed film (fig. 6b) shows superconducting transition temperature T_c of about 70 K. This is comparably low with the observed transition temperature with the bulk target of about 89 K.

The a-axis outgrowths are typically observed in the c-axis oriented thick films. The a-axis outgrowth inhibits transport of the superconducting currents in the films resulting into a low value of T_c [17]. The films are also granular introducing weak links at the grain boundaries affecting T_c. This is the main reason of having low T_c values for the films. Although the lattice parameters indicate fully oxygenated Y-123 lattice, the morphology of the films greatly affects superconducting property of the film. Hence, micron thick YBCO films deposited using IR laser needs ex-situ oxygen annealing to improve the T_c of films. Although the a-axis oriented Y-123 films present always a lower critical temperature than that of a c-axis oriented films (about 10 K lower), the study of the a-axis oriented films are useful for sandwich type Josephson device applications [5].

In summary, the heat treament greatly influence the growth and surface morphology of IR PLD films. Since the films are deposited on un-heated substrates, the habit of the material upon heat treatment becomes an important parameter in choosing heat treatment profiles. Layering and grain movement during high temperature melting can be reduced by exposing the film at temperatures closer to the melting temperature and introduction of annealing steps in an environment conducive for uniform coalesnce and intake of oxygen. The temperature

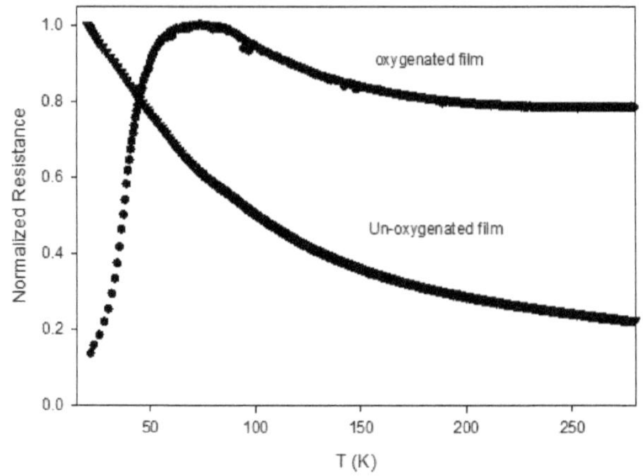

Figure 6. Resistance vs. temperature measurement on Y-123 films heated in (a) ambient air, (b) oxygen ambient. The oxygen annealing helps in forming superconducting films. The low value of T_c is attributed to granular surface morphology of the film and also to the oxygen annealing profile used.

used for Y-Ba-Cu-O is low enough to see the initial stage of growth and will allow us to implement a heat profile that will melt and provide sufficient oxygen on the film.

4. Conclusion

The post heat treatment studies on IR Nd:YAG PLD films is an important stage in developing the technique to be a competitive and alternative technique in producing high T_c superconducting films for electronic applications. The use of non-toxic lasers to deposit coupled with appropriate heat treatment profiles to grow the films is an efficient tool in minimizing the complicated control of parameters in conventional PLD of high T_c superconductor materials. It is envisioned that IR Nd:YAG PLD technique can grow high quality materials with critical current densities useful in the production of films for second generation coated conductors.

Author details

Jeffrey C. De Vero, Rusty A. Lopez, Wilson O. Garcia and Roland V. Sarmago
National Institute of Physics, University of the Philippines, Diliman, Quezon City, Philippines

Acknowledgments

J. C. De Vero acknowledge the Philippine Commission on Higher Education, the Office of the Chancellor in collaboration with the Office of the Vice Chancellor for Research and Development, of the University of the Philippines Diliman, and the National Research Council of the Philippines for funding support.

5. References

[1] Arnold, J., Pfuch, A., Borck, J., Zach, K. & Seidel, P. [1993]. Preparation and characterization of $Bi_2Sr_2CaCu_2O_{8+}$ thin films made by LPVD, *Physica C: Superconductivity* 213.

[2] Awaji, S., Watanabe, K., Badica, P. & Togano, K. [2006]. Growth of superconducting MgB_2 films by pulsed- laser deposition using Nd: YAG laser, *Superconductor Science and Technology* 19: 242–246.

[3] Blanca, G. R. S., Vero, J. C. D., Garcia, W. O. & Sarmago, R. V. [2012]. Enhanced flux pinning in ir pld grown bi-2212 films, *Physica C: Superconductivity, In Press,* doi: 10.1016/j.physc.2012.03.045.

[4] Branescu, M., Vailionis, A., Huh, J., Moldovan, A. & Socol, G. [2007]. Afm and complementary XRD measurements of in situ grown YBCO films obtained by pulsed laser deposition, *Applied Surface Science* 253: 817–8183.

[5] Branescu, M., Ward, I., Huh, J., Matsushita, Y. & Zeltzer, G. [2008]. Scanning electron microscopy and resistive transition of in-situ grown YBCO films by pulsed laser deposition, *Journal of Physics: Conference Series* 94.

[6] Chrisey, D. B. & Hubler, G. K. [1994]. *Pulsed Laser Deposition of Thin Films,* John Wiley and Sons Inc.

[7] Christen, H. M. [2005]. *Pulsed Laser Deposition of $YBa_2Cu_3O_{7-\delta}$for Coated Conductor Applications: Current Status and Cost Issues in Second Generatoin HTS Conductors, A. Goyal (Ed.),* Kluwer Academic Publisher.

[8] Dam, B., Rector, J., Chang, M. F., Kars, S., de Groot, D. G. & Griessen, R. [1997]. The laser ablation threshold of $YBa_2Cu_3O_{6+x}$ as revealed by using projection optics, *Applied Surface Science* 86: 13–17.

[9] De Vero, J., Blanca, G., Vitug, J., Garcia, W. & Sarmago, R. V. [2011]. Stoichiometric transfer of material in the infrared pulsed laser deposition of yttrium doped Bi-2212 films, *Physica C: Superconductivity* 471: 378–383.

[10] De Vero, J., Gabayno, J., Garcia, W. & Sarmago, R. V. [2010]. Growth evolution of $Bi_2Sr_2CaCu_2O_{8+\delta}$ thin films deposited by infrared (1064 nm) pulsed laser deposition, *Physica C: Superconductivity* 470: 149–154.

[11] Eason, R. [2007]. *Pulsed Laser deposition of Thin Films: Applications- Led Growth of Functional Materials,* John Wiley and Sons, Inc.

[12] Ichino, Y., Yoshida, Y., Yoshimura, T., Takai, Y., Yoshizumi, M., Izumi, T. & Shiohara, Y. [2010]. Potential of Nd: YAG pulsed laser deposition method for coated conductor production, *Physica C: Superconductivity* 470: 1234–1237.

[13] Jannah, A. N., Halim, S. A. & Adbullah, H. [2009]. Superconducting properties of bscco thin films by pulsed laser deposition, *European Journal of Scientific Research* 29: 438–446.

[14] Kaneko, S., Shimizu, Y., Yuasa, H. & Ohya, S. [2002]. Change in stoichiometry by laser-induced plasma deposition of high-tc superconducting thin films, *Physica C: Superconductivity* 378–381.

[15] Kastner, G., Schafer, C., Senz, S., Kaiser, T., Hein, M. A., Lorenz, M., Hochmuth, H. & Hesse, D. [1999]. Microstructure and microwave surface resistance of typical YBaCuO thin films on sapphire and $LaAlO_3$, *Superconductor Science and Technology* 12: 366–375.

[16] Kim, C. H., Hong, K. S., Kim, I., Hahn, T. & Choi, S. [2000]. Effects of target microstructure on pulsed laser deposited $YBa_2Cu_3O_{7-\delta}$ thin films, *Thin Solid Films* 358: 223–228.

[17] Kim, C. H., Hong, K. S., Kim, I. T., Hahn, T. S. & Choi, S. S. [1999]. Comparison of microstructures of pulsed laser deposited $YBa_2Cu_3O_{7-\delta}$ thin films using solid-state

sintered and modified melt-textured grown targets, *Physica C: Superconductivity* 325: 127–135.

[18] Kim, S. & Lee, S. [1999]. Characterization of YBCO superconducting films fabricated by pulsed laser deposition, *Thin Solid Films* 355–356: 461–464.

[19] Kume, E., Fukino, H., Zhao, X. & Sakai, S. [2004]. Temperature dependence of composition ratio of $bi_2sr_2cacu_2o_{8+\delta}$ film by pld method, *Physica C: Superconductivity* 412–414: 1354–1357.

[20] Kusumori, T. & Muto, H. [1999]. Fabrication of high quality YBCO epitaxial films by ablation using the fourth harmonic of the Nd: YAG laser, *Physica C: Superconductivity* 321: 247–257.

[21] Kusumori, T. & Muto, H. [2001]. Change in stoichiometry by laser-induced plasma deposition of high-T_c superconducting thin films, *Physica C: Superconductivity* 351: 227–244.

[22] Li, S., Ritzer, A., Arenholz, E., Bauerle, D., Huber, W., Lengfellmer, H. & Prettl, W. [1996]. Step-like growth of$Bi_2Sr_2CaCu_2O_8$ films on off-axis oriented (0 0 1) $SrTiO_3$, *Applied Physics A: Materials Science and Processing* 63.

[23] Marechal, C., Lazaze, E., Seiler, W. & Perriere, J. [1998]. Growth mechanisms of laser deposited BiSrCaCuO films on MgO substrates, *Physica C: Superconductivity* 471.

[24] Mohan, R. & Sighn, K. [2007]. Calcium and oxygen doping in $YBa_2Cu_3O_y$, *Solid State Communication* 141: 605–609.

[25] Morimoto, K., Takezawa, K., Minamikawa, T., Yonezawa, Y. & Shimizu, T. [1998]. Low-temperature growth of YBCO thin films by pulsed laser ablation in reducing environment, *Applied Surface Science* 127–129: 963–967.

[26] Nie, J. C., Yamasaki, H., Develos-Bagarinao, K. & Nakagawa, Y. [2007]. Surface morphology and microstructure of thick $YBa_2Cu_3O_{7-\delta}$ films on vicinal r-cut sapphire buffered with CeO_2, *IEEE Transaction of Applied Superconductivity* 17: 3459–3462.

[27] Park, J. H., Jeong, Y. S. & Lee, S. Y. [1998]. Investigation on the relation between the thickness and the orientation of epitaxially grown YBCO thin films by laser ablation, *Thin Solid Films* 318: 243–246.

[28] Prouteau, C., Verbist, K., Hamet, J., Mervey, B., Hervieu, M., Raveau, B. & Tendeloo, G. B. [1997]. Microstructure of a-axis oriented YBCO films on $SrTiO_3$ substrates using a new template layer $La_4BaCu_5O_{13}$, *Superconductor Science and Technology* 288.

[29] R.Rossler & J.D. Pedarnig, a. C. J. [2001]. $Bi_2Sr_2CaCu_{n-1}O_{(n+2)+\delta}$ thin films on c- axis oriented and vicinal substrates, *Physica C: Superconductivity* 361: 13–21.

[30] Vonk, V., van Reeuwijk, S. J., Dekkers, J. M., Harkema, S., Rijnders, A. J. H. M. & Graafsma, H. [2004]. Strain-induced structural changes in thin $YBa_2Cu_3O_{7-x}$ films on $SrTiO_3$ substrates, *Thin Solid Films* 449: 133–137.

[31] Yamada, Y., Kamashima, J., Wen, J. G., Niiori, Y. & Hirabayashi, I. [2000]. Critical thickness and effective thermal expansion coefficient of YBCO crystalline film, *Japanese Journal of Applied Physics* 39: 1111–1115.

[32] Yavuz, M., Uprety, K. K., Subramanian, G. & Paliwal, P. [2003]. Preparation and characterization of BSCCO 2212 thin films, *IEEE Transaction of Applied Superconductivity* 13: 3295–3297.

[33] Zheng, R., Campbell, M., Ledingham, K., Jia, W., Scott, J. & Singhal, R. [1997]. Diagnostic study of laser ablated $YBa_2Cu_3O_{7-\delta}$ plumes, *Spectrochimica Acta Part B: Atomic Spectroscopy* 52: 339–352.

[34] Zhu, D., Huang, J., H. Li, M. S. & Su, X. [2009]. A new method to preserve the c-axis growth of thick $YBa_2Cu_3O_{7-\delta}$ films grown by pulsed laser deposition, *Physica C: Superconductivity* 469: 1977–1982.

Chapter 10

Novel Pt and Pd Based Core-Shell Catalysts with Critical New Issues of Heat Treatment, Stability and Durability for Proton Exchange Membrane Fuel Cells and Direct Methanol Fuel Cells

Nguyen Viet Long, Cao Minh Thi, Masayuki Nogami and Michitaka Ohtaki

Additional information is available at the end of the chapter

1. Introduction

Traditionally, Pt and Pd based catalysts are widely studied in the continuous developments of next fuel cells (FCs) with the critical issues of energy and environment technologies. So far, Pt and Pd based catalysts have been mainly used in the anodes and the cathodes in FCs by a electrode-membrane technology. In spite of the large advantages of Pt based catalysts in electro-catalysis for FCs, many problems of high cost remain. In addition, so far Pt and Pd catalysts have still exhibited very good catalytic activity and selectivity of hydrogen and oxygen adsorption as well as hydrogen evolution reaction (HER), hydrogen oxidation reaction (HOR) for the dissociation of hydrogen into protons (H^+) and electrons (e^-), and oxygen reduction reaction (ORR). At present, FC technologies and applications are polymer electrolyte fuel cell (PEFC) or also known as proton exchange membrane FC (PEMFC), phosphoric acid FC (PAFC), alkaline FC (AFC), molten carbonate FC (MCFC), solid oxide FC (SOFC). The typical features include operating temperature (°C) for low-temperature PEMFC and DMFC of about 50-80 °C, power density ~350 Mw/cm^2, fuel efficiency ~40-65%, lifetime >40,000 hr, capital cost >200$/kW [5,49,52,173], and other practical applications. According to hydrogen and oxygen reaction, electro-oxidation of carbon monoxide (CO) is intensively studied in low temperature FCs. In DMFCs, methanol oxidation reaction (MOR) in catalytic activity of Pt catalyst is very crucial to improve the whole performance. Therefore, scientists have considerably focused on the various ways of improving HOR, ORR, and MOR in the catalyst layers of various FCs, PEMFCs, and DMFCs [1-3]. So far, ORR has become an important mechanism investigated in PEMFCs and DMFCs for their large-scale commercialization. Recently, U.S. Department of Energy Fuel Cell Technologies

Program (DOE Program), and New Energy and Industrial Technology Development Organization (NEDO Program) in Japan have supported large Research and Development programs (R&D) of FCs and FC systems for stationary, portable and transportation applications, such as FC vehicles. In addition, FCs become promising technology to address global environmental challenges in energy, science and nature issues [4-8]. Now, various DMFCs can work at low and intermediate temperatures up to 150 °C [9]. Thus, next fuel cells also can meet the urgent demands for green energy.

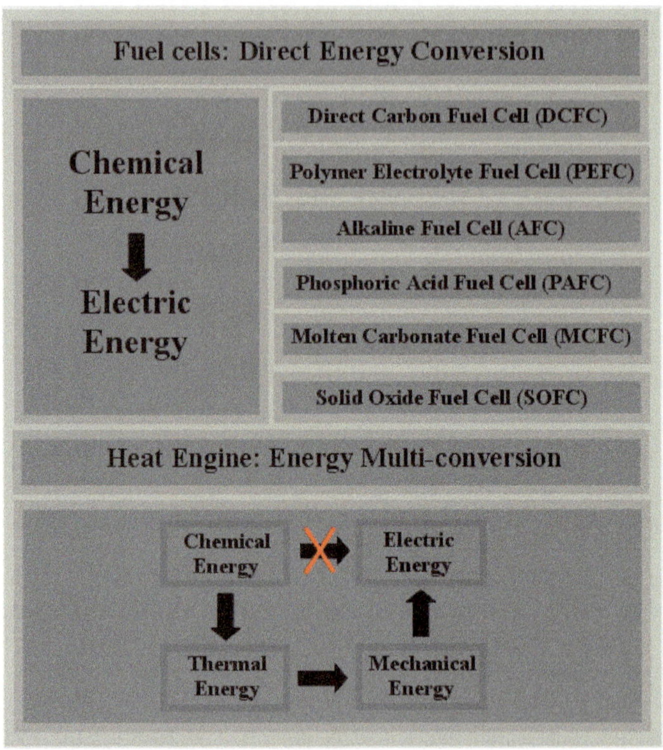

Figure 1. Features of fuel cells and heat engine in the direct or indirect conversion processes from chemical energy into electric energy. Excellent advantage of fuel cells is direct energy conversion

Today, the proton exchange membranes, typically such as perfluorosulfonic acid (PFSA) membranes or Nafion® for FC applications are presented [51-53]. Interestingly, charge carriers in FCs are various kinds of H^+ (PEMFC), H^+ (DMFC), OH^- (AFC), H^+ (PAFC), $CO_3^=$ (MCFC), and $O^=$ (SOFC) [4-9,49,52,173]. Figure 1 shows various energy conversion processes from chemical energy into electric energy through both FCs and heat engine. The operation principle of simple low-temperature FCs mainly depends on the chemical reactions of hydrogen and oxygen with direct conversion into electricity without mediate conversions of thermal energy and mechanical energy.

Fuel cells	Applications
AFC, PAFC, MCFC, SOFC PEMFC, DMFC	AFC: Space, mobile; PAFC: Distributed power; MCFC: Distributed power generation; SOFC: Power generation; PEMFC and DMFC: Portable, mobile, stationary
PEMFC, DMFC	Special use in compact mobile devices and handphones in future

Table 1. Potential applications for various fuel cells [4-9,49,52,173]

2. Pt and Pd based catalysts

In recent years, novel Pt and Pd metals have been known as the best electrocatalysts to important chemical reactions for synthesis of new chemicals as well as the FC reactions. The electrocatalytic properties are typically characterized by hydrogen evolution reaction/hydrogen oxidation reaction (HER/HOR), ORR, and electro-oxidation of CO at the surfaces of Pt(hkl) facets of the as-prepared catalysts as well as the typical oxidations of methanol and formic acid on the surfaces of Pt (hkl) facets of the prepared catalysts. The (111), (100) and (110) low-index facets were proved in high stability and durability in catalysis, and good re-construction in the catalytic FC reactions [10,81,82]. The HER on the Pt catalyst is known by the important Volmer, Tafel, and Heyrovsky mechanisms. In addition, Volmer-Tafel and Volmer-Heyrovsky mechanisms can occur in the complex combinations of the above basic mechanisms [10]. The surface kinetics and chemical activity occurring at the electrode surface containing Pt/support catalyst are characterized as follows [10-14,140].

$$\text{Pt-H}_{ads} \rightarrow \text{Pt} + \text{H}^+ + e^- \quad (1)$$

$$Q_{DL}\ (\text{Charge}) \leftrightarrow Q_{DL}\ (\text{Discharge}) \quad (2)$$

$$\text{Pt} + \text{H}_2\text{O} \rightarrow \text{Pt-OH} + \text{H}^+ + e^- \quad (3)$$

$$\text{PtOH} + \text{H}_2\text{O} \rightarrow \text{Pt(OH)}_2 + \text{H}^+ + e^- \quad (4)$$

$$\text{Pt-(OH)}_2 \rightarrow \text{PtO} + \text{H}_2\text{O} \quad (5)$$

$$2\text{PtO} + 4\text{H}^+ + 4e^- \rightarrow \text{Pt-Pt} + 2\text{H}_2\text{O} \quad (6)$$

$$\text{Pt} + \text{H}^+ + e^- \rightarrow \text{Pt-H}_{ads} \quad (7)$$

To evaluate catalytic activity of the pure Pt catalysts or Pt based catalysts, the electrochemical active surface area (ECSA) is used as ECSA = $Q_H/(0.21 \times L_{Pt})$ [10-14,140]. Therefore, ECSA can be significantly enhanced by the use of a low L_{Pt} loading, and a low content of CO intermediates generated, and new discoveries of highly strong hydrogen reactions in the improvements of the Pt based catalysts. Clearly, the particle size of Pt NPs of 10 nm is crucial in catalysis and FCs, PEMFCs, and DMFCs because the metal NPs showed very large quantum and size effects in the size range of around 10 nm. The ORR is observed in two main pathways in acidic electrolytes as follows [10-15].

$$O_2 + 4H^+ + 4e^- \rightarrow 2H_2O \quad (8)$$

$$O_2 + 2H^+ + 2e^- \rightarrow H_2O_2 + 2H^+ + 2e^- \rightarrow 2H_2O \quad (9)$$

For the ORR, the relationship between kinetic current (i) and potential (E) can be investigated as rate expression

$$i = \underline{n}Fkc(1 - \theta_{ad})^x \exp(-\beta FE/RT)\exp(-\gamma\Delta G_{ad}/RT) \quad (10)$$

Where n, F, K, c, x, β, and γ are constants. In addition, n, F, c, and θ_{ad} indicated the mole (n), Faraday's constant (F), the concentration of O_2 (c), and coverage of adsorbed species (θ_{ad}), respectively. Here, ΔG_{ad} indicated the weak or strong adsorption. We can choose x=1 and γ=1 in the simplification. According to the adsorption degree, the rate may be changed. Thus, it may be change from positive (Weak adsorption) to negative (Strong adsorption). It means that the reaction rate declines when the coverage of intermediates (θ_{ad}) rises [13,173].

At present, the phenomena of ORR kinetics and mechanisms occurring on the Pt catalysts are intensively investigated but a very high overpotential loss observed. Thus, the very high loadings of Pt must be supplied in the high requirements of the FCs operation with large current. It is known that the Pt catalyst has showed the highest activity to the ORR mechanism. Most of research has led to understand ORR on catalytic systems of Pt designed catalysts using the ultra-low Pt loading at minimal level. The issues of the low Pt-catalyst loading, high performance, durability and effective-cost design in FCs systems are very crucial for their large-scale commercialization. The CV results of various Pt NPs (sphere, cube, hexagonal and tetrahedral-octahedral morphology …) in H_2SO_4 showed the strong structural sensitivity of the as-prepared Pt NPs. The most basic (111), (100) and (110) planes were confirmed in the active sites of catalytic activity such as in the edges, corners, and terraces [15]. In particular, monolayer bimetallic surfaces were investigated in the experimental and theoretical studies of the surface monolayer, subsurface monolayer, and inter-mixed bimetallic structures, especially by DFT theoretical approaches [16,17]. So far, the characterization of size, structure, surface structure, internal structure, shape, and morphology has been discussed in various the as-prepared metal NPs by various strategies of syntheses. The noble NPs (Pt, Pd, Ru, Ir, Os, Rh, Au, Ag), and their combinations with cheaper metals (Ni, Co, Cu, Fe …) as alloy and core-shell nanostructures can be used as potential Pt based catalysts for further studies of ORR and CO oxidation reaction in various FCs for long-term physical-chemical stability and durability, such as PEMFCs and DMFCs [18-21]. Besides, the investigations of both theory and applications of alloy clusters and nanoparticles showed potential applications in catalysis and FCs [22]. In various FCs, noble Pt metal is the key to large-scale commercialization of PEMFCs and DMFCs because of its unusually high catalytic properties. Therefore, scientists and researchers try to create highly active and stable catalysts with a low Pt loading. To enhance its catalytic activity, Pt catalyst NPs were supported on various high-surface-area carbon materials, such as carbon black (e.g. Vulcan XC-72) [23-27]. The Pt based catalysts of various nanostructures are discussed in the developments of PEMFCs and DMFCs. Interestingly, electricity is directly generated in PEMFCs by hydrogen oxidation and oxygen reduction reactions through membrane-electrode assembly (MEA) [1-14].

2.1. Preparation methods of Pt and Pd based nanoparticles

Essentially, various top-down physical or bottom-up chemical methods, such as polyol method, and chemical-physical combined methods are widely used for making Pt and Pd based catalysts for homogeneous and heterogeneous catalysis, FCs, PEMFCs, and DMFCs involving in both theory and practice [28-33,173-188]. The core-shell nanostructures of bimetallic nanoparticles can be synthesized by phase-transfer protocol method [34]. The relatively facile method with microwave and ultrasound supports or sonochemical method in the synthesis of nanoparticles without focusing on much consideration of the basic issues of the homogeneity of typical size and morphology for catalysis and FCs has been very attractive to researchers and scientists [35,36]. In the methods, the nanosized ranges of the as-prepared nanoparticles are crucial to practical applications in catalysis, biology and medicine. So far, no comprehensive survey of the effects of heat treatments to achieve the significant enhancements of catalytic activity of Pt and Pd based catalysts has been presented in detail.

2.2. Size, shape and morphology

At present, it is known that the as-prepared Pt nanostructures show a variety of particle morphologies and shapes in homogeneous and heterogeneous characterizations. The main morphology and shape were prepared in the broad forms of cube, octahedra, cubo-octahedra, tetrahedra, prisim, sphere, icosahedra, decahedra, rod, tube, wire, fiber, dendrite, flower, plate, twin, belt, disk etc... in the non-polyhedral and non-polyhedral or irregular shapes and morphologies in the near same size range. The spherical and non-spherical morphologies and shapes are observed in the near same range of particle size. The particle size is discovered in various nanosized ranges of from 10 nm, 20 nm, 30 nm, 40 nm, 50 nm, and 100 nm ... to 1 μm, 10 μm, up to 100 μm ..., especially particle size of around 10 nm for potential promising applications in catalysis, biology, and medicine. In our proposals, catalytic activity and selectivity of interesting homogeneous and heterogeneous morphologies and shapes of the Pt based nanoparticles in the nanosized ranges of 10 nm and 20 nm become important topics for scientific research because their structural transformations in that certain ranges of 1-30 nm are difficult to understand transparently unknown phenomena and properties [37,38]. When the size of Pt nanoparticles is decreased into the range of 10 nm, the total fraction of Pt atoms on the Pt catalyst surfaces is very large. This leads a very significant enhancement of electro-catalytic activity. Most of sizes, shapes, morphologies, nanostructures of the as-prepared Pt nanoparticles are significantly changed in normal conditions after the *in situ* TEM and HRTEM measurements. Clearly, the important effects of temperature on characterization of the pure Pt NPs and Pt/supports need to be studied at different temperatures for one optimum temperature range while keeping their good catalytic characterization. The new discoveries of surface-structure changes of polyhedral Pt shapes and morphologies are crucial in the further catalysis investigations [39-41]. The influence of hydrogen on the morphology of Pt or Pd nanoparticles was found in the structural transformations [42,43]. Therefore, alloy and core-shell nanoparticles with various Pt metal compositions are crucial to practical applications in catalysis and FCs. Nevertheless, most of the as-prepared metal nanoparticles possibly change their certain good shapes and morphologies (e.g. cube, tetrahedra, octahedra …) into hetero-shapes and hetero-morphologies. In fact, the issues of catalytic activity, durability, and stability of the as-prepared metal nanoparticles in various media have become very important to most of current scientific research. For example, non-platinum anode catalysts or without the use of Pt metal for DMFC and PEMFC applications were developed [44]. It has been known that they are transition metal carbides (e.g. WC and W_2C ...) and the promoted transition metal oxides (e.g. TiO_2, SiO_2, CeO_2, Zr_2O_3, CeO_2-Zr_2O_3 ...) that have the advantages of low prices and strong resistance to poisonous substances such as carbon dioxide (CO) poisoning or CO adsorption on the catalysts.

2.3. Structure and composition

Among metal noble nanoparticles (Pt, Pd, Ru, Rh, Ir, Os, Au, and Ag NPs) as well as various cheap metal nanoparticles, metals show a face centered cubic (fcc) structure. The strong emphasis is that they can be used as metal catalysts for catalysis and FCs. Thus, Pt and Pd based alloy and core-shell nanoparticles can be engineered in a variety of composition using various metals (Co, Ni, Fe, Cu …) or oxides, ceramics, and glasses in the next significant efforts of researches according to the discoveries and improvements of catalytic activity, selectivity, durability, and stability in catalysis.

3. Proton exchange membrane fuel cell

At present, PEMFCs are used for mobile, portable, and automobile applications because of generated high power densities. For instance, they can operate at low and high temperatures of 60-100 °C or up to 200 °C [45-50]. In addition, the PEMFC is used for transportation applications when pure hydrogen as fuel can be used in PEMFCs for their operation. The conventional fuels are used as liquid, natural gas or gasoline. Therefore, the direct use of methanol can lead to develop PEMFCs into DMFCs. In particular, DMFCs proved that they can offer potential applications, such as cameras, notebook computers, and portable electronic applications [45-50]. The nanostructured membranes have been extensively reviewed in potential FC applications [50]. In addition, proton exchange membranes for PEMFCs operated at medium temperatures are discussed [51-53]. It is likely that the fast developments of new membrane technology can be realized in FCs, PEMFCs, and DMFCs.

3.1. Operation principle

A simple hydrogen and oxygen PEMFC includes the catalytic anode, membrane electrode assembly (MEA), and the catalytic cathode. Fuel is hydrogen fed to the anode that generates protons (H^+). They travel through proton exchange membrane and combine with electrons (e^-) and oxygen at the cathode to form water (H_2O). Electrons travel through an external circuit. This leads to that electricity is generated by a FC. Figure 2 shows chemical reactions on the anode and the cathode of a PEMFC. The electrochemical reactions typically occur in a PEMFC as follows.

$$\text{At the anode: } H_2 \rightarrow 2H^+ + 2e^- \quad (11)$$

$$\text{At the cathode: } \tfrac{1}{2} O_2 + 2H^+ + 2e^- \rightarrow H_2O \quad (12)$$

$$\text{The overall reaction is } \tfrac{1}{2} O_2 + H_2 \rightarrow H_2O \quad (13)$$

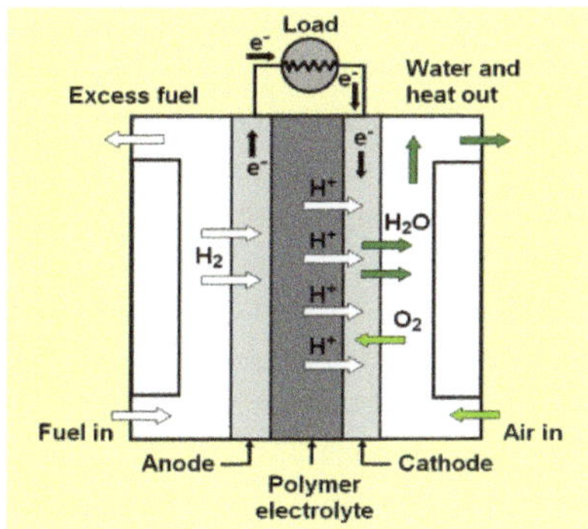

Figure 2. Basic configuration and chemical reaction of PEMFC

3.2. Catalysts in proton exchange membrane fuel cell

Of great interest is the study of the Pt nanoparticles with controlled size and shape around 10 nm because of its importance in electro-catalysis. So far, the Pt and Pt catalysts have showed the best catalytic activities in the HOR and ORR mechanisms for PEMFCs comparable to other metal catalysts. Therefore, new Pt and Pd based catalysts are developed by using various metals combined in alloy nanostructures. Now, various kinds of Pt and Pd core-shell nanoparticles or nanostructures are prepared in the proofs of improving catalytic activities of HOR and ORR. The cost of Pt and Pd catalysts is very high for the large-scale commercialization of FCs. Therefore, cheaper metals such Cu, Co, Fe, Ni ... can be studied in the uses in the alloy and core-shell catalysts with the Pt shells for reducing the Pt loading [54,55]. The thermal cathodic treatments on Pt/C and Pt-Ru/C catalysts were used to enhance methanol electro-oxidation in sulfuric acid solution in electrochemical activation [56-58]. Clearly, Pt/support catalysts are preferred in many applications. Bimetallic catalysts such as PtNi, PtCo, and PtCu have been very important to the ORR activity at cathode in PEMFC. The as-prepared nanoparticles were studied in the de-alloying phenomena of Pt binary alloys with the different nanostructures [59-62]. The nanostructured catalysts were reviewed in various FCs. The next catalysts with the Pt loadings for low-temperature PEMFCs and DMFCs will be proposed in various alloy, multi-composition, and core-shell structures. Here, Pt bimetallic catalysts were prepared by impregnating a commercial Pt/C with various transition metals (Pt/M = 3, M: V, Cr, Mn, Fe, Co, Ni, Cu, Zr, Ag and W), and sintering at 900 °C. An increase in electrode activity in ORR tests using a half cell in phosphoric acid solution at 190 °C at the initial stage was observed for PtCr, PtFe and PtAg catalysts due to the surface roughening effect [63]. So far, the Pt catalysts of high cost have been successfully used in both the anode and the cathode in PEMFCs with high cost. Therefore, binary alloy catalysts with Pt are developed for PEMFCs. Thus, Pt-Ni, Pt-Co or Pt-Cr bimetallic nanoparticles can be used as binary alloy catalysts for the significant enhancement of ORR. In particular, the Pt_3Ni catalysts of the very high ORR activity are investigated [64-66]. A thin-film rotating disc electrode (TF-RDE) method was used for investigating the electrocatalytic activity of high surface area catalysts, and the highest catalytic activity on Pt/C catalysts towards ORR [67]. In PEMFCs, the catalytic activity and stability of the Pt based catalysts are very important. In the cathode of PEMFCs, Pt based alloy catalysts are developed for PEMFCs. Consequently, the issues of sizes, internal nanostructures, surface nanostructures, shapes and morphologies are also important to the stable operations of PEMFCs for the long periods. In addition, the pt based catalysts with the core-shell nanostructures of the thin Pt shell of around several nanometers have been developed for the next PEMFCs. In particular, the non-Pt catalysts without noble Pt metal are developed. Because noble Pt and Pd metals have the near same catalytic activity of HOR, the Pd based catalysts are studied for the improvements of ORR of PEMFCs [68]. The electro-active sites on carbon nanomaterials electro-catalyzed the reduction of peroxide intermediated from ORR on Pt [69]. The Pd based catalysts can be also used in a combination of Fe, Co, and Cu metals for the improvements of ORR, especially methanol tolerant ORR catalysts. It is clear that stability and durability of the catalysts in the cathodes have importance of the operations of PEMFCs. Therefore, PEMCs can be improved in high stability in the uses of the special supports such as various kinds of carbon supports (e.g. CNTs …). It is clearly admitted that Pt-Ru, Pt-Mo, Pt-Ni catalysts … are used as reformate-tolerant catalysts or impure-hydrogen tolerant catalysts for good stability of next PEMFCs. In addition, Pt based ternary catalysts were discussed for the development of various low and high temperature FCs, PEMFCs, and DMFCs involving in their performance, durability and cost [70-73]. Now, an emergence of the hugely urgent demands of the Pt or Pd based catalysts after high heat treatment processes in both low and high temperatures less than 1000 °C, or up to more than 2000 °C offering better characterizations of catalytic activity and stability can be predicted in future due to the catalysts exhibiting the pristine surfaces, shapes and morphologies in the electrode catalysts. The re-constructions, e.g. (111), (110), and (100) planes [10-12], or collapses of the Pt or Pd based nanoparticles and nanostructures with or without heat treatment in the preparation will be very attractive to research and development of new catalysts for PEMFCs and DMFCs [10-12,41]. There are little evidences of the size, structure and morphology of Pt or Pd based catalysts after high heat treatments in media of H_2 or N_2/H_2. These are major challenges in catalysis science, and Pt or Pd based catalysts for FCs, PEMFCs, and DMFCs. Our catalyst preparation gave a better catalytic activity and stability of HER, ORR, and MOR in an environment of mixture H_2/N_2 avoiding the formation of PtO by heat treatment. In future, the Pt catalysts can be treated at high temperature but their size, nanostructure and morphology in the 10 nm range kept [128-130]. Therefore, the pure Pt or Pd based catalysts used in the electrodes of FCs, PEMFCs, and DMFCs can give better catalytic activity, stability, and good performance of the whole FC systems.

4. Direct methanol fuel cell

4.1. Operation principle

In general, PEMFCs can be categorized into various kinds of hydrogen/oxygen FCs, DMFCs, and direct formic acid fuel cells (DFAFC) according to the use of liquid or gas fuels etc. [74-76]. So far, the economic uses of Pt, Pt-Pd, and Pt-Ru based catalysts as well as catalyst supports have been very crucial to MOR at the anode in DMFCs. To date, methanol can be used in direct chemical-electrical energy conversion in a DMFC in figure 3. The electrochemical reactions occurring in a DMFC are:

$$\text{Anode: } CH_3OH + H_2O \rightarrow CO_2 + 6H^+ + 6e^- \quad (14)$$

$$\text{Cathode: } \tfrac{3}{2}O_2 + 6H^+ + 6e^- \rightarrow 3H_2O \quad (15)$$

$$\text{Overall: } CH_3OH + \tfrac{3}{2}O_2 \rightarrow CO_2 + 2H_2O \quad (16)$$

The DMFC is attractive because methanol, being a liquid fuel, is easy to transport and handle. In DMFCs, the MOR mechanism at the anode is crucial. Due to the low operating temperature ~60-150 °C, the Pt based catalysts are sensitive to poisoning. Since CO is formed during electro-oxidation of methanol, CO-tolerant catalysts are used for DMFCs. Nevertheless, these have a much lower power density despite the typically high noble metal loading of the electrodes. In addition, the energy efficiency of DMFCs suffers from high electrode over-potential (voltage losses), and from methanol losses by transfer (by permeation) through the membrane used in DMFCs [77-80].

4.2. Catalysts in direct methanol fuel cell

So far, Pt and Pd catalysts have been known as the most important catalysts for the direct methanol oxidation in electrodes, such as anodes and cathodes. The interesting hydrogen adsorption on Pt or Pd catalyst was studied [81-83]. In this context, PVP and TTAB polymer-Pt nanoparticles were synthesized with the same cubic shape and similar particle size (8.1 and 8.6 nm, respectively). They can be used as the potential Pt catalysts for chemical synthesis [84]. The PVP-Pt nanoparticles of good cubic, tetrahedral, and octahedral morphology in the size range of 10 nm were prepared by polyol method with the use of commercial chemicals for potential catalysts for FCs [85-87]. At such very small scale, the well-controlled synthesis of Pd nanoparticles by polyol synthesis routes using PVP polymer proved that the shapes of Pd NPs provide a good opportunity of investigating their catalytic property [88]. The catalytic reactions of Pt and Pd catalyst have been studied in the different combinations of various metals such as Ru, Rh, and Sn etc. with support materials such as carbon nanomaterials or oxides and glasses in both homogeneous and heterogeneous catalysis [89-97]. The new Pt-monolayer shell electrocatalysts of high stability were developed for the FC cathodes. The role of the Pt shell is to reduce the Pt loading significantly in PEMFCs and DMFCs but synergic effects for enhancing catalytic activity and stability. However, it is very difficult to make the Pt monolayers on the core nanoparticles.

Therefore, scientists and researchers have tries to study the easy-to-use chemical and physical methods for highly homogeneous core-shell nanosystems with the use of the thin Pt shells of several nanometers in the size range of 10 nm for PEMFCs and DMFCs. In addition, the good characterization of the engineered nanostructure of the Pt based core-shell nanosystems need to be stabilized after the high heat treatments and preparation processes for obtaining the best catalysts for PEMFCs and DMFCs [98-103], leading to the FC durability up to 200,000 cycles with core-shell catalysts with the Pt-monolayers shells [99]. In all cases, the as-prepared Pt nanoparticles are supported on various carbon nanomaterials for catalytic enhancement. The improvements of Pt catalyst can be performed through by the use of the core-shell nanoparticles or core-shell catalysts that have been proposed. The systems of Pt based, binary, ternary, and quaternary catalysts are proposed in both homogeneous and heterogeneous catalysis [104,105]. Therefore, the Pt based multi-metal catalysts (PtRu or PtRuIr/C) exhibiting high catalytic activities are developed for the MOR [106,107]. A comparison of catalytic activity of Pt, Ru, and PtRu catalysts using $Pt(NH_3)_2(NO_2)_2$ and $Ru(NO_3)_3$ precursors for DMFCs was performed. The $Pt_{50}Ru_{50}/C$ catalyst prepared at 200 °C has the maximum electrocatalytic activity toward MOR. The Pt-Ru catalysts have shown the excellent MOR in the stable operations of DMFCs but very high cost [108]. For methanol electro-oxidation, binary, ternary, and quaternary Pt alloy catalysts were prepared. As a result, quaternary alloy catalysts (Pt-Ru-Os-Ir) were considered to be far superior to Pt-Ru in DMFCs [109]. In particular, the successful process of Pt based cubic nanoparticles (Pt-Fe NPs, Pt-Fe-Co NPs), Pt-Fe-Co branched nano cubes, Pt-Fe-Co NPs of low and high Co content was presented. Overall, Pt-Fe-Co branched cubes were used as a good catalyst to show best activity and durability for electrocatalytic MOR [110]. In one study, the evidences of linear sweep voltammetry (LSV) tests indicated that the high peak current density on Pt_2Rh/C is about 2.4 times higher than that of Pt/C in alkaline media direct methanol FCs [111]. So far, the Pt-Au catalysts engineering has become very important to the developments of PEMFCs and DMFCs [112]. For the continuous developments of DMFCs, the Pt-Ru alloy and core-shell electrocatalysts for FCs are discussed in a critical survey [113]. The as-prepared catalysts of PtRuMo NPs supported on graphene-carbon nanotubes (G-CNTs) nanocomposites were developed for DMFCs. The results showed that the catalytic activity and stability of the PtRuMo/G-CNTs catalyst are higher than those of PtRuMo/G and PtRuMo/CNTs catalysts. However, the new trends of using Pt based multi-component catalysts without considering the structural issues lead to the complexity of preparation and synthesis [114]. The Pd-Pt catalysts are recognized as potential candidates for PEMFCs and DMFCs operating at low temperatures [115]. Electrocatalytic activity for ORR and oxygen binding energy was intensively studied in the metal catalysts among the different combinations of W, Fe, Mo, Co, Ru, Ni, Rh, Cu, Ir, Pd, Pt, Ag, and Au. According to the oxygen binding energy by density functional theory (DFT), the volcano-type dependence of ORR was discovered in a high catalytic activity of Pt, Pt, Ir, and Ag metal, especially for the very thin Pt and Pd layers as the monolayers. The structural investigation of ORR and oxygen binding energy by the *d*-band model or calculation method of *d*-band center is very crucial to the issues of engineering a novel catalyst [116-118]. In particular, a X-ray absorption spectroscopy (XAS) method was employed in the characterization of a number of catalysts for low temperature FCs to determine the existent oxidation state of metal atoms in the catalyst, or in the case of Pt, the d-band vacancy per atom. It is a good tool for confirming the structures of the catalyst *in situ* [119]. There is an important discovery of a synergistic effect to being studied. In theory and practice, synergistic effect needs to be intensively investigated in quantum property in respective to the development of bimetallic catalysts of alloy and core-shell nanostructures. In this context, the synergistic effect of core-shell bimetallic nanoparticles gives an excellent catalytic enhancement. In catalytic activity, the shell provides strongly catalytic sites. The core element gives an electronic effect (a ligand effect) on the shell element because the surface atoms of the shell are coordinated to the core in their catalytic reactions. Therefore, the shell is an important factor to control the catalytic properties. In addition, the core-shell bimetallic structures cause better suppression of adsorbed poisonous species [120,140-143]. This effect is discussed in the metallic NPs of hetero-morphologies and hetero-structures that can be as new types of important catalyst [121]. Of all catalysts, the bimetallic nanoparticles have played important roles in promising applications of catalysis, and FCs [122]. The core-shell nanoparticles and nanostructures were intensively investigated [123-124]. The Pt-Ni-graphene catalysts were prepared for the high MOR activity observed [125]. In particular, an important role on the MOR activity of Pt-Co and Pt-Ni alloy electrocatalysts for DMFCs has to be ascribed to the degree of alloying [126]. The interaction of Pd NPs and Pd(111) with CH_3OH and CH_3OH/O_2 mixtures was examined from ultrahigh vacuum conditions up to ambient pressures [127]. Here, the particle size and structure dependent effects in methanol oxidation and decomposition are very crucial to the use of the supported Pd catalysts as well as the good incorporation of the pure Pt into various supports with high homogeneous distribution. In addition, we should find suitable heat treatment in good experimental conditions, and Pt based catalyst engineering to increase catalytic activity and durability.

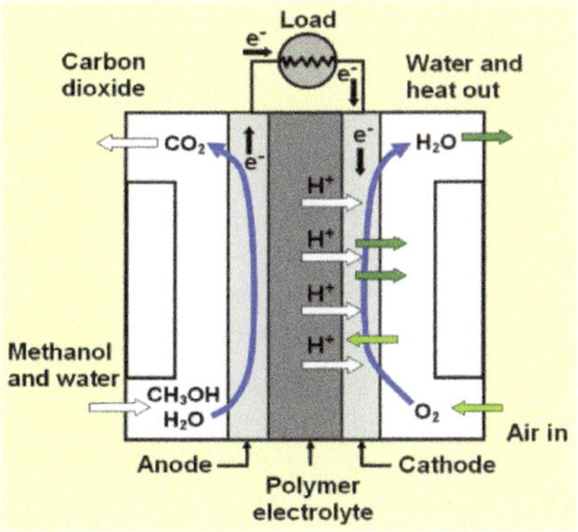

Figure 3. Chemical reactions of direct methanol fuel cell (DMFC)

5. Catalysts and heat treatment

Of interest to scientists, at present the Pt catalysts are used in both the anode and the cathode in DMFCs. However, the Pt NPs showed a variety of sizes, shapes, and morphologies in the different size ranges. The shapes and morphologies of Pt NPs are synthesized in the forms of cube, octahedra, tetrahedra, plate, wire, flower, rod, fiber etc. The certain issues of their particle sizes are studied in the nanosized ranges of 10 nm, 20 nm, 50 nm etc as well as the homogeneous nucleation, growth, and formation of polyhedral nanoparticles. In particular, the range of the Pt NPs of about 10 nm exhibit the highest catalytic activity with the good morphology and shape such as sphere, cube, octahedra, tetrahedra, and polyhedra etc. The high stability and durability of the shape-dependent catalytic activity are needed to be confirmed in the homogeneity of the particle size in a whole nano-system. In the heat treatments of the Pt NPs, the characterization of size, surface, structure, shape, and morphology of Pt NPs need to be preserved to obtain the Pt catalysts of good catalytic activity, stability, and durability for a long time. The new generations of the Pt-Pd core-shell catalysts with the thin Pd shells of less than several nanometers were developed. They showed the excellent catalytic activity. In our recent research, the Pt NPs of less than 10 nm in size were used as the Pt catalysts, which can be used as the standards for any comparison of catalytic activity of new Pt based catalysts. So far, the heat treatment procedures and methods of heat treatments of the as-prepared metal nanoparticles for catalyst engineering in catalysis have not been considered in their very crucial issues of the size, surface, structure, shape, and morphology. The transformations of structure and property of the as-prepared metal nanoparticles in the ranges of 10 nm and 20 nm through heat treatments or sintering of Pt nanoparticles are very critical to nanoparticles as electro-catalysts or catalytic nanosystems for PEMFCs and DMFCs [128]. We need to confirm that Pt nanoparticles with the rough curved surfaces exhibit catalytic activity much better than Pt nanoparticles with flat and smooth surface through the measurement results. The important effects of heat treatment and the removal of poly(vinylpyrrolidone) (PVP) polymer on electrocatalytic activity of polyhedral Pt NPs towards the ORR mechanism have been investigated. The methods of keeping the size, surface structure, internal structure, shape, and morphology were proposed in our catalyst engineering processes, especially heat treatments of the as-prepared PVP-Pt nanoparticles at 300 ^{o}C. The loaded electrodes were carefully dried in air for 3 h at 25 °C and heated with the heating rate of 1 °C/min up to 450 °C in air and a keeping time of 2 h in order to remove any organic species. The pure Pt nanoparticles need to be kept in their good characterization of size and morphology in the size range of 10 nm. We proposed that the polyhedral Pt NPs of around 10 nm can be used the standard Pt catalyst in all research of electrocatalysts in PEMFCs and DMFCs due to their higher concentration of surface steps, kinks, islands, terraces, and corners [129-137]. In our research, Pt-Au NPs were prepared by polyol method, which can be used as Pt-Au catalysts for DMFCs [138]. In particular, Pt-Pd alloy and core-shell NPs were synthesized by polyol method. The core-shell NPs can be used for low temperature PEMFCs and DMFCs for the excellent advantages of reducing the Pt total metal weight. Therefore, it is an economic solution of the suitable use of Pt based core-shell catalysts for next FCs. The new

Pt monolayer Pd-Au catalyst of a core-shell structure with double shells was found in a good stability and activity of ORR. The double shells have an outermost shell of Pt monolayer and a sub-layer shell of Pd-Au alloy [139,140]. In our new results and findings, new evidence of fast enhancement of ORR on the electrode of the new Pt-Pd core-shell catalyst in the size range of 25 nm is clearly observed in a comparison with the prepared Pt catalyst in the size range of 10 nm. Therefore, it is important to use core-shell bimetallic catalysts to increase the ORR rate in the electrode catalyst. For the case of Pt-Pd core-shell catalysts, the fast hydrogen-desorption response and high sensitivity in our results after reaching the stable characterization after only the first CV cycle in a comparison to Pt catalyst. This enables the realization of robust and efficient Pt- or Pd-based core-shell catalysts that are extremely sensitive to the fast hydrogen desorption. Most of the alternative method of improving the hydrogen reaction by the core-shell nanostructures can be clearly realized. During the measurement, the electrodes are swept from -0.2 to 1.0 V with respective to the kinds of saturated standard electrodes. There are the specific regions in the cyclic voltammogram (Figures 4 and 5). They show highly and good catalytic activity and surface kinetics for the case of both the Pt catalysts of 10 nm and the Pt-Pd core-shell catalysts of 25 nm. It is clear that the high heat treatments of our catalyst preparation in H_2/N_2 offer the good characterization of the size, surface, structure, and morphology. Therefore, the highly long-term catalytic activity, stability, and durability in chemical and physical Pt or Pd based catalysts are needed in the catalyst layers of FCs, DMFCs, and PEMFCs. The effects of using a suitable temperature range in heat treatment should be suitable to various FCs, for example the better ORR activity. This depends on the operating temperature of various FCs. They were characterized by the chemical activity occurring at the electrode surface. In the forward sweep, the first region assigned to hydrogen desorption is crucial to confirm catalytic activity of the Pt catalysts. The slow kinetics of hydrogen desorption of the case of Pt catalyst was confirmed in the cell before the stabilization of CV was achieved from the first cycle to the twentieth cycle, and the fast kinetics of hydrogen desorption of the case of Pt-Pd core-shell catalyst. Indeed, the results proved the good desorption and adsorption of hydrogen of both Pt catalysts and Pt-Pd core-shell catalyst as evidences of good catalytic activity of two kinds of important catalysts in the preparation process and heat treatment in figures 4 and 5. To evaluate the catalytic activity of the prepared catalysts, the electrochemical active surface area (ECSA) of the Pt catalyst is calculated to be (10.5 m^2g^{-1}) in a comparison with that of the Pt-Pd core-shell catalysts (27.7 m^2g^{-1}) in our catalytic investigations. Thus, suitable heat treatments or sintering processes of the Pt or Pd based catalysts for obtaining good catalytic activity and stability in the desirable nano and micro structures are very crucial to enhance, and improve the continuous operation of direct chemical-into-electrical energy conversion, high stability and durability of various FCs, PEMFCs and DMFCs for the urgent global challenges of energy and environment. Obviously, the heat treatments can lead to significantly reduce the effects of CO poisoning to the electrodes in PEMFCs and DMFCs. One of the most challenging goals in the heat treatment is to develop successful protocols for keeping the good characterization of the as-preparation Pt NPs such as surface, structure, size, and shape in their nanosized ranges. In our research, the polyol method was used for our synthesis of the Pt and Pd bimetallic nanoparticles with alloy and core-shell structure. In comparison, we also can control the time and temperature of convenient heat-treatments to the pure Pt or Pt/support catalysts for their better catalytic activity. Clearly, the temperature plays an important role in heat treatments for making the better Pt based catalysts. The shape and morphology of polyhedral-like or spherical-like Pt nanoparticles are crucial in the further study of electro-catalytic activity.

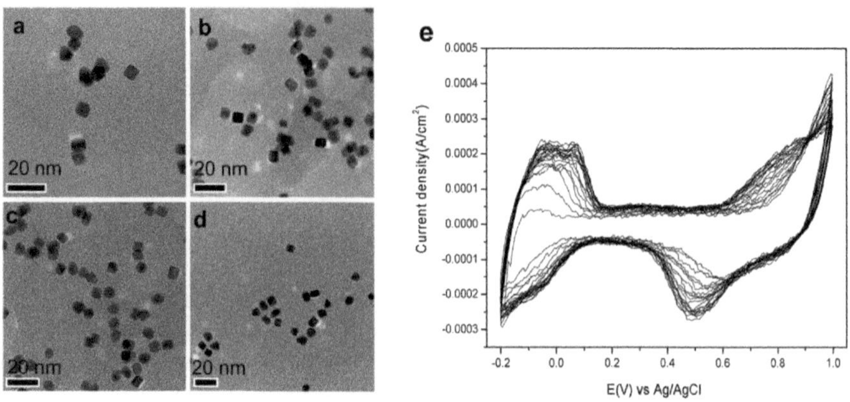

Figure 4. TEM images of the polyhedral Pt nanoparticles. Scale bars: (a)-(d) 20 nm. (e) Cyclic voltammograms of the pure Pt catalysts in 0.5 M H_2SO_4 from -0.2 V to 1.0 V. Reprinted from: Long NV, Ohtaki M, Hien TD, Randy J, Nogami M, Electrochim. Acta. 56:9133-9143. Copyright 2011 with permission from Elsevier [140]

It means that the high heat treatment to the as-prepared Pt nanoparticles for the good catalyst can be performed in the ways with various solvents or pure water during their preparation. However, the desirable characterization of size, shape, morphology, and structure need to be kept for the better catalytic activity. Therefore, scientists need provide more research results of Pt based catalysts with heat treatment effects in electro-catalysis. Thus, new Pt based catalysts, electrodes, and membranes need to be considerably studied for the reduction of their high costs but the high whole performance. The effects of heat treatments on size, shape, surface, and structure of the pure Pt nanoparticles (pure Pt catalyst) or Pt based catalysts for catalytic activity, durability and stability can be intensively analyzed by *in situ* TEM and electrochemical measurements as well as the chemical reactions and their kinetics at the surfaces of the only Pt catalyst or Pt/support catalysts in fuel cells.

In electrochemical measurements, we can use electrolyte solution of using 0.5 M H_2SO_4+1.0 M CH_3OH (a scan rate of 50 mV s^{-1}). Our comparisons of cyclic voltammograms were done between the pure Pt catalysts in the nanosized range of 10 nm, and the pure Pt-Pd core-shell catalysts in the nanosized range of 25 nm in the mixture of 0.5 M H_2SO_4 and 1 M CH_3OH in figure 6. A stable voltammogram was attained after 20 cycles of sweeping between 0 and 1 V. Two oxidation peaks are observed. The good MOR was confirmed in the evidences of the typical peaks at 0.6 V and 0.8 V in the forward sweep, and the other peaks at 0.4 and 0.5 V in

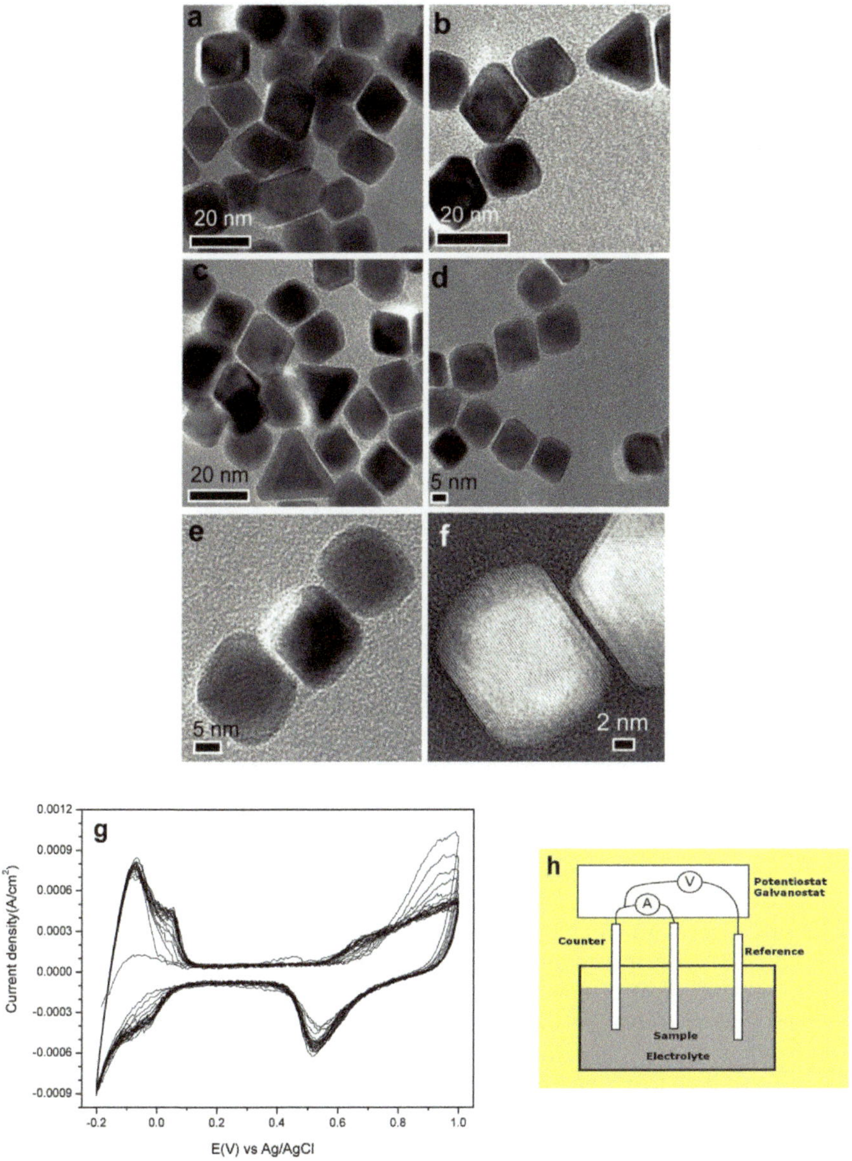

Figure 5. (a)-(f) TEM and HRTEM images of the as-prepared Pt-Pd core-shell. The thin Pd shells protect polyhedral Pt cores. The nucleation and growth of Pd shells are controlled by a chemical synthesis. Scale bars: (a)-(c) 20 nm. (d) 5 nm. (e) 5 nm. (f) 2 nm. (h) Schematic of a standard three-electrode electrochemical cell. Reprinted from: Long NV, Ohtaki M, Hien TD, Randy J, Nogami M, Electrochim. Acta. 56:9133-9143. Copyright 2011 with permission from Elsevier [140]

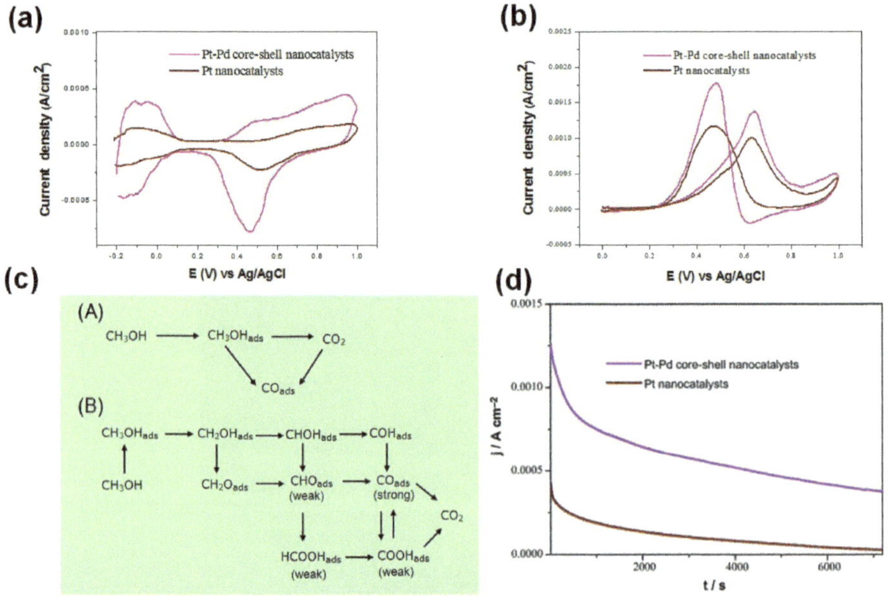

Figure 6. (a) Cyclic voltammogram of Pt catalysts, and Pt-Pd core-shell catalysts on glassy carbon electrode in N_2-bubbled 0.5 M H_2SO_4 electrolyte (scan rate: 50 mV s^{-1}). (b) Cyclic voltammogram towards methanol electro-oxidation of Pt catalyst and Pt-Pd core-shell catalyst. (c) Simple pathways of MOR (A). Possible reaction pathways of MOR (B). (d) Chronoamperometry data of Pt and Pt-Pd catalysts in 0.5 M H_2SO_4+1.0 M CH_3OH and polarization potential about 0.5 V. Reprinted from: Long NV, Ohtaki M, Hien TD, Randy J, Nogami M, Electrochim. Acta. 56:9133-9143. Copyright 2011 with permission from Elsevier [140]

the backward sweep. Our results of the MOR mechanisms showed that the two peaks are directly related to methanol oxidation and the associated intermediates. The fascinating high stable characterization was observed for Pt-Pd core-shell catalysts. They showed a very high initial current about 1.29×10^{-3} A cm^{-2} in comparison with the Pt catalyst of initial current about 4.33×10^{-4} A cm^{-2}. In all our research, the samples with the Pt or Pt based catalysts were carried out in the heat treatment procedures at 300 °C and up to 450 °C. The loaded electrodes were dried in air for 3 h at 25 °C and heated with the heating rate of 1 °C/min up to 450 °C in air and a keeping time of about 2 h for the complete removal of any organic species. After heat treatment, the specific size, shape, structure and morphology of Pt nanoparticles were retained in the good conditions of the Pt based catalyst for the CV measurements [141-143]. New hydrogen absorption was found in the Pt-Pd bimetallic nanoparticles [144]. In our research, we suggested that there are the relationships and dependences of electro-catalytic properties on issues of size, internal structure and surface structure, shape and morphology, and composition of Pt or Pd based catalysts in catalysis, biology, and medicine. The homogeneous properties of size, internal structure and surface structure, shape and morphology, and composition of nanoparticles can create novel and promising properties in catalysis, biology and medicine. According to our research, the effects of the nanostructures can be confirmed in the better catalytic activity. In general, the Frank-van der Merwe (FM), Volmer-Weber (VW), and Stranski-Krastanov (SK) growth modes show the formation of various nanostructures of core-shell nanoparticles [141], such as core-shell bimetallic NPs and oxide-metal NPs [104,141]. It is clear that the formations of novel Pt based core-shell bimetallic nanoparticles the thin-based metal or alloy shells can be controlled in both the FM and SK growth modes for the utilization of Pt metal. The FM overgrowth mode or the epitaxial overgrowth of the thin Pt shells on the metal or oxide cores due to the layer-by-layer mechanism are very crucial to discover most of new Pt based core-shell catalysts for FCs, such as PEMFCs or DMFCs. The core-shell engineered nanoparticles in the nanosized ranges of about 10 nm and 20 nm with the Pt metal shell or the Pt based bimetallic shells or the Pt based multi-metallic shells become an important topic of next research and scientific investigations for FCs. The thin shells can be noble metals such as Pt, Pt-Au, Pt-Pd, Pt based alloys with the use of Ag, Au, Rh, Ru, and Pd metal. The thick cores can be cheap metals such as Ni, Co, Fe, Cu .. and alloys, even and ceramics and glasses. With regards to low-cost issues, we suggest that typical core-shell configurations of bimetallic nanoparticles are proposed in the main goals of designing Pt and Pd based catalysts with the shells of using ultra-low mass of Pt metal or Pt-noble (Pd, Ru, Rh, Au, Ag) bimetals [140-143, 190]. Thus, a variety of the core-shell configurations with the thin Pt shell or the thin Pt based bimetallic shells controllably created in chemical and physical engineering is one of the best ways of the realization of large-scale commercialization and development of FCs, such as PEMFCs and DMFCs. We suggest that noble Au NPs can be used the good core for the Pt shells for DMFCs and PEMFCs. In most cases, Pt-based bimetallic NPs were supported on carbon nanomaterials such as MWCNTs for the catalytic enhancement in the operations of various DMFCs due to their large surface area and good conductivity [145,146]. In addition, the Co-Pt core-shell nanoparticles can be used as potential catalysts in cathode of PEMFCS because of their high stability and durability [147-149]. In many efforts, the catalysts such as Pt-Pd, Pt-Ru, Pt-Rh and Pt-Sn/C exhibiting an improved performance in MOR as anode materials were prepared by ultrasonic method [150,151]. Importantly, the catalytic properties of well-characterized Ru-Pt core-shell NPs were demonstrated for preferential CO oxidation [152]. The metal-oxide nanostructures of Pd or Pt metals and TiO_2 or Zr_2O_3 oxides by self-assembly are proposed for catalysis and FCs. The core-shell nanoparticles and nanostructures will offer potentially promising applications in homogeneous and heterogeneous catalysis, biology and medicine [153,154]. The recent developments of polymer electrolyte membranes for FCs were discussed for improving the long-term stability [155]. The new methanol-tolerant catalysts with the use of Pd nano cubes exhibiting high electro-activity for ORR in the 0.5 M H_2SO_4 solution were confirmed [156,157]. Thus far, the catalysts with the use of the Pt-Pd nanoparticles or the Pt-Pd nanostructures of various size, internal structure, surface structure, shape, morphology, and composition by various preparation methods have been intensively studied for practical applications in various FCs, PEMFCs, and DMFCs. There are the very huge demands of high-surface area catalysts of Pt-Pd NPs supported on carbon nanomaterials that have been used in various FCs, PEMFCs and PEMCs [158-170,188,189]. The nano-porous SiO_2 solids with Pt or Pd metal nanoparticles were also used [171-172], and they offered higher electro-oxidation current densities for both methanol and ethanol.

5.1. Heat treatment

Before the TEM and HRTEM measurements, copper grids with the prepared nanoparticles were annealed in the heat treatment at 300 °C for 4 h, 6 h or even a half day. The evidences and results of the characterization of size, structure, shape, morphology with the only Pt composition by TEM and HRTEM show the uniform polyhedral Pt nanoparticles synthesized by polyol method with the controlled size in the nanosized range of 10 nm and 20 nm, and sharp shapes and morphologies through an introduction of low contents of $AgNO_3$ as size, structure, morphology and shape-controlling reagent and the gradual addition of the precursors of PVP and H_2PtCl_6 in the suitable volume ratio. In the characterizations of homogeneous polyhedral morphology and shape, the appearances of the main sharp cubic, octahedral, and tetrahedral shapes of Pt NPs in the controlled growth of (100), (110), and (111) selective surfaces are very good for their applications in catalytic activity, e.g. the ORR and MOR chemical reactions because the polyhedral Pt nanoparticles have at a higher concentration of surface steps, kinks, islands, terraces, and corners [10-12,188]. In addition, the TEM and HRTEM images of Pt NPs by polyol method show interestingly important phenomena of particle-particle surface attachment, self-aggregation, and assembly in the case of the use of the pure Pt NPs or the PVP-Pt NPs. Most of Pt NPs showed various large morphologies and shape after heat treatment at high temperature from 20 to 60 nm for the case of the PVP-Pt NPs [128-130]. The clear overgrowths and structural transformation in the nanosized range of polyhedral Pt nanoparticles annealed at 300 °C for 4 h were observed. Clearly, the morphology of PVP-Pt nanoparticles was significantly changed by heat treatment. During their synthesis, PVP polymer is used to stabilize and control the size and morphology of Pt nanoparticles against their aggregation. In addition, the amount of PVP polymer can bind the nanoparticle surface after the catalyst synthesis. Therefore, PVP or other polymers for the protections of the as-prepared nanoparticles should be removed in the minimal content before the catalytic reactions. In our research, PVP polymer plays an important role to stabilize their morphology and size of these Pt nanoparticles. Despite the fact that these Pt nanoparticles were put on copper grids, they still have their interfacial interactions to their particle-particle surface attachments, aggregation, and self-assembly leading to form the larger and irregular Pt particles or the larger particles in the forms of hetero-morphologies and hetero-nanostructures in the final formation by the heat treatment. It is known that Pt nanostructures with the issues of size, shape and morphology under heat treatment procedures become important to further investigation in the confirmation of catalytic activity in catalysis. The critical issues of size, shape and morphology of colloidal nanoparticles need to be intensively studied in different media (condense, liquid, and gas) for the certain confirmations of high and long-term stability, high durability, and safety in their practical applications in catalysis, biology and medicine. In our experimental methods, in order to obtain the pure Pt catalyst for the standard catalyst of the standard nanosized range of 10 nm in investigations of catalytic activity, PVP polymer of the as-prepared product of PVP-Pt nanoparticles was removed. Therefore, the pure Pt nanoparticles for electrochemical measurements can be obtained by using centrifuge. The resultant solution of as-prepared nanoparticle was washed by using the suitable mixtures of acetone and followed by centrifugation at 5,000 rpm up to 15,000 rpm. Next, the black solid product was re-dispersed in the ethanol/hexane mixtures with a suitable volume rate. The resultant mixture was centrifuged in order to obtain the fresh product. Experimentally, the procedures of washing and clean Pt nanoparticles in the mixture of ethanol and hexane were done many times. After washing PVP polymer and contaminations in the as-prepared product, the nanoparticles were dispersed in milli-Q water in order to achieve the fixed density of colloidal Pt nanoparticles at 1 mg/mL with the aid of ICPS analyzer. The working electrode was a glassy carbon rod (RA-5, Tokai Carbon Co., Ltd.) with a diameter of 5.2 mm. Then, the electrode surface was cleaned and activated by using a kind of polishing-cloth (Buehler Textmet) with alumina slurry (Aldrich, particle size of 50 nm), followed by washing any contaminations with milli-Q water. The step-by-step procedures were repeated until the surface looked like a mirror. Then, a fixed weight of some μg (e.g. 5 μg or 10 μg) of the Pt loading was set onto the surface of the polished electrode for the electrochemical measurements. The loaded electrodes were dried in air for 3 h at 25 °C and heated with the heating rate of 1 °C/min up to 450 °C or 723 K in air and a keeping time of 2 h in order to remove organic species [128-130]. We used a very slow heating rate to avoid the serious problem of sintering of the pure nanoparticles so that the layer catalyst is the nanoparticles. The electrodes were allowed to cool normally and then exposed into the flow of the mixture of H_2/N_2 gases (20%,80%) at 100 °C for 3 h to reduce the existence of PtO and ensure a pristine catalyst surface. In order to improve the mechanical stability of electrode surfaces, Nafion® solution, e.g. 10 μL of 5 wt.%, was added onto the electrode and followed by drying in air for a long time, e.g. overnight, before the electrochemical measurements. The cyclic voltammetry experiment was performed at room temperature using a typical setup of three-electrode electrochemical system in figure 5(h) connected to Potentiostat (SI-1287 Electrochemical Interface, Solartron). The cell was a 50-mL glass vial, which was carefully treated with the mixture of H_2SO_4 and HNO_3, and then washed generously with milli-Q water. A leak-free AgCl/Ag/NaCl electrode (RE-1B,ALS) served as the reference and all the potentials were reported vs. Ag/AgCl. The counter electrode was a Pt coil (002234,ALS) [128-130, 140-143]. The electrolyte solution was bubbled with N_2 gas for 30 min before every measurement. This N_2 blanket was kept during the actual course of potential sweeping. For the base voltammetry, the electrolyte was a solution of 0.1 M $HClO_4$ that was diluted from 70% concentrated solution (Aldrich) using milli-Q water. Additionally, 0.5 M H_2SO4, H_2SO_4 and 1 M CH_3OH are prepared for electrochemical measurements. The potential window between -0.2 to 1.0 V with a sweep rate of 50 mV/s was used. For the methanol oxidation, the electrolyte was added with 1.0 M methanol in milli-Q water. The system measurements were cycled until the stable voltammograms were achieved. The electrochemical surface area (ECA) or ECSA was estimated by considering the area under the curve in the hydrogen desorption region of the forward scan and using 0.21 mC/cm^2 for the monolayer of hydrogen adsorbed on the surface of Pt catalyst [13,174]. The CO poisoning issues or phenomena of CO-stripping voltammetry of the Pt catalysts cause a significant decrease in the overall efficiency of DMFCs. In all CV research, we did not observe CO poisoning in the CV data. The evidences proved that heat treatments are very important to engineer the best Pt catalyst or the better Pt based catalyst for FCs, PEMFCs, and DMFCs. In the developments of FCs, PEMFCs, DMFCs, novel Pt and Pd based core-shell bimetallic nanosystems or Pt and Pd core-shell catalysts will be the next catalysts with use of the thin Pt or Pt-Pd shells on the thick cores (metals, oxides, glasses, and ceramics) in the nanosized ranges of 10 nm, 20 nm and 30 nm. The thin Pt metal shells or Pt based noble bimetal (noble Pt-Pd, Pt-Ir, Pt-Ru, Pt-Rh, Pt-Au, Pt-Ag) thin shells can be tuned to be several nanometers on the metal or alloy cores in the ranges of 10 nm, 20 nm and 30 nm, and so on through the controlled synthesis and preparation processes. Unfortunately, quantum-size, structural and surface effects of metal nanoparticles around 10 nm in electro-catalysis are not fully understood in their chemical and structural changes.

5.2. Stability and durability

With respect to the heat treatment, electrocatalytic characterizations of the pure catalyst with our as-prepare products of Pt nanoparticles showed highly good quality, long-term stability and durability. Thus, the significant effects of time and temperature during heat treatment for the pure Pt catalyst or Pt/support catalysts are crucial. In our research, the findings and results involve in the cyclic voltammograms of polyhedral Pt nanoparticles with the different removal of PVP polymer acquired at 50 mV/s in 0.1 M $HClO_4$ solution (or in 0.5 H_2SO_4 or in 0.5 H_2SO_4 and 1 M CH_3OH). The voltammograms data proved the typical shape for the base voltammetry for the high activity, stability and durability of our catalysts used [98,99,129]. The very high activity, durability and stability in FCs described were achieved up to 200,000 cycles with the use of core-shell catalysts with the Pt-monolayers shells in the cathode as important scientific evidences [99]. In our interesting research, the structural effects of low and high-index planes of the fcc structure of Pt catalyst were confirmed in electrochemical measurements, typical low-index planes of (111), (110), and (100) or more high index planes (hkl). The Volmer-Tafel and Volmer-Heyrovsky mechanisms are observed in hydrogen reactions. Most of ORR evidences showed good oxygen reduction. Therefore, our results showed the important evidences of the catalytic activity of the Pt catalyst or Pt based catalysts with alloy structures or core-shell structures. The ECA of the Pt catalyst the use of the Pt nanoparticles washed and heated showed the best catalytic activity (ECA=10.53 m^2/g), better than Pt catalyst the use of the Pt nanoparticles heated only-Pt nanoparticles (8.56 m^2/g), and better than Pt catalyst the use of the Pt nanoparticles of washed-only Pt nanoparticles (6.75 m^2/g). The values proved that the high stability and durability of the catalyst preparation, processes of heat treatment, methods of washing and clean the as-prepared Pt nanoparticles for the pure Pt catalyst. However, we should find a good process of heat treatment to avoid particle sintering in the electrodes at high temperature more than 450 °C and keep the good characterizations of size, internal structure, surface structure, shape and morphology of the Pt nanoparticles or the Pt based nanoparticles for the pure Pt based catalysts for FCs, PEMFCs and DMFCs. Therefore, we propose that homogeneous polyhedral Pt nanoparticles under control in the size, shape and morphology in the nanosized range of 10 nm should be used the standard catalyst in the scientific investigations of catalysis for FCs, PEMFCs and DMFCs. Our electrochemical measurements were performed by the use of the pure Pt catalysts of polyhedral Pt nanoparticles or the Pt based catalysts in alloy and core-shell structure in the mixture of 0.1 M $HClO_4$/1 M methanol (or 0.5 H_2SO_4/1 M methanol). In these CV measurements, the stable voltammograms are attained after about 10-20 cycles of sweeping between -0.2 to 1.0 V potential range. The typical two oxidation peaks are observed in the evidences of methanol oxidation. The first one is between 0.6 and 0.7 V in the forward scan, and the other at around 0.4 V and 0.5 V in the reverse scan [129]. The two peaks showed the good methanol oxidation and its associated intermediate species. Our results of methanol oxidation with the use of the pure Pt catalyst prepared are agreement with other scientific reports. The peak current density in the forward scan serves as benchmark for the catalytic activity of Pt nanoparticles during methanol dehydrogenation. For the prepared catalyst samples, its values are 7.62×10^{-4} A/cm^2 (washed-only samples), 8.75×10^{-4} (heated-only samples), and 9.90×10^{-4} (washed and heated samples), respectively [129]. Therefore, the as-prepared Pt nanoparticles should be washed with organic solvents before heating them at a specific temperature. It is known that PVP can be polymer that only protects the as-prepared Pt nanoparticles in the solution products that seriously decrease the catalytic activity of Pt nanoparticles in MOR. Therefore, PVP or other polymers should be completely removed in the centrifugation processes by centrifuge systems for the pure Pt and Pd based catalyst before high heat treatments for to enhance catalytic activity of Pt nanoparticles for FCs, PEMFCs, and DMFCs, primarily towards methanol electro-oxidation (MOR) in DMFCs.

6. Prospects and conclusion

In the development and commercialization of various FCs, PEMFCs, and DMFCs, the next Pt and Pt based bimetal nanoparticles of internal structure, surface structure, shape and morphology in the nanosized ranges, typically 10 nm, 20 nm, and 30 nm are the best potential catalysts for a significant reduction of the very high costs of various FCs, PEMFCs, and DMFCs. The aim of this chapter is to demonstrate that the urgent demands of studying and synthesizing for the next Pt based catalysts of highly long-term stability and durability are crucial to create low-cost products of PEMFCs and DMFCs. The low cost of low and high temperature FCs, PEMFCs and DMFCs is an extremely important factor for large-scale commercialization. The low-cost Pt and Pd based catalysts can be the good solution to the big challenges of the costs of low and high temperature FCs, PEMFCs, and DMFCs. The modifications and improvements of the catalyst layer materials, and their synthesis become important factors. It should be stressed that next low and high temperature FCs for direct conversion from chemical energy into electrical energy will be very crucial to offer potential applications in portable power supply for mobile use or portable devices or transportation vehicles. Thus, large-efficiency FCs for direct energy conversion via chemical reaction into electricity can be commercially realized in near future.

Author details

Nguyen Viet Long

Department of Molecular and Material Sciences, Interdisciplinary Graduate School of Engineering Sciences, Kyushu University, 6-1 Kasugakouen, Kasuga, Fukuoka, Japan

Department of Education and Training, Posts and Telecommunications Institute of Technology, Nguyen Trai, Ha Dong, Hanoi, Vietnam

Department of Materials Science and Engineering, Nagoya Institute of Technology, Gokiso-cho, Showa-ku, Nagoya, Japan

Laboratory for Nanotechnology, Ho Chi Minh Vietnam National University, Linh-Trung, Thu-Duc, Ho Chi Minh, Vietnam

Cao Minh Thi

Ho Chi Minh City University of Technology, Dien-Bien-Phu, Binh-Thach, Ho Chi Minh City, Vietnam

Masayuki Nogami

Department of Materials Science and Engineering, Nagoya Institute of Technology, Gokiso-cho, Showa-ku, Nagoya, Japan

Michitaka Ohtaki

Department of Molecular and Material Sciences, Interdisciplinary Graduate School of Engineering Sciences, Kyushu University, 6-1 Kasugakouen, Kasuga, Fukuoka, Japan

Acknowledgement

This work was supported by NAFOSTED Grant No. 104.03-2011.33, 2011. This work was supported by Laboratory for nanotechnology, Vietnam National University, Ho Chi Minh in Viet Nam. We greatly thank Kyushu University in the Global COE Program, Novel Carbon Resource Sciences for the financial support in our research and development of science and nanotechnology in Kyushu University, Japan. We greatly thank Nagoya Institute of Technology for kind supports in Japan.

7. References

[1] Mehta V, Cooper JS (2003) Review and analysis of PEM fuel cell design and manufacturing. J. Power Sources. 114:32-53.

[2] Wang Y, Chen KS, Mishler J, Cho SC, Adroher XC (2011) A review of polymer electrolyte membrane fuel cells: Technology, applications, and needs on fundamental research. Appl. Energy. 88:981-1007.

[3] Gewirth AA, Thorum MS (2010) Electroreduction of dioxygen for fuel-cell applications: materials and challenges. Inorg. Chem. 49:3557-3566.

[4] Borup R, Meyers J, Pivovar B,Kim YS, Mukundan R, Garland N, Myers D, Wilson M, Garzon F, Wood D, Zelenay P, More K, Stroh K, Zawodzinski T, Boncella J, McGrath JE,

Inaba M, Miyatake K, Hori M, Ota K, Ogumi Z, Miyata S, Nishikata A, Siroma Z, Uchimoto Y, Yasuda K, Kimijima K, Iwashita N (2007) Scientific Aspects of Polymer Electrolyte Fuel Cell Durability and Degradation. Chem. Rev. 107:3899-4435.

[5] Bullinger H (2009) Technology Guide: Principles-Applications-Trends. In Bullinger H, editor. Fuel cells and hydrogen technology. Germany: Springer. pp.368-373.

[6] Spendelow JS, Papageorgopoulos DC (2011) Progress in PEMFC MEA Component R&D at the DOE Fuel Cell Technologies Program. Fuel Cells. 11:775-786.

[7] Litster S, McLean G (2004) PEM fuel cell electrodes. J. Power Sources. 130(1-2):61-76.

[8] Okada O, Yokoyama K (2001) Development of Polymer Electrolyte Fuel Cell Cogeneration Systems for Residential Applications. Fuel Cells. 1:72-77.

[9] Aricò AS, Srinivasan S, Antonucci V (2001) DMFCs: From Fundamental Aspects to Technology Development. Fuel Cells. 1:133-161.

[10] Markoví NM, Ross Jr PN (2002) Surface science studies of model fuel cell electrocatalysts. Surf. Sci. Rep. 45(4-6):117-229.

[11] Schmidt TJ, Stamenkovic V, Arenz M, Markovic NM, Ross PN Jr (2002) Oxygen electrocatalysis in alkaline electrolyte: Pt(hkl), Au(hkl) and the effect of Pd-modification. Electrochim. Acta. 47:3765-3776.

[12] Somorjai GA, Contreras AM, Montano M, Rioux RM (2006) Clusters, surfaces, and catalysis. PNAS. 103: 10577-10583.

[13] Basu S (2007) Recent Trends in Fuel Cell Science and Technology. New Delhi, India: Anamaya Publishers 375 p.

[14] Shao-Horn Y, Sheng WC, Chen S, Ferreira PJ, Holby EF, Morgan D (2007) Instability of supported platinum nanoparticles in low-temperature fuel cells. Top. Catal. 46(3-4):285-305.

[15] Sánchez-Sánchez CM, Solla-Gullón J, Vidal-lglesias FJ, Aldaz A, Montiel V, Herrero E (2010) Imaging structure sensitive catalysis on different shape-controlled platinum nanoparticles. J. Am. Chem. Soc. 132:5622-5624.

[16] Chen J, Menning C, Zellner M (2008) Monolayer bimetallic surfaces: Experimental and theoretical studies of trends in electronic and chemical properties.Surf. Sci. Rep. 63(5):201-254.

[17] Antolini E (2009) Palladium in fuel cell catalysis. Energy Environ. Sci. 2(9):915-931.

[18] Cheong S, Watt JD, Tilley RD (2010) Shape control of platinum and palladium nanoparticles for catalysis. Nanoscale. 2:2045-2053.

[19] Lee K, Kim M, Kim H (2010) Catalytic nanoparticles being facet-controlled. J. Mater. Chem. 20:3791-3798.

[20] Solla-Gullón J, Vidal-Iglesias FJ, Feliu JM (2011) Shape dependent electrocatalysis. Annu. Rep. Prog. Chem., Sect. C: Phys. Chem.107:263-297.

[21] Subhramannia M, Pillai VK (2008) Shape-dependent electrocatalytic activity of platinum nanostructures. J. Mater. Chem. 18:5858-5870.

[22] Ferrando R, Jellinek J, Johnston RL (2008) Nanoalloys: From theory to applications of alloy clusters and nanoparticles. Chem. Rev. 108(3):845-910.

[23] Service RF (2007) Platinum in fuel cells gets a helping hand. Science. 315:172.

[24] Gellman AJ, Shukla N (2009) Nanocatalysis: More than speed. Nat. Mater. 8:87-88.

[25] Rao CV, Viswanathan B (2010) Monodispersed Platinum Nanoparticle Supported Carbon Electrodes for Hydrogen Oxidation and Oxygen Reduction in Proton Exchange Membrane Fuel Cells. J Phys Chem C.114(18):8661-8667.

[26] Qiao Y, Li CM (2011) Nanostructured catalysts in fuel cells. J. Mater. Chem. 21:4027-4036.

[27] Takasu Y, Fujiwara T, Murakami Y, Sasaki K, Oguri M, Asaki T, Sugimoto W (2000) Effect of structure of carbon-supported PtRu electrocatalysts on the electrochemical oxidation of methanol. J. Electrochem. Soc. 147(12):4421-4427.

[28] Burda C, Chen X, Narayanan R, El-Sayed MA (2005) Chemistry and properties of nanocrystals of different shapes. Chem. Rev. 105(4):1025-1102.

[29] Olenin AY, Lisichkin GV (2011) Preparation and bulk and surface structural dynamics of metallic nanoparticles in condensed media. Russ. Chem. Rev. 80(7):605-630.

[30] Toshima N, Yonezawa T (1998) Bimetallic nanoparticles-novel materials for chemical and physical applications. New. J. Chem. 22:1179-1201.

[31] Moshfegh AZ (2009) Nanoparticle catalysts. J. Phys. D: Appl. Phys. 42:233001-233030.

[32] Rao CN, Ramakrishna Matte HS, Voggu R, Govindaraj A (2012) Recent progress in the synthesis of inorganic nanoparticles. Dalton Trans. 41(17):5089-120.

[33] Qiang Q, Ostafin AE (2004) Metal nanoparticles in catalysis. In: Nalwa HS, editor. Encyclopedia of nanoscience and nanotechnology. American Scientific Publishers. Vol. 5. pp.475-505.

[34] Yang J, Sargent E, Kelley S, Ying JY (2009) A general phase-transfer protocol for metal ions and its application in nanocrystal synthesis. Nat. Mater. 8:683-689.

[35] Baghbanzadeh M, Carbone L, Cozzoli PD, Kappe CO (2011) Microwave-assisted synthesis of colloidal inorganic nanocrystals. Angew. Chem. Inter. Ed. 50(48):11312-11359.

[36] Pollet BG. The use of ultrasound for the fabrication of fuel cell materials (2010) Int. J. Hydrogen Energy. 35(21):11986-12004.

[37] Cuenya BR (2010) Synthesis and catalytic properties of metalnanoparticles: Size, shape, support, composition, and oxidation state effects. Thin Solid Films. 518(12):3127-3150.

[38] Guo S, Wang E (2011) Noble metal nanomaterials: Controllable synthesis and application in fuel cells and analytical sensors. Nano Today. 6(3):240-264.

[39] Tao A, Habas S, Yang P (2008) Shape control of colloidal metal nanocrystals. Small. 4:310-325.

[40] Habas S, Lee H, Radmilovic V, Somorjai G, Yang P (2007) Shaping binary metal nanocrystals through epitaxial seeded growth. Nat. Mater. 6:692-697

[41] Long NV, Thi CM, Nogami M, Ohtaki M (2012) Novel issues of morphology, size, and structure of Pt nanoparticles in chemical engineering: surface attachment, aggregation or agglomeration, assembly, and structural changes. New J. Chem. Advance Article. DOI: 10.1039/C2NJ40027H.

[42] Salzemann C, Petit C (2012) Influence of Hydrogen on the Morphology of Platinum and Palladium Nanocrystals. Langmuir. 28:4835-4841.

[43] Lim B, Xia Y (2011) Metal Nanocrystals with Highly Branched Morphologies. Angew. Chem. Int. Ed. 50:76-85.

[44] Serov A, Kwak C (2009) Review of non-platinum anode catalysts for DMFC and PEMFC application. Appl. Catal., B. 90(3-4):313-320.

[45] Ralph TR, Hogarth MP (2002) Catalysis for Low Temperature Fuel Cells - Part I. Platinum Metals Rev. 46:3-14.

[46] Ralph TR, Hogarth MP (2002) Catalysis for Low Temperature Fuel Cells - Part II. Platinum Metals Rev. 46:117-135.

[47] Ralph TR, Hogarth MP (2002) Catalysis for Low Temperature Fuel Cells - Part III. Platinum Metals Rev. 46:146-164.

[48] Sharma S, Pollet BG (2012) Support materials for PEMFC and DMFC electrocatalysts-A review. J. Power Sources. 208:96-119.

[49] Shekhawat D, Spivey JJ, Berry DA (2011) Fuel cells: Technologies for fuel processing. Amsterdam: Elsevier BV. 555 p.

[50] Thiam HS, Daud WRW, Kamarudin SK, Mohammad AB, Kadhum AAH, Loh KS, Majlan EH (2011) Overview on nanostructured membrane in fuel cell applications. Int. J. Hydrogen Energy. 36:3187-3205.

[51] Dupuis A (2011) Proton exchange membranes for fuel cells operated at medium temperatures: Materials and experimental techniques. Prog. Mater. Sci. 56:289-327.

[52] Peighambardousta SJ, Rowshanzamira S, Amjadia M (2010) Review of the proton exchange membranes for fuelcell applications. Int. J. Hydrogen Energy. 35:9349-9384.

[53] Balgis R, Anilkumar GM, Sago S, Ogi T, Okuyama K (2012) Nanostructured design of electrocatalyst support materials for high-performance PEM fuel cell application. J. Power Sources. 203:26-33.

[54] Zhang J (2011) Recent advances in cathode electrocatalysts for PEM fuel cells. Front. Energy. 5:137-148.

[55] Bing Y, Liu H, Zhang L, Ghosh D, Zhang J (2010) Nanostructured Pt-alloy electrocatalysts for PEM fuel cell oxygen reduction reaction. Chem. Soc. Rev. 39:2184-2202.

[56] Díaz V, Ohanian M, Zinola CF (2010) Kinetics of methanol electrooxidation on Pt/C and PtRu/C catalysts. Int. J. Hydrogen Energy. 35(19):10539-10546.

[57] Shao Y, Liu J, Wang Y, Lin Y (2009) Novel catalyst support materials for PEM fuel cells: current status and future prospects. J. Mater. Chem. 19:46-59.

[58] McConnell VP (2009) High-temperature PEM fuel cells: Hotter, simpler, cheaper. Fuel Cells Bull. 2009(12):12-16.

[59] Hasché F, Oezaslan M, Strasser P (2012) Activity, Structure and Degradation of Dealloyed $PtNi_3$ Nanoparticle Electrocatalyst for the Oxygen Reduction Reaction in PEMFC. J. Electrochem. Soc. 159(1):B25-B34.

[60] Oezaslan M, Hasché F, Strasser P (2012) Oxygen electroreduction on $PtCo_3$, PtCo and Pt_3Co alloy nanoparticles for alkaline and acidic PEM fuel cells. J. Electrochem. Soc. 159:B394-B405.

[61] Oezaslan M, Hasché F, Strasser P (2012) $PtCu_3$, PtCu and Pt_3Cu Alloy Nanoparticle Electrocatalysts for Oxygen Reduction Reaction in Alkaline and Acidic Media. J. Electrochem. Soc. 159(4):B444-B454.

[62] Oezaslan M, Heggen M, Strasser P (2012) Size-Dependent Morphology of Dealloyed Bimetallic Catalysts: Linking the Nano to the Macro Scale. J. Am. Chem. Soc. 134(1):514-524.

[63] Sung Y, Hwang J, Chung JS (2011) Characterization and activity correlations of Pt bimetallic catalysts for low temperature fuel cells. Int. J. Hydrogen Energy. 36(6):4007-4014.

[64] Zhong C, Luo J, Fang B, Wanjala BN, Njoki PN, Loukrakpam R (2010) Nanostructured catalysts in fuel cells. Nanotechnology. 21:2010.

[65] Loukrakpam R, Luo J, He T, Chen Y, Xu Z, Njoki PN, Wanjala BN, Fang B, Mott D, Yin J, Klar J, Powell B, Zhong C (2011) Nanoengineered PtCo and PtNi Catalysts for Oxygen Reduction Reaction: An Assessment of the Structural and Electrocatalytic Properties. J. Phys. Chem. C. 115:1682-1694.

[66] Wanjala BN, Loukrakpam R, Luo J, Njoki PN, Mott D, Zhong C, Shao M, Protsailo L, Kawamura T (2010) Thermal Treatment of PtNiCo Electrocatalysts: Effects of Nanoscale Strain and Structure on the Activity and Stability for the Oxygen Reduction Reaction. J. Phys. Chem. C. 114:17580-17590.

[67] Mayrhofer K, Strmcnik D, Blizanac B, Stamenkovic V, Arenz M, Markovic N (2008) Measurement of oxygen reduction activities via the rotating disc electrode method: From Pt model surfaces to carbon-supported high surface area catalysts. Electrochim. Acta. 53:3181-3188.

[68] Othman R, Dicks AL, Zhu (2012) Non precious metal catalysts for the PEM fuel cell cathode. J. Power Sources. 130(1-2):61-76.

[69] Xu W, Zhou X, Liu C, Xing W, Lu T (2007) The real role of carbon in Pt/C catalysts for oxygen reduction reaction. Electrochem. Commun. 9(5):1002-1006.

[70] Antolini E (2007) Platinum-based ternary catalysts for low temperature fuel cells. Part I. Preparation methods and structural characteristics. Appl. Catal., B. 74(3-4):324-336.

[71] Antolini E (2007) Platinum-based ternary catalysts for low temperature fuel cells. Part II. Electrochemical properties. Appl. Catal., B. 74(3-4):337-350.

[72] Antolini E (2009) Carbon supports for low-temperature fuel cell catalysts. Appl. Catal., B. 88 (1-2):1-24.

[73] Antolini E, Perez J (2011) The renaissance of unsupported nanostructured catalysts for low-temperature fuel cells: from the size to the shape of metal nanostructures. J. Mater. Sci. 46:4435-4457.

[74] Wang X, Wang S (2011) Nanomaterials for proton exchange membrane fuel cells. In: Zang L, editor. Energy efficiency and renewable energy through nanotechnology. London: Springer. pp. 393-424.
[75] Apanel G, Johnson E (2004) Direct methanol fuel cells - Ready to go commercial?. Fuel Cells Bull. 11:12-17.
[76] Zhao X, Yin M, Ma L, Liang L, Liu C, Liao J, Lu T, Xing W (2011) Recent advances in catalysts for direct methanol fuel cells. Energy Environ. Sci. 4:2736-2753.
[77] Hamnett A (1997) Mechanism and electrocatalysis in the direct methanol fuel cell. Catal. Today. 38(4):445-457.
[78] Basri S, Kamarudin SK, Daud WRW, Yaakub Z (2010) Nanocatalyst for direct methanol fuel cell (DMFC). Int. J. Hydrogen Energy. 35:7957-7970.
[79] Álvarez GF, Mamlouk M, Scott K (2011) An Investigation of Palladium Oxygen Reduction Catalysts for the Direct Methanol Fuel Cell. Int. J. Electrochem. 1-12.
[80] Dillon R, Srinivasan S, Aricò AS, Antonucci V (2004) International activities in DMFC R&D: Status of technologies and potential applications. J. Power Sources. 127(1-2):112-126.
[81] Furuya N, Koide S (1989) Hydrogen adsorption on platinum single-crystal surfaces. Surf. Sci. 220:18-28.
[82] Gómez R, Fernández-Vega A, Feliu JM, Aldaz A (1993) Hydrogen evolution on Pt single crystal surfaces. Effects of irreversibly adsorbed bismuth and antimony on hydrogen adsorption and evolution on Pt(100). J. Phys. Chem. 97:4769-4776.
[83] Ogi T, Honda R, Tamaoki K, Saitoh N, Konishi Y (2011) Direct room-temperature synthesis of a highly dispersed Pd nanoparticle catalyst and its electrical properties in a fuel cell. Powder Technol. 205:143-148.
[84] Kim C, Lee H (2009) Change in the catalytic reactivity of Pt nanocubes in the presence of different surface-capping agents. Catal. Commun. 10(9):1305-1309.
[85] Teranishi T, Hosoe M, Tanaka T, Miyake M (1999) Size control of monodispersed Pt nanoparticles and their 2D organization by electrophoretic deposition. J. Phys. Chem. B. 103(19):3818-3827.
[86] Song H, Kim F, Connor S, Somorjai SA, Yang P (2005) Pt nanocrystals: shape control and Langmuir-Blodgett monolayer formation. J. Phys. Chem. B. 109:188-193.
[87] Koebel MM, Jones LC, Somorjai GA (2008) Preparation of size-tunable, highly monodisperse PVP-protected Pt-nanoparticles by seed-mediated growth. J. Nanopart. Res. 10(6):1063-1069.
[88] Lim B, Jiang M, Tao J, Camargo P, Zhu Y, Xia Y (2009) Shape-controlled synthesis of Pd nanocrystals in aqueous solutions. Adv. Funct. Mater. 19(2):189-200.
[89] Antolini E (2009) Palladium in fuel cell catalysis. Energy Environ. Sci. 2(9):915-931.
[90] Adams BD, Chen A (2011) The role of palladium in a hydrogen economy. Mater. Today. 14(6):282-289.
[91] Shao M (2011) Palladium-based electrocatalysts for hydrogen oxidation and oxygen reduction reactions. J. Power Sources. 196(5):2433-2444.

[92] Kim I, Han OH, Chae SA, Paik Y, Kwon S, Lee K, Sung Y, Kim H (2011) Catalytic Reactions in Direct Ethanol Fuel Cells. Angew. Chem. Int. Ed. 50:2270-2274.
[93] Peng Z, Yang H (2009) Designer platinum nanoparticles: Control of shape, composition in alloy, nanostructure and electrocatalytic property. Nano Today. 4(2):143-164.
[94] Moshfegh AZ (2009) Nanoparticle catalysts. J. Phys. D. Appl. Phys. 42:233001.
[95] Jia C, Schüth F (2011) Colloidal metal nanoparticles as a component of designed catalyst. Phys. Chem. Chem. Phys. 13:2457-2487.
[96] Bianchini C, Shen PK (2009) Palladium-based electrocatalysts for alcohol oxidation in half cells and in direct alcohol fuel cells. Chem. Rev. 109:4183-206.
[97] Kamarudin SK, Hashima N (2012) Materials, morphologies and structures of MEAs in DMFCs. Renewable Sustainable Energy Rev. 16:2494-2515.
[98] Adzic R, Zhang J, Sasaki K, Vukmirovic M, Shao M, Wang J, Nilekar A, Mavrikakis M, Valerio JA, Uribe F (2007) Platinum monolayer fuel cell electrocatalysts. Top. Catal. 46:249-262.
[99] Sasaki K, Naohara H, Cai Y, Choi YM, Liu P, Vukmirovic MB, Wang JX, Adzic RR (2010) Core-Protected Platinum Monolayer Shell High-Stability Electrocatalysts for Fuel-Cell Cathodes. Angew. Chem. Int. Ed. 49:8602-8607.
[100] Cai Y, Adzic RR (2011) Platinum Monolayer Electrocatalysts for the Oxygen Reduction Reaction: Improvements Induced by Surface and Subsurface Modifications of Cores. Adv. Phys. Chem. 2011:1-16.
[101] Chen A, Holt-Hindle P (2010) Platinum-based nanostructured materials: synthesis, properties, and applications. Chem. Rev. 110:3767-3804.
[102] Zhong C, Maye M (2001) Core-Shell Assembled Nanoparticles as Catalysts. Adv. Mater. 13(19):1507-1511.
[103] Kulp C, Chen X, Puschhof A, Schwamborn S, Somsen C, Schuhmann W, Bron M (2010) Electrochemical synthesis of core-shell catalysts for electrocatalytic applications. Chem. Phys. Chem. 11(13):2854-2861.
[104] Carbone L, Cozzoli PD (2010) Colloidal heterostructured nanocrystals: Synthesis and growth mechanisms. Nano Today. 5(5):449-493.
[105] Chen J, Lim B, Lee EP, Xia Y (2009) Shape-controlled synthesis of platinum nanocrystals for catalytic and electrocatalytic applications. Nano Today. 4(1):81-95.
[106] Saida T, Ogiwara N, Takasu Y, Sugimoto W (2010) Titanium oxide nanosheet modified PtRu/C electrocatalyst for direct methanol fuel cell anodes. J. Phys. Chem. C. 114:13390-13396.
[107] Geng D, Matsuki D, Wang J, Kawaguchi T, Sugimoto W, Takasu Y (2009) Activity and Durability of Ternary PtRuIr/C for Methanol Electro-oxidation. J. Electrochem. Soc. 156(3):B397-B402.
[108] Kawaguchi T, Sugimoto W, Murakami Y, Takasu Y (2005) Particle growth behavior of carbon-supported Pt, Ru, PtRu catalysts prepared by an impregnation reductive-pyrolysis method for direct methanol fuel cell anodes. J. Catal. 229(1):176-184.

[109] Gurau B, Viswanathan R, Liu R, Lafrenz TJ, Ley KL, Smotkin ES, Reddington E, Sarangapani S (1998) Structural and electrochemical characterization of binary, ternary, and quaternary platinum alloy catalysts for methanol electro-oxidation. J. Phys. Chem. B. 102(49):9997-10003.

[110] Kim SS, Kim C, Lee H (2010) Shape- and Composition-Controlled Pt-Fe-Co Nanoparticles for Electrocatalytic Methanol Oxidation. Top. Catal. 53(7-10):686-693.

[111] Shen SY, Zhao TS, Xu JB, Carbon supported PtRh catalysts for ethanol oxidation in alkaline direct ethanol fuel cell. Int. J. Hydrogen Energy. 35(23):12911-12917.

[112] Wang J, Yin G, Wang G, Wang Z, Gao Y (2008) A novel Pt/Au/C cathode catalyst for direct methanol fuel cells with simultaneous methanol tolerance and oxygen promotion. Electrochem. Commun. 10:831-834.

[113] Petrii OA (2008) Pt-Ru electrocatalysts for fuel cells: A representative review. J. Solid State Electrochem. 12(5):609-642.

[114] Ye F, Cao X, Yu L, Chen S, Lin W (2012) Synthesis and catalytic performance of PtRuMo nanoparticles supported on graphene-carbon nanotubes nanocomposites for methanol electro-oxidation. Int. J. Electrochem. Sci. 7(2):1251-1265.

[115] Wang H, Xu C, Cheng F, Zhang M, Wang S, Jiang SP (2008) Pd/Pt core-shell nanowire arrays as highly effective electrocatalysts for methanol electrooxidation in direct methanol fuel cells. Electrochem. Commun. 10(10):1575-1578.

[116] Nørskov JK , Rossmeisl J, Logadottir A, Lindqvist L, Kitchin J, Bligaard T, Jonsson H (2004) The origin of the overpotential for oxygen reduction at a fuel cell cathode. J. Phys. Chem. B. 108:17887-17892.

[117] Norskov JK, Bligaard T, Rossmeisl J, Christensen CH (2009) Towards the computational design of solid catalysts. Nat. Chem. 1:37-46.

[118] Nørskova JK, Abild-Pedersen F, Studt F, Bligaard T (2011) Density functional theory in surface chemistry and catalysis. PNAS. 108(3):937-943.

[119] Russell AE, Rose A (2004) X-ray Absorption Spectroscopy of Low Temperature Fuel Cell Catalysts. Chem. Rev. 104(10):4613-4636.

[120] Toshima N, Yan H, Shiraishi Y (2008) Recent progress in bimetallic nanoparticles: their preparation, structures and functions. In: Corain B, Schmid G, Toshima N, editors. Metal nanoclusters in catalysis and materials science: the issue of size control. Amsterdam: Elsevier BV. pp.49-75.

[121] Jiang H, Xu Q (2011) Recent progress in synergistic catalysis over heterometallic nanoparticles. J. Mater. Chem. 21:13705-13725.

[122] Liu X, Liu X (2012) Bimetallic Nanoparticles: Kinetic Control Matters. Angew. Chem. Int. Ed. 51:3311-3313.

[123] Kalidindi SB, Jagirdar BR (2012) Nanocatalysis and Prospects of Green Chemistry. ChemSusChem. 5:65-75.

[124] Carlton CE, Chen S, Ferreira PJ, Allard LF, Shao-Horn Y (2012) Sub-Nanometer-Resolution Elemental Mapping of "Pt_3Co" Nanoparticle Catalyst Degradation in Proton-Exchange Membrane Fuel Cells. J. Phys. Chem. Lett. 3:161-166.

[125] Hu Y, Wu P,Yin Y, Zhang H, Chenxin Cai (2012) Effects of structure, composition, and carbon support properties on the electrocatalytic activity of Pt-Ni-graphene nanocatalysts for the methanol oxidation. Appl. Catal., B. (111-112):208-217.

[126] Antolini E, Salgado JRC, Gonzalez ER (2006) The methanol oxidation reaction on platinum alloys with the first row transition metals: The case of Pt-Co and -Ni alloy electrocatalysts for DMFCs: A short review. Appl. Catal., B. 63(1-2):137-149.

[127] Bäumer M, Libuda J, Neyman KM, Rösch N, Rupprechter G, Freund H (2007) Adsorption and reaction of methanol on supported palladium catalysts: microscopic-level studies from ultrahigh vacuum to ambient pressure conditions. Phys. Chem. Chem. Phys. 9:3541-3558.

[128] Long NV, Ohtaki M, Hien TD, Randy J, Nogami M (2011) Synthesis and characterization of polyhedral and quasi-sphere non-polyhedral Pt nanoparticles: Effects of their various surface morphologies and sizes on electrocatalytic activity for fuel cell applications. J. Nanopart. Res. 13:5177-5191.

[129] Long NV, Ohtaki M, Nogami M, Hien TD (2011) Effects of heat treatment and poly(vinylpyrrolidone) (PVP) polymer on electrocatalytic activity of polyhedral Pt nanoparticles towards their methanol oxidation. Colloid Polym. Sci. 289:1373-1386.

[130] Long NV, Ohtaki M, Uchida M, Jalem R, Hirata H, Chien ND, Nogami M (2011) Synthesis and characterization of polyhedral Pt nanoparticles: Their catalytic property, surface attachment, self-aggregation and assembly. J. Colloid Interface Sci. 359:339-350.

[131] Long NV, Chien ND, Hirata H, Matsubara T, Ohtaki M, Nogami M (2011) Highly monodisperse cubic and octahedral rhodium nanocrystals: Their evolutions from sharp polyhedrons into branched nanostructures and surface-enhanced Raman scattering. J. Cryst. Growth. 320:78-89.

[132] Long NV, Chien ND, Matsubara T, Hirata H, Lakshminarayana G, Nogami M (2011) The synthesis and characterization of platinum nanoparticles: A method of controlling the size and morphology. Nanotechnology. 21:035605.

[133] Long NV, Ohtaki M, Ngo VN, Thi CM, Nogami M (2012) Structure and morphology of platinum nanoparticles with critical issues of low and high-index facets. Adv. Nat. Sci. Nanosci. Nanotechnol. 3: 025005.

[134] Nguyen VL, Hayakawa T, Matsubara T, Chien ND, Ohtaki M, Nogami M (2012) Controlled Synthesis and properties of palladium nanoparticles. J. Exp. Nanosci. 7:426-439.

[135] Nguyen VL, Chien ND, Hayakawa T, Matsubara T, Ohtaki M, Nogami M (2012) Sharp cubic and octahedral morphologies of poly(vinylpyrrolidone)-stabilised platinum nanoparticles by polyol method in ethylene glycol: their nucleation, growth and formation mechanisms. J. Exp. Nanosci. 7:133-149.

[136] Nguyen VL, Ohtaki M, Nogami M (2011) Control of morphology of Pt nanoparticles and Pt-Pd core-shell nanoparticles. Journal of novel carbon resource sciences, Kuyshu university. 3:40-44.

[137] Nguyen VL, Chien ND, Hirata H, Ohtaki M, Hayakawa T, Nogami M (2010) Chemical synthesis and characterization of palladium nanoparticles. Adv. Nat. Sci. Nanosci. Nanotechnol. 1:035012.

[138] Long NV, Chien ND, Uchida M, Matsubara T, Jalem R, Nogami M (2010) Directed and random self-assembly of Pt-Au nanoparticles. Mater. Chem. Phys. 124:1193-1197.

[139] Xing Y, Cai Y, Vukmirovic MB, Zhou WP, Karan H, Wang JX, Adzic RR (2010) Enhancing oxygen reduction reaction activity via Pd-Au alloy sublayer mediation of Pt monolayer electrocatalysts. J. Phys. Chem. Lett. 1(21):3238-3242.

[140] Long NV, Ohtaki M, Hien TD, Randy J, Nogami M (2011) A comparative study of Pt and Pt-Pd core-shell nanocatalysts. Electrochim. Acta. 56:9133-9143.

[141] Long NV, Asaka T, Matsubara T, Ohtaki M, Nogami M (2011) Shape-controlled synthesis of Pt-Pd core-shell nanoparticles exhibiting polyhedral morphologies by modified polyol method. Acta Mater. 59:2901-2907.

[142] Long NV, Hien TD, Asaka T, Ohtaki M, Nogami M (2011) Synthesis and characterization of Pt-Pd nanoparticles with core-shell morphology: Nucleation and overgrowth of the Pd shells on the as-prepared and defined Pt seeds. J. Alloys Compd. 509:7702-7709.

[143] Long NV, Hien TD, Asaka T, Ohtaki M, Nogami M (2011) Synthesis and characterization of Pt-Pd alloy and core-shell bimetallic nanoparticles for direct methanol fuel cells (DMFCs): Enhanced electrocatalytic properties of well-shaped core-shell morphologies and nanostructures. Int. J. Hydrogen Energy. 36:8478-8491.

[144] Yamauchi M,Kobayashi H, Kitagawa H (2009) Hydrogen Storage Mediated by Pd and Pt Nanoparticles. Chem. Phys. Chem. 10:2566-2576.

[145] Yen CH, Shimizu K, Lin YY, Bailey F, Cheng IF, Wai CM (2007) Chemical fluid deposition of Pt-based bimetallic nanoparticles on multiwalled carbon nanotubes for direct methanol fuel cell application. Energy Fuels. 21(4):2268-2271.

[146] Chen M, Wu B, Yang J, Zheng N (2012) Small Adsorbate-Assisted Shape Control of Pd and Pt Nanocrystals. Adv. Mater. 24:862-879.

[147] Wu H, Wexler D, Wang G, Liu H (2012) Cocore-Ptshell nanoparticles as cathode catalyst for PEM fuel cells. J. Solid State Electrochem. 16:1105-1110.

[148] Sato K, Yanajima K, Konno TJ (2012) Structure and compositional evolution in epitaxial Co/Pt core-shell nanoparticles on annealing. Thin Solid Films. 520:3544-3552.

[149] Qiu H, Zou F (2012) Nanoporous PtCo Surface Alloy Architecture with Enhanced Properties for Methanol Electrooxidation. ACS Appl. Mater. Interfaces. 4(3):1404-1410.

[150] Avila-Garcia I, Plata-Torres M, Dominguez-Crespo MA, Ramirez-Rodriguez C, Arce-Estrada EM (2007) Electrochemical study of Pt-Pd, Pt-Ru, Pt-Rh and Pt-Sn/C in acid media for hydrogen adsorption-desorption reaction. J. Alloys Compd. 434-435:764-767.

[151] Ávila-García I, Ramírez C, Hallen López JM, Arce-Estrada EM (2010) Electrocatalytic activity of nanosized Pt alloys in the methanol oxidation reaction. Journal of Alloys and Compounds. 495(2):462-465.

[152] Alayoglu S, Nilekar AU, Mavrikakis M, Eichhorn B (2008) Ru-Pt core-shell nanoparticles for preferential oxidation of carbon monoxide in hydrogen. Nat. Mater. 7:333-338.

[153] Bakhmutsky K, Wieder NL, Cargnello M, Galloway B, Fornasiero P, Gorte RJ (2012) A Versatile Route to Core-Shell Catalysts: Synthesis of Dispersible M@Oxide (M=Pd, Pt; Oxide=TiO_2, ZrO_2) Nanostructures by Self-Assembly. ChemSusChem. 5:140-148.

[154] Chaudhuri RG, Paria S (2012) Core/Shell Nanoparticles: Classes, Properties, Synthesis Mechanisms, Characterization, and Applications. Chem. Rev. 112(4):2373-433.

[155] Zhang H, Shen PK (2012) Recent Development of Polymer Electrolyte Membranes for Fuel Cells. Chem. Rev. 112(5):780-2832.

[156] Lee C, Chiou H (2012) Methanol-tolerant Pd nanocubes for catalyzing oxygen reduction reaction in H_2SO_4 electrolyte. Appl. Catal., B. 117-118:204-211.

[157] Luo L, Futamata M (2006) Competitive adsorption of water and CO on Pd modified Pt electrode from CH3OH solution. Electrochem. Commun. 8(2):231-237.

[158] Koenigsmann C, Zhou W, Adzic RR, Sutter E, Wong SS (2010) Size-dependent enhancement of electrocatalytic performance in relatively defect-free, processed ultrathin platinum nanowires. Nano Lett. 10:2806-2811.

[159] Tao F, Grass ME, Zhang YW, Butcher DR, Renzas JR, Liu Z, Chung JY, Mun BS, Salmeron M, Somorjai GA (2008) Reaction-driven restructuring of Rh Pd and Pt Pd core-shell nanoparticles. Science. 322:932-934.

[160] Lim B, Jiang M, Camargo PHC, Cho EC, Tao J, Lu X, Zhu Y, Xia Y (2009) Pd-Pt Bimetallic Nanodendrites with High Activity for Oxygen Reduction. Science. 324(5932):1302-1305.

[161] Jiang M, Lim B, Tao J, Camargo PHC, Ma C, Zhu Y, Xia Y (2010) Epitaxial overgrowth of platinum on palladium nanocrystals. Nanoscale. 2:2406-2411.

[162] Lim B, Wang J, Camargo P, Jiang M, Kim M, Xia Y (2008) Facile Synthesis of Bimetallic Nanoplates Consisting of Pd Cores and Pt Shells through Seeded Epitaxial Growth. Nano. Lett. 8:2535-2540.

[163] Hong J, Kang S, Choi B, Kim D, Lee SB, Han SW (2012) Controlled synthesis of Pd-Pt alloy hollow nanostructures with enhanced catalytic activities for oxygen reduction. ACS Nano. 6(3):2410-2419.

[164] Chen, X, Wang H, He J, Cao Y, Cui Z, Liang M (2010) Preparation, Structure and catalytic activity of Pt-Pd bimetallic nanoparticles on multi-walled carbon nanotubes. J. Nanosci. Nanotechnol. 10(5):3138-3144.

[165] Kim I, Lee H, Shim J (2008) Synthesis and characterization of Pt-Pd catalysts for methanol oxidation and oygen reduction. J. Nanosci. Nanotechnol. 8(10):5302-5305.

[166] Cho YH, Choi B, Cho YH, Park HS, Sung YE (2007) Pd-based PdPt(19:1)/C electrocatalyst as an electrode in PEM fuel cell. Electrochem. Commun. 9:378-381.

[167] Zhang H, Jin M, Wang J, Kim MJ, Yang D, Xia Y (2011) Nanocrystals composed of alternating shells of Pd and Pt can be obtained by sequentially adding different precursors. J. Am. Chem. Soc. 133(27):10422-10425.

[168] Jung D, Park S, Ahn C, Kim J, Oh E (2012) Performance comparison of $Pt_{1-x}Pd_x$/carbon nanotubes catalysts in both electrodes of polymer electrolyte membrane fuel cells. Fuel Cells. 12:398-405.

[169] Guo S, Dong S, Wang E (2010) Three-dimensional Pt-on-Pd bimetallic nanodendrites supported on graphene nanosheet: Facile synthesis and used as an advanced nanoelectrocatalyst for methanol oxidation. ACS Nano:4(1):547-555.

[170] Bernardi F, Alves MCM, Traverse A, Silva DO, Scheeren CW, Dupont J, Morais J (2009) Monitoring atomic rearrangement in Pt_xPd_{1-x}(x = 1, 0.7, or 0.5) nanoparticles driven by reduction and sulfidation processes. J. Phys. Chem. C. 113:3909-3916.

[171] Dimos MM, Blanchard GJ, Evaluating the role of Pt and Pd catalyst morphology on electrocatalytic methanol and ethanol oxidation, J. Phys. Chem. C 2010, 114, 6019-6026.

[172] Shiju NR, Guliants VV (2009) Recent developments in catalysis using nanostructured materials. Appl. Catal., A. 356(1)1-17.

[173] Vielstich W, Gasteiger HA, Yokokawa H (2009) Handbook of Fuel Cells: Advances in Electrocatalysis, Materials, Diagnostics and Durability, Volumes 5&6. United Kingdom: John Wiley & Sons. 1090 p.

[174] Cudero A, Gullón J, Herrero E, Aldaz A, Feliu J (2010) CO electrooxidation on carbon supported platinum nanoparticles: Effect of aggregation. J. Electroanal. Chem. 644:117-126.

[175] Mizukoshi Y, Fujimoto Y, Nagata Y, Oshima R, Maeda Y (2002) Characterization and Catalytic Activity of Core-Shell Structured Gold/Palladium Bimetallic Nanoparticles Synthesized by the Sonochemical Method. J. Phys. Chem. B. 104(25):6028-6032.

[176] Teng X, Yang H (2003) Synthesis of face-centered tetragonal FePt nanoparticles and cranular films from Pt@Fe_2O_3 Core-Shell Nanoparticles. J. Am. Chem. Soc. 125:14559-14563.

[177] Chen Y, Yang F, Dai Y, Wang W, Chen S (2008) Ni@Pt Core-Shell Nanoparticles: Synthesis, Structural and Electrochemical Properties. J. Phys. Chem. C. 112:1645-1649.

[178] Huang Y, Zheng S, Lin X, Su L, Guo Y (2012) Microwave synthesis and electrochemical performance of a PtPb alloy catalyst for methanol and formic acid oxidation. Electrochim. Acta. 63:346-353.

[179] Debe MK (2012) Effect of Electrode Surface Area Distribution on High Current Density Performance of PEM Fuel Cells. J. Electrochem. Soc. 159(1):B54-B67.

[180] Li G, Lu W, Luo Y, Xia M, Chai C, Wang X (2012) Synthesis and characterization of cendrimer-encapsulated bimetallic core-shell PdPt nanoparticles. Chin. J. Chem. 30:541-546.

[181] Marcu A, Totha G, Srivastava R, Strasser P (2012) Preparation, characterization and degradation mechanisms of PtCu alloy nanoparticles for automotive fuel cells. J. Power Sources. 208:288-295.

[182] Maillard F, Dubau L, Durst J, Chatenet M, André J, Rossinot E (2010) Durability of Pt_3Co/C nanoparticles in a proton-exchange membrane fuel cell: Direct evidence of bulk Co segregation to the surface. Electrochem. Commun. 12:1161-1164.

[183] Du S (2012) Pt-based nanowires as electrocatalysts in proton exchange fuel cells. Int. J. Low-Carbon Tech. 7(1):44-54.

[184] Huaman J, Fukao S, Shinoda K, Jeyadevan B (2011) Novel standing Ni-Pt alloy nanocubes. CrystEngComm.13:3364-3369.

[185] Gong K, Vukmirovic M, Ma C, Zhu Y, Adzic R (2011) Synthesis and catalytic activity of Pt monolayer on Pd tetrahedral nanocrystals with CO-adsorption-induced removal of surfactants. J. Electroanal. Chem. 662:213-218.

[186] Taufany F, Pan C, Rick J, Chou H, Tsai M, Hwang B, Liu D, Lee J, Tang M, Lee Y, Chen C (2011) Kinetically controlled autocatalytic chemical process for bulk production of bimetallic core-shell structured nanoparticles. ACS Nano. 5(12):9370-9381.

[187] Wei Z, Feng Y, Li L, Liao M, Fu Y, Suna C, Shao Z, Shen P (2008) Electrochemically synthesized Cu/Pt core-shell catalysts on a porous carbon electrode for polymer electrolyte membrane fuel cells. J. Power Sources. 180:84-91.

[188] Wang ZL (2000) Transmission electron microscopy of shape-controlled nanocrystals and their assemblies. J. Phys. Chem. B 104:1153-1175.

[189] Wang Y, Toshima N (1997) Preparation of Pd-Pt bimetallic colloids with controllable core/shell structures. J. Phys. Chem. B. 101:5301-5306.

[190] Nguyen VL, Ohtaki M, Matsubara T, Cao MT, Nogami M (2012) New experimental evidences of Pt-Pd bimetallic nanoparticles with core-shell configuration and highly fine-ordered nanostructures by high-resolution electron transmission microscopy. J. Phys. Chem. C 116(22):12265-12274.

Chapter 11

Monitoring the Effects of Thermal Treatment on Properties and Performance During Battery Material Synthesis

Wesley M. Dose and Scott W. Donne

Additional information is available at the end of the chapter

1. Introduction

1.1. Energy supply

Utilizing an inexpensive, clean and sustainable supply of energy is one of the world's foremost challenges heading into the future. The present energy supply scenario is dominated by fossil fuels, which are both a finite resource and a substantial contributor greenhouse gas emissions. Many renewable sources of energy (e.g., nuclear, solar, wind, etc.) exist for consumer use, although they all have associated pros and cons which means they cannot be used ubiquitously across the planet [1]. As a result, the future supply of energy is not likely to be centralized in a limited number of large power stations, but rather much more distributed as smaller scale renewable energy sources are utilized.

Much has been made in the literature concerning solar energy harvesting, both in terms of photovoltaics and solar thermal. Of all the renewable forms of energy, solar energy has the capacity to completely replace society's dependence on fossil fuels. However, of course, the challenge remains to make this a reality, particularly so with the high cost of photovoltaics [1]. Another perceived problem with the use of photovoltaics, and indeed with many other renewable energy sources, is their intermittency. Whether this be over short time frames, such as with cloud cover, or extended periods of time, such as overnight, steps need to be taken to ensure the consistency of power supply. In other words, some form of energy storage must be present to complement the primary source of energy.

1.2. Energy storage

Energy can be stored in many different ways, some examples of which include [2]:

i. Thermal energy storage as heat;
ii. Chemical energy storage in the form of a fuel;
iii. Electrical energy storage in the form of charge separation; and
iv. Mechanical storage in the form of kinetic energy.

Each of these types of energy storage has its own set of performance characteristics in terms of the energy and power that they can deliver, ultimately meaning that they will be best suited in specific applications. Of course cyclability is also a key performance characteristic.

Chemical energy, stored as a fuel, can deliver very high specific power and energy, with reasonable efficiency, particularly when it is used in an internal combustion engine [3]. However, such a combustion reaction does little to abate the demand for fossil fuels (which are used most commonly in this domain) or the contributions to greenhouse gas emissions.

Chemical energy storage has an added advantage in the sense that it can also be released electrochemically (rather than thermally), through devices such as batteries, supercapacitors and fuel cells [4]. While the specific energy and power performance of these devices is much less than that of the internal combustion engine, their efficiency is much higher, in some instances approaching 100%, meaning that they ultimately utilize any fuel much better.

1.3. Batteries and battery materials

There are many battery systems available for consumer use, and over the years since commercial introduction, their performance has improved due to a combination of advances in material design and cell engineering [5]. The various battery chemistries that are available can be categorized as either primary (single use) or secondary (rechargeable) systems. The battery system that is currently receiving the most attention in the literature is based on the Li-ion chemistry. Here Li^+ ions are reversibly shuttled between the positive and negative electrodes in the cell, which with the use of an appropriate non-aqueous electrolyte, can achieve a potential of over 4 V [5,6]. Some of the commonly found materials in Li-ion batteries include:

i. Negative electrode materials: Li metal, carbon, $Li_4Ti_5O_{12}$, Sn and its alloys, Si and its alloys;
ii. Electrolytes: organic carbonate mixtures; e.g., 1:1 ethylene carbonate:dimethyl carbonate, ionic liquids; and
iii. Positive electrode materials: $LiMO_2$ (where M is a combination of Ni, Mn and/or Co), $LiMn_2O_4$ (and doped varieties thereof), $LiFePO_4$, MnO_2.

Whatever the combination of positive and negative electrode that is used, the performance of the resultant cell is quite significantly determined by the properties of the electroactive materials used, which in turn are determined by the way in which they were prepared.

Material properties such as phase purity, crystallinity and particle size (or extent of agglomeration) all affect performance.

1.4. Common synthetic routes

Many different synthetic routes have been used to prepare the materials mentioned in the previous section, so much so that to list them here would be excessive (note the recent review in reference [6]). Nevertheless, the more common approaches can be categorized as being based on either (i) thermal methods, (ii) solvothermal methods, (iii) mechanical methods, and (iv) electrochemical methods. It is also reasonably common to find that a combination of these methods has been used to produce the resultant material. As an example, $LiFePO_4$ can be made by first using a solvothermal process to intimately mix the precursors, with the resultant mixture then subjected to a thermal treatment to make the final product [7]. Overall, the majority of the positive electroactive materials listed above involve a thermal processing step as the last step in their synthesis.

1.5. Pitfalls of thermal processing methods

The overall objective of any synthesis method is to obtain the final product in the desired form for immediate use. This is particularly true for thermal synthesis methods where there is a delicate balance between heat treatment temperature and duration so as to produce the desired material. Of course the choice of these thermal parameters is also dependent on the effectiveness of precursor mixing, with various solvothermal methods being used to ensure appropriate mixing on the molecular level. Contrast this with some of the initial solid state mixing methods (grinding) used in some synthetic efforts, and the implications it has on the thermal conditions necessary [8].

Let us begin by assuming that we have sufficient mixing of our precursors, since the focus of the discussion here is on the actual thermal conditions to be used. Under these circumstances if we were to thermally treat this mixture the temperature and duration of heat treatment would determine the phase purity and crystallinity of the resultant material. Of course a higher heat treatment temperature, and a longer heat treatment duration would ensure phase purity, as well as lead to a more crystalline material. The question at this time then becomes: What is the preferred material crystallinity?

Many positive electroactive materials in Li-ion batteries require very small crystallite sizes so as to minimize Li^+ ion diffusion paths, which is commonly regarded as a key limiting factor in performance [7]. Therefore, excessive heat treatment temperatures and durations, while they may ensure phase purity, also lead to excessive crystallization, which is detrimental. Additionally from a commercial perspective, excessive material heating leads to a waste of energy, which can be costly. What is required, therefore, is a method for predicting the optimum heat treatment temperature and duration so as to ensure phase purity and small crystallite size.

1.6. This work

What we will describe here in this chapter is a method for identifying the optimum thermal synthesis conditions, and then explore their effects on the resultant material properties. The system we will use to demonstrate the approach is the heat treatment process used to remove water from the structure of γ-MnO_2 prior to its use in non-aqueous Li-MnO_2 cells. While this will be used as a representative example, the general method is applicable to any other thermally based synthetic method.

2. Methods

2.1. Preparation of starting material

The sample of γ-MnO_2 used in this work was prepared by anodic electrodeposition, and hence given the designation electrolytic manganese dioxide, or EMD. The cell used for electrolysis was based on a temperature controlled 2 L glass beaker in which two 144 cm^2 (72 cm^2 on either side) titanium sheets were used as the anode substrate, and three similarly sized copper sheets were used as the cathode substrate. The electrodes were arranged alternately so that each anode was surrounded on both sides by a cathode. The electrolyte was an aqueous mixture of 1.0 M $MnSO_4$ and 0.25 M H_2SO_4 maintained at 97°C. Electrodeposition of the manganese dioxide was conducted with an anodic current density of 65 A/m^2 according to the reactions:

$$\text{Anode (Ti)}: \; Mn^{2+} + 2H_2O \longrightarrow MnO_2 + 4H^+ + 2e^- \quad (1)$$

$$\text{Cathode (Cu)}: \quad 2H^+ + 2e^- \longrightarrow H_2 \quad (2)$$

$$\text{Overall}: Mn^{2+} + 2H_2O \longrightarrow MnO_2 + 2H^+ + H_2 \quad (3)$$

The overall process was carried out for three days, during which time the electrolyte Mn^{2+} concentration was of course depleted, while the H^+ concentration increased. To counteract this, and hence maintain a constant electrolyte concentration over the duration of the deposition, a concentrated (1.5 M) $MnSO_4$ solution was added continually at a suitable rate to replenish Mn^{2+} and dilute any excess H_2SO_4 produced. Under these conditions control of the solution conditions was typically maintained to within ±2%.

After deposition was complete, the solid EMD deposit was mechanically removed from the anode and broken into chunks ~0.5 cm in diameter, and then immersed in 500 mL DI water to assist in the removal of entrained plating electrolyte. The pH of this chunk suspension was adjusted to pH 7 with the addition of 0.1 M NaOH. After ~24 h at a pH of 7 the suspension was filtered and the chunks then dried at 110°C. After drying the chunks were then milled to a -105 μm powder (mean particle size ~45 μm) using an orbital zirconia mill.

The powder was then suspended in ~500 mL of DI water and its pH again adjusted to 7 with the further addition of 0.1 M NaOH. When the pH had stabilized, the suspension was filtered and the collected solids dried at 110°C. When dry the powdered EMD was removed from the oven, allowed to cool to ambient temperature in a dessicator and then transferred to an airtight container for storage.

2.2. Thermogravimetric (TG) analysis

TG analysis was conducted using a Perkin Elmer Diamond TG/DTG controlled by Pyris software. Approximately 10 mg of γ-MnO_2 sample was added to an aluminium sample pan and placed into the analyser. The same mass of α-Al_2O_3 in a similar aluminium pan was used as the reference material for DTA measurements. With the sample and reference materials loaded, and the furnace closed, dry nitrogen gas was passed over the sample at 20 mL/min for 30 minutes prior and during the heating profile. The heating profile applied to the sample was essentially a linear ramp at rates ranging from 0.25°-10°C/min.

2.3. Material characterization

The structure of the materials generated in this work were identified by X-ray diffraction using a Phillips 1710 diffractometer equipped with a Cu Kα radiation source (λ=1.5418 Å) and operated at 40 kV and 30 mA. The scan range was from 10° to 80° 2θ, with a step size of 0.05° and a count time of 2.5 s. Analysis of the diffraction patterns was carried out by fitting a Lorentzian lineshape to individual peaks, with the fitting parameters then used to calculate structural properties such as the fraction of pyrolusite (P_r; as described by Chabre and Pannetier [9]) and the unit cell parameters.

Morphology was examined by gas adsorption using a Micromeritics ASAP 2020 Surface Area and Porosity Analyser. A representative 0.10 g sample of the manganese dioxide material was degassed under vacuum at 110°C for 2 h prior to analysis. An adsorption isotherm was then determined over the partial pressure (P/P_0) range of 10^{-7}-1 using N_2 gas as the adsorbate at 77 K. The specific surface area was extracted from the gas adsorption data using the linearized BET isotherm [10] in the range $0.05<P/P_0<0.30$, while the pore size distribution was determined using a Density Functional Theory-based approach (Micromeritics DFTPlus V2.00).

The composition of the materials was determined using two consecutive potentiometic titrations (Pt indicator and SCE reference electrode) as outlined in Vogel [11]. A blank titration was carried out first in which 10 mL of acidified 0.25 M ferrous ammonium sulfate ($NH_4FeSO_4.6H_2O$, BDH Chemicals, 99%) was titrated with a standardised 0.03 M potassium permanganate solution ($KMnO_4$, Ajax Finechem, 99%; standardized using the oxalate method [11]), and the volume of permanganate added to reach the end point denoted as V_0. Sample analysis was conducted by digesting 0.050 g of the manganese dioxide being studied into another 10 mL aliquot of the acidified 0.25 M ferrous ammonium sulfate solution. After complete dissolution the resultant solution was then titrated using the same 0.03 M permanganate solution, with the volume to reach the end point recorded as V_1. For this titration it is important to stop at or just after the end point has been attained. To the solution resulting from the first titration ~6 g of tetra-sodium pyrophosphate ($Na_4P_2O_7.10H_2O$, Ajax Finechem, 99%) was added and allowed to dissolve. The pH of this solution was then adjusted to lie within the range 6-7 by the drop-wise addition of ~0.20 M sulfuric acid. The second potentiometic titration was performed using the same 0.03 M permanganate solution, and the volume to reach the end point recorded as V_2. The value for x in MnO_x was then calculated using:

$$x=1+\frac{5(V_0-V_1)}{2(4V_2-V_1)} \tag{4}$$

The total manganese content in the sample can be found from the second titration by taking into account the amount of manganese added through the addition of permanganate in the first titration. Using this result, and the dry mass of the manganese dioxide sample, found by subtracting the mass of surface water lost from the sample after heating at 110°C (%H_2O(<110°C)) from the original mass, the total manganese content (%Mn) can be found using:

$$\%Mn=\frac{n_{Mn}}{n_{MnO_2(dry)}}\times 100 \tag{5}$$

To calculate the relative proportion of manganese (III) and (IV) species (%Mn(III) and %Mn(IV) respectively), we have:

$$\%Mn(III)=(4-2x)\times\%Mn \tag{6}$$

$$\%Mn(IV)=(2x-3)\times\%Mn \tag{7}$$

Finally, the cation vacancy fraction (CVF) can be found by taking into account the percentage structural water (i.e., water removed after heating at 400°C, but above 110°C, (%H_2O(>110°C)), found by considering the difference in mass after heating the sample at 400°C for 2 h, and using:

$$CVF=\frac{m}{m+2} \tag{8}$$

where

$$m=(2-x)+\frac{M_{Mn}\times\%H_2O(>110°C)}{M_{H_2O}\times\%Mn} \tag{9}$$

and M_{Mn} and M_{H_2O} are the molar masses of manganese and water, respectively.

2.4. Thermal treatment of EMD

Approximately 10 g of EMD was heated in an alumina boat crucible by a Eurotherm HTC1400 furnace with a static air atmosphere set at the required temperature. After the elapsed isothermal heating time, the sample was removed from the oven and allowed to cool to room temperature.

2.5. Electrochemical performance

To evaluate the electrochemical performance the heat treated materials prepared were first thoroughly mixed with graphite and polyvinylidene fluoride, in a 1:8:1 ratio. Around 0.30 g of this mixture was compressed in a 10 mm die press under 1 t into a disk electrode ~1 mm thick. The electrodes were dried at 110°C under vacuum and accurately weighed prior to introduction into an Ar-filled dry box, where cell construction took place.

CR2032 size coin cells were constructed for electrochemical testing. The coin cells were comprised of a heat treated EMD (HEMD) cathode, lithium metal anode, with electrolyte made up from 1 M $LiPF_6$ (Sigma-Aldrich (≥99.99%) in 1:1 w/w of ethylene carbonate (EC, Sigma-Aldrich 99%) and dimethyl carbonate (DMC, Sigma-Aldrich 99+%). A Celgard 2400 micro-porous separator was used in these cells. After construction, cells were left to equilibrate for 3-4 days before being used for electrochemical testing.

The electrochemical characteristics of the cells prepared were assessed using a Perkin-Elmer VMP multichannel potentiostat/galvanostat on which a modular galvanostatic discharge program was performed at rates of 2, 5, 10 and 20 mA/g of active material.

3. Data analysis

3.1. Starting material properties

The compositional, morphological and structural data for the starting EMD sample are shown in the first row of Table 1. While the details of these initial properties and the resulting changes to the measured parameters as a result of heat treatment will be discussed in detail later, we note here that the EMD chosen for this work is a typical EMD sample. The composition of samples prepared via electrolysis can vary considerably depending on the experimental deposition conditions. We find that the compositional data collected for our starting EMD fit comfortably within the typical range for samples termed EMD [12]. The structure of the starting EMD, as measured by XRD, is shown in Figure 1. The Miller indicies for the peaks in the starting EMD pattern are labelled assuming an orthorhombic unit cell.

	Composition					
Temp (°C)	Mn(T) (%)	Mn(IV) (%)	Mn(III) (%)	CVF	%H_2O (>110°C)	%H_2O (<110°C)
25	59.45	55.34	4.11	0.081	2.13	4.11
200	59.00	55.03	3.98	0.051	1.67	2.73
250	61.47	57.28	4.20	0.027	1.60	1.81
300	61.82	55.22	6.60	0.008	1.66	1.41
350	60.94	54.94	6.00	0.000	0.94	0.84
400	60.41	53.79	6.61	0.000	1.15	0.90

	Morphology			Structure				
Temp (°C)	BET SA (m^2/g)	Micro-pore Volume (cm^3/g)	Meso-pore Volume (cm^3/g)	Temp (°C)	P_r	a_0 (Å)	b_0 (Å)	c_0 (Å)
25	37.22	0.0074	0.0293	25	0.34	4.47	9.55	2.83
200	36.68	0.0071	0.0358	200	0.50	4.41	9.35	2.85
250	32.46	0.0047	0.0327	250	0.52	4.43	9.33	2.85
300	30.67	0.0038	0.0348	300	0.73	4.42	9.22	2.86
350	24.90	0.0021	0.0401	350	0.83	4.42	9.11	2.87
400	27.49	0.0034	0.0392	400	0.84	4.42	9.20	2.87

Table 1. Composition, morphology and structure of the starting and heat treated EMD samples

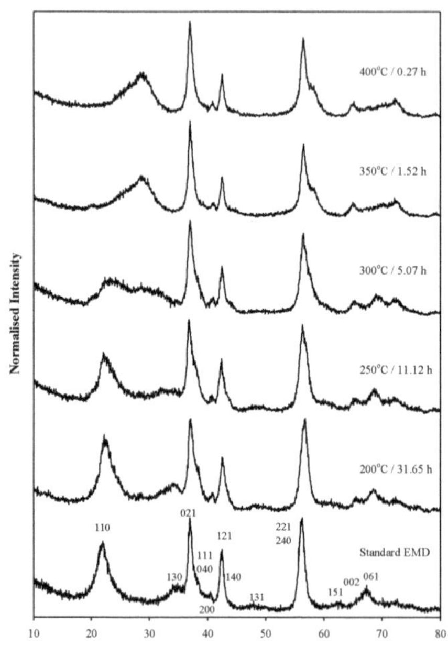

Figure 1. XRD pattern of the starting EMD showing the γ-MnO_2 structure, with the corresponding Miller indices indexed using an orthorhombic unit cell, and XRD patterns for the heat treated EMD.

3.2. Thermogravimetric analysis

The thermogravimetric (TG) and differential thermogravimetric (DTG) data for the thermal decomposition of the EMD sample at the various heating rates are shown in Figures 2 and 3, respectively. The initial loss in mass up to ~120°C is due to the removal of physisorbed water from the EMD surface. Manganese oxides are well known for their ability to adsorb water [13],

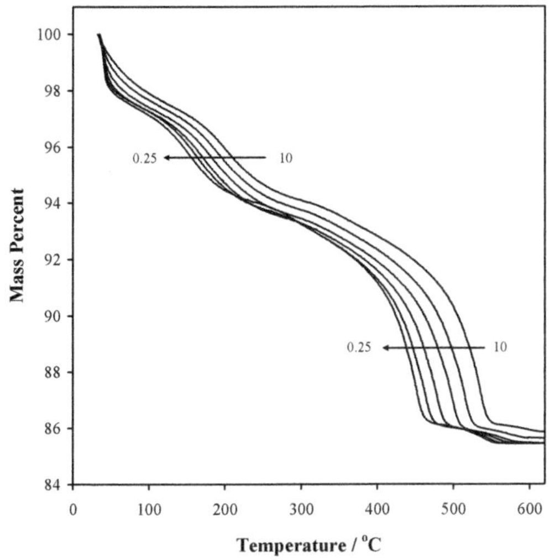

Figure 2. TGA data for the EMD sample used in this work recorded at rates of 0.25, 0.5, 1.0, 2.5, 5.0 and 10.0°C/min.

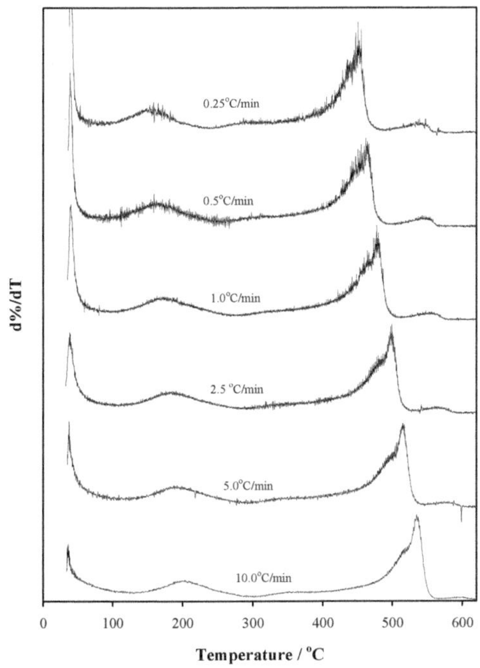

Figure 3. Differential thermogram (DTG) for the EMD sample heated at different heating rates.

so it is not surprising that 2-3% of the initial sample mass is physisorbed water. The broad peak at ~200°C in the DTG relates to the removal of structural water; i.e., protons associated with cation vacancies and Mn(III) ions within the manganese dioxide structure. The sharper peak at ~500°C relates to the thermal reduction of the MnO_2 to form Mn_2O_3. The use of faster heating rates has shifted the decomposition temperature to higher values, possibly as a result of slow reaction kinetics and/or since less time is allowed for the equivalent reaction. It is also possible that the thermal conductivity of the EMD contributed to this effect, although with the use of a relatively small sample size (~10 mg) its contribution is expected to be minor.

3.3. Multiple curve isoconversional analysis

In this analysis, we will be considering the first step in the EMD thermal decomposition; i.e., the process of removing water from the structure beginning at ~175°C, since this is most important when the material is to be used in a non-aqueous battery system.

The first step in the analysis was background correction of the DTG data. To do this an exponential background curve was fitted to the data surrounding the peak for each heating rate used. The resulting curve after background correction describes the processes occurring in this region, as shown in Figure 4 for a heating rate of 10°C/min. The normalized extent of conversion (α) was then found by numerical integration of the background-corrected DTG data, with the normalization being carried out by expressing each point relative to the maximum area determined. A plot of α as a function of temperature for the range of heating rates considered is shown in Figure 5. From this data the temperatures corresponding to a pre-defined set of α values can be found for each heating rate, as shown in Figure 6.

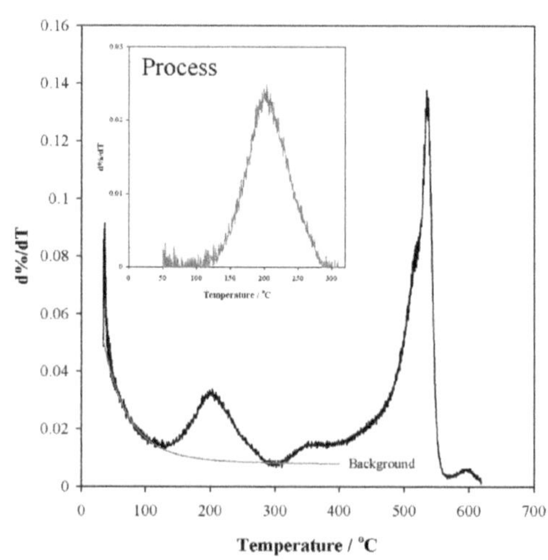

Figure 4. Fitting of DTG data for heating rate 10°C/min with background curve and resulting process.

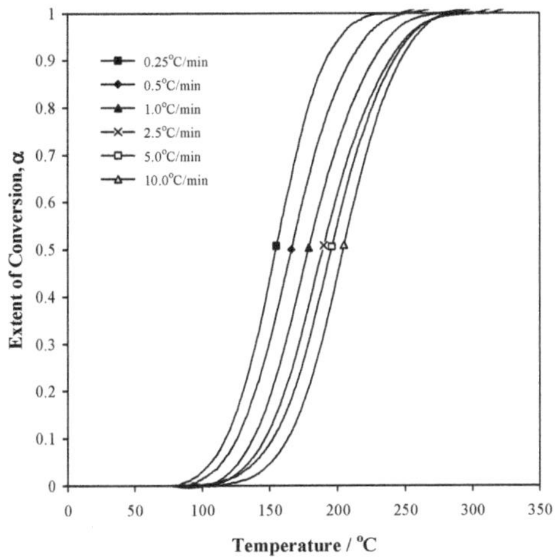

Figure 5. Extent of conversion versus temperature for the different heating rates.

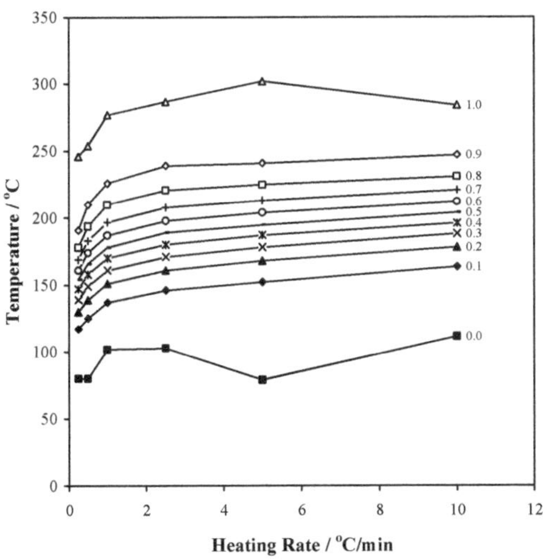

Figure 6. Variation of temperature with heating rate for given extent of conversion.

Kinetic analysis using this method is conventionally applied to the data corresponding to α=0 for the different heating rates. However, since we have access here to the temperature corresponding to various extents of conversion for different heating rates, and the fact that we will apply the method used for α=0 to all α values, then it will be interesting to observe how the resultant kinetic parameters change. Kinetic analysis is based on the rate equation:

$$\beta\frac{d\alpha}{dT}=Af(\alpha)\exp\left(-\frac{E_A}{RT}\right) \tag{10}$$

where A is the pre-exponential factor (min^{-1}), β is the heating rate (°C/min), E_A is the activation energy (J/mol), the term $f(\alpha)$ represents the model chosen to represent the mechanism of thermal decomposition, and all other symbols have their usual significance. In this case we will use $f(\alpha)=1-\alpha$ as our decomposition model [14]. Separation of variables leads to:

$$\frac{d\alpha}{f(\alpha)}=\frac{A}{\beta}\exp\left(-\frac{E_A}{RT}\right)dT \tag{11}$$

and then integration of the left-hand side from 0 to α_i in Eqn. 11, gives:

$$F(\alpha_i)-F(0)=\int_0^{T_i}\frac{A}{\beta}\exp\left(-\frac{E_A}{RT}\right)dT \tag{12}$$

where F represents the integrated form of $f(\alpha)$, and T_i corresponds to the temperature at α_i. Upon rearrangement we can write:

$$\beta=\int_0^{T_i}\frac{A}{F(\alpha_i)-F(0)}\exp\left(-\frac{E_A}{RT}\right)dT \tag{13}$$

This integration was performed numerically using the trapezium method and optimised to fit the experimental data using a linear least squares regression. Consequently, the dependence of activation energy on the extent of conversion and the pre-exponential factor was found, as shown in Figure 7. With the exception of the first point at α=0 (which was likely due to the noisy data for these conditions, cf. Figure 6), the activation energy clearly increases with extent of conversion, ranging from 109-250 kJ/mol.

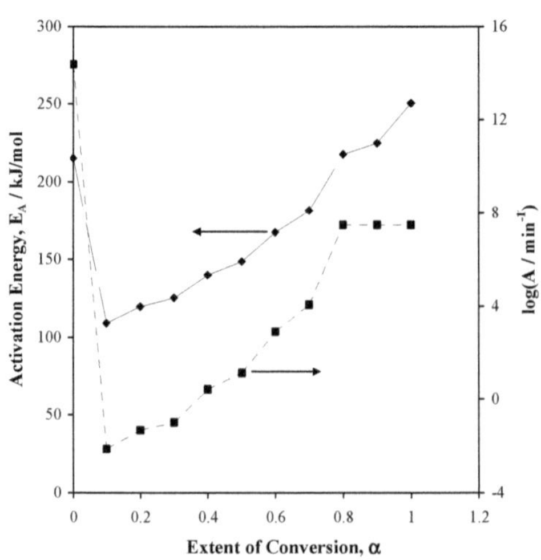

Figure 7. Activation energy and pre-exponential factor for process occurring against extent of conversion calculated using the first order kinetic analysis.

Before going further, an additional comment needs to be made regarding the choice of thermal decomposition model ($f(\alpha)$) used in the analysis. Each of the models listed in reference [14] were examined here using this approach. Firstly, despite the broad range of curve shapes (α vs. T) that these models generate, no single one was able to fit satisfactorily to the experimental data – hence the use of the incremental approach in our analysis. As we will discuss later, this provides strong supporting evidence for multiple weight loss processes occurring. Additionally, with the application of the incremental approach, it was noted that for each thermal decomposition function ($f(\alpha)$) used, the measured activation energy was similar to that reported in Figure 7, with similar variation in the activation energy across the extent of conversion (α). Again, this suggests the presence of multiple weight loss processes, as well as providing us with some confidence for using the first order $f(\alpha)=1-\alpha$ expression over the range used. Finally, the use of this model enables us to quite easily calculate the required isothermal time necessary to achieve a specified extent of conversion.

3.4. Single curve incremental isoconversional analysis

Another approach to solving the rate expression in Eqn. 10 is the incremental integral method [14,15]. This method can also be used to take into account the dependence of the kinetic parameters on the extent of conversion, focussing instead on an individual TG experiment rather than a range of different heating rate experiments. In this case, the Runge-Kutta method, an iterative technique for the approximation of ordinary differential equations, was used to solve Eqn. 10. This method uses the previous point (α_n, T_n) to approximate the next (α_{n+1}, T_{n+1}), by using the size of the interval between the points (h) and an estimated average of the slopes. To begin we have the initial condition:

$$\alpha(T_0)=a_0=0 \tag{14}$$

Then, using the Runge-Kutta method, α_{n+1} and T_{n+1} are given by:

$$\alpha_{n+1}=\alpha_n+\frac{1}{6}h(k_1+2k_2+2k_3+k_4) \tag{15}$$

$$T_{n+1}=T_n+h \tag{16}$$

where h is the size of the interval (1°C was used in this analysis), and:

$$k_1=f(T_n,\alpha_n) \tag{17}$$

$$k_2=f\left(T_n+\frac{h}{2},\alpha_n+\frac{hk_1}{2}\right) \tag{18}$$

$$k_3=f\left(T_n+\frac{h}{2},\alpha_n+\frac{hk_2}{2}\right) \tag{19}$$

$$k_4=f(T_n+h,\alpha_n+hk_3) \tag{20}$$

where the function $f(T,\alpha)$ is implied by Eqn. 10. This gives rise to a theoretical curve for the extent of conversion against temperature, which can be fitted to the experimental curve using linear least squares regression in a restricted range of α (hence employing the incremental integral method), by varying values for the activation energy, E_A, and the pre-exponential factor, *A*. Figure 8 demonstrates how the activation energy changes as a function of extent of conversion and heating rate for this analysis method. Because the focus here is on just one data set, particular attention was paid to the fitting procedure to ensure that a global minimum was determined. This was achieved by repeating the fitting multiple times, from different starting points. In each case, across the complete α range and for each heating rate used, the same result was achieved.

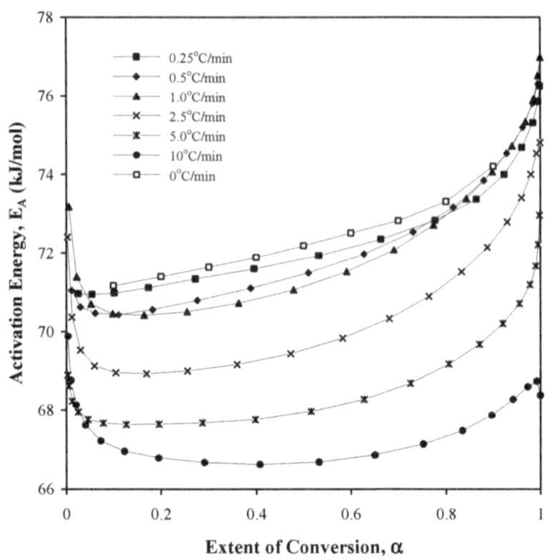

Figure 8. Activation energy with respect to extent of conversion for the different heating rates.

The pre-exponential factor for all analyses remained essentially constant at a value of $(5.3\pm0.8)\times10^{6}$ min^{-1}, where the error analysis here takes into account the variation in the calculated *A* value.

Overall, the calculated activation energy for the mass loss associated with the thermal decomposition of γ-MnO_2 fell within the rather narrow range 66-77 kJ/mol. Nevertheless, within this range there were some systematic changes observed. For all heating rates there was a minimum, or for lower heating rates, a plateau in the activation energy within the extent of conversion range $0.1<\alpha<0.7$. At both higher and lower α values the activation energy increased. This increase is interesting in that it tells us something about the availability of energy or heat, as well as reactants, to affect the thermal transformation. At low α values, corresponding to lower temperatures, there is insufficient heat to activate the reaction, so for all intents and purposes, the activation energy is much larger than normal because very little reaction is occurring. Conversely, at higher α values there is a relatively low concentration of unreacted species available to actually undergo the thermal transformation, and as such the rate of the thermal transformation here is also inhibited (manifested as an increase in activation energy) since reactant concentration is also a limiting factor in chemical reactions. What is also interesting about the data in Figure 8 is the general decrease in calculated activation energy as the heating rate was increased. This is quite clearly demonstrated in Figure 9, which shows how the activation energy at α=0.5 changes with heating rate. The data in this figure indicates that there are two heating rate regions for which there is an exponential decrease in calculated activation energy with heating rate. The cause of this may lie in the apparently less than ideal thermal transfer of heat from the furnace to the sample during the TG experiment. Whether there is a thermal gradient within the powdered sample in the TG pan, or within individual sample particles, it does mean that the thermal decomposition reaction will be occurring at different rates within the sample and/or individual particles. Certainly at higher heating rates this thermal gradient will be much more pronounced, meaning that there is expected to be a greater error in the estimated activation energy when determined at faster heating rates. While the thermal conductivity of manganese dioxide is, to the best of our knowledge, not available in the literature, other similar metal oxides have relatively high thermal conductivities, as shown in Table 2 [16]. Typical thermal conductivity values lie within the range 2-30 W m^{-1} K^{-1}. Of these, the value for titanium dioxide (3.8 W m^{-1} K^{-1}) is most likely very similar to manganese dioxide given the proximity of the metals to each other in the periodic table, and the iso-structural nature of the corresponding oxides. This relatively low value does imply that there will be a reasonable thermal gradient across each particle. To eliminate the effect of thermal conductivity, the exponential relationship between heating rate and calculated activation energy was extrapolated to predict the activation energy under the hypothetical condition of a 0°C/min heating rate. This extrapolated data is also shown in Figure 8, and is expected to more closely represent the true activation energy for this thermal decomposition process.

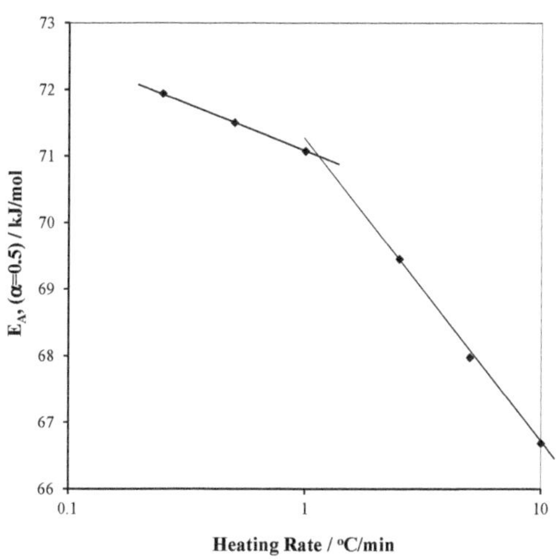

Figure 9. Activation energy (for α=0.5) as a function of heating rate.

Metal Oxide	Thermal Conductivity (W m^{-1} K^{-1})
Al_2O_3 (sintered)	26 (373 K)
$BaTiO_3$	6.2 (300 K)
Fe_3O_4 (magnetite)	7.0 (304 K)
MnO	3.5 (573 K)
SiO_2 (fused silica)	1.6 (373 K)
$SrTiO_3$	11.2 (300 K)
TiO_2	3.8 (400 K)

Table 2. Thermal conductivity of selected metal oxides [16].

As was mentioned previously, the systematic variation in the activation energy, as shown in Figure 8 and highlighted specifically in Figure 9, quite nicely validates the use of this method for the thermal analysis of kinetic parameters. The International Confederation for Thermal Analysis and Calorimetry (ICTAC) has recommended that single scan methods of analysis be avoided where possible (see for example reference [17]). However, in this work the activation energies obtained were not just based on one analysis. In fact, what we have shown here is that the systematic variation within the activation energy over two orders of magnitude change in heating rate (Figure 9) not only provide us with considerable confidence in the resultant data, but also allows us to extrapolate to what the theoretical activation energy would be at a 0°C/min heating rate. This latter outcome has not been reported previously, at least for this system, and in actual fact represents a novel approach to determining an activation energy that is free of experimental artefacts, such as thermal transfer of heat to the sample.

Finally, the kinetic parameters determined for the water loss process can be used to calculate the required isothermal time to achieve complete conversion of the material (α=1), i.e., completely remove water from the material, at a range of temperatures. This value can be found by employing the Arrhenius equation to calculate the rate constant (k) and subsequently, assuming first order kinetics, the heating time can be determined. This was performed for the heating rate 1°C/min, and is shown in Table 3.

Temperature (°C)	Heating time (h)
200	31.65
250	11.12
300	5.07
350	1.52
400	0.27

Table 3. Isothermal heating regimes to achieve complete conversion of the water loss process.

3.5. Comparison of methods

The first order analysis of the kinetics has shown that the activation energy for the loss of water from the manganese dioxide structure increased relatively linearly with extent of conversion, varying from ~109 kJ/mol at α=0.1 to ~250 kJ/mol at α=1.0. On the other hand the incremental integral method led to an activation energy of '66—76 kJ/mol throughout the majority of the thermal transformation, with slightly higher values at both lower and higher extents of the conversion. Clearly there are significant differences in the results obtained from the application of the two kinetic analysis methods, and these differences need to be addressed.

From a purely statistical perspective, the use of multiple experiments to produce data for analysis is bound to contain more variation than just using a single experiment. Under these circumstances, therefore, we might expect that the incremental integral approach should inherently be more reliable than the first order analysis method. Nevertheless, the contribution to the total variation in the analysis made by the individual TG experiments is expected to be quite small, certainly not enough to account for the significant difference between the two methods.

As part of the analysis, a background correction of the DTG data was employed to focus specifically on the mass loss process of interest; i.e., the loss of structural water from the manganese dioxide. While this background correction was applied to all of the data reported here, it was in no way constant between experiments. Furthermore, the shape of the background correction curve was arbitrarily chosen to be exponential. The point being made is that the background correction being made could have easily over- or under-compensated its contribution to the total response, thus inducing some variability between experiments. This would certainly suggest that the incremental integral approach should be the preferred method.

Another likely contributor to variability in the analysis is the thermal conductivity of the manganese dioxide in relation to the heating rate used. Despite the fact that only a small quantity of material (~10 mg) was used in each experiment, the rate with which heat is transferred through the sample is very critical in determining the validity of the resultant information, particularly so since the kinetic analysis model assumes that the sample temperature is uniform throughout. As has already been mentioned, the thermal conductivity of manganese dioxide is not available in the literature; however, the thermal conductivity of similar materials (e.g., titanium dioxide) does suggest that there may be thermal gradients within the manganese dioxide, particularly with the use of fast heating rates. Therefore, those experiments that make use of the higher heating rates could be judged as having a higher relative error compared to those using slower heating rates. Therefore, the incremental integral approach, particularly for those experiments employing a slower heating rate, is most likely to be the preferred method.

As a final comment, the first order kinetic analysis involved the use of separation of variables to solve Eqn. 10. An inherent assumption made with this approach is that the extent of conversion is independent of temperature, when in fact this is not the case given the thermal gradients within the manganese dioxide sample and the fact that the conversion process we are examining covers a very broad temperature range. Under these circumstances, while the data we have collected shows a nice asymptotic change with heating rate (Figure 9), the assumptions made in the numerical analysis do not lead to an accurate estimate of the activation energy, again implying that the incremental integral approach is superior.

4. Material effects

4.1. Introduction

In the preceding discussion we found the incremental iso-conversional approach for analysing thermogravimetric data to be the favourable method to investigate the kinetics of water loss during the heat treatment of a γ-MnO_2 sample. Using the kinetic parameters from this analysis, particular heating regimes at a selection of temperatures were devised to prepare materials with a theoretical complete water loss (α=1), thereby avoiding the use of excessive temperatures and/or times. We now consider the effects of these optimised thermal treatment regimes on the heat treated material structure, composition and morphology. The electrochemical characteristics of these heat treated EMD samples are examined and the observed changes in the material properties used to relate the electrochemical performance to the material properties.

4.2. Heat treated material properties

The XRD patterns for the heat treated MnO_2 samples are shown in Figure 1. The most obvious changes in these XRD patterns as a result of heat treatment are (i) the merging of the (110) (~22° 2θ) and the (130) (~36° 2θ) lines in the starting γ-MnO_2 to form a single peak at ~29° 2θ in the sample heat treated at 400°C; (ii) clearer separation of the (221) and (240) peaks at ~56° 2θ by 400°C; and (iii) disappearance of the peak at ~68° 2θ and the emergence of two peaks at ~66° 2θ and ~73° 2θ. The extent of conversion from the γ-MnO_2 phase, which predominantly displays a ramsdellite composition, to the more thermodynamically stable pyrolusite structure can be determined quantitatively by calculation of the fraction of pyrolusite (P_r) in the samples using the method outlined by Chabre and Pannetier [9]. These results are listed in Table 1. Interestingly, these changes are consistent with literature investigating γ-MnO_2 heat treated at various temperatures for 24 hours [18,19], indicating a progressive structural conversion with increasing heat treatment temperature. This is despite the largely different experimental heating times used. The similarities indicate that this conversion is mostly influenced by the thermodynamics of the heating process rather than the kinetics. For instance, samples heated at lower temperatures (i.e., 200°C and 250°C) have begun the conversion from $\gamma \rightarrow \beta$-MnO_2, although this process is clearly retarded by insufficient thermal energy to drive this process to completion. This is further supported by comparing the data recorded at 400°C using this analysis method (16 mins heating time) with a sample heated at the same temperature for 7 days. In both cases, a P_r value of 0.84 was calculated from the XRD data suggesting few or no kinetic limitations for this process at the elevated temperature.

The structural changes can be further elucidated by consideration of the unit cell parameters (determined from the XRD patterns of these materials assuming an orthorhombic unit cell) as a function of heat treatment temperature. The unit cell parameters for the starting EMD were a_0 = 4.468 Å, b_0 = 9.554 Å and c_0 = 2.833 Å, which is an expansion in the *a-b* plane, but a slight contraction in the *c* direction, compared with ramsdellite [20]. The unit cell parameters for the heat treated materials are shown in Table 1. The decrease shown for both the a_0 and b_0 parameters represents a structural contraction in these directions, while the steady increase in the c_0 parameter indicates lattice expansion in this direction. By considering the differences in the crystal structures of γ-MnO_2 and pyrolusite, it is clear that the main differences are found in the *a-b* plane. Since the unit cell is found to contract along both these directions, this suggests that ion (Mn(IV)) movement predominates in these directions during heat treatment. The excess of edge sharing octahedra in a uniform array in the *c* direction, without any vacancies present to compensate or provide a buffer for the close proximity of the Mn(IV) ions, results in expansion in this direction [18].

The changes in BET surface area for the HEMD samples are shown in Table 1. The relatively high surface area for these materials indicates that they are quite porous. Most evident from this data is the general decrease in surface area as temperature increases, barring a slight increase between the 350°C and 400°C samples. Given the structural changes occurring during heat treatment, the decreasing surface area suggests that the pores in EMD are removed as Mn(IV) ions diffuse through the structure, creating a more defect free and crystalline material [18]. The slight increase in surface area for the 400°C material is likely to be caused by slow kinetics for this process (relating to the much shorter heating period applied to this material), thus not allowing for the completion of pore collapse. Noticeably, at the lower temperatures (e.g., 200°C), the surface area has not decreased significantly from the original value. This is likely to be connected to insufficient activation energy at the relatively low temperature to drive Mn(IV) diffusion, a factor responsible for pore closure. Table 1 also lists the changes in micro- (<2 nm) and meso-pore (2-50 nm) structures as a result of heat treatment. Clearly, heat treatment causes the collapse of micro-pores, while an increase in the meso-pore volume was observed. The increase in the micro-pore volume for the 400°C material with respect to the 350°C sample indicates that kinetic limitations in the collapse of these pores is responsible for the increase in BET surface area for this material.

A comparison of how the structural changes relate to morphological changes clearly portrays the key differences in the HEMD materials prepared. Figure 10 compares the changes in the orthorhombic unit cell volume with BET surface area. From this data, it is evident that between the temperatures tested, small changes in BET surface area can relate to large structural changes (e.g., between the standard EMD and 200°C material), or vice versa. Generally, however, the interplay of the thermodynamics and kinetics influencing the variation in these parameters leads to an approximately exponential decrease in unit cell volume with respect to BET surface area. Decreases in the unit cell volume can be attributed to manganese ions within the structure having sufficient thermal energy to move to positions consistent with pyrolusite, thus causing intra-crystallite rearrangement within the material. Conversely, changes in BET surface area relate to either the sintering of crystallites (both intra- and inter-) by joining together across the pores thus resulting in pore closure, or the opening of existing pores.

The underlying assumption in the preparation of these samples is that structural water, which is associated with defects (i.e., Mn(III) and cation vacancies), has been completely removed from the material. There is an expectation then that each of the HEMD samples will have no Mn(III), cation vacancies or water removed above 110°C (%H_2O(>110°C)). However, the compositional data for the HEMD samples (Table 1) shows that this is not the case. In light of this, the trends evident in these parameters provide important insights into the kinetic and thermodynamic limitations of the heat treatment under the set conditions and at these temperatures. The data for these samples show that the total manganese content (%Mn(T)) increases with heating temperature to an optimum at 300°C. This increase relates to the rearrangement of Mn(IV) ions to form a more defect free structure as cation vacancies are annealed via loss of structural protons. As a consequence, a steady decrease in the cation vacancy fraction is observed with heating temperature. The gradual decline in this parameter suggests that as the thermal energy required for this process is met by the higher temperatures, vacancy removal is able to proceed to a greater degree. After heat treatment at 350°C and 400°C, vacancy defects and associated water have been completely removed from the structure, indicating that these heating regimes have provided sufficient thermal energy to complete this process, without any kinetic limitations.

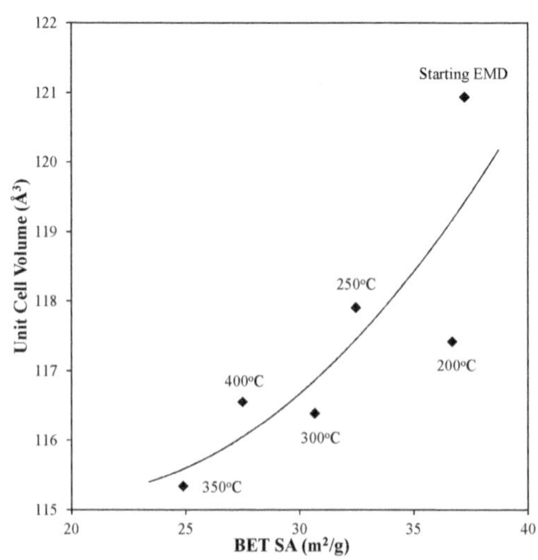

Figure 10. Relationship between BET surface area and unit cell volume for heat treated EMD samples.

The trends in Mn(III) and Mn(IV) percentages are confused to some degree due to a number of processes taking place during the heating experiment. First, the oxidation of Mn(III) to Mn(IV) takes place as a consequence of the oxidising air atmosphere present, thus lowering Mn(III) and increasing Mn(IV). Further, both Mn(III) and Mn(IV) experience a proportionate increase due to the increase in the total manganese content. Additionally, although limited, some thermal reduction of the material is possible at the higher temperatures considered, thereby causing slight increases to Mn(III) at the expense of Mn(IV). These processes superimpose to give the relatively steady Mn(IV) and slightly increasing Mn(III) content with respect to heat treatment temperature.

The proportions of water removed above and below 110°C '(%H_2O(>110°C)) and (%H_2O(<110°C)) respectively, further support these findings. The %H_2O(<110°C), which is water adsorbed on the surface of the material, generally decreases as heat treatment temperature increases, barring slightly higher values for the 300°C and 400°C materials relative to those around them. Clearly, the amount of water adsorbed to the surface of the material is largely determined by the available surface area and therefore it is no surprise that this result reflects the relative BET surface area for these samples (Table 1). The steady decrease in %H_2O(>110°C) (i.e., structural water associated with Mn(III) and vacancy defects) with heat treatment temperature provides a second, independent measure of the extent of structural water removal from the HEMD samples. By 350°C it appears that all structural defects that will be removed in the heat treatment have been, evidenced by the approximately constant %H_2O(>110°C) value for the 350°C and 400°C materials.

The trends in material properties discussed above are specific to a single EMD heat treated under the temperature/time determined from kinetic analysis. However, our investigation of the influence of the starting EMD properties on the resultant HEMD properties has shown these relationships to hold for a broad range of EMD materials [21].

4.3. Electrochemical performance

Due to the large number of cells tested (i.e., five HEMD materials each discharged at four currents), only the discharge characteristics of each HEMD at 2 mA/g are shown in Figure 11. The typically flat discharge curve of Li/MnO_2 at 2.8-3.0 V is evident from this plot. The primary discharge capacity (in mAh/g of MnO_2) for the cells tested was calculated by using a 2.0 V cut-off point. This result is shown in Table 4. As expected, the capacity of a given material decreases with higher discharge currents. On average the capacity for unheated EMD was 104 mAh/g (or 34% utilisation), which was less than half that of HEMD, with an average of 238 mAh/g (77% utilisation) at the 2 mA/g discharge rate. This clearly demonstrates the importance of the heat treatment process, as has been previously reported in the literature [22-24]. This vast difference in performance also suggests that EMD based cells are failing in a different way to HEMD cells. The high water content of the EMD likely leads to destructive side reactions with the electrolyte and anode causing cell failure. Conversely, HEMD is limited by the intrinsic material properties.

Considering the HEMD materials in greater detail, we note that manganese dioxide heat treated at 250°C and above maintained relatively good capacity at the higher discharge rates. Also of note are the excellent performance characteristics of the 300°C and 350°C materials particularly at the lower discharge rates. It was also observed that of the various HEMD samples, the material treated at 200°C exhibited relatively poor discharge capacities, this being especially noticeable at the higher discharge rates.

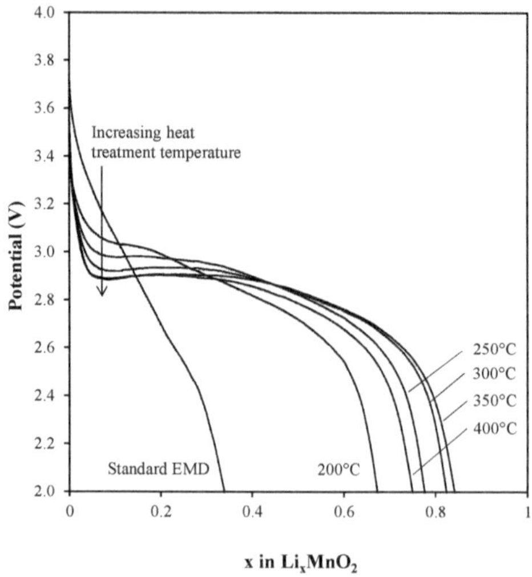

Figure 11. Discharge curves for heat treated EMD samples discharged at 2 mA/g.

Temperature (°C)	Capacity (mAh/g)			
	2 mA/g	5 mA/g	10 mA/g	20 mA/g
25	104.1	62.9	59.8	-
200	207.5	185.0	169.4	103.8
250	242.9	208.4	215.1	183.8
300	252.9	229.7	195.4	178.4
350	252.0	221.5	218.9	170.5
400	233.4	226.0	223.7	185.4

Table 4. Discharge capacity for heat treated EMD samples at various discharge currents (2.0 V cut-off).

Normalising the capacity of the cell in terms of the proportion of electrochemically active material (i.e., Mn(IV) percentage) reveals the extent to which the available material is being used. This will have a significant effect on the relative capacity of these HEMD materials since the proportion of Mn(IV) to Mn(III) has been seen to vary depending on the heat treatment conditions. The normalised capacities are listed in Table 5, and demonstrate that heat treatment at 300°C and 350°C has resulted in the highest proportion of electrochemically active Mn(IV) being utilised at the at 2 mA/g rate (94%).

Another useful way in which to present the data from these tests is through the use of a Ragone diagram, comparing the relative specific energy and power output of the materials. Due to the number of cells tested, only the points corresponding to the energy delivered by the cell by the cut-off voltage, and the corresponding power value, are shown in Figure 12. This diagram clearly shows a greater differentiation in energy delivered at the low discharge rates, with the 300°C and 350°C materials exhibiting superior performance. Further, at high rates heat treatment at 250°C and above has resulted in materials exhibiting virtually equivalent performance. Comparison with literature Ragone diagrams [4] reveals that the performance of our materials is superior in terms of specific energy, and comparable in terms of specific power.

Temperature (°C)	Utilization of Active Material			
	2 mA/g	5 mA/g	10 mA/g	20 mA/g
25	0.39	0.23	0.22	-
200	0.77	0.69	0.63	0.39
250	0.87	0.75	0.77	0.66
300	0.94	0.85	0.73	0.66
350	0.94	0.83	0.82	0.64
400	0.89	0.86	0.85	0.71

Table 5. Utilization of active material (Mn(IV) percentage) for heat treated EMD samples at various discharge currents (2.0 V cut-off).

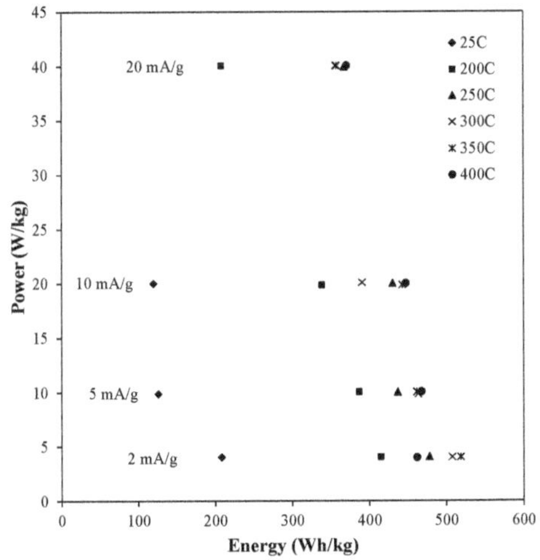

Figure 12. Ragone diagram showing points corresponding to energy delivered by the cell by the cut-off voltage and the corresponding power value for the heat treated EMD samples with respect to discharge rates.

As the electrochemical behaviour of manganese dioxide is largely dependent on the structure, composition and morphology of the material, the changes in these parameters with heat treatment temperature are invoked in explaining their relative electrochemical performance. Since the performance is not directly proportional to the heating temperature, it is immediately evident that there are a number of factors influencing material performance. From the current understanding of the Li/MnO_2 system, these factors are outlined as follows:

i. It is well known that the removal of both surface and structural water from the manganese dioxide structure is paramount, and that higher heating temperatures perform this function most completely and effectively [22].
ii. A greater amount of the more electrochemically active Mn(IV) species (compared to Mn(III)) is advantageous, and therefore materials in which oxidation of the Mn(III) to Mn(IV) has occurred will exhibit better performance.
iii. Samples retaining the γ-MnO_2 structure, with larger [1×2] tunnels in ramsdellite regions of the structure, are thought to provide both a greater degree of mobility and a larger number of insertion sites for ions introduced into the structure during discharge [25]. These relationships certainly hold for the aqueous system, although as yet there is currently no clear indication in the literature relating these structural features to the ease of lithium diffusion through materials used in the non-aqueous Li/MnO_2 system.
iv. The effect of sample porosity on the discharge characteristics of Li/MnO_2 cells remains relatively unknown. However, it has been previously determined that during heat treatment micro-pores within the EMD are sintered shut as a result of manganese ion movement, while the structural contraction connected with the phase transition from $\gamma\rightarrow\beta$-MnO_2 and the increase in material density, induces cracking [18]. This results in an increase in the number of larger pores. These factors will be especially influential on the rate capabilities and total capacity of the cell [26].

In explaining the relative performance of the HEMD tested, each material will be considered separately and the above factors used to explain the resulting electrochemical performance. Some comparison will also be made highlighting differences between materials heat treated at consecutive temperatures and the effect these differences have had on the electrochemical performance. In the proceeding discussion, those materials exhibiting poorest performance will be discussed first, followed by the HEMD samples with better electrochemical characteristics.

From the physical characterisation of the 200°C material, it is clear that it has retained much of its original BET surface area, experienced some removal of cation vacancies, lost only a small amount of structural water, and maintained to a large extent the γ-MnO_2 parent structure. Of these features, the presence of structural water associated with cation vacancies and Mn(III) will be a limiting factor on the electrochemical performance. The water present evidently makes for poor performance at all discharge rates (on average 15% lower capacity than the 250°C material), highlighting the importance of its removal prior to use in Li/MnO_2 system. Perhaps an additional feature that limits the performance of this material is the relative ease of lithium ion movement in γ/β-MnO_2. While fundamental work has been performed analysing the mechanism and activation energy for proton diffusion in MnO_2 structures [25], these relationships are currently unknown for lithium diffusion in EMD/HEMD. The data presented here may indicate slower lithium ion diffusion through less pyrolusitic structures.

The properties of the 400°C material are very similar to the sample heat treated at 350°C; i.e., both samples have had basically all cation vacancies removed, with minor, but similar, amounts of structural water remaining, and both have experienced a high structural conversion to pyrolusite. However, an interesting difference is apparent in Figure 10, when comparing the structural and morphological properties. Here, the sample heated at 400°C has maintained a larger unit cell and higher surface area compared to the 350°C material, most likely as a consequence of the shorter heating time applied to this material. This difference is likely responsible for its slightly lower performance at 2 mA/g (233 mA/g compared to 252-3 mA/g for the 300°C and 350°C materials). It is also possible that the conversion of γ-MnO_2 to a more pyrolusite-like structure as a consequence of the high heating temperature has been limiting on the performance. In conjunction with the low performance for the 200°C material, this may suggest that the optimum γ/β-MnO_2 structure for lithium insertion is not highly ramsdellitic or pyrolusitic structures, but rather some intermediate structure. From this, we propose an optimum value of P_r for HEMD at low discharge rates, lying somewhere between 0.5 and 0.84, which leads to superior performance. At high rate discharge however, the 400°C material capacity is similar to the other materials, suggesting the differences in material properties have become more or less immaterial in determining the electrochemical performance.

The capacity of the material heat treated at 250°C (243 mAh/g) is higher the 400°C material (233 mAh/g) and slightly below the 300°C and 350°C materials (252-3 mAh/g) at the 2 mA/g discharge rate, but is essentially equal to them at high rates (~180 mAh/g at 20 mA/g). This is despite the fact that this material still contains reasonable amounts of structural water associated with both cation vacancies and Mn(III). The decreased capacity at the low rate can likely be attributed to the moderate retention of the γ-MnO_2 structure (P_r=0.52) which may limit the diffusion of lithium through the structure, although evidently to a lesser degree than in the 200°C material or the highly pyrolusitic 400°C structure. Additional differences between the 250°C material and those around it can be clearly seen in Figure 10. Noticeably, although the unit cell volume has not varied significantly between the 200°C and 250°C materials, the surface area has decreased by around 10%. This is largely due to a lower micropore volume in the 250°C material (Table 1). This feature would likely aid the discharge of this material since the small micro-pores can only accommodate a limited number of lithium ions. Hence, when the cell is discharged, the few ions in these pores are soon inserted into the structure. The subsequent deficiency of lithium ions in this locality limits the discharge capabilities of the cell, especially at high discharge rates.

The materials heat treated at 300°C and 350°C demonstrated the highest capacity over the range of discharge rates tested. Relating this back to the properties, the 300°C material had a moderate amount of defects remaining in the structure, with a pyrolusite to ramsdellite ratio of 0.73. The particular structural arrangement is likely of particular significance in light of the poorer performance of materials with low or high P_r values (e.g., 200°C and 400°C materials, respectively). Also of interest are the relationships between the structural and morphological features of the 300°C sample (Figure 10). This data suggests that this material has undergone significant amounts of structural rearrangement (shown by a smaller unit cell), but maintained a relatively high surface area (31 m^2/g). The combination of the structural arrangement, composition and surface area has resulted in this material delivering one of the highest capacities at low discharge rates, with comparable performance at high rates.

Finally, as we have already noted the material properties of the 350°C material are similar to that of the 400°C sample in terms of composition, although key differences are noted in the structure and morphology, as highlighted in Figure 10. These differences are significant in light of the superior electrochemical characteristics of the 350°C material. Figure 10 shows that heat treatment at 350°C has resulted in the greatest contraction of the unit cell and also a lowering of the BET surface area relative to the other HEMD materials. The smaller unit cell is the main structural difference between the 350°C and 400°C materials, which otherwise have very similar P_r values (0.83 and 0.84, respectively). This suggests that the proportion of each phase is not the only structural feature influencing the movement of lithium ions through the material, and that the unit cell volume is also significant. The high performance of the 350°C material is an indication that lower unit cell volumes are favoured for lithium diffusion in HEMD. The morphology differences between these materials are also likely to play a significant role in determining their relative electrochemical performance. As previously noted, the lower BET surface area for the 250°C sample due to a lower volume of micropores, was found to enhance its capacity over the 200°C material. Similarly, the change in morphology brought about by the removal of these micropores in the 350°C material has clearly improved the material capacity across discharge rates.

5. Conclusion

In conclusion, this work has shown that kinetic analysis can be employed in optimising the thermal processing of positive electroactive materials. As an example, kinetic analysis using a first-order kinetic analysis method and an incremental integral approach was performed for the water loss process from electrolytic manganese dioxide, the precursor material for use in non-aqueous Li/MnO_2 cells. Both methods determined the pre-exponential factor and the activation energy for this process as a function of the extent of conversion, hence allowing for variation in the values over the process studied, while the incremental integral method also varied with heating rate.

Comparison and investigation of the two methods revealed the incremental integral method as the preferred means for obtaining the kinetic parameters describing this process. This result stems back to the use of a single TG experiment and the exclusive use of a slower scan rate using the incremental integral method (multiple scan rates, and thus experiments, are required in the first order kinetic analysis method). A single experiment avoids inherent variability between multiple experiments and limits errors induced in the somewhat arbitrary application of the background correction. Using slower scan rates reduces the possible undesired effects arising from thermal gradients across the sample as a result of the thermal conductivity of the material.

Using the kinetic parameters determined for this process, and assuming first order kinetics, a family of thermal treatment regimes based on heating temperature and duration were calculated to, in theory, completely remove water from the manganese dioxide structure.

Characterization of the resulting HEMD materials revealed considerable differences in their properties. It was found that materials undergo a mainly kinetically limited thermal conversion from γ-MnO_2 towards β-MnO_2. The material surface area generally decreases with heating temperature, although noticeably, the pore collapse process is thermally limited at lower temperatures, while at the higher temperatures the reaction kinetics dictate the extent to which this process proceeds. Despite the theoretical expectation of complete water removal from the EMD, experimental results demonstrate that thermal and kinetic limitations present under these heating conditions prevent the completion of the water loss process. In general, however, as heating temperature increases there is a decrease in cation vacancy fraction and amount of structural water, while Mn(III) and Mn(IV) increase and remain relatively steady, respectively.

Electrochemical performance of the HEMD materials generated in this work revealed they are superior in terms of specific energy, and comparable in terms of specific power, to recent reports in the literature for the primary Li/MnO_2 system. Greatest differentiation of materials was noted at the low discharge rates, where materials prepared at 300°C and 350°C demonstrated best performance.

Finally, it is important to note that the method of kinetic analysis outlined here is not limited to only mass loss processes. It can in fact be employed to monitor the progress of any thermally based reaction which undergoes changes measurable by TG, DTA, DSC, etc. Therefore, the method of kinetic analysis outlined here could be applied in optimising the preparation and electrochemical performance of numerous other positive electroactive materials with similar improvements to performance anticipated.

Author details

Wesley M. Dose and Scott W. Donne
Discipline of Chemistry, University of Newcastle, Callaghan, Australia

Acknowledgement

The authors would like to acknowledge the University of Newcastle EM-X-ray Unit and Ms. Jenny Zobec for assistance in obtaining the XRD data. WMD acknowledges the University of Newcastle for the provision of an APA PhD scholarship.

6. References

[1] Lewis N.S (2005) Scientific Challenges in Sustainable Energy Technology, Keynote address at 208th meeting elecrochem. soc., Los Angeles.
[2] Ristinen R.A, Kraushaar J.P (2005) Energy and the Environment. 2nd Ed. USA: Wiley.
[3] Ragone D (1968) Review of Battery Systems for Electrically Powered Vehicles. SAE Technical Paper 680453. doi:10.4271/680453.
[4] Simon P, Gogotsi Y (2008) Materials for Electrochemical Capacitors. Nat. mater. 7: 845-854.
[5] Linden D, Reddy T.B, editors (2001) Handbook of Batteries. 3rd ed. USA: McGraw-Hill.
[6] Whittingham M.S (2004) Lithium Batteries and Cathode Materials. Chem. rev. 104: 4271-4301.

[7] Kim D.H, Kim J (2006) Synthesis of $LiFePO_4$ Nanoparticles in Polyol Medium and Their Electrochemical Properties. Electrochem. solid-state lett. 9: A439-A442.
[8] Wickham D.G, Croft W.J (1958) Crystallographic and Magnetic Properties of Several Spinels Containing Trivalent JA-1044 Manganese. J. phys. chem. solids. 7: 351-360.
[9] Chabre Y, Pannetier J (1995) Structural and Electrochemical Properties of the Proton/γ-MnO_2 System. Prog solid state chem. 23: 1-130.
[10] Brunauer S, Emmett P.H, Teller E (1938) Adsorption of Gases in Multimolecular Layers. J am chem soc. 60: 309-319.
[11] Vogel A.I (1989) Quantitative Chemical Analysis. New York: John Wiley & Sons.
[12] Williams R.P (1995) Characterisation and Production of High Performance Electrolytic Manganese Dioxide for use in Primary Alkaline Cells. PhD Thesis, University of Newcastle, Australia.
[13] Fraioli A.V (1985) Investigation of Manganese Dioxide as an Improved Solid Desiccant. Proc. electrochem. soc. 85-4: 342-368.
[14] Budrugeac P, Segal E (2001) Some Methodological Problems Concerning Nonisothermal Kinetic Analysis of Heterogeneous Solid-Gas Reactions. Int. j. chem. kinet. 33: 564-73.
[15] Simon P, Thomas P.S, Okiliar J, Ray A.S (2003) An Incremental Integral Isoconversional Method. J. therm. anal. calorim. 72: 867-874.
[16] Haynes W.M, editor (2009) CRC Handbook of Physics and Chemistry. 91st ed. USA: CRC Press.
[17] Maciejewski M (2000) Computational Aspects of Kinetic Analysis. Part B: the ICTAC Kinetics Project B - the Decomposition Kinetics of Calcium Carbonate Revisited, or Some Tips on Survival in the Kinetic Minefield. Thermochim. acta. 355: 145-154.
[18] Arnott J.B, Williams R.P, Pandolfo A.G, Donne S.W (2007) Microporosity of Heat-Treated Manganese Dioxide. J. power sources. 165: 581-590.
[19] Malloy A.P, Browning G.J, Donne S.W (2005) Surface Characterization of Heat-Treated Electrolytic Manganese Dioxide. J. colloid. interface. sci. 285: 653-664.
[20] Bystrom A.M (1949) Crystal Structure of Ramsdellite, an Orthorhombic Modification of Manganese Dioxide. Acta chem. scand. 3: 163-173.
[21] Dose W.M, Donne S.W (2011) Heat Treated Electrolytic Manganese Dioxide for Li/MnO_2 Batteries: Effect of Precursor Properties. J. electrochem. soc. 158: A1036-A1041.
[22] Ikeda H, Saito T, Tamura H (1975) Manganese Dioxide as Cathodes for Lithium Batteries. Manganese dioxide symp. [proc.]. 1: 384-401.
[23] Ilchev N, Manev V, Hampartzumian K (1989) The Lithium-Manganese Dioxide Cell II. Behaviour of Manganese Dioxide in Nonaqueous Electrolyte. J. power sources. 25: 177-185.
[24] Ohzuku T, Kitagawa M, Hirai T (1989) Electrochemistry of Manganese Dioxide in Lithium Nonaqueous Cell. J. electrochem. soc. 136: 3169-3174.
[25] Balachandran D, Morgan D, Ceder G (2002) First Principles Study of H-Insertion into MnO_2. J. solid state chem. 166: 91-103.
[26] Lee T.J, Cheng T.T, Juang H.K, Chen S.Y (1993) Relationship of Cathode Pore-Size Distribution and Rated Capacity in Lithium/Manganese Dioxide Cells. J. power sources. 43-44: 709-712.

Chapter 12

Synthesis, Characterization and Properties of Zirconium Oxide (ZrO_2)-Doped Titanium Oxide (TiO_2) Thin Films Obtained via Sol-Gel Process

Rabah Bensaha and Hanene Bensouyad

Additional information is available at the end of the chapter

1. Introduction

Sol–gel process [1-3] is an attractive alternative to other methods for synthesis of ceramics and glasses for many reasons: for example, low temperature synthesis, simple equipment's to be used, thin film formability and so on. Particularly, sol–gel process is very useful for thin film deposition because of the capability to coat materials of various shapes and/or large area, to control the composition easily for obtaining solutions of homogeneity and controlled concentration without using expensive equipment.

Historically, metal alkoxides have been employed in sol–gel process, which readily undergo catalyzed hydrolysis and condensation to form nanoscale oxide or hydroxide particles. Still in general, metal alkoxides are often used as raw materials in sol–gel process, but many of the alkoxides are very difficult to be obtained because of the high sensitivity to the atmospheric moisture [4-9]. In ordinary sol–gel processing, starting compositions as well as reaction conditions are selected so as to maintain the mixture in a homogeneous state throughout the processes including mixing of starting compounds, gelation, aging, drying and heat-treatment.

Titanium and zirconium oxides are very promising candidates for future technology of thin layers because of their interesting mechanical, thermal and chemical properties. Titanium oxide (TiO_2) is a cheap, non-toxic, and non-biodegradable material, besides their widely uses in various industries [10]. Moreover, it is a semiconductor under the form of thin films. Its insensitivity to visible light due to its band gap (3.2 eV) enables it to absorb in the near ultraviolet region [11], even though its low efficiency. Hence, it can be sensitized by a great number of dyes; some of them allow a conversion rate incident photon–electron approaching unity. Thus, these various applications arouse great interest in the study of titanium oxide thin films. The significant uses of TiO_2 thin films are in solar cells [12], photo-catalytic [13] and electro-chromic systems [14], in other words, they are mainly found their use in optics.

TiO_2 thin films are extensively studied because of their interesting chemical, electrical and optical properties [15,16]. TiO_2 film in anatase phase could accomplish the photocatalytic degradation of organic compounds under the radiation of UV. So, it has a variety of application prospects in the field of environmental protection [17,18]. TiO_2 thin film in rutile phase is known as a good blood compatibility material and can be used as artificial heart valves [19]. In addition, TiO_2 films are important optical films due to their high reflective index and transparency over a wide spectral range [16].

During the two last decades, several methods have been used for the TiO_2 thin films preparation, such as chemical vapor deposition [20], chemical spray pyrolysis [21], pulsed laser deposition [22] and sol–gel method [23]. In comparison with other methods, the sol–gel method has some advantages such as controllability, reliability, reproducibility and can be selected for the preparation of nano-structured thin films [23,24]. Sol–gel coating has been classified as two different methods such as dip and spin coating.

The dip-coating has considerably been used for preparation TiO_2 nanostructured thin films [25–27]. Experimental results have shown that the preparation of high transparent TiO_2 thin film by dip-coating method needs to control morphology, thickness of the film and the anatase-brookite-rutile phase transformation [26,28].

Additions of another semiconductor have been used to improve the properties of titanium dioxide. In principle, the coupling of different semiconductor oxides seems useful in order to achieve a higher photocatalytic activity [29]. Various composites formed by TiO_2 and other inorganic oxides such as SiO_2 [30], ZrO_2 [31] SnO_2 [32], Cu_2O [33], MgO [34], WO_3 [35], In_2O_3 [36], ZnO [37], MoO_3 [38], CdS [39], PbS [40], and so on, have been reported.

Zirconium oxide (ZrO_2) has good dielectric and optical properties [41,42] it has a high refraction index [43]. Additionally, it has a very good transparency on a broad spectral field [44], a great chemical stability and a threshold of resistance to high laser flow. All these properties led to miscellaneous applications such as optical filters, laser mirrors [45] or barriers layers from the heat [46]. ZrO_2 films are also employed as plug layer for super-conducting ceramics [47,48], like biomaterial for prostheses [49,50], as gas sensor [51] or like component in combustible batteries [52]. Basically, ZrO_2 itself is an insulating direct wide gap metal oxide, with an optical band gap in the range 5.0-5.85 eV [53].

The aim of the present work is to investigate the transformation behaviors and the effect; of a smaller ratio range of ZrO_2; doping on the surface area of TiO_2 thin films, light absorption, band gap energy, variations of crystal granularity, phase composition and especially on the evolution of the crystallite size and defects concentration with annealing treatments (heat treatments) and layers thickness of the samples produced. So that in this chapter, we report the study of structural, thermal and optical properties of ZrO_2-doped TiO_2 thin films deposited by the sol–gel process. Several experimental techniques were used to characterize structural and optical properties resulting from different annealing treatments and different layer thicknesses: X-ray powder diffraction, Fourier transforms infrared (FTIR), Raman spectroscopy, Scanning electron microscopy (SEM), differential scanning calorimetric (DSC), Scanning electron microscopy (SEM), the energy dispersive X-ray spectrometry (EDX) and UV spectroscopy.

To obtain nanomaterial's with controlled properties, it is most often involves the use of mineral additives (dyes, semiconductors, metal particles, rare earth, etc.) in small quantities. These additions can promote densification or control the phenomenon of grain growth, or to change the structural, physical and optical properties. So the presence of impurities in a matrix can stabilize, improve or modify the various properties of a material. Generally, thin layers of doped TiO_2 give hope of significant performance gains and new applications.

2. Experiments

Our 5% ZrO_2-doped TiO_2, thin films were prepared by dip coating, in three steps. The first step: the dissolution of 1 mol of butanol (C_4H_9OH) as solvent, 4 mol of acetic acid ($C_2H_4O_2$), 1mol of distilled water and 1 mol of tetrabutylorthotitanate $(C_4H_9O)_4Ti$. In the second step, the solution of ZrO_2 was prepared from the dissolution of 1 mol of zirconium oxychloride salt ($ZrOCl_2 \bullet 8H_2O$) in distilled water and 2 mol of ethanol (95%) as catalyst. Finally, the solution of TiO_2 was doped with ZrO_2. Then, the resultant yellowish transparent solutions were ready for use. The substrates were dip-coated in the solutions at a constant rate of 6.25 cm.s-1. After each dipping, thin films were dried for 30 min at a distance of 40 cm from a 500 W light source. The drying temperature of the light source is approximately equal to 100 °C. Subsequently, thin films were heat treated in the temperature range 350–450 °C, with a temperature increase rate of 5°C.min^{-1}, for 2 h in the furnace. The powders obtained from the xerogel were prepared in room temperature and under air atmosphere.

After each dipping, the thin films were dried for 30 min, at a distance of 40 cm from a 500 Wight source. The drying temperature of the light source is approximately equal to 100 °C. Subsequently, thin films were heat treated in the temperature range 350–450 °C, with a temperature increase rate of 5 °C min^{-1}, for 2 h in the furnace. The powders obtained from the xerogel were prepared with an annealing till three months in room temperature and under air atmosphere.

To investigate the transformation behaviors and the effect; of a smaller ratio range of ZrO_2; doping on the surface area of TiO_2 thin films, light absorption, band gap energy, variations of crystal granularity, phase composition and especially on the evolution of the crystallite size and defects concentration with annealing treatments (heat treatments) and layers thickness of the samples produced.

So that in this chapter, we report the study of structural, thermal and optical properties of ZrO_2-doped TiO_2 thin films deposited by the sol–gel process.

Several experimental techniques were used to characterize structural and optical properties resulting from different annealing treatments and different layer thicknesses: X-ray powder and films diffraction, Fourier transforms infrared (FTIR), Scanning electron microscopy (SEM), Raman spectroscopy, differential scanning calorimetric (DSC), Scanning electron microscopy (SEM), the energy dispersive X-ray spectrometry (EDX) and UV spectroscopy.

To determine the transformation points, the obtained powdered xerogels were analyzed by Differential Scanning Calorimetry (DSC) using a SETARAM DSC–92 analyzer equipped with a processor and a measuring cell. The thermal cycle applied consists of heating from room temperature to 520°C, holding for 5min at this temperature and finally cooling back to room temperature with the same rate (5°C/min). X-ray powder diffraction was performed by Siemens D5005 diffractometer using a Cu $K_{\alpha 1}$ radiation. The patterns were scanned at room temperature, over the angular range 10-70° 2θ, with a step length of 0.1° 2θ and counting time of 1 $s.step^{-1}$. The UV absorption studies were carried out using UV-VIS double–beam spectrophotometer SHIMADZU (UV3101PC). Its useful range is between 190 and 3200 nm. The treatment of the spectra was performed using the UVPC software. A surface profiler DEKTAK 3ST AUTO1 (VEECO) was used to determine film thicknesses. Raman spectra were recorded in a back scattering configuration with a Jobbin Yvon micro Raman spectrometer coupled to a DX40 Olympus microscope. The samples of doped and undoped TiO_2 thin films were excited with a 632.8 nm wavelength with an output of 20 mw.

3. Results

3.1. Solution properties

3.1.1. Viscosity

This part is devoted to study the effect of solution aging and its viscosity on the films thickness. To do this study, five samples are developed, successively on the day, the next day, after seven days, after 10 days, and after fourteen days of solution synthesis. The conservation of the sol during the 14 days is made at room temperature.

Table 1 shows an important change of layers thickness with the increasing of the solution viscosity. After a repose period of 14 days of synthesis solution doped with ZrO_2, the thickness of layer changes from 32 nm, with a viscosity 10 mPa.s, to 81 nm when the viscosity is 180 mPa.s. We notice that the ZrO_2-doped TiO_2 solution becomes more viscous over time (Figure 1). This reflects the rate of polycondensation reaction progress.

solution age	η (mPa.s)	Measured thicknesses d (nm)
0 day	10	32
1 day	20	33
7 days	60	39
10 days	110	61
14 days	180	81

Table 1. Variation of film thicknesses d (nm) with solution viscosity.

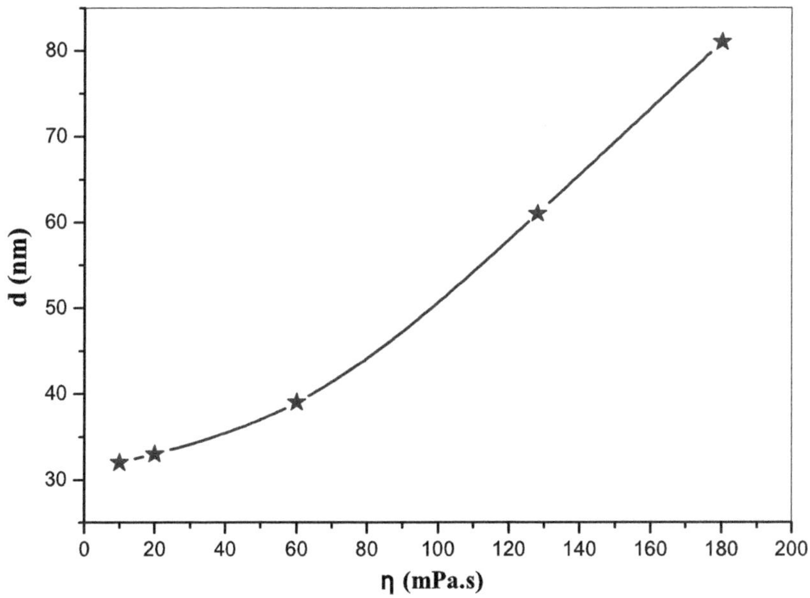

Figure 1. Variation of film thicknesses d (nm) with solution viscosity η(mPas.s)

3.2. Powder Properties (Xerogel)

3.2.1. X-Ray Diffraction (XRD)

Figure 2a and b shows the X-ray diffraction (XRD) patterns of TiO_2 xerogels of undoped (Figure 2a) and 5% ZrO_2-doped TiO_2 (Figure 2b). The XRD pattern evolution of titanium xerogel obtained after the evaporation of the organic compounds during 3 months of aging at ambient temperature shows that it is an amorphous phase as reported in [54].

It has been reported that the used acid catalyst, during sol–gel preparation, plays a crucial role for determining the TiO_2 phase, in literature [54, 55], they found that powder is amorphous when they use acetic acid as catalyst. However, when using formic acid they found that, in addition to amorphous phase, there is an amount of the anatase nanoparticles. This analysis of the doped TiO_2 xerogel exhibits that the addition of 5% ZrO_2 (Figure 2b) would be largely sufficient to form nanoparticles of anatase which crystallizes with (101) plane. It is interesting to note that the addition of a minor amount of ZrO_2 starts crystallization of anatase. Whereas, A. Kitiyanan et al. [56] B. Neppolian et al. [57] reported that addition of ZrO_2 has no effect on TiO_2 oxide morphology..

3.2.2. Thermal analysis

The differential scanning calorimetric (DSC) curves of undoped TiO_2 and 5% ZrO_2-doped TiO_2 xerogels are shown in Figure 3 It is interesting to note that both doped and undoped xerogels showed a similar thermal behavior in the temperature range 20–250 °C. Generally, weight loss corresponds to the evaporation of water, thermal decomposition of butanol as well as carbonization or combustion of acetic acid and other organic compounds [58-60] which constitute metal alkoxides. Hence, the above thermal events were represented by an endothermic peak spreading from 50 to 250 °C.

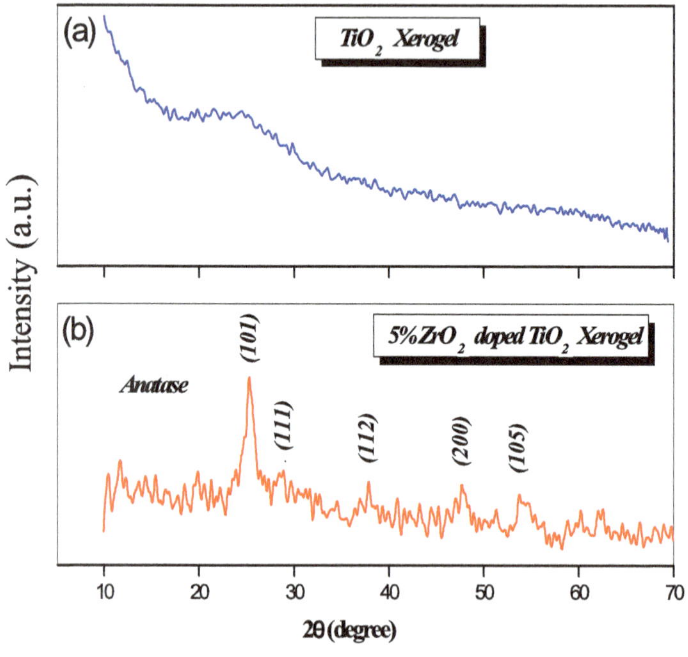

Figure 2. Evolution of XRD patterns of xerogels (a) undoped TiO_2, and (b) 5% ZrO_2-doped TiO_2.

A double exothermic peak in the 260–450 °C temperature range of TiO_2 xerogel can be attributed to the crystallization of titanium oxide [61].

The addition of 5% of zirconium oxide led to a shift of exothermic peak phase towards lower temperatures. This may be due to the speeding up of the crystallization of titanium oxide compared to the undoped one.

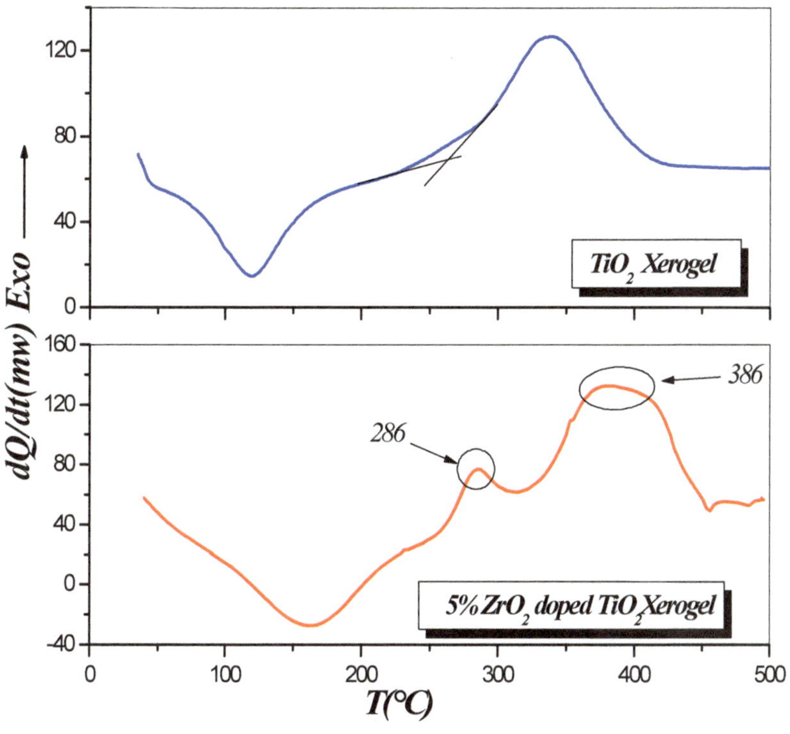

Figure 3. Differential scanning calorimetric curves of xerogels: (a) undoped TiO_2 and (b) 5 % ZrO_2-doped TiO_2.

3.3. Thin films properties

3.3.1. Structural Properties

3.3.1.1. Thin films thickness

The measured values of thin films thickness, given in Figure 4, were determined with a surface profiler for various layers and at different annealing temperatures. It is clearly observed that the film thickness increases with the number of dipping and annealing temperatures.

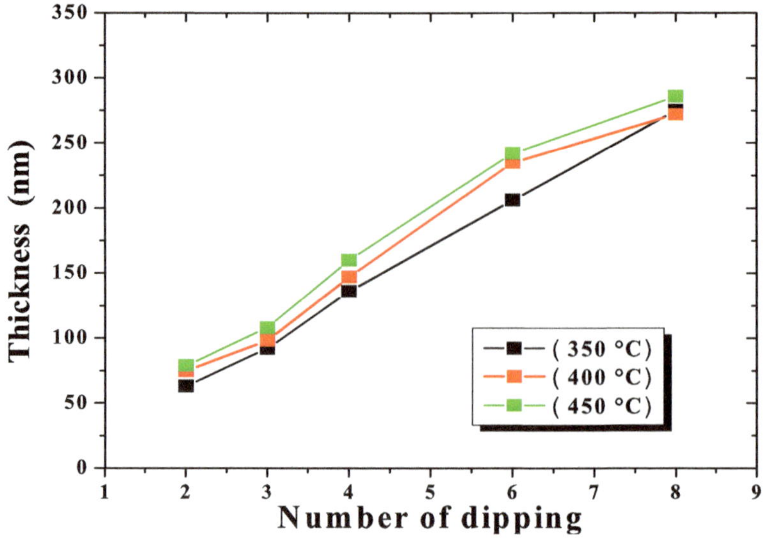

Figure 4. Variation of film thicknesses d (nm) for different annealing temperatures and different dipping

3.3.1.2. Study of deposition rate

A thin film deposited by dip-coating method has a thickness which can be controlled by the deposition rate. To simplify, d thickness depends on the speed according to the following relationship:

$$d = k.V_d^a$$

Where:

V_d is the deposition rate, k is the empirical factor depending on the viscosity, surface tension and density of the solution used and a is the exponent with 2/3 value according to Landau and Levich [62], and 1/2 according to Michels et al. [63] or even proportionately at the speed of dipping Hewak et al. [64].

We then set the objective of determining this factor to validate one of these different models. So, we have prepared six samples with different deposition rate from 0.6 $cm.s^{-1}$ to 2 $cm.s^{-1}$ in the same conditions (deposited in solution with a viscosity 40 mPa.s at 21°C and treated at 400°C); we measured their thicknesses by profilometer and the results are grouped in table 2 and represented in figure 5.

This curve shows a linear increase in ln(d) versus ln(V_d). For the exponent a, values of 0,965 have been obtained, which is in good agreement with the exponent obtained by Hewak et al. [64]. This simple comparison shows that the increase in the speed of dipping, results an increase in the thickness of doped thin films.

Deposition rate ($cm.s^{-1}$)	0.6	0.8	1	1.4	1.8	2
Thickness (± 0.1 nm)	38	56	74	93	116	143

Table 2. Variation of film thickness d (nm) for different deposition rate.

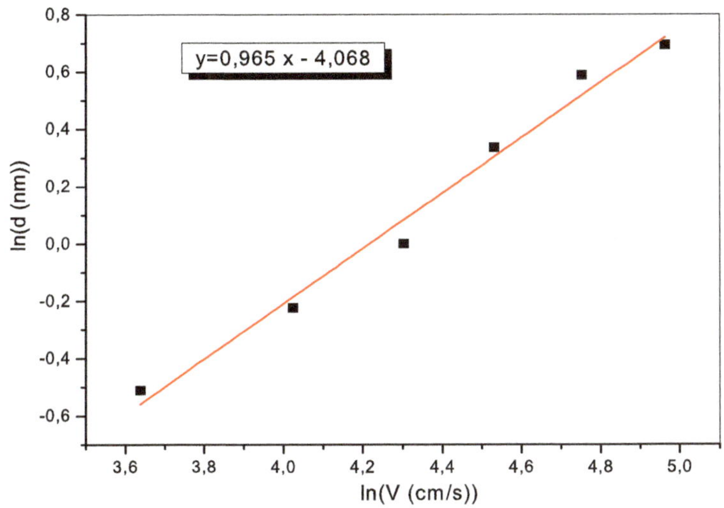

Figure 5. Variation of thickness logarithm with deposition rate logarithm.

3.3.1.3. X-Ray Diffraction (XRD)

3.3.1.3.1. Crystallization

a. Influence of annealing temperature

Our samples were analyzed using X-ray diffraction (XRD) to investigate the transformation behaviors. We have studied the structural properties of undoped TiO_2, and doped with 5% ZrO_2 thin films and deposited by the sol–gel method. The dip-coated thin films have been examined at different annealing temperatures (350 to 450 °C).

Figure 6 shows XRD patterns of 5% ZrO_2-doped TiO_2 thin film obtained after different dipping and treated at 450 °C. However, Figure 7 shows XRD patterns of 5% ZrO_2-doped TiO_2 thin film obtained after 2 dipping and treated at various annealing temperatures at 350 °C, 400 °C and 450 °C. Clearly, titanium oxide starts to crystallize starting from annealing at 350 °C.

This analysis of the doped TiO_2 thin films exhibits that the addition of a minor amount of ZrO_2 would be largely sufficient to form 4 sharp diffraction peaks at 25.35°, 35.51°, 50.90°

and 60.76°, these are assigned to (101), (112), (200), and (105) planes which are attributed to anatase nanoparticles (crystalline) phase of TiO_2.

Furthermore, all XRD patterns show a peak at 30.41° corresponding to (111) plane which is attributed to the brookite formation whatever the annealing temperature.

Peak intensities corresponding to characteristic planes of anatase (101) and brookite (111) phases are obviously increased with the increase of annealing temperature and number of dipping. This latter is interpreted as due not only to increase in proportion of titanium oxide but also to the improvement of the crystalline quality.

However, in the same conditions Kitiyanan et *al.* [56] B. Neppolian et *al.* [57] showed that titanium oxide, crystallizes in anatase phases and they found that addition of minor amount of ZrO_2 has not contributed to any change in the TiO_2 morphology.

b. Influence of doping with ZrO_2

Comparison between XRD patterns (Figure 8) of both undoped and doped with 5% ZrO_2 thin films obtained after 2 dippings and annealed at 400 °C temperature showed a similar behavior, so characteristic peaks correspond to the crystallization of anatase and brookite phases of the doped state is shifted to larger angles compared to the undoped one.

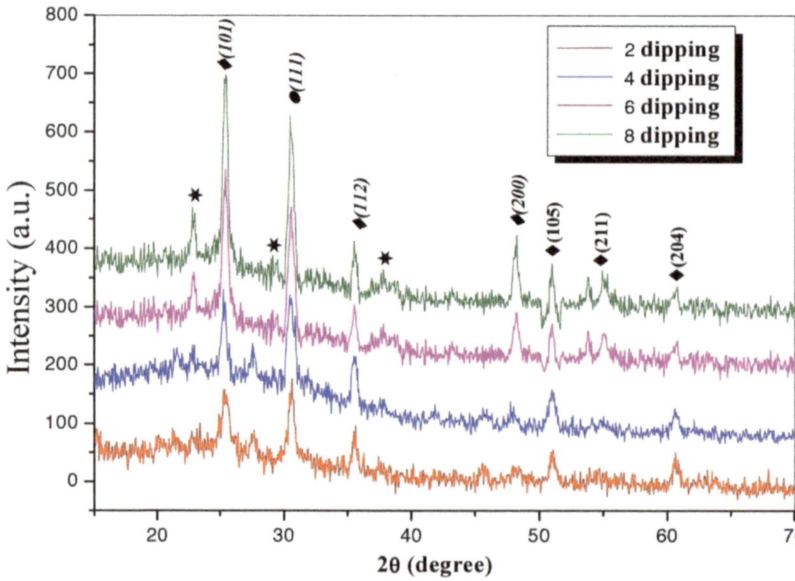

Figure 6. Evolution of diffraction patterns of 5% ZrO_2-doped TiO_2 thin films; obtained after various dipping (2, 4, 6, 8) annealed at 450°C. ★: substrat, ♦: anatase, ●: brookite.

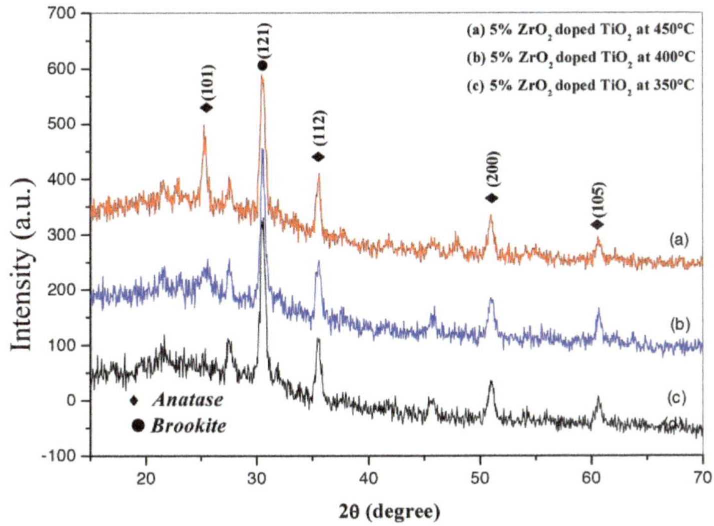

Figure 7. Evolution of diffraction patterns of 5% ZrO_2-doped TiO_2 thin films; obtained at various annealing temperatures (350, 400, 450 °C) for the same thickness.

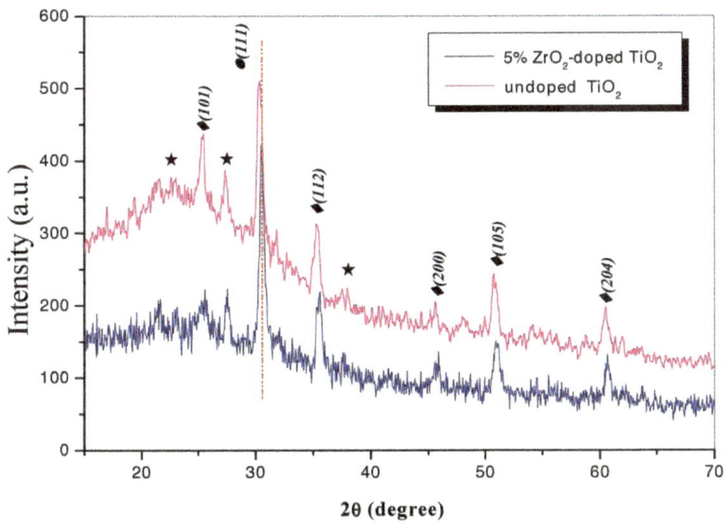

Figure 8. Comparison between undoped and 5% ZrO_2-doped TiO_2 diffraction patterns.

3.3.1.3.2. Surface morphology and grain size

The crystallite size D of TiO_2 doped with ZrO_2 thin films can be deduced from XRD line broadening using Scherrer equation [65]:

$$D = \frac{0.94 \times \lambda}{\sqrt{\left(\Delta_{hkl}^2 - \Delta_{instr}^2\right)}} \frac{1}{\cos\theta}$$

λ is the wavelength of X-ray beam (Cu K_α=1.5406 Å), Δ_{hkl} is the full width at half maximum (FWHM) of (hkl) diffraction peak, Δ_{instr} is the FWHM corresponding to the instrumental limit, and θ is the Bragg angle.

Using the size of the crystallites, the dislocation density (δ) [66], the number of crystallites per unit surface area (N) and strain in the films (ε), which are newly introduced by Ray et al. [66], has been determined:

$$\delta = \frac{1}{D^2}$$

$$N = \frac{d}{D^3}$$

$$\varepsilon = \frac{\Delta(2\theta) \cdot \cos\theta}{4}$$

Where d is the film thickness.

The calculated structural parameters are presented in Table 3.

		Annealed at	Phase	(hkl)	L (nm)	δ (10^{-4} traits/nm^2)	N (10^{-3} nm^{-2})	ε (10-4)
Undoped TiO_2	Xerogel	3 months at T ambient	Amorphous	-	-	-	-	-
5 %ZrO_2-doped TiO_2	Xerogel	3 months at T ambient	Anatase	(101)	14,80	-	-	-
	Same thickness	350 °C	Anatase	(101)	8,58	135,84	99,74	3,11
			Brookite	(111)	17,50	32,65	11,76	3,89
			Anatase	(112)	16,66	36,03	13,62	4,52
			Anatase	(200)	14,74	46,03	19,67	6,69
			Anatase	(105)	16,33	37,50	14,47	7,53
		400 °C	Anatase	(101)	10,09	98,22	73,01	3,07
			Brookite	(111)	17,61	32,25	13,73	3,82
			Anatase	(112)	17,27	33,53	14,56	4,44
			Anatase	(200)	15,57	41,25	19,87	6,37
			Anatase	(105)	18,71	28,57	11,45	7,58
		450 °C	Anatase	(101)	13,92	51,61	29,29	2,98
			Brookite	(111)	18,06	30,66	13,41	3,74
			Anatase	(112)	19,09	27,44	11,36	4,38
			Anatase	(200)	18,63	28,81	12,22	6,34
			Anatase	(105)	20,56	23,66	9,09	7,47

Table 3. Structural parameters of xerogels and thin films, for different annealing temperatures and same thickness.

The computed values of grain sizes, given in Table 3, were calculated for different annealing temperatures with the same thickness. Thus, the obtained grain sizes of anatase and brookite increase from 8.58 nm to 20.56 nm and from 17.50 nm to 18.06 nm, respectively. In fact, as annealing temperature increases grain sizes also increases.

It is interesting to note that the grain size improves and the defects like dislocation density and strain in the films decrease with film thickness. This may be due to the improvement in crystallinity in the films with film thickness. As we also note that the variation of the strain is perfectly correlated with that of the dislocation density δ. When these increase, they cause the decrease in grain size and leads to recrystallization of the nanoparticles. Furthermore, the stages of nucleation, growth and coalescence become stable, which causing the reduction of constraints in the film formed.

The evolution of the grain size D according the annealing temperature can be interpreted by the Arrhenius law (figure 9):

$$D = D_0 exp\left(-E_a / k_b T\right),$$

Where:

- E_a is the energy activation of crystallization;
- K_B the Boltzmann constant;
- D_0 pre-exponential factor.
- The size D tends towards the infinite for high temperatures [67].

The values of activation energies of crystallization corresponding respectively to anatase and brookite phases are calculated from the curve showed in figure 9, we note that the activation energy of the anatase E_a=0,096 eV crystallization is lower than that of the brookite E_a=0,012 eV. This means that the formation of anatase requires more energy than that of brookite.

3.3.1.4. RAMAN

The Raman spectra of undoped and 5% ZrO_2-doped TiO_2 thin films annealed at 450°C for different dipping (figure 10) display various peaks related to titanium oxide as anatase and brookite phases. These spectra exhibit bands at around 138 (strong), 235 (weak), 514 (weak) and 632 cm-1 (medium)) for the thin layers of ZrO_2-doped TiO_2 corresponding to the Eg modes of vibration. The above bands can be assigned to anatase phase except the band 235 cm^{-1} corresponding to the B_{1g} modes of vibration, which is due to the crystallization of brookite phase. While bands of 144, 188 and 651 cm-1 can be assigned to both anatase and brookite phases [68,69].

A slight shift of the most intense peak, Eg, to smaller wavenumber is observed for all thin films doped with ZrO_2 by comparison with anatase of undoped phase (figure 11). Similar displacements have been previously reported in XRD patterns and they can be correlated with the confinement effects in nano-structured anatase crystallites.

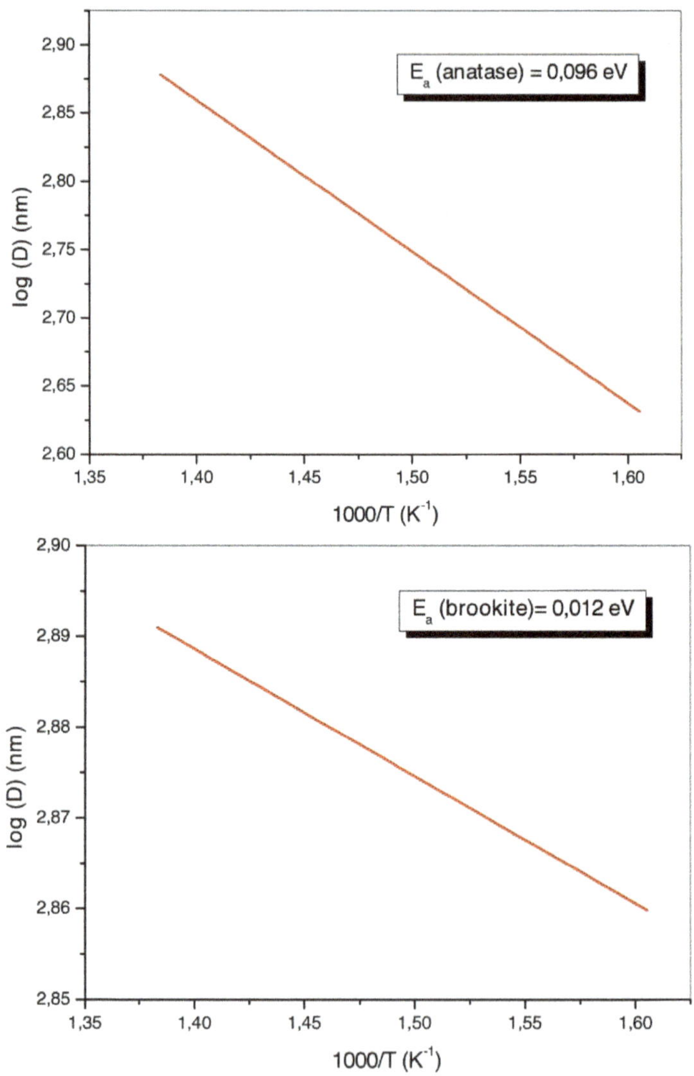

Figure 9. Plot of log (D) versus (1000/T) for determination of activation energies of anatase and brookite

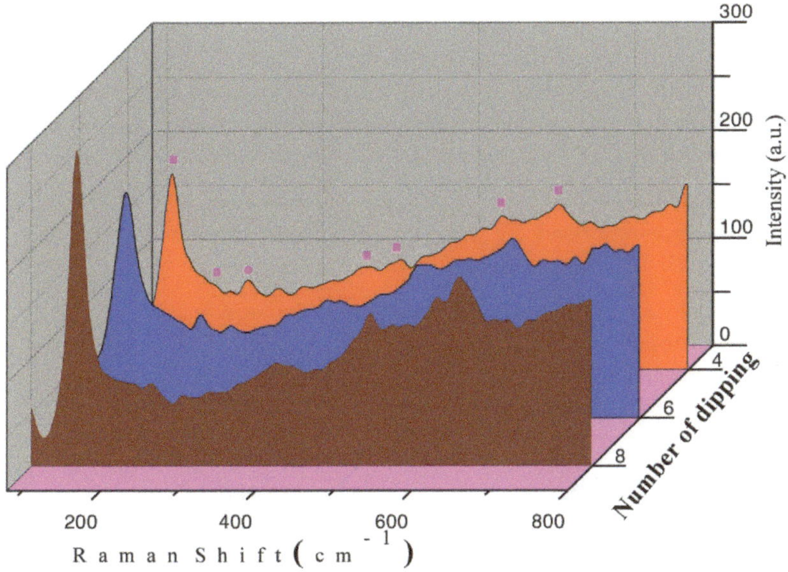

Figure 10. Raman spectrum of 5% ZrO_2-doped TiO_2 thin films annealed at 450°C for different dipping; ■=anatase, •=brookite.

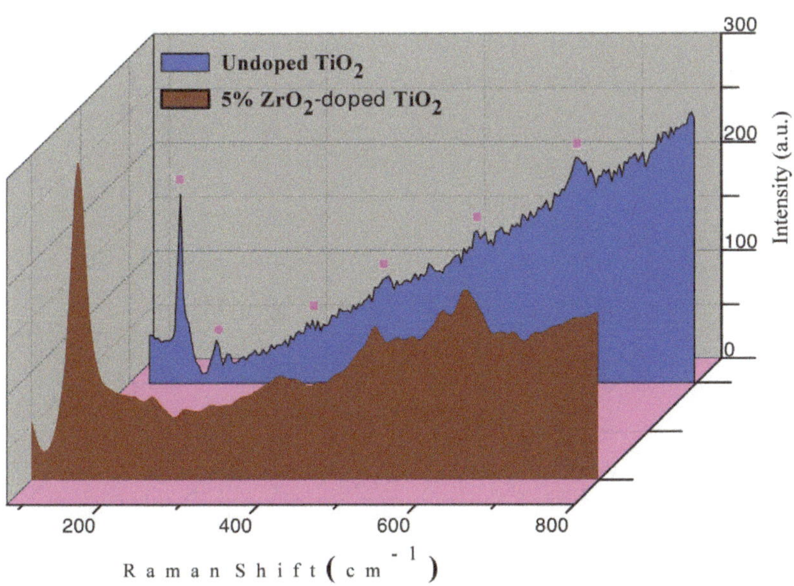

Figure 11. Comparison between Raman spectrum of undoped and 5% ZrO_2-doped TiO_2 thin films; ■=anatase, •=brookite.

3.3.1.5. FTIR

Figure 12 shows the infrared absorption spectrum of the 5% ZrO_2 -doped TiO_2 films annealed at different temperature. The peak at 2360 cm^{-1} resulted from the adsorbed H_2O molecules, which were not removed completely after sol–gel coating. The peaks at 1242 cm^{-1}, 1111 cm^{-1}, 1035 cm^{-1} and 860 cm^{-1} correspond to the vibration mode of Ti-OH [70, 71].

The band around 665 cm^{-1} was attributed to the vibration mode of Ti-O-Ti bond [72] and another bond appears around 455 cm^{-1}, this is the O-Ti-O band corresponding to the crystalline titania in the anatase form [73-76].

We find that the vibration bands intensity located in the vicinity of 665 cm^{-1} and 455 cm^{-1} increase when annealing temperature increases. This indicates that the number of Ti-O-Ti and O-Ti-O links of titanium dioxide crystallization is also growing.

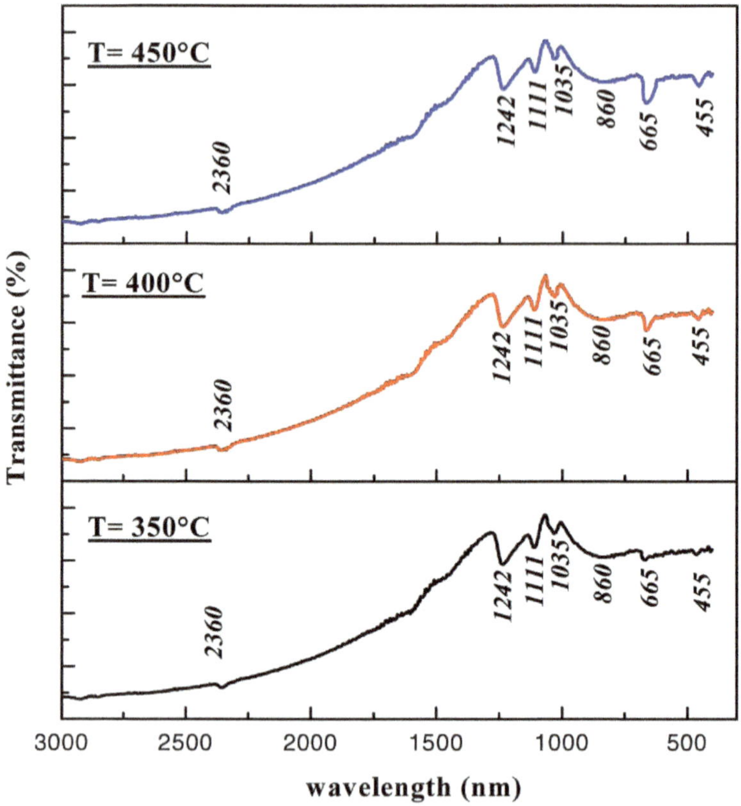

Figure 12. FTIR spectra of 5% ZrO_2-doped TiO_2 thin films, obtained at various annealing temperatures (350, 400, 450 °C).

3.3.1.6. Scanning electron microscopy (SEM) and EDX

5% ZrO_2-doped TiO_2 thin films deposited on ITO substrates and obtained after various annealing temperature at 350°C and 450°C were coated and examined in a scanning electron microscope (SEM) to investigate their structure and surface characteristics. We observed that the coating was homogeneous without any visual cracking over a wide area. The increase in the treatment temperature, did not affect the uniformity of the film.

The surface composition of films is further identified by EDX measurement. EDX result shown in Figure 13 demonstrates that the peaks of Ti, O and Zr can be clearly seen in the survey spectrum. While the other elements as Si, In, Ca, Na and Mg are the components of ITO substrate.

The chemical compositions of thin film analyzed are given in table 4.

	O	Na	Mg	Si	Ca	In	Zr	Ti	Total
at.%	40,13	2,03	3,15	37,83	3,07	4,39	1,23	8,17	100

Table 4. Elemental composition (at. %) of 5% ZrO_2 -doped TiO_2 thin films, treated at 450 ° C.

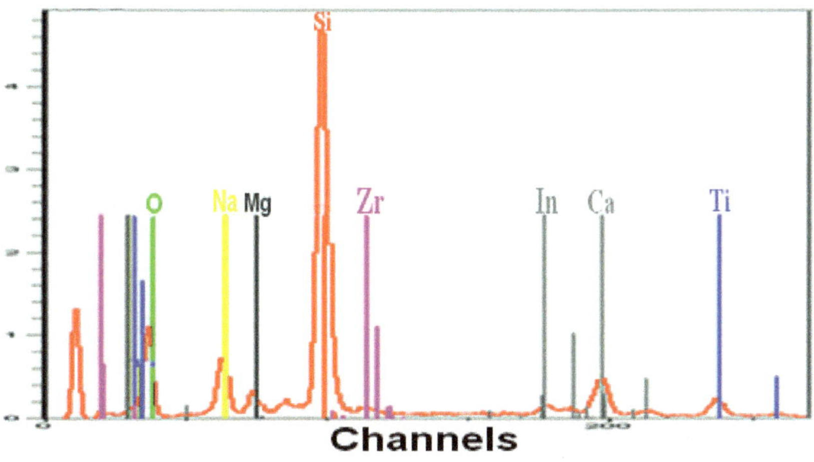

Figure 13. EDX spectra of 5% ZrO_2-doped TiO_2 thin films annealed at 450°C.

3.3.2. Optical properties

3.3.2.1. UV absorption analysis

Figure 14 display diffused scattering UV-VIS transmittance spectra of TiO_2 thin films undoped and 5% ZrO_2-doped TiO_2, for different annealing temperatures from 350°C to 450°C and different numbers of dipping (3, 4, 6, 8 dipping) in the wavelength range 300–800 nm, where the film due to interference phenomena between the wave fronts generated at

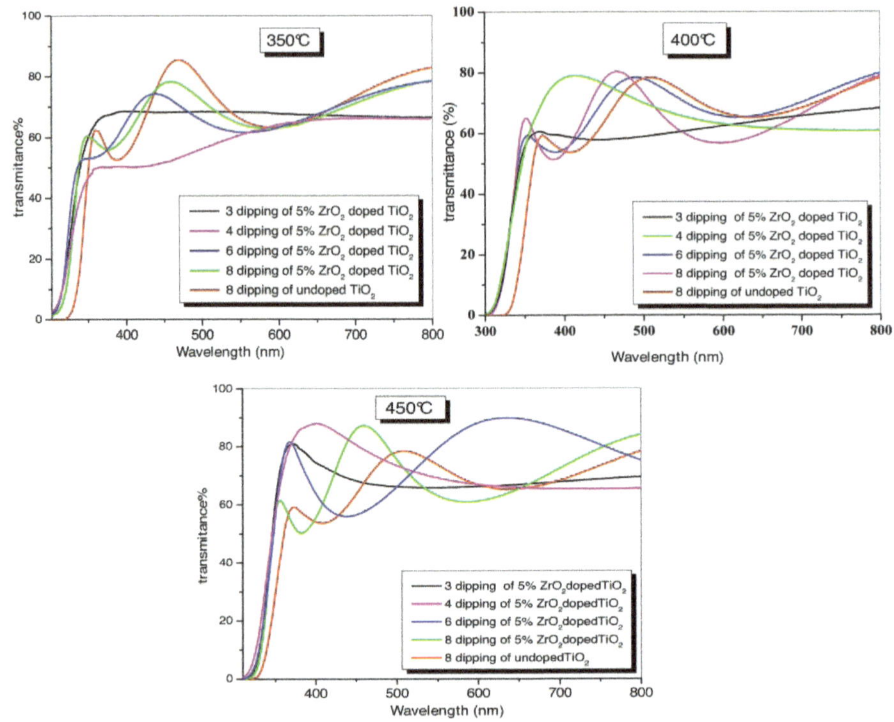

Figure 14. UV–VIS spectra of undoped TiO_2 and 5% ZrO_2-doped TiO_2 thin films, for various dipping and at different annealing temperatures.

the two interfaces (air and substrate) defines the sinusoidal behavior of the curves' transmittance versus wavelength of light. The curves showed a similar behavior in the temperature range:

- A **region** characterized by a strong absorption at $\lambda < 380$ nm, this absorption is due to the transition electronic inter-band.
- High transmittance region, from 60 to 95% on a wide range of wavelength in the visible region (from 380 to 800 nm) has been observed which may be used in applications in solar cells. High transparency is one of the most important properties that explain the interest in undoped or doped TiO_2 thin films.

As can be seen, all the spectrums exhibit interference fringes, which are due to the multiple reflections at the two film edges, i.e. at the film/air and the film/substrate interfaces. This indicates that the top film surface is smooth and uniform and exhibits a good transparency in the visible region. So that the excellent surface quality and homogeneity of the film were confirmed from the appearance of interference fringes in the transmission spectra. This occurs when the film surface is reflecting without much scattering/absorption in the bulk of the film.

If the film surface is rough, the radiation in film/air interface undergo scattering in all directions instead of a reflection. Oh et al. [28], Kim et al. [77] show that the interference fringes are due the increase in thin films thickness. The occurrence of such fringes means that our films are sufficiently thick

Analysis of UV– VIS transmission spectra shows that the 5% ZrO_2-doped TiO_2 thin films are transparent in the visible range and opaque in the UV region, whatever are the annealing temperature and number of dipping

The amplitude of interference spectra increased with increasing calcination temperature. These results show that the refractive index of TiO_2 thin films is increased while the film thickness is decreased. This can be due to the formation stage of anatase and with the increase in the grain size [28].

A slight shift of transmission curves to lower wavelengths is observed for curves of ZrO_2-doped thin films in comparison with those undoped (figure 14) This shift is ascribed to the increase in band gap energy.

3.3.2.2. Refractive index, density, thickness and porosity

The refractive index of TiO_2 thin films was calculated from measured UV–VIS transmittance spectrum. The evaluation method used in this work is based on the analysis of UV–VIS transmittance spectrum of a weakly absorbing film deposited on a non-absorbing substrate [78]. The refractive index n (λ) over the spectral range is calculated by using the envelopes that are fitted to the measured extreme:

$$n(\lambda) = \sqrt{S + \sqrt{S^2 - n_0^2(\lambda) n_S^2(\lambda)}}$$

$$S = \frac{1}{2}\left(n_0^2(\lambda) + n_S^2(\lambda)\right) + 2n_0 n_S \frac{T_{max}(\lambda) - T_{min}(\lambda)}{T_{max}(\lambda) \times T_{min}(\lambda)}$$

Where n_0 is the refractive index of air, n_s is the refractive index of the film, T_{max} is the maximum envelope, and T_{min} is the minimum envelope. The thickness of the films was adjusted to provide the best fits to the measured spectra. In this study, all the deposited films are assumed to be homogeneous.

The porosity of the thin films is calculated using the following equation [79]

$$Porosit\acute{e} = \left(1 - \frac{n^2 - 1}{n_d^2 - 1}\right) \times 100(\%)$$

Where n_d is refractive index of pore-free anatase (n_d = 2.52) [80], and n is refractive index of porous thin films.

The relationship between density, and refractive index for the polymorphs was proposed by Gladstone-Dale [81]. This relationship is as follows:

$$n = 1 + 0{,}4\rho$$

Where :

n: mean index of refraction,
d: density
0,40: Gladstone-Dale constant for TiO_2
The thickness of the films was calculated using the equation:

$$d = \lambda_1\lambda_2/2\ (\lambda_1 n_2 - \lambda_2 n_1)$$

Where n_1 and n_2 are the refractive indices corresponding to the wavelengths λ_1 and λ_2, respectively [82].

The results of the computed refractive index (n) (figure 15), density (ρ) and porosity (p) (figure 16) are shown in Table 5. It is noted that the refractive index and the density of thin films of doped titanium oxide increases with increasing annealing temperature and number of dipping; due to phase transition (anatase, anatase–brookite), which increases grain sizes and/or the density of layers.

This phenomenon is related to crystallization, pores destruction and densification of associated film, as well as the elimination of organic compounds.

However, the porosity decreases with increasing annealing temperature and film thickness.

		4 Dipping			*6 Dipping*			*8 Dipping*		
T (°C)	*Films of*	*n*	*ρ*	*P(%)*	*n*	*ρ*	*P(%)*	*n*	*ρ*	*P(%)*
350°C	TiO_2	1,92	2,30	49,8	2,15	2,88	36,3	2,19	2,98	29,1
	$TiO_2 : ZrO_2$	1,62	1,55	69,3	2,13	2,83	33,9	2,18	2,95	29,7
400°C	TiO_2	2,11	2,78	35,5	2,21	3,03	27,4	2,25	3,13	24,1
	$TiO_2 : ZrO_2$	1,91	2,28	50,3	2,18	2,95	29,9	2,21	3,03	27,4
450°C	TiO_2	2,15	2,88	36,3	2,29	3,23	20,7	2,37	3,43	13,7
	$TiO_2 : ZrO_2$	2,17	2,93	30,7	2,23	3,08	25,7	2,29	3,23	20,7

Table 5. Variation of refractive index (n), density (ϱ), porosity (p) of undoped and 5% ZrO_2-doped TiO_2 for different annealing temperatures and different thickness.

The calculated values of thin films thickness are given in table 6. It is clearly observed that the film thickness increases with the number of dipping and annealing temperatures, which is in good agreement with results obtained previously of thickness determined with a surface profiler.

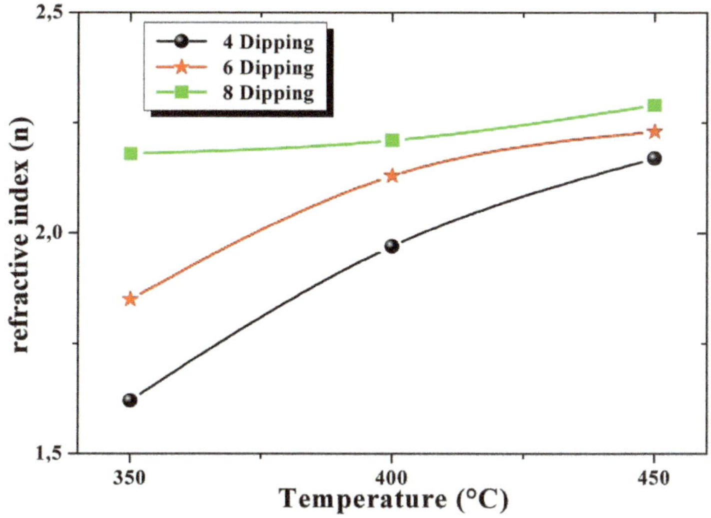

Figure 15. Variation of refractive index (n) of 5% ZrO_2-doped TiO_2 for different annealing temperatures and different thickness

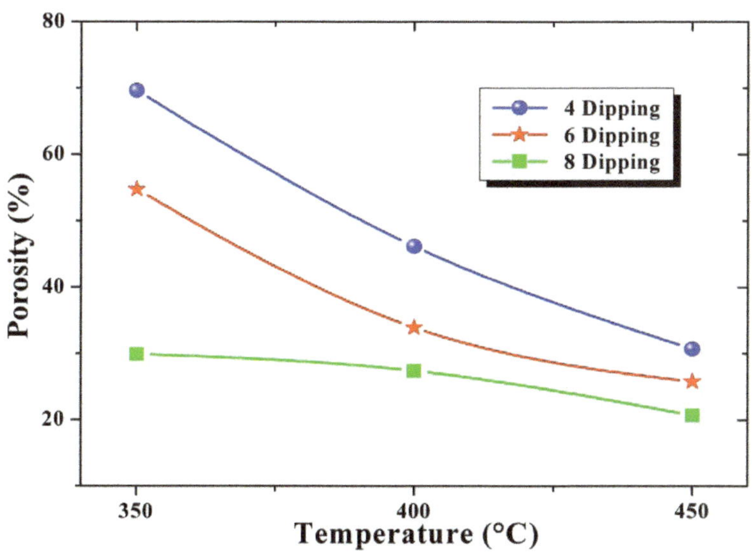

Figure 16. Variation of porosity (p) of 5% ZrO_2-doped TiO_2 for different annealing temperatures and different thickness.

	Thickness d (nm)		
T (°C)	*4 Dipping*	*6 Dipping*	*8 Dipping*
350°C	127	194	268
400°C	139	216	274
450°C	158	233	289

Table 6. Variation of calculated film thicknesses d (nm) for different annealing temperatures and different dipping.

3.3.2.3. Optical band gap:

The band gap is then found as the intercept of the linear portion of the plot. For a direct band gap semi-conductor, the absorption near the band edge can be estimated from the following equation known as the Tauc plot [83]:

$$(ah\nu) = C\left(h\nu - E_g\right)^n$$

Where C is a constant, E_g the optical band, α is the optical absorption coefficient, $h\nu$ is the photon energy gap, h the Plank's constant and the exponent n characterizes the nature of band transition; the values of n = 1/2 and 3/2 correspond to direct allowed and direct forbidden transitions, n = 2 and 3 are related to indirect allowed and indirect forbidden transitions [83] and in the cases for a direct band gap semi-conductor like TiO_2 the relation become [84, 85]:

$$(ah\nu)^2 = C\left(h\nu - E_g\right)$$

The energy band gap (E_g) of the films can be estimated by plotting $(\alpha h\nu)^2$ versus $h\nu$ (Figure 17), then extrapolating the straight-line part of the plot to the photon energy axis. The energy band gap of 5% ZrO_2-doped TiO_2 films, given in table 7, decrease owing to an increase in annealing temperatures and the number of dipping. The values are 3.65 and 3.54 eV at 350°C and 450°C respectively.

This decrease was correlated with grains size increases with temperature, when the latter increases the defects and impurities tend to disappear causing a reorganization of the structure. We find that doping with ZrO_2 causes an increase in the band gap by contrast to that of undoped TiO_2 (3.50 eV).

	Band gap (eV)			
T (°C)	*3 Dipping*	*4 Dipping*	*6 Dipping*	*8 Dipping*
350°C	3,79	3,74	3,71	3,65
400°C	3,73	3,70	3,68	3,59
450°C	3,67	3,63	3,61	3,54

Table 7. Variation of band gap of 5% ZrO_2-doped TiO_2 thin films for different annealing temperatures and different thickness.

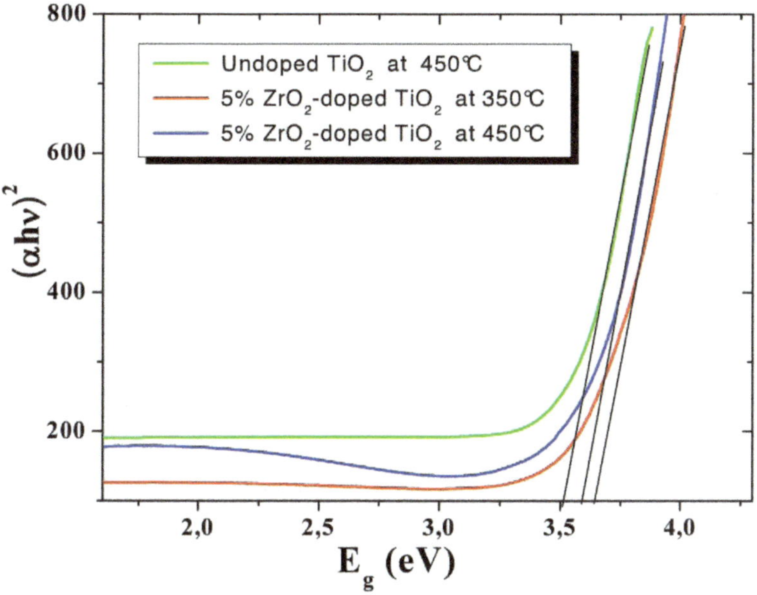

Figure 17. Plot of $(\alpha h\nu)^2$ versus ($h\nu$) for determination of band gap of undoped and 5% ZrO_2-doped TiO_2.

4. Conclusion

In this study, we investigated the transformation behaviors and the effect; of a smaller ratio range of ZrO_2; doping on the surface area of TiO_2 thin films, band gap energy, variations of crystal granularity, phase composition and especially on the evolution of the crystallite size and defects concentration with annealing treatments and layers thickness of the samples produced. So that in this chapter, we report the study of structural, thermal and optical properties of ZrO_2-doped TiO_2 thin films deposited by the sol–gel process

Analyses of doped TiO_2 xerogel show that addition of 5% ZrO_2 would be largely sufficient to form nanoparticles of anatase (size of grain of 14.78 nm) by contrast to that of undoped TiO_2. X-ray diffraction and Raman spectroscopy analyses exhibit that doped thin films obtained starting from annealing at 350°C crystallize in both anatase and brookite phases. Calculation of grain sizes by Scherrer's formula, gives sizes ranging from 8.58 to 20.56 nm and we note an increase in grain sizes by increasing the annealing temperature for all structures. Raman spectroscopy studies confirms the results found by XRD and reveal that the films annealed from 350 to 450◦C crystallizes in anatase and brookite structure. From the DSC analysis, we have demonstrated that an annealing temperature equal or higher than 340 °C for undoped and 260 ◦C for 5% ZrO_2-doped would be sufficient to form titanium oxide. The addition of 5% of zirconium oxide led to a shift of exothermic peak phase towards lower temperatures, due to the speeding up of the crystallization of titanium oxide compared to the undoped one.

Analysis of UV-VIS transmission spectra shows that the 5% ZrO_2-doped TiO_2 thin films are transparent in the visible range and opaque in the UV region, whatever the annealing temperature and the number of dipping. Refractive index of the thin films of titanium oxide increases with increasing annealing temperature and number of dipping, but the porosity decreases, due to phase transition (anatase, anatase–brookite), which increases grain sizes and/or density of layers. Energy band gap of 5% ZrO_2-doped TiO_2 films decrease owing to an increase in annealing temperatures, also we find that doping with ZrO_2 causes an increase in the band gap by contrast to that of undoped TiO_2.

The optical properties of the films are found to be closely related to the microstructure and crystallographic structure which depend on the annealing temperature. In summary, In this study, we successfully fabricated ZrO_2-doped TiO_2 thin films, with desired structural and optical properties by sol–gel dip coating using the titanium alkoxide (tetrabutyl-orthotitanate) as a starting material.

Author details

Rabah Bensaha and Hanene Bensouyad
Laboratory of ceramics, University of Constantine 1, Constantine, Algeria

5. References

[1] Hu L, Yoko T, Kozuka H, Sakka S (1992) Thin Solid Films j. 18: 219.
[2] Brinker C.J, Scherer G.W (1990) Sol–Gel Science, Academic Press, San Diego, CA.
[3] Brinker C.J, Frye G.L, Hurd A.J, Ashlet C.S (1991) Thin Solid Films j. 97: 201.
[4] Nishio K, Sei T, Tsuchiya T (1996a) Preparation and electrical properties of ITO thin films by dip-coating process. Mater. Sci. Tech. J. 31: 1761–1766
[5] Nishio K, Miyake S, Sei T, Watanabe Y, Tsuchiya T (1996b) Preparation of highly oriented thin film exhibiting transparent conduction by the sol–gel process. Mater. Sci. J. 31: 3651–3656
[6] Nishio K, Seki N, Thongrueng J, Watanabe Y, Tsuchiya T (1999a) Preparation and properties of highly oriented thin films by a sol–gel process. Sol–Gel Sci. Tech J. 16: 37–45

[7] Nishio K, Kudo C, Nagahama T, Manabe T, Yamaguchi I, Watanabe Y, Tsuchiya T (2000b) Preparation and characterzation of epitaxial thin films prepared by sol–gel process. Mater. Res. Soc. J. 623: 377–382
[8] Nishio K, Watanabe Y, Tsuchiya T (2003) Epitaxial growth of thin films prepared from sol–gel process. Sol–Gel Sci. Tech. J. 26: 245–250
[9] Kodaira T, Nishio K, Yamaguchi I, Suzuki S, Tsukada K, Tsuchiya T (2003) Synthesis and properties of highly conductive thin films as buffer layer from sol–gel process. Sol–Gel Sci. Tech. J. 26: 1049–1053
[10] Larson S.A, Falconer J.L (1994) Appl. Catal. B: Environ. j. 4: 149.
[11] Pas Y, Heller A (1997) Mater. Res. J. 12: 2759.
[12] Hara K, T. Hariguchi, T. Kinoshita, K. Sayama, H. Arakawa, Solar Energy Mater. Solar Cell. 70 (2001) 151.
[13] Sakha V.A, Arabatzis I.M, Konstantinou I.K, Dimou A.D, Albanis T.A, Falaras P (2004) Appl. Catal. B: Environ. J. 49: 195.
[14] Tachibana Y, Ohsaki H, Hayashi A, Mitsui A, Hayashi Y (2000) Vacuum j. 59: 836.
[15] Zheng S.K, Wang T.M, Xiang G, Wang C (2001) Vacuum j. 62: 361.
[16] Radecka M, Zakrzewska K, Czternastek H, Stapinéski T, Debrus S (1993) Appl. Surf. Sci. j. 65: 227.
[17] Dumitriu D, Bally A.R, Ballif C, Hones P, Schmid P.E, Sanjinés R, Lévy F, Pârvulescu V.I (2000) Appl. Catal. B: Environ. J. 25: 83.
[18] Takeda S, Suzuki S, Odaka H (2001) Thin Solid Films j. 392: 338.
[19] Zhang F, Huang N, Yang P (1996) Surf. Coat. Technol. J. 84: 476.
[20] Aidla A, Uustare T, Kiisler A.A, Aarik J, Sammelselg V (1997) Thin Solid Films j. 305: 270.
[21] Abou-Helal M.O, Seeber W.T (2002) Appl. Surf. Sci. j. 195: 53.
[22] Choi Y, Yamamoto S, Umebayashi T, Yoshikawa M (2004) Solid State Ionics j. 172: 105.
[23] Hu L, Yoko T, Kozuka H (1992) Thin Solid Films j. 219: 18.
[24] Chrysicopoulou P, Davazoglou D, Trapalis C, Kordas G (1998) Thin Solid Films j. 323: 188.
[25] Kim D.J, Hahn S.H, Oh S.H, Kim E.J (2002) Mater. Lett. j. 57: 355.
[26] Yang J, Li D, Wang H, Wang X, Yang X, Lu L (2001) Mater. Lett. j. 50: 230.
[27] Huber B, Brodyanski A, Scheib M, Orendorz A, Ziegler C, Gnaser H (2005) Thin Solid Films j. 472: 114.
[28] Oh S.H, Kim D.J, Hahn S.H, Kim E.J (2003) Mater. Lett. j. 57: 4151.
[29] Wang C, Xu BQ, Wang XM, Zhao JC (2005) Solid State Chem J. 178:3500.
[30] Hu C, Tang YC, Yu JC, Wong PK (2003) J photocatalyst Appl Catal B Environ 40:131.
[31] Bensouyad H, Sedrati H, Dehdouh H, Brahimi M, Abbas F, Akkari H, Bensaha R (2010) Thin Solid Films J. 519:96.
[32] Yang J, Li D, Wang X, Yang XJ, Lu LD (2002) J Solid State Chem 165:193.

[33] Li JL, Liu L, Yu Y, Tang YW, Li HL, Du FP (2004) Electrochem Commun J. 6:940.
[34] Bandara J, Hadapangoda CC, Jayasekera WG (2004) J Appl Catal B Environ 50:83.
[35] Li XZ, Li FB, Yang CL, Ge WK (2001) Photochem Photobiol A Chem J. 141:209.
[36] Shchukin D, Poznyak S, Kulak A, Pichat P (2004) Photochem Photobiol A Chem J. 162:423.
[37] Bensouyad H, Adnane D, Dehdouh H, Toubal B, Brahimi M, Sedrati H, Bensaha R (2011) *Sol-Gel Sci Technol J.* 59: 546.
[38] Takahashi YK, Ngaotrakanwiwat P, Tatsuma T (2004) photocatalysts Electrochim Acta J. 49: 2025.
[39] Wu L, Yu JC, Fu XZ (2006) Mol Catal A Chem J. 244:25.
[40] Su H, Xie Y, Gao P, Xiong Y, Qian Y (2001) J Mater Chem 11:684.
[41] Zelner M, Minti H, Reisfeld R, Cohen H, Tenne R (1997) Preparation and characterization of CdS films synthesized in situ in zirconia sol-gel matrix, Chem. Mater. J. 9: 2541.
[42] Sashchiuk A, Lifshitz E, Reisfeld R, Saraidarov T, Zelner M, Willenz A (2001) Optical and conductivity properties of PbS nanocrystals in amorphous Zirconia sol-gel films, of Sol-Gel Science and Technology J. 24: 31.
[43] Wang M.T, Wang T.H, Lee J (2005) Electrical conduction mechanism in high-dielectric-constant ZrO2 thin films, Microelecton. Reliability j. 45: 969.
[44] Harrison H.D.E, McLamed N.T, Subbaro E.C (1963) A new family of self-activated phosphors, Electrochem. Soc J. 110: 23.
[45] Lowdermilk W.H, Milam D, Rainer F (1980) Optical coatings for laser fusion applications, Thin Solid Films j. 73: 155.
[46] Meier S.M, Gupta D.K (1994) The evolution of thermal barrier coatings in gas turbine engine applications, Eng. Gas Turbines Power Trans. ASME J. 116: 250.
[47] Wendel H, Holzschuh H, Suhr H, Erker G, Dehnicke S, Mena M (1990) Thin zirconium dioxide and yttrium oxide-stabilized zirconium dioxide films prepared by plasma-CVD, Mod. Phys. Lett. B j. 4: 1215.
[48] Fork D.K, Fenner D.B, Connell G.A.N, Phillips J.M, Geballe T.H (1990) Epitaxial yttria-stabilized zirconia on hydrogen-terminated Si by pulsed laser deposition, Appl. Phys. Lett. j. 57: 1137.
[49] Patel A.M, Spector M (1997) Tribological evaluation of oxidized zirconium using an articular cartilage counterface: a novel material for potential use in hemiarthroplasty, Biomaterials j. 18: 441.
[50] Balamurugan A, Kannan S, Rajeswari S (2003) Structural and electrochemical behaviour of sol-gel zirconia films on 316L stainless-steel in simulated body fluid environment, Mater. Lett. j. 57: 4202.
[51] Ritala M, Leskela M (1994) Zirconium dioxide thin films deposited by ALE using zirconium tetrachloride as precursor, Appl. Surf. Sci. j. 75: 333.
[52] Cao G.Z, Brinkman H.W, Meijerink J, De Vries K.J, Burgraaf A.J (1993) Pore narrowing and formation of ultrathin yttria-stabilized zirconia layers in ceramic membranes by

chemical vapor deposition/electrochemical vapor deposition, Am. Ceram. Society. J. 76: 2201.

[53] Emeline A, Kataeva G.V, Litke A.S, Rudakova A.V, Serpone N (1998) Langmuir j. 14: 5011.

[54] Mechiakh R, Bensaha R (2006) C.R. Phys j. 7: 464.

[55] Ivanda M, Music S, Popovic S, Gotic M (1999) J Mol Struct 480: 645.

[56] Kitiyanan A, Yoshikawa S (2005) Materials Letters J. 59: 4038.

[57] Neppolian B (2007) Applied Catalysis A: General j. 333: 264.

[58] Pérez-Hernandez R, Gomez-Cortes A, Arenas-Alatorre J, Rojas S, Mariscal R, Fierro J.L.G, Diaz G (2005) Catal. Today j. 107: 149.

[59] Perdew J.P, Wang Y (1992) Phys. Rev. B j. 45: 13244.

[60] Montoya I.A, Viveros T, Dominguez J.M, Channels L.A, Schifter I (1992) Catal. Lett. J. 15: 207.

[61] Mechiakh R, Meriche F, Kremer R, Bensaha R, Boudine B, Boudrioua A (2007) Opt. Mater. J. 30: 645.

[62] Landau L.D, Levich B (1942) Acta Physicochim. J. 17: 42.

[63] Michels A. F, Manegotto T, Horowitz F (2004) Appl. Opt. j. 43: 820.

[64] Hewak D. W, Lit J. W. Y (1988) Can. J. Phys, 66: 861.

[65] Cullity B.D (1978) Elements of X-Ray Diffraction, 2nd ed.Addison-Wesley, Reading, MA.

[66] Ray S, Banerjee R, Baraua A.K (1980) Jpn J. of Appl Phys 19:1889.

[67] Yang H, Huang C, Tang A, Zhang X, Yang W, *Materials Research Bulletin,* 40 (2005) 1690.

[68] Lottici P.P, Bersani D, Braghini M, Montenero A (1993) Mater. Sci. J. 28: 177.

[69] Djaoued Y, Badilescu S, Ashrit, Bersani D, Lottici P.P, Robichaud J (2002) *Sol-Gel Sci Technol J.* 24: 255.

[70] Liao M.H, Hsu C.H, Chen D.H (2006) J Solid State Chem 179: 2020.

[71] Xie Y, Liu X, Huang A, Ding C, Chu P.K (2005) J Biomaterials 26: 6129.

[72] McDevitt N.T, Baun W.L (1964) J Spectrochim Acta 20: 799.

[73] Music S, Gotic M, Ivanda M, Popovic S, Turkovic A, Trojko R, Sekulic A, Furic K (1997) Mater Sci Eng B 47: 33.

[74] Ocana M, Fornes V, Serna J.V (1988) J Solid State Chem 75: 364.

[75] Djaoued Y, Badilescu S, Ashrit P.V (2002) *Sol-Gel Sci Technol J.* 24: 247.

[76] Tian J (2009) Surface & Coatings Technology j. 204: 723.

[77] Kim S, Yum J, Sung Y (2005) Photochem Photobiol A: Chem J. 171: 269.

[78] Manifacier J.C, Gasiot J, Fillard J.P (1976) Phys. E J. 9: 1002.

[79] Yoldas B.E, Partlow P.W (1985) Thin Solid Films j. 129: 1.

[80] Kingery W.D, Bowen H.K, Uhlmann D.R (1976) Introduction to Ceramics, Wiley, New York.

[81] Gladstone, Dale (1863), Phil. Trans. J. 153: 317

[82] Swanepoel R (1983) Phys. E: Sci. Instrum. J. 16: 1214.

[83] Pankove J.I (1971) *Optical Process in Semiconductors*; Prentice Hall; Englewood; Cliffs; NJ; 34. p.
[84] Zhao X. K, Fendler J. H (1991) Phys. Chem. J. 95: 3716.
[85] Tang H, Prasad K, Sanjines R, Schimid P.E, Levy F (1994) App. Phys. J. 75: 2042.

Chapter 13

Lithium Niobiosilicate Glasses Thermally Treated

Manuel Pedro Fernandes Graça and Manuel Almeida Valente

Additional information is available at the end of the chapter

1. Introduction

The term glass comes from the latin word, vitrium, and refers to one of the oldest known material. It is defined in accordance with the ASTM C-169-92 norm, as an inorganic product obtained by quenching a melt until hardness conditions, without crystallization. This definition is however too restrictive because it only applies to glasses prepared by fusion method. A broader definition is that proposed by A. Paul: the glass shows the elastic behavior characteristic of the crystalline state and the behavior of the viscous liquid state. The most common properties of the glasses are the transparency to visible radiation, mechanical stability, inert at the biological level and dielectric material. However, due to the possibility of controlling the microstructure by changing the composition or by applying heat treatments, controlling the process of nucleation and crystallization, the properties of glasses can be modified.

The initial chemical composition is a factor, controllable, which allows to shape some of the properties of the glasses. In any glass, the units that define its structure can be divided into three categories, defined according to their structural function, network former; modifier and intermediate species. The formers are units that, without the addition of other elements, can form glass such as SiO_2, B_2O_3, GeO_2 and P_2O_5. The network modifiers, do not form glass by itself, but are often combined with a former. Examples of modifying elements are the alkaline elements (Li, Na, K, etc.) and alkaline-earth metals (Mg, Ca, etc.). The intermediate species are elements that can both play a role in forming or modifying the network (Al, Nb, etc.).

The formation of glass ceramics, for example by heat treatment of the as-prepared glass, shows at the technological level great advantages, for the single crystals and sintered ceramics, as the possibility of their properties (optical, electrical, mechanical, chemical, etc.) be controlled via the volume fraction of the active phase dispersed in the matrix. For example, to maintain optical transparency, the process of nucleation and crystal growth requires a great degree of control being achieved when the size of the crystals dispersed in the glassy matrix is not high enough to cause light scattering. However, for most electrical applications it is necessary that the crystals are large enough to present, for example, a ferroelectric response. This commitment is not easy to perform. Another condition that can maintain the optical transparency of a glass ceramic is the small difference between the refractive indices of the crystals and of the glassy matrix. If this difference is insignificant it allows, regardless of the size of the crystals, to keep the optical transparency. In recent years there has been a growing interest in the preparation, characterization and technology implementation of glasses and glass ceramics in different type of devices. However, it is important to note that, in general, the electrical and optical properties of glass ceramics are not as good as their single crystal when not embedded in a matrix. This is due to the fact that the glass ceramics present at least two distinct phases, the crystal (considered the active phase) and the amorphous (support). The electric polarization of the crystals embedded in a glassy matrix is more difficult due to the low value of the dielectric constant of the glassy phase. On the other hand and because of the single crystal growth processes present a high economic cost, have now start to been replaced by glass ceramics. Some glass ceramics have the additional advantage of being dense and not porous materials.

The growth and crystal orientation can generally be achieved through different processes, such as: mechanical deformation, thermodynamic control, kinetic control (electrochemical induced nucleation). In glasses the use of thermodynamic control is the most common. However, control of crystallization, with the desired crystalline phase is usually difficult because crystallization is a complex process affected by various factors such as composition, surface conditions and heat treatment parameters.

The crystallization process, in a glass, of a particular crystalline phase oriented in a pre-defined direction is usually a desirable objective but difficult to implement. One way that can induce the crystallization of oriented particles in a glass is the application, together with the thermal process of external fields (magnetic / electric). In the presence of an electric field polar nuclei should be oriented parallel to the field and the existing nuclei and / or crystallites can rotate until reach the same direction. However, this can only occur if the value of the electric field is high enough to allow the resultant force to overcome the viscous medium, enabling the rotation. Currently, glass ceramics containing ferroelectric crystals are a class of materials with high technological interest because of the ferroelectric crystals presenting a structural anisotropy results in the formation of electric dipoles and therefore a spontaneous electric polarization.

A large number of ferroelectric materials is presented in the form of crystalline ceramic. The scientific and technological development that photonics has been presenting, has required, particularly in terms of applications, new materials which exhibit characteristics such as optical transparency and are optically active so they can be used as amplifiers, switches, sensors, transducers, filters (...) , that is because there is a need for materials that use light to perform functions already implemented in electronic devices (mainly in the sectors of communications, energy, instrumentation, etc.), but more efficiently or giving rise to new devices. Thus the preparation, structural, electrical and optical characterization of glass ceramic showing ferroelectric properties is of great importance for potential technological applications.

Of the various ferroelectric known materials, lithium niobate ($LiNbO_3$), first synthesized in 1949 by Matthias and Remeika, had attracted attention from many researchers due to their excellent piezoelectric properties, electro optics, electroacoustic, pyroelectric and photorefractive. In late 1960, due to the appearance and development of various applications of fiber optics, several research centers, including Bell Laboratories, studied in detail the structural characteristics and properties of $LiNbO_3$ crystal, especially its electro optics properties.

The fact that the preparation of $LiNbO_3$ single crystals, by the traditional growth techniques is technically difficult and economically costly, the scientific research about the preparation methods of inorganic glasses containing $LiNbO_3$ crystals is the main topic of this chapter. Their structural, morphological, optical and electrical properties will be discussed in function of the glass treatment parameters.

The as-prepared glasses were prepared by sol-gel method. This method allowed to prepare glasses of molar composition $(100-2x)SiO_2$-xLi_2O-xNb_2O_5, with $x < 10$, that through the melt-quenching method are of extreme difficulty to prepare. The synthesis of the glass ceramics was performed by heat treatment of the as-prepared glass and the nucleation and growth of crystals in the glass matrix controlled by the heating rate, temperature and treatment time. It was introduced for the first time, a new variable in this type of glass treatments that was the presence of a external dc electric field, during the heat-treatment cycle.. These heat treatments, with external electrical field applied treatments were called thermoelectric (TET). The main objective of the TET is the hypothesis of promote the precipitation of nano-size crystals in a crystalline preferred orientation. In all prepared compositions, $LiNbO_3$ particles were precipitated due to the thermal treatments, in the glass surfaces. The electric analysis showed the presence of conduction, relaxation and polarization mechanisms.

2. Preparation of glasses by the Sol-Gel method

In the literature there is a considerable number of works on the preparation of glasses by melt-quenching. However, it does not exist, to our knowledge, a comparable number of publications on the same type of glasses prepared by sol-gel method. The preparation of materials by sol-gel, with relevant scientific and technological characteristics only began after 1930. However this method was discovered in late 1800. It was, only, after 1970, and with the preparation of inorganic monolithic gels and their subsequent thermal treatment, resulting in glasses, that this method start to developed [1-2].

The sol-gel method has opened a new path for the preparation of glasses of high technological interest, presenting a high degree of chemical purity and very high homogeneity. The preparation of a glass according to this method is performed by mixing, in the liquid state, its constituents until a homogeneous solution, at the molecular level, is reached. It is then subjected to polymerization and gelling (gel point is defined as the point, in time, where the mixture forms a rigid substance and can be removed from the original container [3]). The resulting gel is transformed into glass by heat treatment during which the volatile species are eliminated and the material undergoes a densification [4]. This method of preparation is known by the generic name sol-gel process. In this process there are two variants, which have the common passage through a gel phase (forming a 3D network [1]) but which differ in the starting products and the first steps of reaction. In one, it starts with a colloidal suspension and in the other with metal-organic compounds, which are dissolved in alcohols. Indeed, the scientific name sol-gel process can only be applied to the first case but it is also used to the second [4;5].

The method using solutions of polymerizable species, which was used in the preparation of the gels presented in this chapter, starts with liquids or alcoholic solutions of an organometallic compound such as the metal alkoxides M(OR)n, where M is a metal and R an alkyl group, exposing them to reactions of hydrolysis, polymerization and dehydration [3;6]. In this method, the alkoxides usually used are the tetraethylorthosilicate (TEOS) and the tetrametoxisilicate (TMOS). However, in several papers other type of alkoxides can be found, such as boron, aluminum or titanium but, usually, mixed with the TEOS. In the case of the gels prepared in this work, the TEOS solution was the only used and the remained components were introduced in the form of nitrate ($LiNO_3$) and chloride ($NbCl_5$), mainly due to their high solubility in water and/or alcohol.

Regarding the chemical formation of a polymeric gel from metal alkoxides (M (OR) n), it is typically described by two consecutive reactions:

- hydrolysis ($M(OR)_n + xH_2O \rightarrow M(OH)_x(OR)_{n-x} + xROH$)

and the condensation, which can be subdivided into two phases:

- alcohol condensation ($-M-OH+R-O-M \rightarrow -M-O-M- \quad +ROH$)
- and water condensation ($-M-OH+H-O-M \rightarrow -M-O-M- \quad +H_2O$) [1;4;5].

Briefly, the hydrolysis reaction is the substitution of the alkoxy group (OR) with a hydroxide group (OH). The subsequent condensation reactions involves, in the case of TEOS ($Si(OC_2H_5)_4$), the silanol group (Si-OH) that gives rise to siloxane bonds Si-O-Si), water and alcohol. It must be noticed that these reactions are, however, a simplified version of which occurs during the hydrolysis and condensation of the alkoxide solutions [4;5].

The diagram showed in figure 1 represents, according to Sakka et al. [6], the preparation of a glass from metal alkoxides. This diagram shows the preparation of a binary glass of the SiO_2-TiO_2 system. The treatment temperatures found in the flowchart are only examples, because they depend directly on the glass composition.

According to figure 1, the preparation process can be divided into three steps. The first involves the mixing of the alkoxide in the amounts corresponding to the final composition of the glass, yielding a clear solution. In the second step it is added to the alkoxide solution water, alcohol and acid to induce the hydrolysis. Parameters such as the pH, the molar ratio between H_2O and the alkoxide and the presence of catalysts (acids) may, as desired, that the onset of condensation is given only after the end of the hydrolysis of the metal alkoxide. In addition to this fact and due to the immiscibility between the water and the alkoxide it is necessary to use a mutual solvent, such as ethanol (C_2H_5OH). The reactions of hydrolysis and polymerization induce an increase of the solution viscosity until it becomes a gel. In the third stage the gel is heat treated until be converted into a glass [2;6].

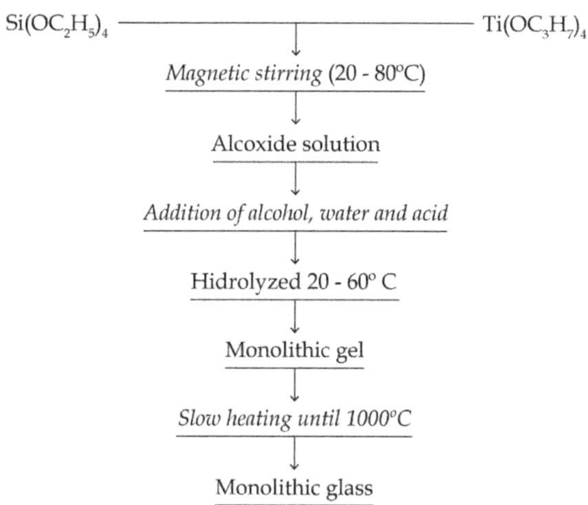

Figure 1. Flowchart of the sol-gel process used in the formation of a SiO_2-TiO_2 glass [6].

With this sol-gel process it is possible to have a high control in the interaction between the liquid precursors, minimizing the energy required for the process, obtaining a final product with high homogeneity and with a high control of its morphology [7]. In summary, the major advantages of this sol-gel process are the high purity of the end product, the low temperature processing (these characteristics are related to the nature of the starting materials), high homogeneity, novel compositions, which are very difficult or even impossible to synthesize by other processes [7;8] and the ability of the shape of final products to be wide (thin film, monolithic blocks, fibers, beads and powders) [6;9;10;11;12].

The major disadvantage of the sol-gel method, for the preparation of glasses, when compared to the melt quenching method (traditional method for glass preparation), is the high cost of starting materials, namely the metal alkoxides [7;8].

A major problem related with the preparation of monolithic blocks is the shrinkage of the gel during the drying process, leading to very high internal mechanical stresses and therefore fracture of the block can occur [13]. This shrinkage is related to the removal of fluids which are inside the pores, resulting in a stress on the capillary walls, which is inversely proportional to the pore diameter [14]. One way to minimize the gel fracture is to control the drying process. For example, Chou et al. [13] present a thermal process in which includes: (a) removal of ethanol (50 °C), water (90 °C, 4h), formamide (170 °C, 4h) and glycerol (230 °C, 14-18h); (2) burning of organic waste; (3) elimination of pores.

The heating rate, usually less than or equal to 5 °C/min, the treatment temperature and the treatment duration time are critical factors that must be control, to prevent fracture. Furthermore, the use of long treatment times, at or near room temperature (20-50 °C), promote poly-condensation, which is an advantage to produce gels with well-defined microstructure, reducing the stresses [13]. Another important factor is to control the thickness of the gel. The greater the thickness of the gel greater the time required to complete the reaction [3-5]. The transition from gel to glass is usually accompanied through drying and sintering by an appropriate heat treatment process [2;13]. This process depends on the composition.

3. Preparation of SiO_2-Li_2O-Nb_2O_5 glasses by sol-gel – Experimental description

The sol-gel method was used to prepare clear glasses of the ternary system SiO_2-Li_2O-Nb_2O_5. The choice of the molar compositions was based on the following criteria: equal molar amounts of lithium oxide and niobium oxide; obtain a clear gel. In this method, the starting materials were the lithium nitrate ($LiNO_3$), niobium chloride ($NbCl_5$), hydrogen peroxide (H_2O_2 - 30% V/V), tetraethylorthosilicate (TEOS) and ethanol (C_2H_5OH) as the mutual solvent.

The preparation steps are presented in the diagram of figure 2.

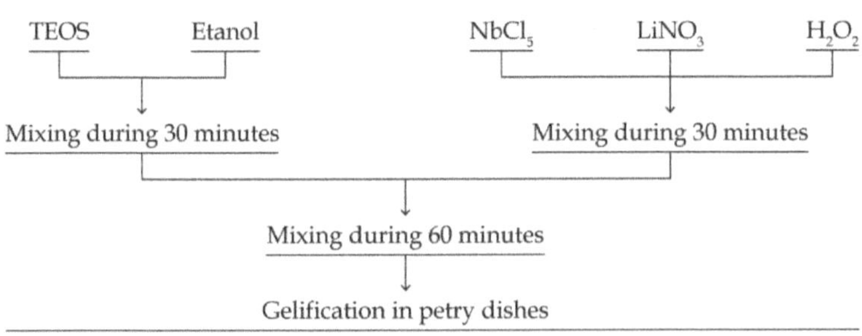

Figure 2. Diagram of the method used to prepare the glasses [16;17].

All samples were prepared using a molar ratio between $(C_2H_5O)_4Si$: C_2H_5OH : H_2O_2 of 1:3:8. Hydrogen peroxide (H_2O_2) was used in the form of aqueous H_2O_2 (3% V/V). This dilution was carried out using deionized water. All solutions, placed in petri boxes were left to gelling at a constant temperature of 30 °C, during more than 1 week. The gel was submitted to a heat treatment which gave rise to the as-prepared sample. This treatment (Fig.3) consists of two steps. First, at a temperature of 120 °C for 48 °C, with the purpose of release the maximum number of free H_2O groups. The second stage has two heat levels.

The first at 250 °C, with the purpose of releasing some H_2O groups, that probably still exist [4;5] and the second, at 500 °C, whose main aim is the liberation of the CO_2 groups from the oxidation of organic radicals [nav91; sil90] . For this reason the heating rate must be as slow as possible.

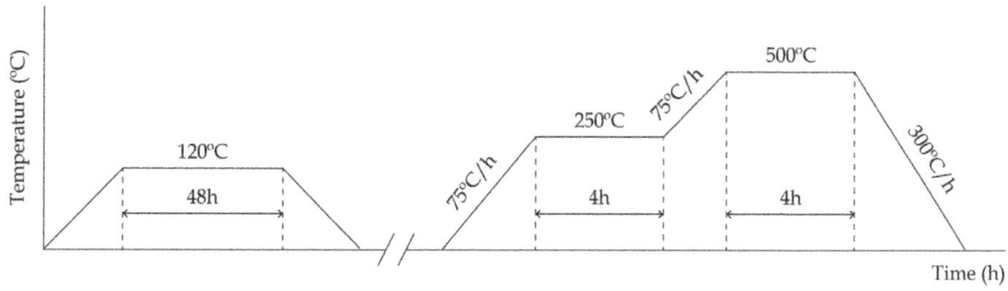

Figure 3. Diagram of the drying process.

4. Glass-ceramics preparation

4.1. Thermal treatments (TT)

The dried gels were submitted to heat treatments in order to obtain glass-ceramics with the $LiNbO_3$ crystalline phase. It is important to refer that the samples prepared by this method present a thickness of 1 mm, approximately. Figure 4 depicts the profile of the heat treatment used. These treatments were performed in a horizontal tube furnace. The value of the threshold temperature parameter (T_P) was chosen based on the information obtained from the thermal analysis of each composition. This thermal analysis was performed using a Linseis Aparatus [18].

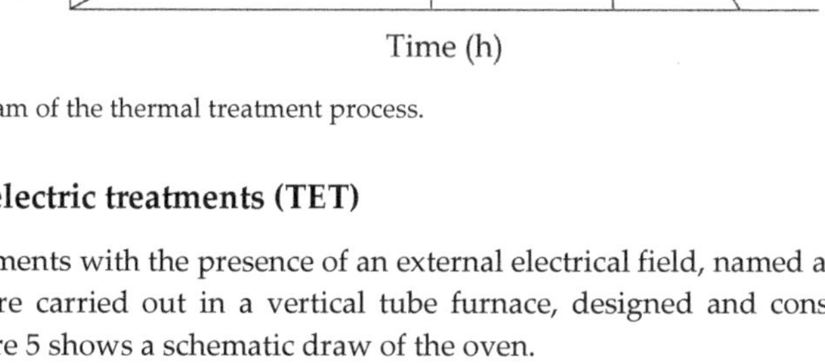

Figure 4. Diagram of the thermal treatment process.

4.2. Thermoelectric treatments (TET)

The heat treatments with the presence of an external electrical field, named as thermoelectric treatment, were carried out in a vertical tube furnace, designed and constructed for this purpose. Figure 5 shows a schematic draw of the oven.

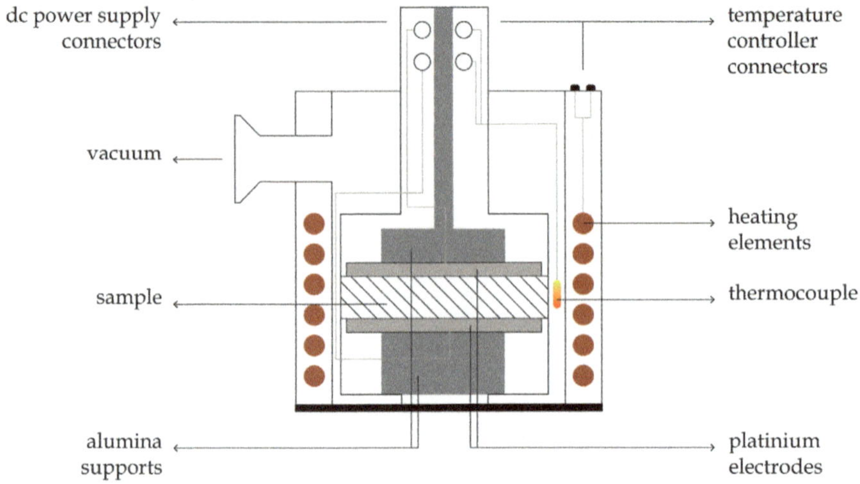

Figure 5. Schematic draw of the oven used for the thermoelectric treatments [19].

The dc external electric field was produced using a high dc voltage source (PS325 Stanford Research System), which could apply a potential difference between 25 V and 2500 V, with a maximum current of 10.5 mA. The temperature was controlled by a Digi-Sense Temperature Controller R/S. All the treatment process is controlled by computer. In these treatments, the thermoelectric cycles used (heating ramp, threshold temperature, treatment time and cooling ramp) were equal to those used in the treatments without the presence of an external field (horizontal tube furnace). In the TET treatments the dc electric field was applied during the periods of heating and dwell, and switched off at the beginning of the cooling step. The parameters: temperature level (T_p) and electric field amplitude are specified and justified in following sections.

5. Samples composition

By the sol-gel method the following two compositions were prepared:

- $92SiO_2$-$4Li_2O$-$4Nb_2O_5$ (mole%);
- $88SiO_2$-$6Li_2O$-$6Nb_2O_5$ (mole%);
- $84SiO_2$-$8Li_2O$-$8Nb_2O_5$ (mole%);

referenced from here by 92Si, 88Si and 84Si, respectively. The 84Si composition did not form a transparent and amorphous gel and glass, indicating the composition limit for those characteristics. This composition was therefore not full characterized.

The preparation process of the glasses and glass ceramics, the results of structural and electrical analyzes and their discussion, are the following sections.

6. Samples preparation

The preparation of the based glass with molar composition $92SiO_2$-$4Li_2O$-$4Nb_2O_5$ and $88SiO_2$-$6Li_2O$-$6Nb_2O_5$ followed the procedure described in figure 2. With the aim of obtain glass ceramics containing $LiNbO_3$ crystallites, heat treatments (TT) were carried out on the as-prepared glass samples (treated at 120 ° C for 48 h and subsequently at 500 ° C for 4 h), which present a thickness between 0.6 and 1.0 mm. For the TT, carried out in a horizontal tubular furnace, the threshold temperature (T_p) choice, differential thermal analyzes (DTA) was performed to the base glass of each composition. The temperatures at the observed exothermic effects, which can indicate the occurrence of crystallization, lead to the definition of the threshold temperatures, which in the 92Si composition case were the following: 650, 700, 750 and 800 °C. The 88Si based glass was TT at 600, 650, 700 and 800 ° C. [16;17]

The 92Si based glass was also subjected to thermoelectric treatments (TTE), which followed the same thermal profile of the TT. The 92Si based samples were therefore TTE at 650, 700 and 750 °C, for 4 hours. For each temperature, three different TTE were performed, differing in the amplitude value of the electric field applied: i) 100 kV/m, ii) 500 kV/m and iii) 1000 kV/m. These values were selected based on the thickness of the samples and the characteristics of the dc voltage source.

7. 92Si samples composition results

Figure 6 shows the macroscopic aspect of the 92Si samples, TT at the temperatures of 650, 700 and 750 °C. The based glass, completely transparent, becomes translucent for temperatures above 700 °C.

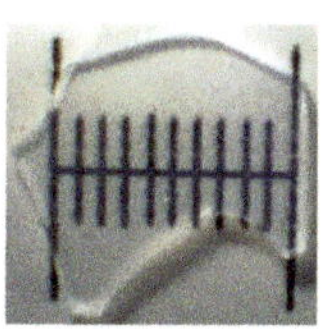
As-prepared
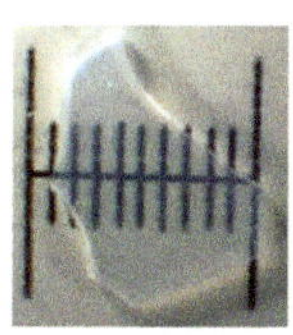
TT700

TT750

Figure 6. Photographs of the 92Si TT glasses (minor division = 0 1 mm).

The 92Si samples, TTE, were named as: 650A (sample TTE at 650 °C with an electric field of 100 kV/m), 650B (sample TTE 650 °C with an electric field of 500 kV/m) and 650C (TTE sample at 650 °C with an electric field of 1000 kV/m). The same designation was used in TTE samples at temperatures of 700 (700A, 700B ...) and 750 °C. In figure 7, photographs of all samples subjected to those treatments can be seen.

The samples 650A, 700A and 750A (samples TTE with a field amplitude of 100 kV/m) have a macroscopic aspect very similar to the sample 650B. With the increase of the TET temperature and applying 500 kV/m and 1000 kV/m, all samples become translucent.

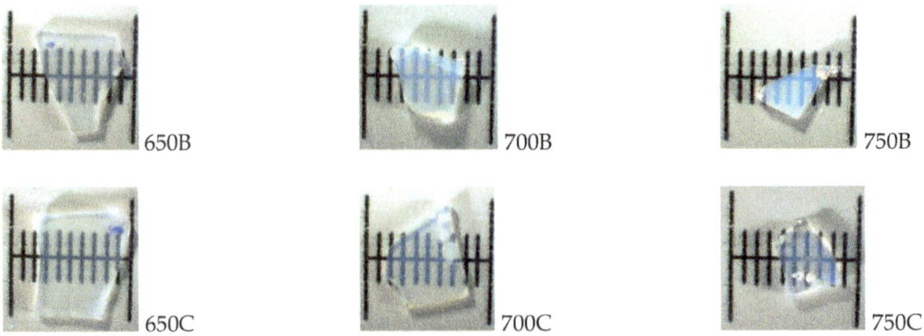

Figure 7. Photographs of the TTE samples at 650, 700 and 750 ºC, with the applied field of (B) 500 kV/m and (C) 1000 kV/m (minor scale division = 1mm).

Figure 8 shows the XRD patterns of the 92Si samples thermal treated without the presence of the external electrical field (TT). It can be observed for the samples TT at temperatures above 700 °C the presence of diffraction peaks associated with the $LiNbO_3$ crystalline phase. The sample TT at 800 °C also presents a second crystalline phase (SiO_2, quartz). The X-ray diffraction was performed at room temperature on a Phillips X'Pert system, where the X-ray production is performed on a Cu ampoule, operating at 40 kV and 30 mA, emitting the monochromatic Kα radiation (λ = 1,54056 Å - graphite monochromator). In this system the sweep is continuous, from 10.025 up to 89.975 º (2θ) with a speed of 1.5 degrees per minute and with a step of 0.02 º. The identification of the crystalline phases was based on the database provided by the JCPDS (Joint Committee on Powder Diffraction Standards). The figures 9 to 11 present the XRD spectra of the samples TET at 650, 700 and 750 ºC, respectively.

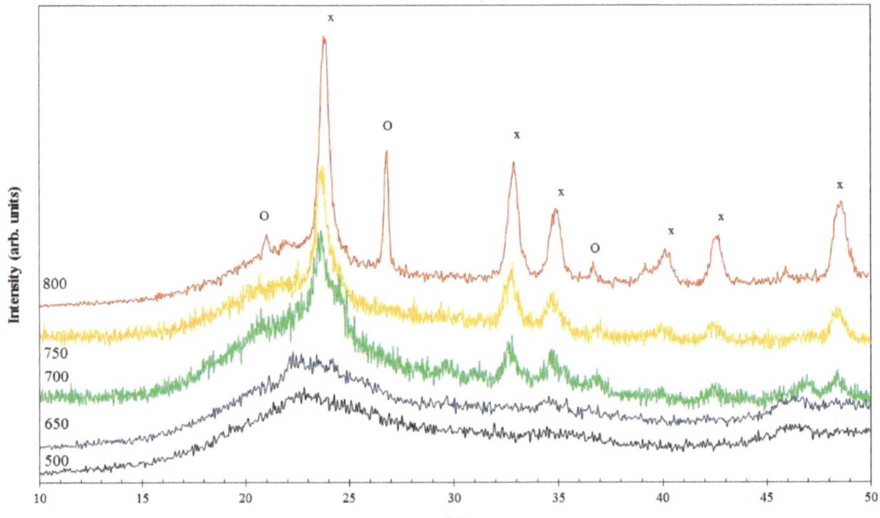

Figure 8. XRD of the 92Si samples TT at 650, 700, 750 and 800 ºC (x $LiNbO_3$; O SiO_2-quartzo).

The XRD spectra of the samples series treated at 650 °C (Fig. 9) shows that the increase in amplitude of the external electrical field favors the formation of the $LiNbO_3$ crystalline phase. In the sample series treated at the temperatures of 700 °C and 750 ºC (Figs. 10 and 11) it was detected the presence of $LiNbO_3$, SiO_2 (quartz) and $Li_2Si_2O_5$ crystalline phases. It is important to refer that the $Li_2Si_2O_5$ phase only appears in the samples thermo-electrically treated.

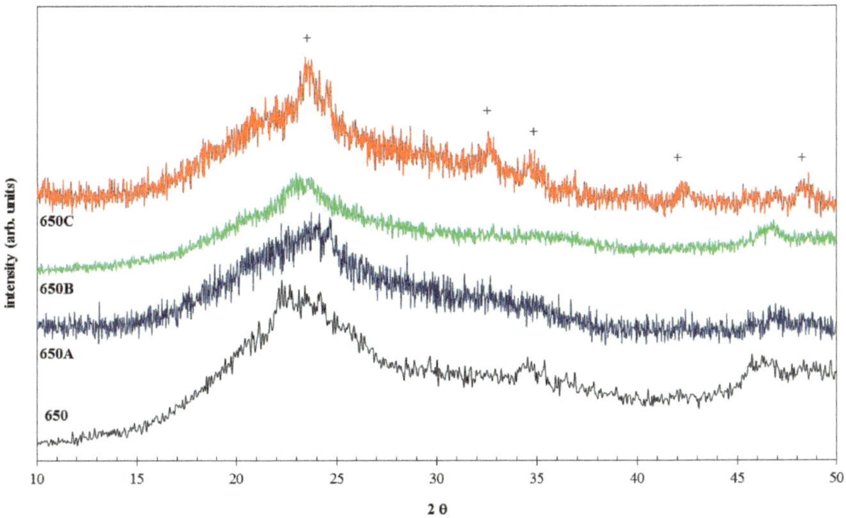

Figure 9. XRD spectra of the 92Si samples TET at 650ºC (+ $LiNbO_3$).

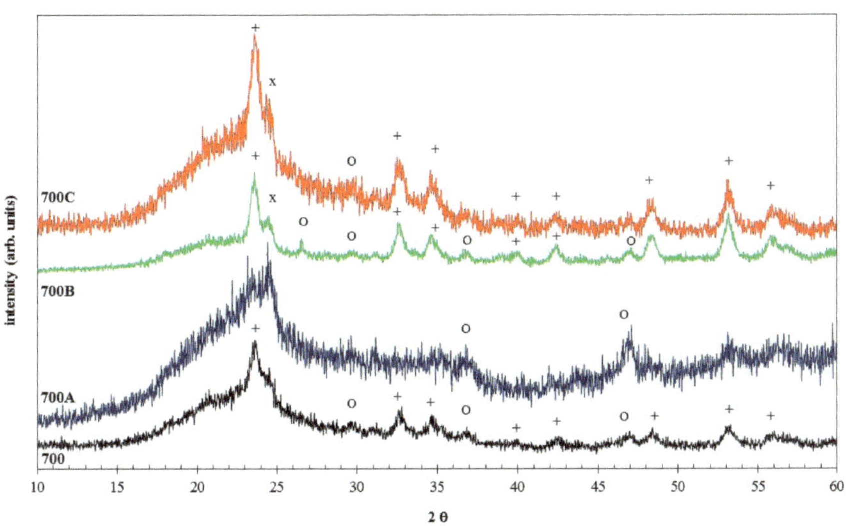

Figure 10. XRD spectra of the 92Si samples TET at 700ºC (+ $LiNbO_3$; O SiO_2 (quartz); x $Li_2Si_2O_5$).

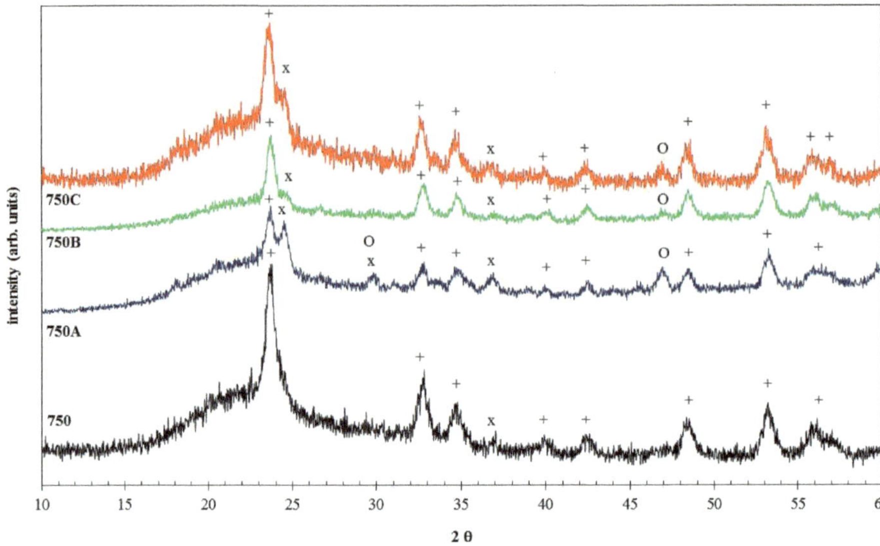

Figure 11. XRD spectra of the 92Si samples TET at 750ºC (+ $LiNbO_3$; O SiO_2 (quartz); x $Li_2Si_2O_5$).

Figures 12 to 15 show the Raman spectra of the free surface of the heat treated samples with or without external electrical field applied. It must be motes that no differences were detected between the spectra obtained on the free surface of the samples and fracture zones (bulk). This analysis was performed on a spectrometer T64000, Jobin Yvon SPEX using an argon laser operating at 514.5 nm. The Raman spectrum was obtained with a back-scattering geometry (back-scattering) between 100 and 2000 cm^{-1}. The amplitude of the lens used was of 50x which allows a laser spot diameter on the sample of about 5 mm.

In all Raman spectra (Figs. 12 to 15), the bands centered at 630, 590, 435, 375, 335, 330, 325, 280, 265, 240 and 155 cm^{-1} are associated to vibrations of the NbO_6 octahedrons [20;21;22]. The bands at 465, 415 and 130 cm^{-1} (sample 700C) are assigned to vibration of Si-O-Si bonds [23]. The broad band centered at 330 cm^{-1}, observed in the samples treated in the presence of an electric field of 1000 kV/m (samples C), seems to be due to the overlapping of the bands centered at 335 and 325 cm^{-1}. The band at 830 cm^{-1}, detected in the samples treated at 650 °C, with no external field applied, are attributed to the vibrations of the Nb-O-Si bonds [20;21;22;23;24].

In the following figures (Figs. 16) SEM micrographs of the free surface of the samples treated with and without the presence of an external electrical field are presented. The scanning electron microscopy was performed in a Hitachi S4100-1, on the surface and fracture of the samples covered with carbon.

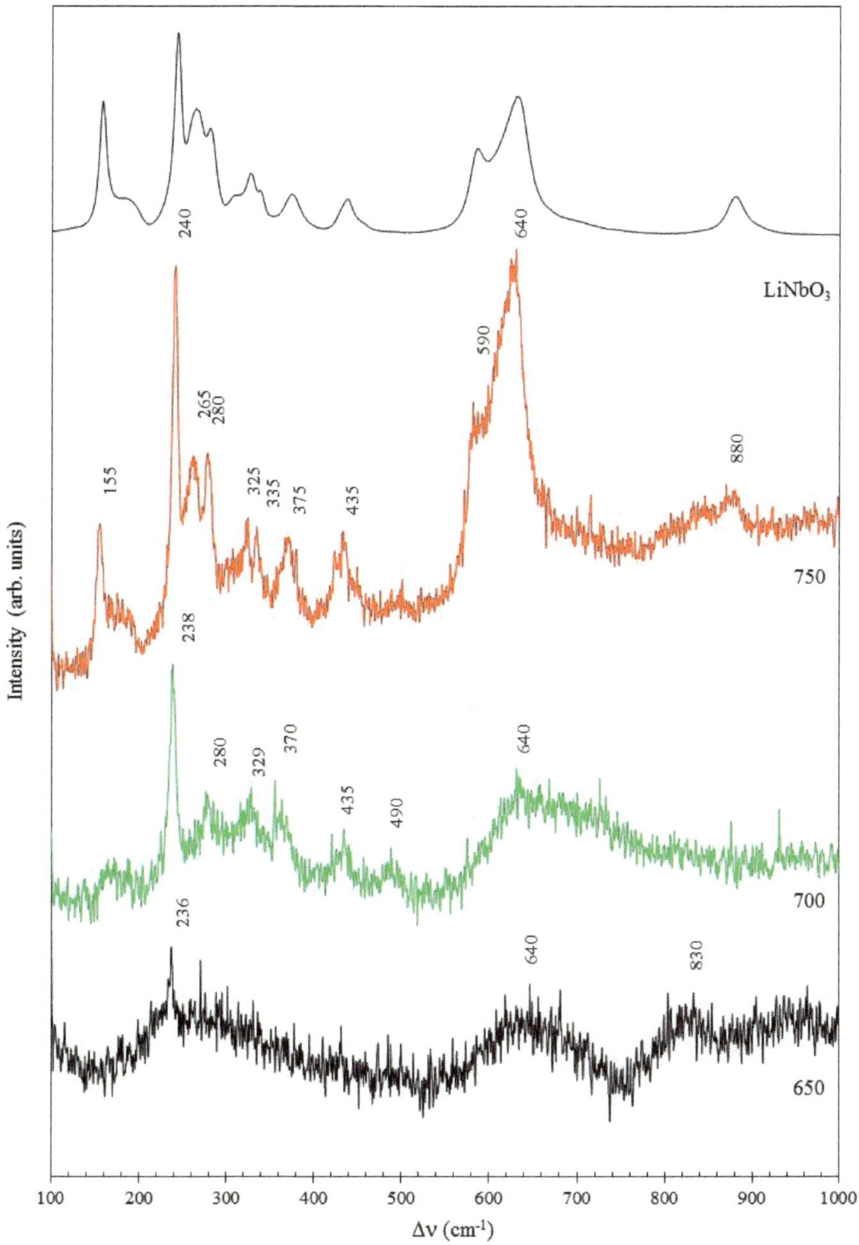

Figure 12. Raman spectra of the 92Si samples TT at 650, 700 and 750 ºC. The Raman spectrum of LiNbO3 crystalline powders is also presented.

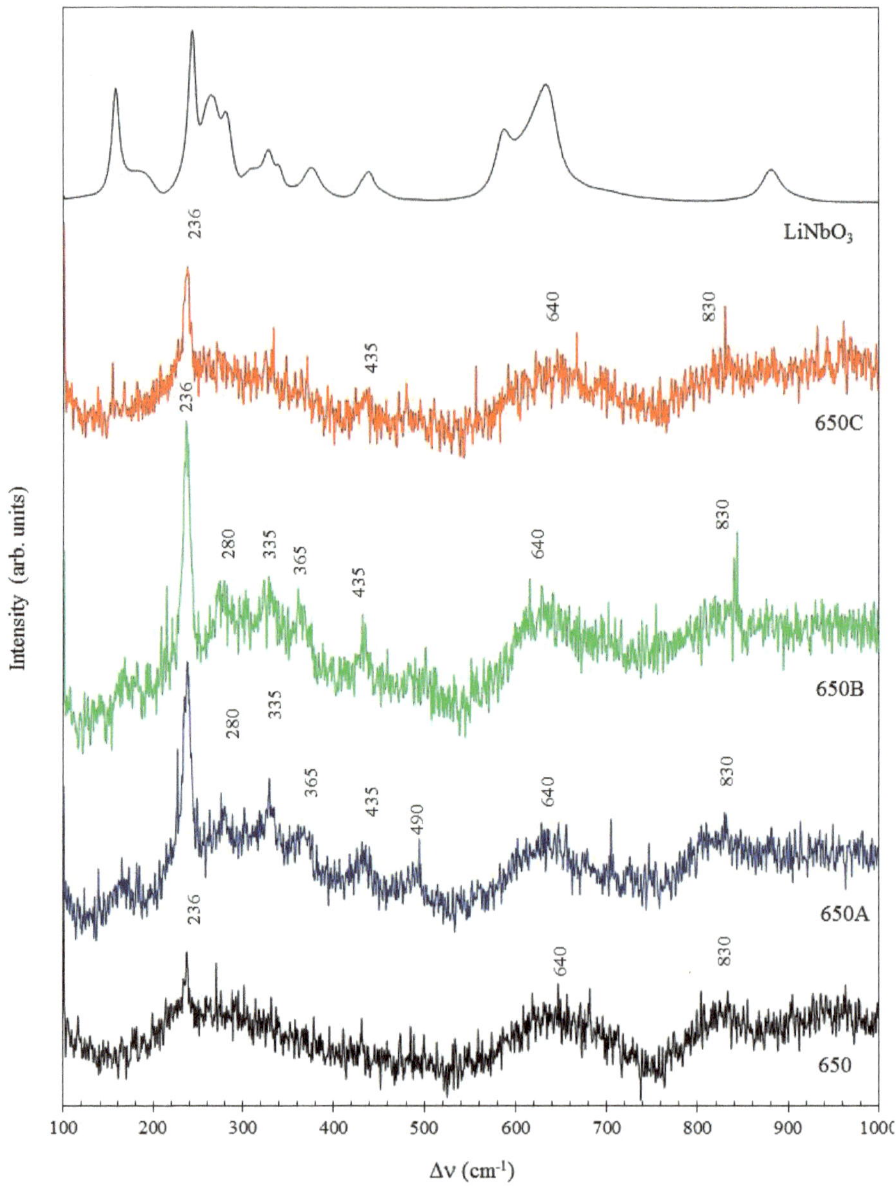

Figure 13. Raman spectra of the 92Si samples TET at 650 ºC. The Raman spectrum of $LiNbO_3$ crystalline powders is also presented.

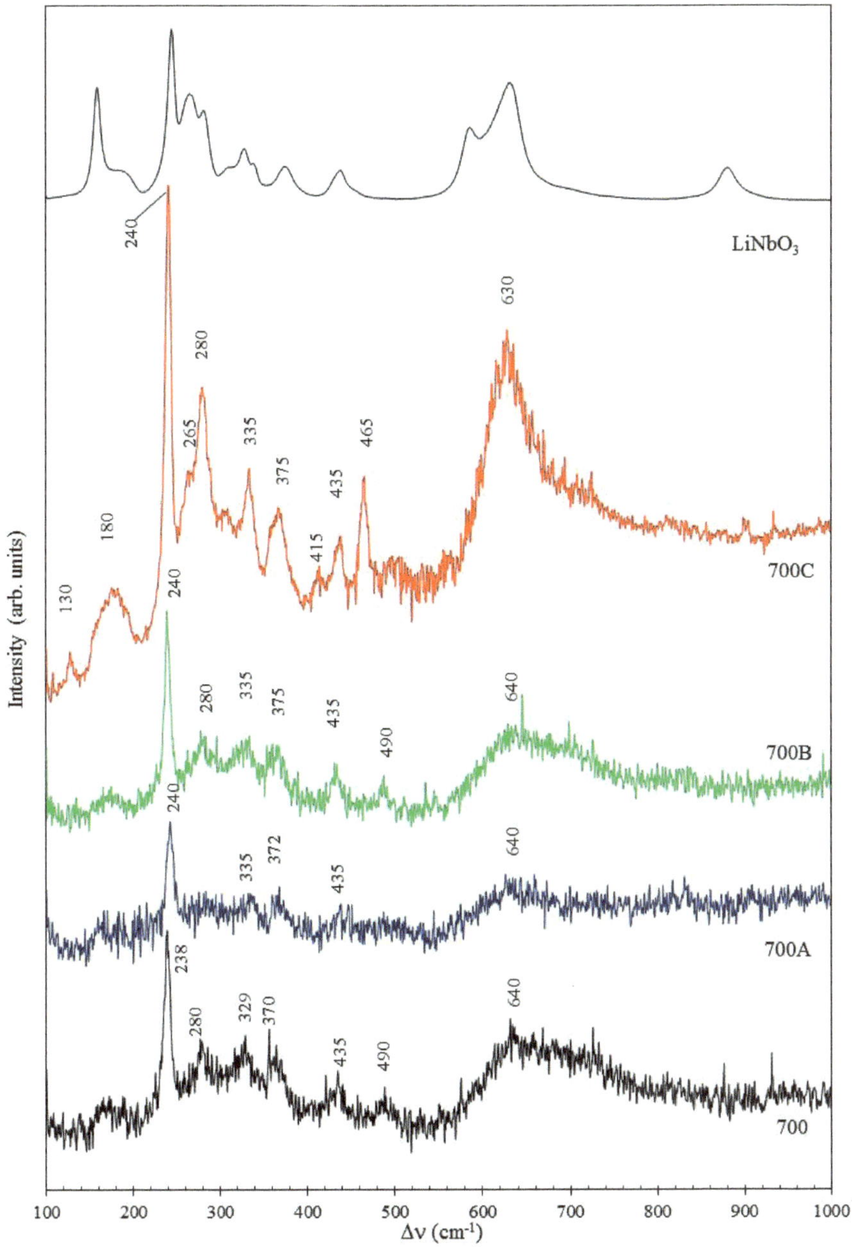

Figure 14. Raman spectra of the 92Si samples TET at 700 ºC. The Raman spectrum of $LiNbO_3$ crystalline powders is also presented.

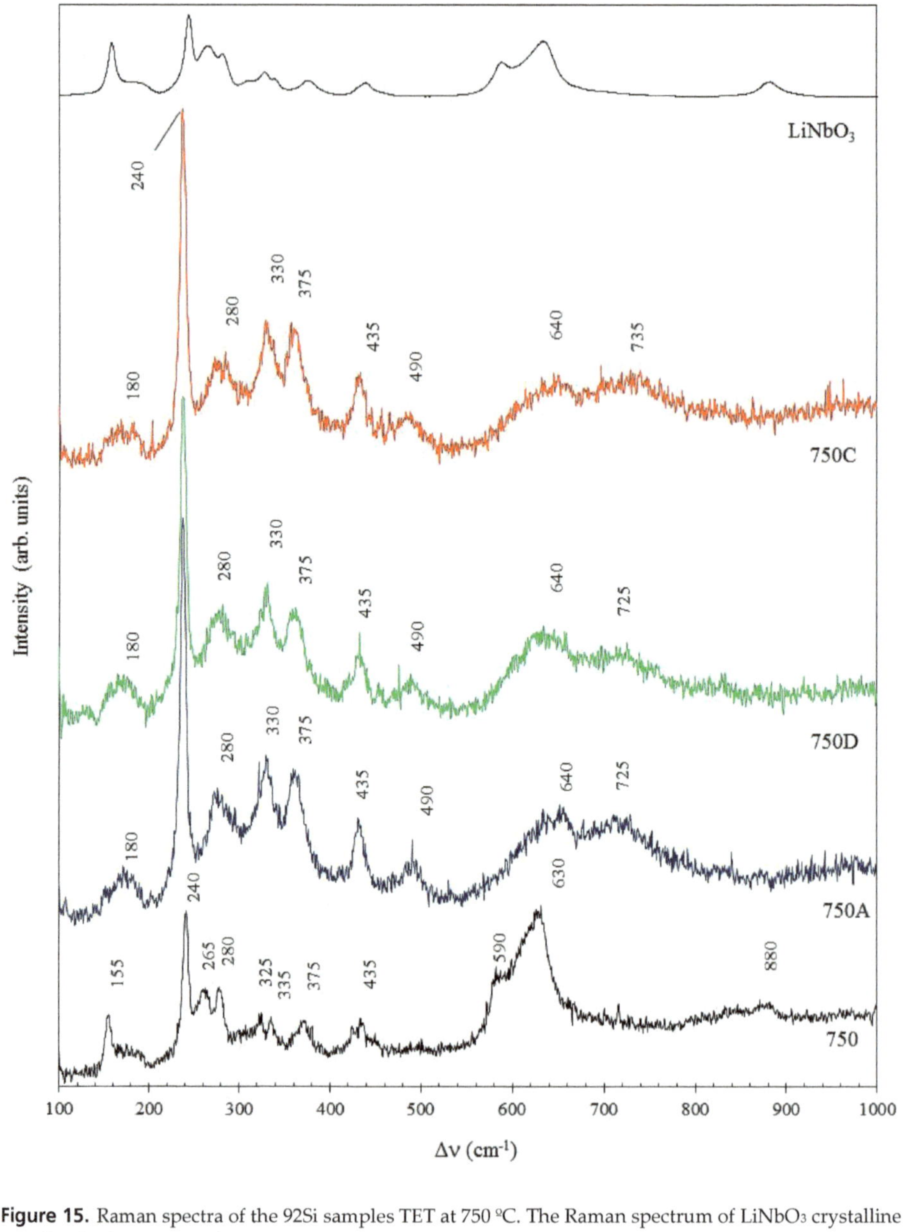

Figure 15. Raman spectra of the 92Si samples TET at 750 ºC. The Raman spectrum of $LiNbO_3$ crystalline powders is also presented.

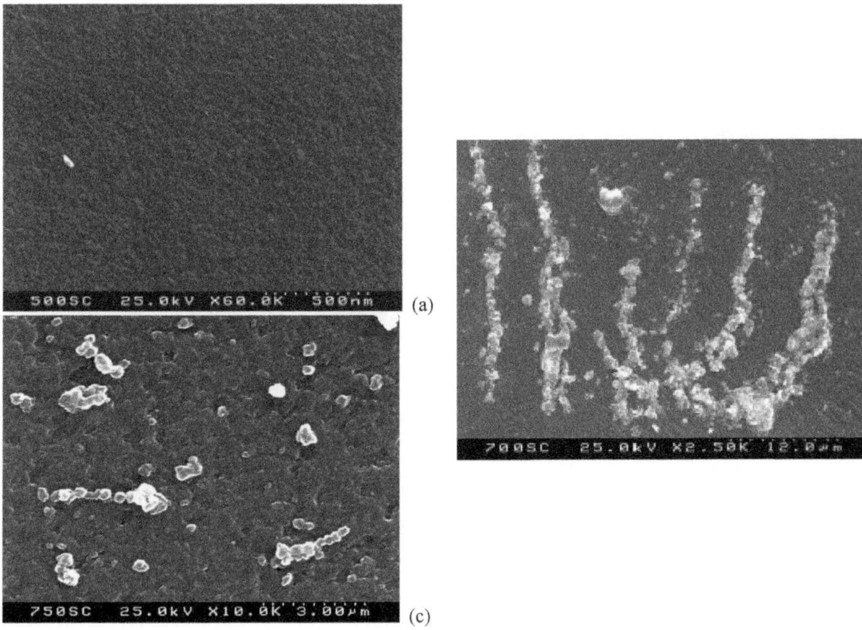

Figure 16. SEM micrographs of the 92Si samples: a) as-prepared; b) TT at 700 ºC; c) TT at 750 ºC.

It was observed in the SEM micrographs of the 92Si based glass (sample TT at 500 °C) the inexistence of particles. Increasing the treatment temperature an increase in the number of surface particles is induced. In the sample heat treated at 700 °C, without applying the external electric field (Fig. 16), particles showing a preferential growth direction were observed. On the surface of the sample TT at 750 °C (Fig. 16c), particle with a size of approximately 500 nm are observed and with a distribution similar to that observed in the sample TT at 700 °C. It must be notice that it was not detected in any sample of the 92Si series, the existence of particles in fracture zone (bulk). In the samples TET at 650 °C it was observed an increase in the size and number of the particles dispersed in the glass matrix, with the increase in amplitude of the applied field (Fig. 17).

In the sample 700A, the number of particles present on the surface, which during the TTE was in contact with the positive electrode, is greater than that the number observed on the opposite surface, but with similar sizes (~ 100nm). The sample 700B registered a particle size distribution similar to the one observed in the sample 700A, but with a larger size (~ 1 µm). Increasing the amplitude of the external electric field up to 1000 kV/m (sample 700C) it was observed the presence of particle aggregation in the two opposite surfaces of the sample. However, the number and size of those aggregates are larger in the surface that was in contact with the positive electrode. The growth of those aggregates seems to have a preferred direction.

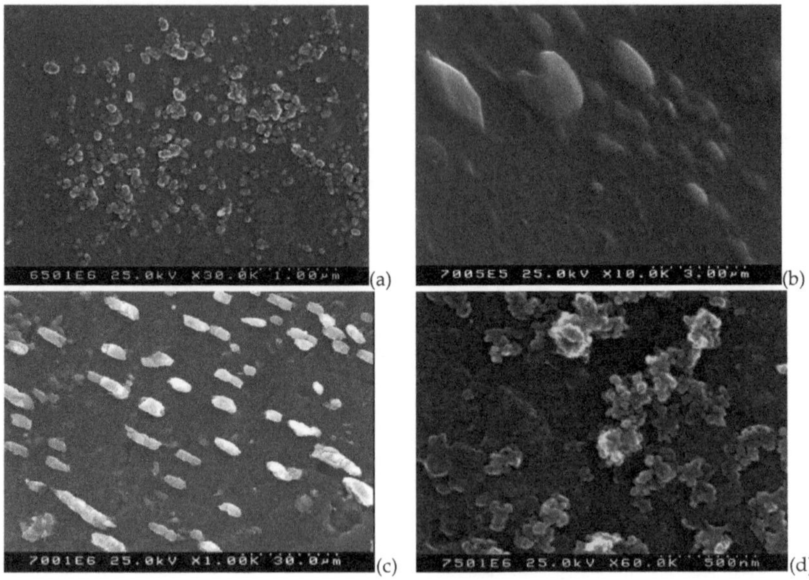

Figure 17. SEM micrographs of the samples 92Si TET: (a) 650C; (b) 700B; (c) 700C; (d) 750C.

Increasing the amplitude of the external electrical field, in the TET at 750 °C, it was observed an increasing in the number of particles but with a reduction in their size. Samples 750A and 750B have particles with an average size of 300 and 250 nm, respectively. The sample 750C presents particles with a maximum size of ~ 50 nm. These particles tend to aggregate themselves.

The dependence of the dc conductivity, in logarithm scale ($\ln(\sigma_{dc})$), with the temperature of measurement, for the samples TT at 650, 700 and 750 °C, with or without external electrical field applied, is shown in figures 18 to 20. All samples exhibit, for temperatures below 270 K, dc conductivity values (σ_{dc}) lower than 10^{-15} Sm^{-1}. The sample 750C shows the highest σdc value (2.63×10^{-9} Sm^{-1}), at the temperature of 370 K. With increasing the treatment temperature, the σ_{dc} measured at 300 K, decreases. These measurements also showed that the σ_{dc} is lower in samples treated with an electric field of 100 kV/m of amplitude (samples 650A, 700A and 750A) than in the samples treated without the presence of the external electrical field. In all sample series (650, 700 and 750), the increase of the amplitude of the electric field applied during the heat treatment induces an increase in σ_{dc} (Table 1).

The dependence of the σ_{dc} with the measurement temperature was adjusted by the Arrhenius equation [25;26;27] -

$$\sigma_{dc} = \sigma_0 e^{\left(-\frac{E_A}{kT}\right)}$$

This adjustment presented graphically in figures 14 to 16 by the lines, allowed the calculation of the dc activation energy ($E_{a(dc)}$ – table 1). It is observed, in all samples, the existence of at least two regions with different activation energies. The first (A), between 270 and 300 K, and the second (B), between 345 and 370 K. The activation energy associated with the conduction mechanism detected at higher temperatures ($E_{a(dc)}$ (B), Table 1) decreases with the increase of the amplitude of the applied electric field, for the sample series treated at 700 and 750 °C. It is observed, with the exception of the sample 700A, that the activation energy of the process A (at lowest temperatures) is always lower than the $E_{a(dc)}$ of the process B.

Sample	σ_{dc} x 10^{-14} ($\Omega^{-1}m^{-1}$)	$E_{a(dc)}$ (A) (kJ/mol)	$E_{a(dc)}$ (B) (kJ/mol)
650	128,48 ± 5,15	24,64 ± 1,19	79,98 ± 1,12
650A	2,04 ± 0,06	27,91 ± 1,09	62,66 ± 2,66
650B	3,03 ± 0,09	27,94 ± 1,40	77,56 ± 0,95
650C	36,46 ± 1,46	48,83 ± 0,72	64,22 ± 1,41
700	13,21 ± 0,55	46,47 ± 1,92	72,98 ± 0,78
700A	4,96 ± 0,21	32,63 ± 3,00	71,38 ± 3,97
700B	8,64 ± 0,29	66,35 ± 1,02	66,79 ± 0,83
700C	15,18 ± 0,47	100,18 ± 3,31	57,15 ± 0,99
750	12,39 ± 0,43	31,48 ± 0,29	51,02 ± 1,55
750A	6,09 ± 0,35	31,63 ± 0,68	82,08 ± 3,19
750B	21,71 ± 1,06	7,71 ± 2,08	79,26 ± 0,82
750C	393,87 ± 12,91	24,47 ± 2,59	62,88 ± 1,22

Table 1. dc conductivity (σ_{dc}), at 300 K, dc activation energy ($E_{a(dc)}$) for the: A - low temperature region (230-300 K); B – high temperature region (310-370 K).

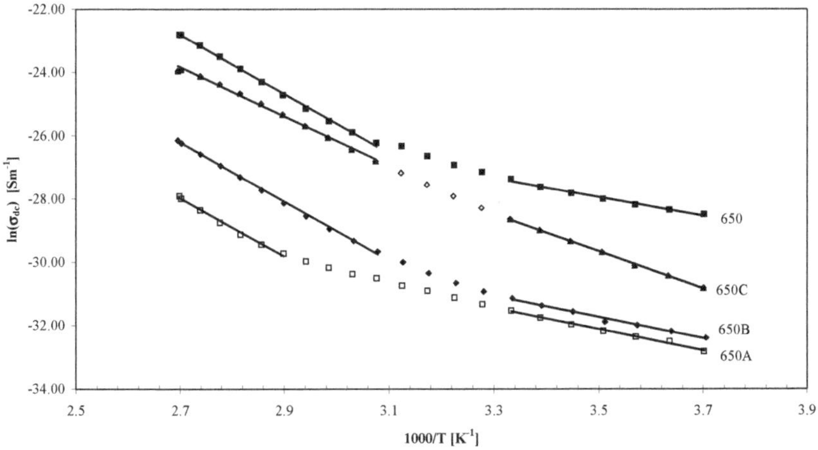

Figure 18. The dc conductivity (σ_{dc}, in logarithm scale) in function of 1000/T, for the 92Si samples treated at 650 ºC.

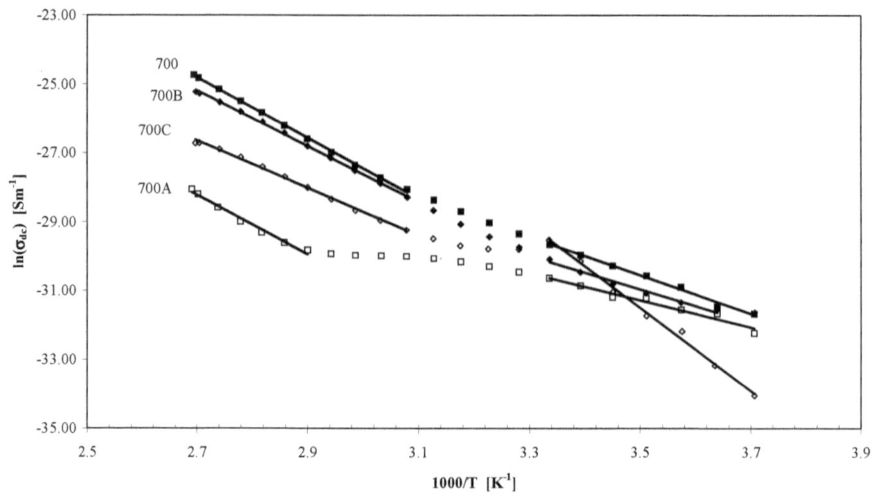

Figure 19. The dc conductivity (σ_{dc}, in logarithm scale) in function of 1000/T, for the 92Si samples treated at 700 ºC.

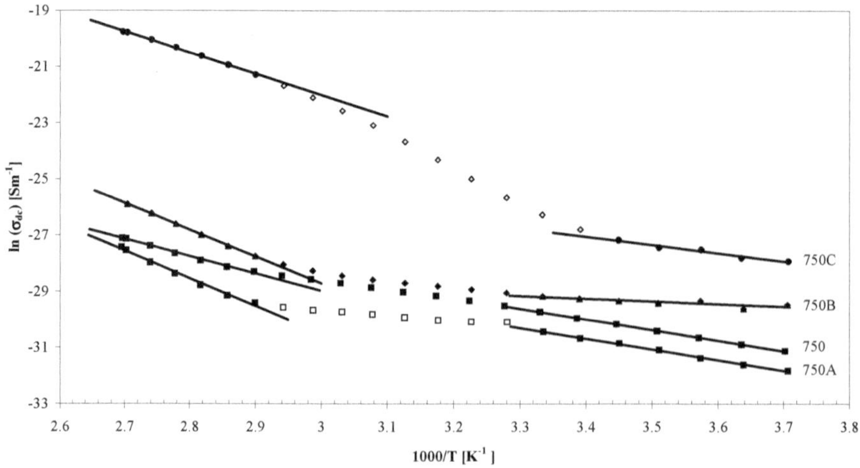

Figure 20. The dc conductivity (σ_{dc}, in logarithm scale) in function of 1000/T, for the 92Si samples treated at 700 ºC.

Figures 21 to 23 show the frequency dependence of the imaginary part of the impedance (Z´´) for the sample series treated at 650, 700 and 750 °C, respectively. It was observed, for the samples series treated at 650 and 700 °C, that Z" decreases with the increase of frequency. In the frequency range used, the impedance (Z*), the admittance (Y*), the permittivity (ε*) and the dielectric modulus (M*) formalisms did not revealed the existence of dielectric relaxation(s). It must be noted that, for frequencies below 100 Hz, there is a high dispersion of the Z′′ values, which is associated with the sensitivity of the measuring apparatus, in this frequency range.

An adjustment of the Z* spectrum was carried out using a complex non-linear least squared deviations method (CNLLS) associated with the electrical equivalent circuit model formed by the parallel between a resistor (R) and a constant phase element (CPE). This constant phase element is characterized by keeping constant the angle of the impedance as a function of frequency, i.e. the ratio between the real and imaginary part of the impedance is constant across the all frequency range. The impedance of this intuitive element (Z_{CPE}) can be represented by

$$Z_{CPE} = \frac{1}{Q_0 (j\omega)^n}$$

where Q_0 and n are frequency independent parameters, but usually are temperature dependent. The parameter n varies between 0 and 1, when n = 1 the CPE is reduced to a capacitance element and when n =0 to a resistive element [18].The lines ("small dots") in the figures 21 to 23 represent the result of this adjustment. The values obtained for the parameters of the equivalent electric circuit are in table 2.

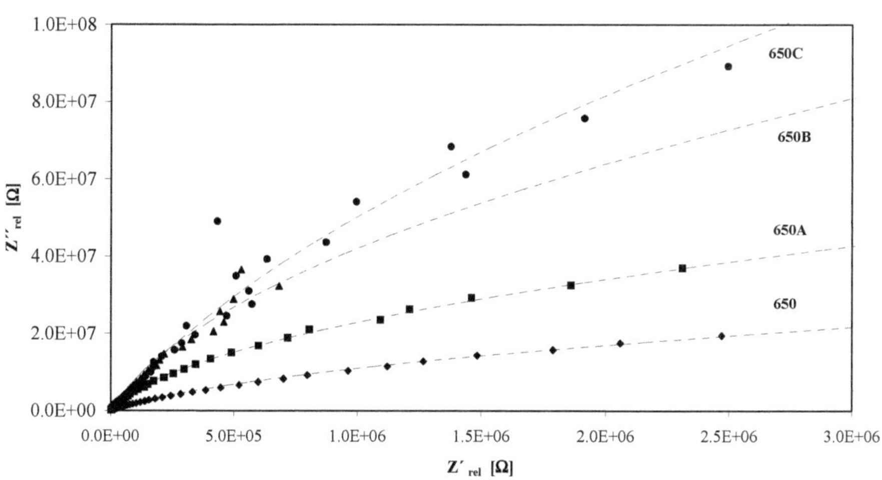

Figure 21. Z′′ versus Z′, for the 92Si samples treated at 650 ºC.

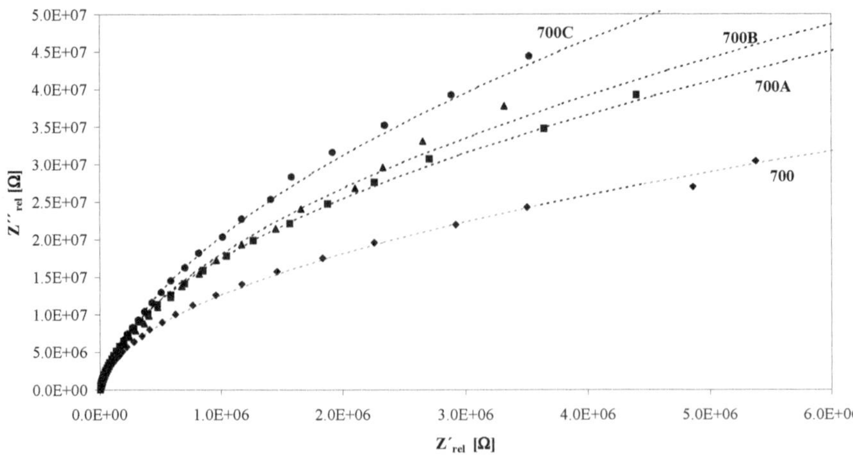

Figure 22. Z'' versus Z', for the 92Si samples treated at 700 ºC.

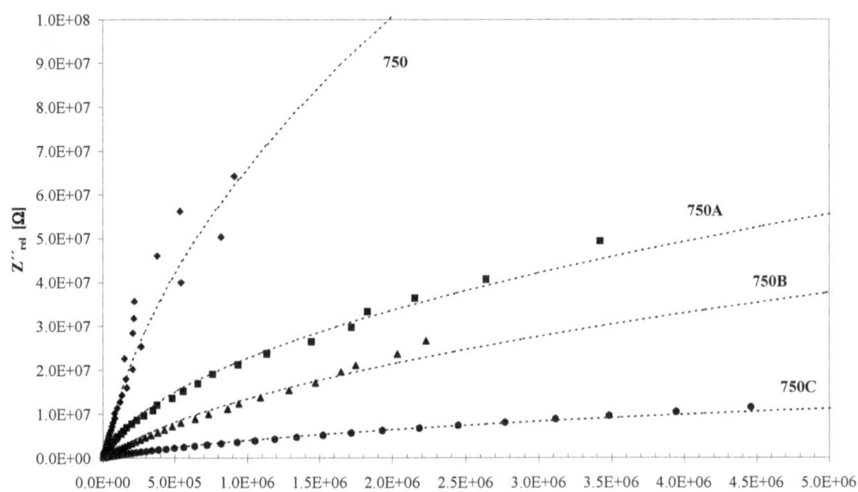

Figure 23. Z'' versus Z', for the 92Si samples treated at 750 ºC.

From the results obtained through the CNLLS fitting process, it is verified that the R parameter, of the samples treated at 650 and 700 °C increases, with the increase of the amplitude of the external electric field. In the sample series treated at 750 °C this parameter has the opposite behavior, i.e., it decreases with the increase of the applied electric field amplitude. However, it should be referred that the sample treated at 750 °C, but without the presence of an external electric field, shows the lowest value of R and the sample 750C (TTE with a field of 1000 kV/m) the largest value of R for all samples. The behavior of the Q_0 parameter, in function of the thermal treatment conditions, is opposite to the observed for parameter R. In all samples, the value of the parameter n is very close to 1. Based on these calculated values a relaxation time, which represents the time average of the relaxation time distribution, was calculated ($\tau_Z = 1/\omega_{Z''_{max}}$) and the value of the capacity nearest of the CPE element value (Table 2). Evaluating the behavior of the parameter τ_Z (Table 2), with the increase of the electric field amplitude, it is verified that τ_Z increases with the increase of that amplitude in the sample series treated at 650 and 700 °C. In the samples treated at 750 °C, the increase of the amplitude of the external electrical field, induces an opposite behavior, i.e., a decrease of τ_Z.

Sample	ε′	tan δ (x10^{-2})	R (x10^{8}) [Ω]	Qo (x10^{-11}) [$\Omega^{-1}m^{-2}s^{n}$]	n (x10^{-1})	τ_Z (x10^{-3}) [s]	C_{CPE} (x10^{-11}) [F]
650	5,21 ± 0,18	6,13 ± 0,34	2,34	6,01	9,72	1,96	5,76
650ª	4,66 ± 0,12	1,81 ± 0,07	7,41	4,47	9,92	5,13	4,46
650B	4,82 ± 0,14	1,29 ± 0,06	23,90	4,54	9,94	20,94	4,54
650C	2,03 ± 0,07	1,61 ± 0,12	68,51	1,94	9,92	20,67	1,94
700	5,57 ± 0,20	2,79 ± 0,16	1,85	5,25	9,93	1,50	5,24
700ª	4,41 ± 0,16	2,39 ± 0,14	3,75	4,17	9,94	2,42	4,16
700B	4,50 ± 0,14	2,68 ± 0,13	4,59	4,41	9,90	3,10	4,38
700C	3,82 ± 0,11	2,92 ± 0,13	7,19	3,79	9,87	4,14	3,77
750	2,79 ± 0,09	8,40 ± 0,05	7,45	2,56	9,96	29,47	2,56
750ª	3,68 ± 0,19	2,03 ± 0,17	7,26	3,52	9,92	3,95	3,51
750B	6,07 ± 0,27	5,86 ± 0,40	4,26	6,91	9,73	4,26	6,69
750C	9,44 ± 0,28	21,20± 0,98	0,47	20,80	9,01	0,95	13,62

Table 2. Dielectric constant (ε′) and dielectric loss (tan δ), at 1kHz and 300 K, parameters of the equivalent electric circuit (R, Q_0 and n), relaxation time (τ_Z) and the C_{CPE} capacitor.

The dependence of the dielectric constant (ε') with the frequency, at the temperature of 300 K, for the sample series of 650, 700 and 750 is represented in figures 24 to 26, respectively. It is observed that the value of ε' decreases with the increase of the frequency. Table 2 shows the values of ε', measured at 300 K and 1 kHz, for all samples. It can be verified that the increase of the amplitude of the external electrical field, for the samples treated at 650 and 700 °C, promotes a decrease of ε'. In the samples series treated at 750 °C, ε′ increases from 2.8 to 9.4, with the increase of the amplitude of the applied external electric field. The C_{CPE} capacitance behavior (Table 2), with the increase of the amplitude of the external electrical field, is similar to the one observed on the ε'. Table 2 contains also the values of the dielectric loss factor (tan δ = ε''/ε'), at room temperature (300 K) and at the frequency of 1 kHz.

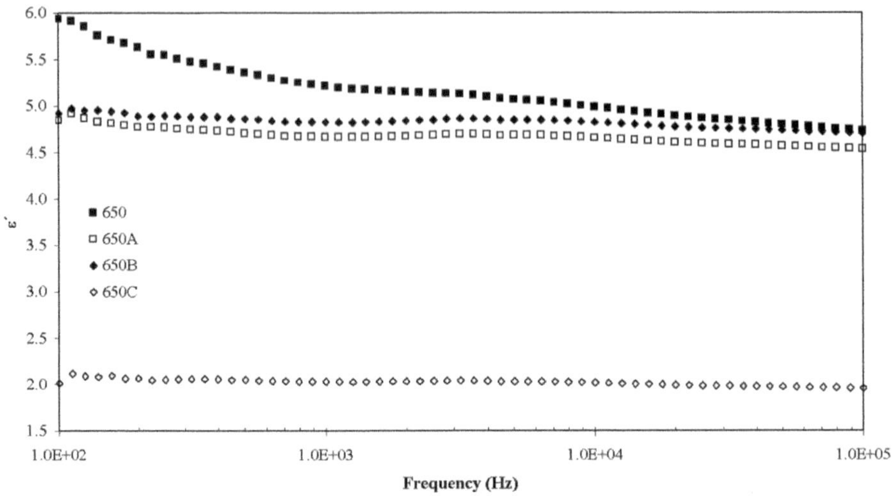

Figure 24. ε´ versus frequency, at 300 K, for the 92Si samples treated at 650 ºC.

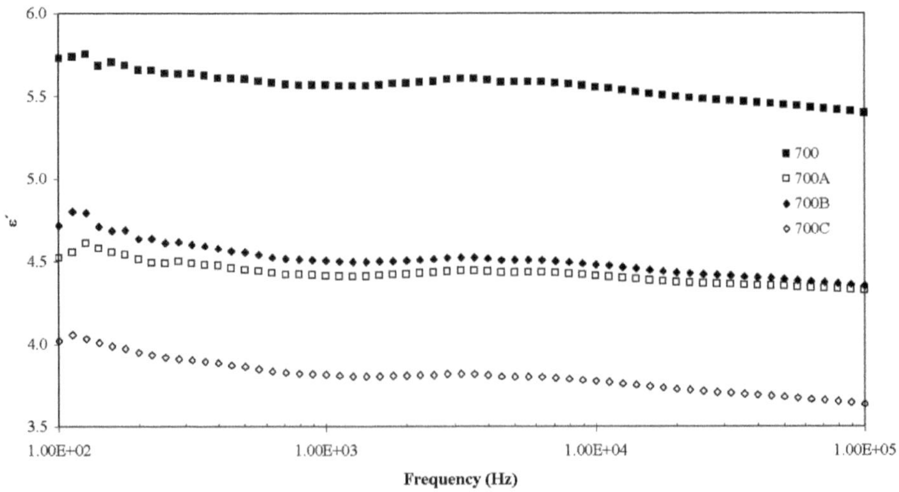

Figure 25. ε´ versus frequency, at 300 K, for the 92Si samples treated at 700 ºC.

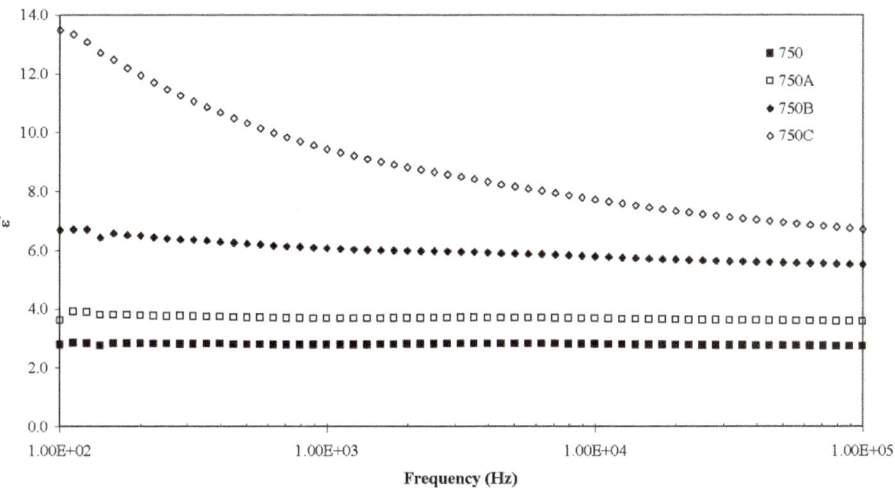

Figure 26. ε′ versus frequency, at 300 K, for the 92Si samples treated at 750 ºC.

8. 88Si samples composition results

Figure 27 shows the macroscopic aspect of the 88Si sample composition, TT at 500, 600, 650, 700 and 800 ºC, during 4 hours. It can be seen that the as-prepared glass (TT at 500 ºC) is translucent and for TT above 700 ºC it becomes opaque.

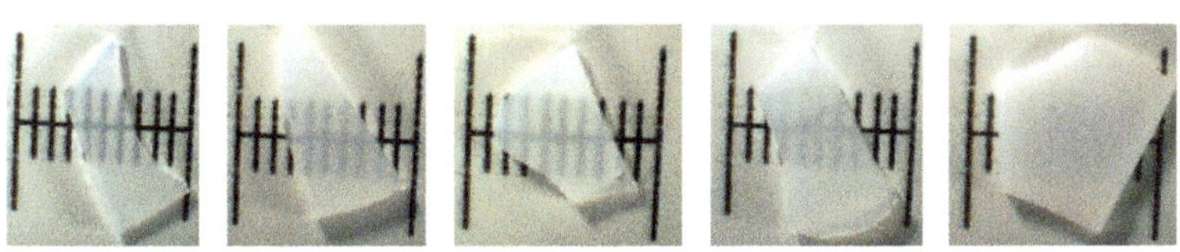

As-prepared TT600 TT650 TT700 TT800

Figure 27. Photographs of the 88 Si samples TT at temperatures between 500 and 800 ºC (the minor scale division = 1mm).

Figure 28 shows the XRD patterns of the samples TT. This spectrum shows the presence of $LiNbO_3$ crystalline phases and cristobalite (SiO_2), in the samples treated at temperatures above 650 °C. With the increase of the TT temperature up to 700 °C it was detected also the lithium silicate ($Li_2Si_2O_5$) crystalline phase. In order to confirm the indexing of some diffraction peaks observed in the sample TT at 800 °C, the XRD was carried out in a new sample, TT at 800 °C but during 8h (sample 800-8h). Analyzing the pattern of this sample it is suggested the presence, in the samples treated at 800 °C, of the Li_3NbO_4 crystalline phase.

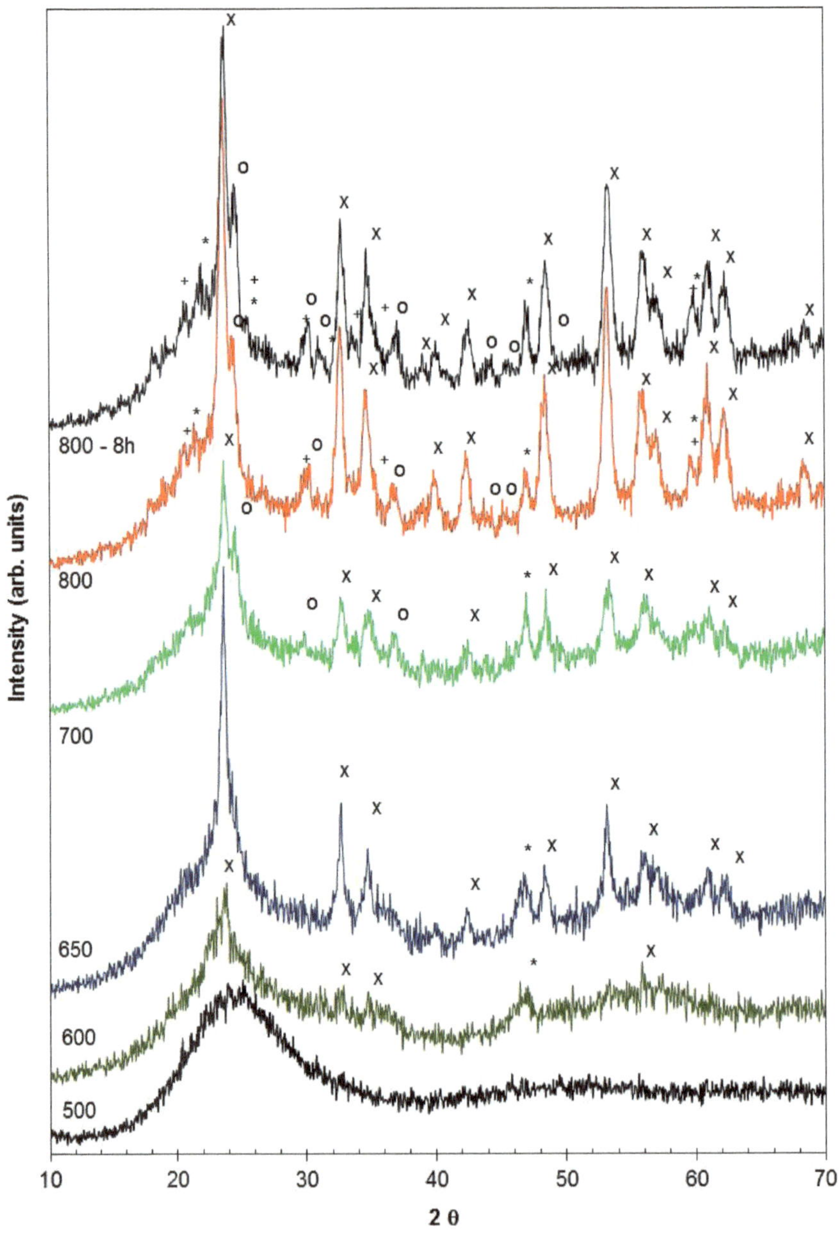

Figure 28. XRD spectra of the 88Si samples TT at temperatures between 500 and 800 ºC (x $LiNbO_3$; O $Li_2Si_2O_5$; * SiO_2 (cristobalite); + Li_3NbO_4).

In figure 29 it is shown the Raman spectra of the surface of all 88Si TT samples. The Raman spectrum of the samples treated at temperatures below 600 °C shows the presence of broad bands centered at ~ 680 and ~ 280 cm^{-1}. With the increase of the thermal treatment temperature, less wide bands but with higher intensity are detected. The bands centered at 630-680 and 240 cm^{-1} observed in samples TT at 500, 600 and 650 °C are due to vibrations of NbO_6 octahedrons [16;17]. The bands at 440, 360, 334 and 280 cm^{-1}, detected in the sample TT at 800 °C, are assigned to the vibration of NbO6 octahedrons, which are associated with the $LiNbO_3$ crystal structure [16;17]. It must be noted the non-detection of Raman bands associated with vibrations of the type Si-O-Si or Nb-O-Si. The non-detection of vibrations associated to Nb-Si-O bonds, which according to Lipovski et al. [24] should be present between 800 and 850 cm^{-1}, suggests that the niobium ions are introduced into the glass matrix only as network modifier.

Micrographs of the sample surface of the 88Si samples TT at 500ºC, 600, 650, 700 and 800 °C are shown in figure 30. The SEM micrographs show, in the surface of the sample heat-treated at 600 °C, particles (Fig. 30b). The size of those particles, also observed in samples TT at 700 and 800 ºC is similar (150-200 nm). However, increasing the temperature of TT leads to an increase in the number of particles. The sample TT at 650 °C has a particle size of around 2 μm (Fig. 30c). The micrograph of this sample also present particle agglomerates (Fig. 30d), not detected in any other sample.

The dependence of the dc conductivity with the temperature of measurement, for all TT samples, is shown in figure 31. The Arrhenius model was used to adjust the $\ln(\sigma_{dc})$ with the inverse of the measurement temperature, allowing the calculation of the activation energy ($E_{a(dc)}$) through the Arrhenius equation. From figure 31 it is verified the existence of two temperature zones, with different activation energies. The first zone (A) is between 230 and 300 K and second (B) is between 310 and 370 K. The as-prepared sample (TT at 500 °C) presents, for measuring temperatures above 300 K, a behavior that is non-adjustable through the Arrhenius equation. The calculated values of $E_{a(dc)}$ are shown in Table 3.

The ac conductivity (σ_{ac}), measured at 300 K and 1 kHz, decreases with the increase of the TT temperature (Table 3). The activation energy, $E_{a(ac)}$, was calculated using the Arrhenius formalism. The lines in figure 31 represent the result of that calculation and the obtained values are in Table 3.

Figures 34 to 41 show the dependence of Z" with the frequency, at various measurement temperatures, for all TT samples. It was observed the existence of two dielectric relaxation mechanisms. The first, in the low frequency region (< 100 Hz) and the second in the high frequency region (> 1 kHz). For frequencies below 1 Hz, a high dispersion of the Z* values is observed and assigned to the sensitivity of the measuring instrument in this low frequency region. The impedance spectra were adjusted to the electrical equivalent circuit model, shown in figure 33, using the CNLLS algorithm. Thus, in the spectra shown in figures 34 to 41, the lines represent the adjustment obtained with this fitting process. It is observed that with the increase of the temperature of measurement, there is a tendency for a better definition of the relaxation curves. Thus, and for the lower measuring temperatures it was found that the adjustment process diverged, not been possible to fit the experimental data with this theoretical model.

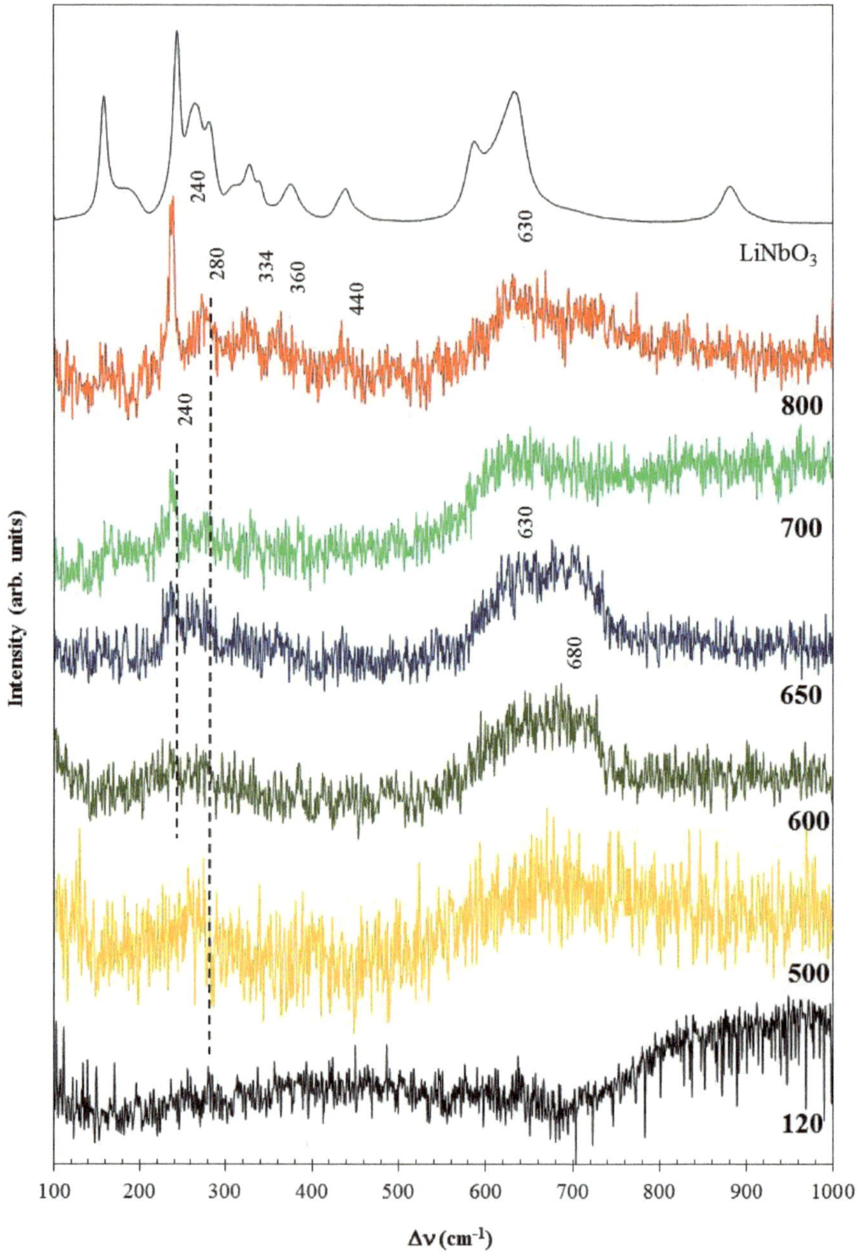

Figure 29. Raman spectra of the 88Si samples TT at 120, 500, 600, 650, 700 and 800 ºC.

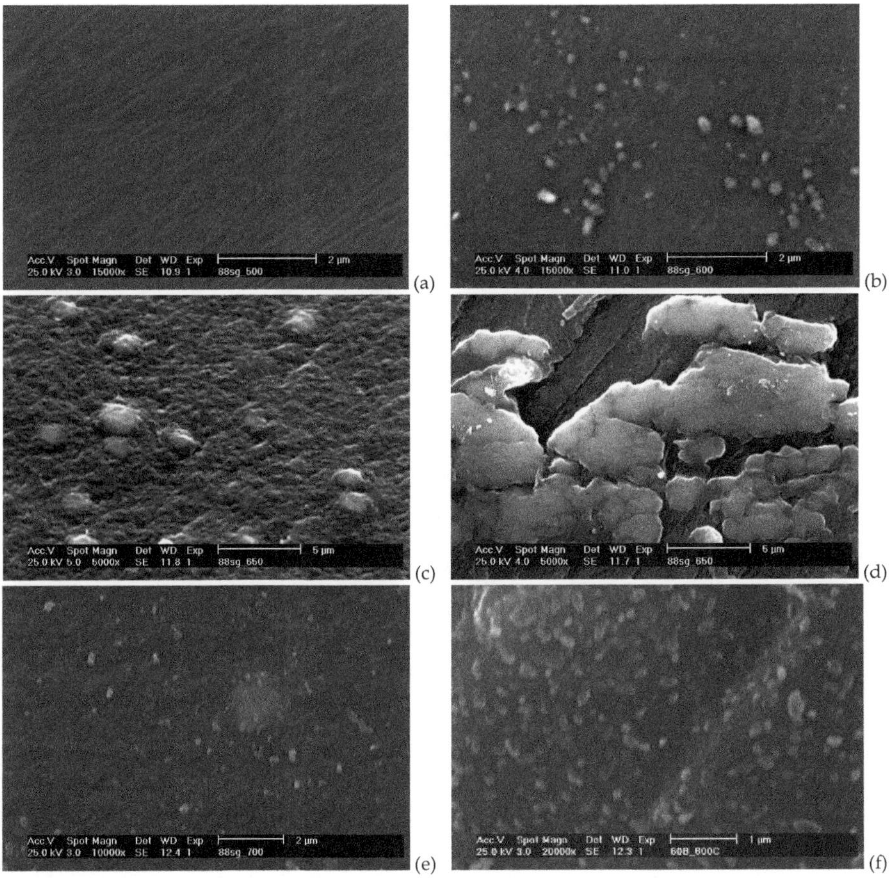

Figure 30. SEM micrographs of the 88Si samples: a) as-prepared (TT at 500 ºC); b) TT600; c) TT650; d) TT650; e) TT700; f) TT800.

Sample	σ_{dc} (x10^{-8}) [$\Omega^{-1}m^{-1}$]	$E_{a(dc)}$ (A) [kJ/mol]	$E_{a(dc)}$ (B) [kJ/mol]	σ_{ac} (x10^{-6}) [$\Omega^{-1}m^{-1}$]	$E_{a(ac)}$ [kJ/mol]
500	379,70 ± 9,86	50,95 ± 1,78	--	20,10 ± 0,02	45,79 ± 0,53
600	163,71 ± 2,41	57,66 ± 1,03	31,66 ± 1,71	15,02 ± 0,62	31,71 ± 0,31
650	55,42 ± 0,74	66,24 ± 0,40	44,00 ± 0,94	2,49 ± 0,09	52,17 ± 2,27
700	120,82 ± 1,66	64,35 ± 0,43	32,76 ± 0,77	2,80 ± 0,12	45,60 ± 2,70
800	3,47 ± 0,05	61,02 ± 0,89	49,96 ± 0,72	0,60 ± 0,02	43,83 ± 2,33

Table 3. dc conductivity (σ_{dc}), at 300 K, dc activation energy ($E_{a(dc)}$) for the: A - low temperature region (230-300 K); B – high temperature region (310-370 K).

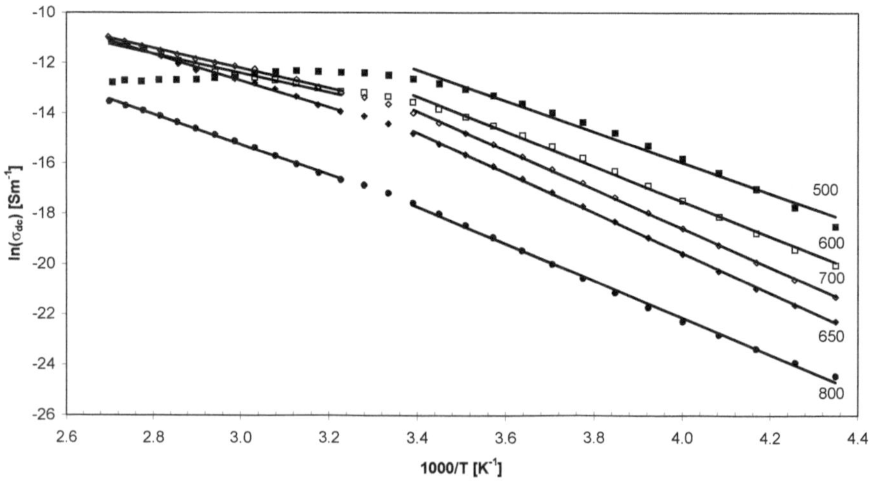

Figure 31. ln(σ_{dc}) versus 1000/T for all 88Si samples TT between 500 and 800 ºC.

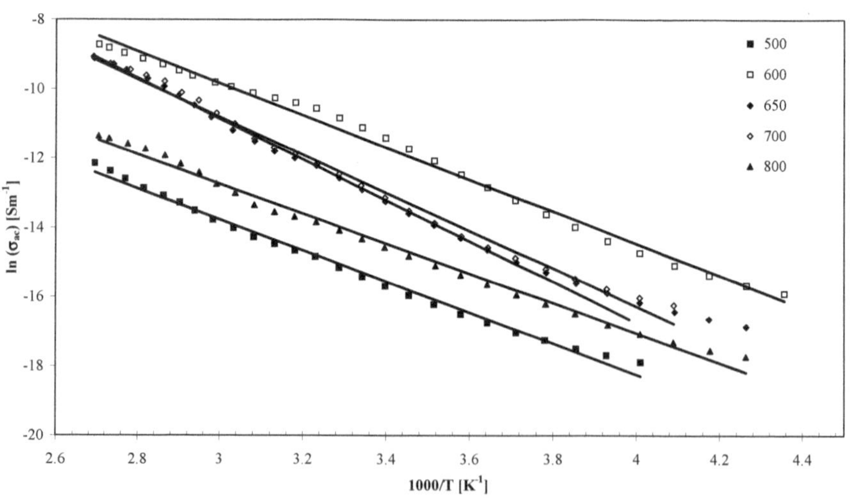

Figure 32. ln(σ_{ac}) versus 1000/T for the samples 88Si TT between 500 and 800 ºC.

Table 4 presents the parameter values of the electrical equivalent circuit for all analyzed samples. Note that the value of the parameter R_0, representing the value of Z' when $\omega \to \infty$ was considered equal to 0 Ω. The parameters R_1 and CPE_1 (CPE represents a constant phase element [18]) are associated with the dielectric relaxation mechanism found at low frequencies (< 100 Hz) and the parameters R_2 and CPE_2 with the relaxation mechanism detected in the high frequency region.

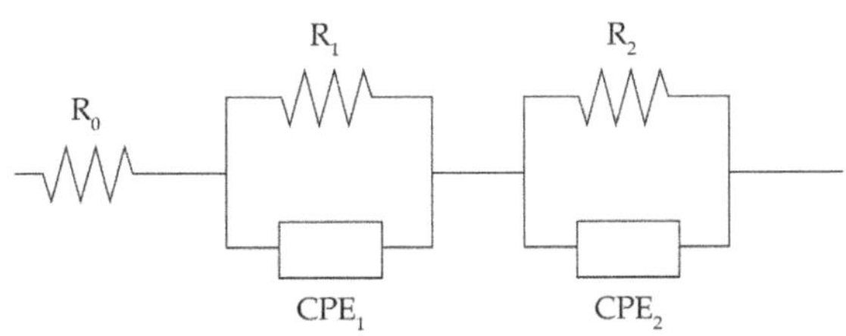

Figure 33. Equivalent electric circuit model used to adjust the impedance data of the 88Si samples ($[R_0(R_1CPE_1)(R_2CPE_2)]$).

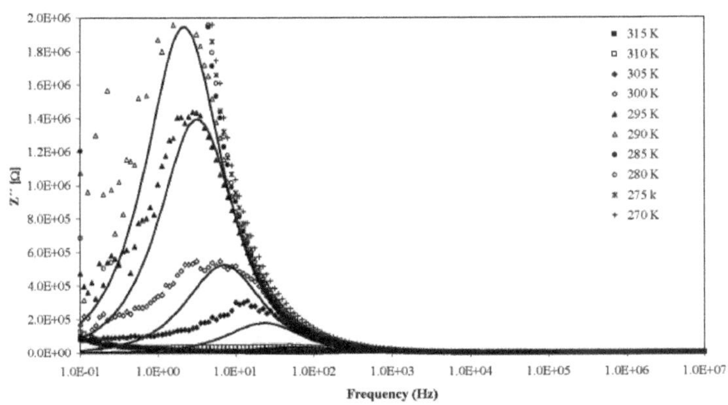

Figure 34. Z'' versus frequency for the 88Si sample TT at 600 ºC (low frequency region). The lines represent the theoretical adjust.

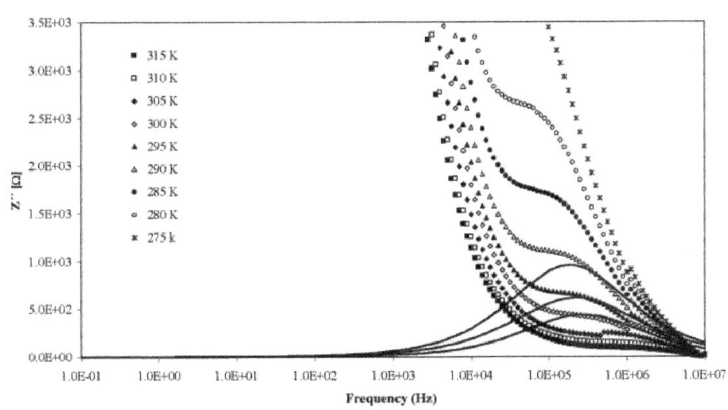

Figure 35. Z'' versus frequency for the 88Si sample TT at 600 ºC (high frequency region). The lines represent the theoretical adjust.

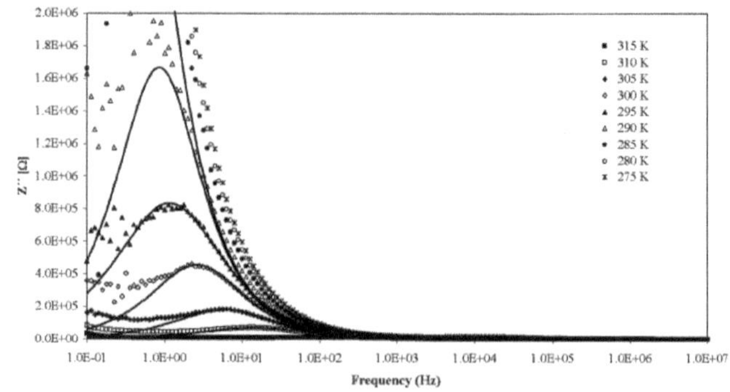

Figure 36. Z′′ versus frequency for the 88Si sample TT at 650 ºC (low frequency region). The lines represent the theoretical adjust.

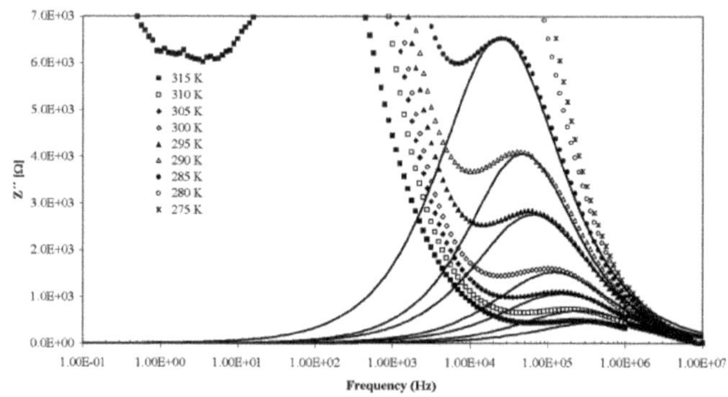

Figure 37. Z′′ versus frequency for the 88Si sample TT at 650 ºC (high frequency region). The lines represent the theoretical adjust.

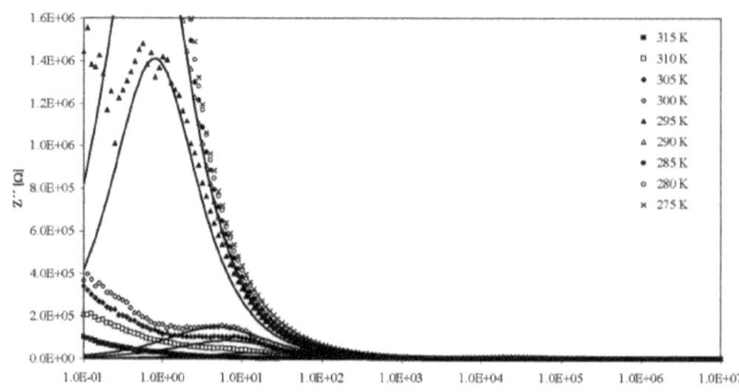

Figure 38. Z′′ versus frequency for the 88Si sample TT at 700 ºC (low frequency region). The lines represent the theoretical adjust.

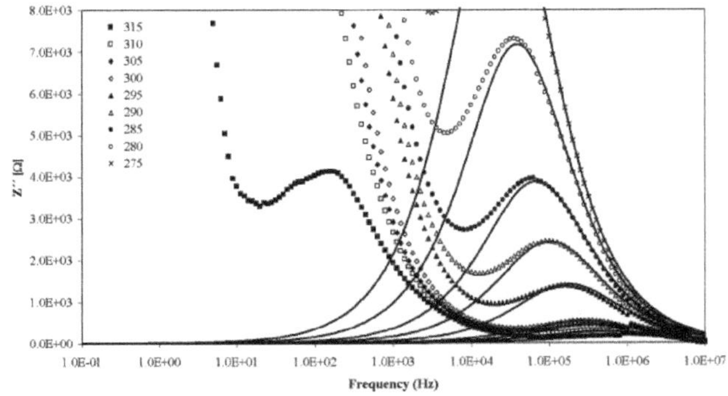

Figure 39. Z′′ versus frequency for the 88Si sample TT at 700 ºC (high frequency region). The lines represent the theoretical adjust.

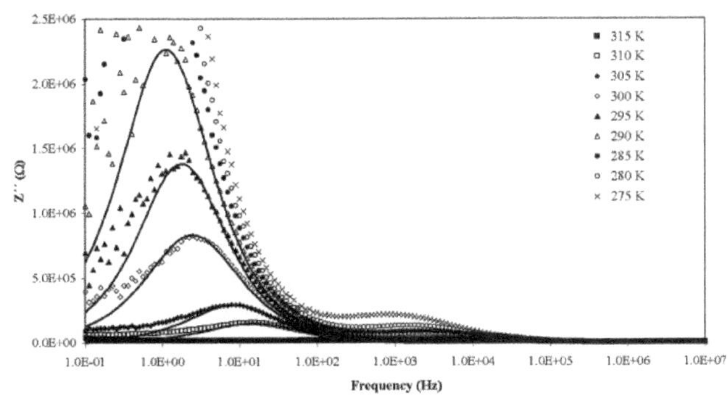

Figure 40. Z′′ versus frequency for the 88Si sample TT at 800 ºC (low frequency region). The lines represent the theoretical adjust.

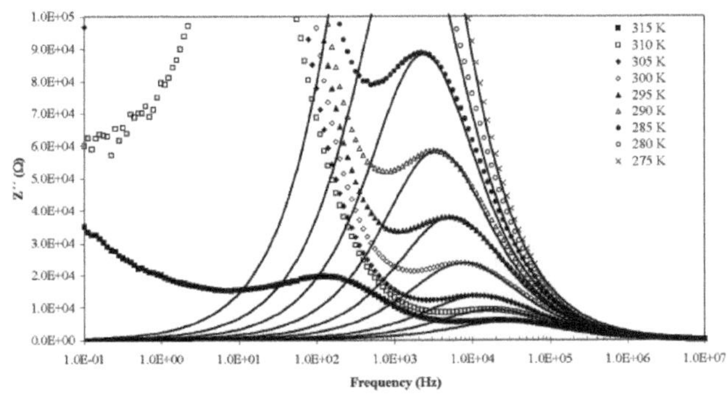

Figure 41. Z′′ versus frequency for the 88Si sample TT at 800 ºC (high frequency region). The lines represent the theoretical adjust.

Sample	Temp. (K)	R_1 (x10^{+5}) [Ω]	Q_{01} (x10^{-7}) [$Ω^{-1}m^{-2}s^n$]	n_1	R_2 (x10^{+4}) [Ω]	Q_{02} (x10^{-8}) [$Ω^{-1}m^{-2}s^n$]	n_2
600	315	0,18	1,20	1,00			
	310	0,90	3,76	0,83			
	305	3,57	1,21	1,00			
	300	10,50	1,34	1,00	0,14	7,60	0,73
	295	27,93	1,11	1,00	0,20	9,66	0,69
	290	39,00	1,18	1,00	0,30	5,12	0,72
	285				0,48	4,40	0,73
	280				0,71	3,43	0,74
	275				1,08	2,44	0,76
650	315	0,29	10,08	0,74	0,15	6,77	0,71
	310	2,01	6,24	0,79	0,22	5,12	0,73
	305	4,86	4,76	0,81	0,34	5,70	0,72
	300	11,21	3,91	0,87	0,47	3,89	0,74
	295	23,12	3,86	0,80	0,85	3,51	0,73
	290	37,10	3,17	0,93	1,06	1,81	0,78
	285	86,23	3,28	0,90	2,01	3,23	0,72
	280				3,07	1,48	0,78
	275				4,84	2,43	0,72
700	315	0,11	13,64	0,84	0,05	6,75	0,75
	310	0,75	14,05	0,82	0,09	5,21	0,77
	305	2,34	9,66	0,79	0,13	3,70	0,77
	300	3,93	7,09	0,83	0,17	6,54	0,73
	295	30,86	4,07	0,94	0,40	1,94	0,78
	290	49,56	3,16	0,95	0,69	1,76	0,78
	285				1,09	1,48	0,78
	280				1,99	1,11	0,80
	275				3,13	0,95	0,80
800	315	0,65	5,78	0,68	1,80	3,38	0,72
	310	3,88	2,60	0,84	2,85	2,53	0,74
	305	7,05	2,25	0,88	4,24	2,29	0,73
	300	21,02	2,19	0,85	7,13	1,73	0,75
	295	32,91	1,81	0,89	11,05	1,25	0,77
	290	55,64	1,63	0,87	16,78	1,07	0,78
	285				25,29	0,95	0,78
	280				37,81	0,86	0,78
	275				60,92	0,70	0,78

Table 4. Equivalent electric circuit parameters (R_1, Q_{01}, n_1, R_2, Q_{02} and n_2) for all samples at several measuring temperatures.

The values obtained by fitting the experimental data using the model of the electrical equivalent circuit shown in figure 33 (Table 4), indicate that the parameters R_1 and R_2 have similar behavior, increasing with the increase of the measurement temperature. With the increase of the TT temperature, the parameters R_1 and R_2 have a similar behavior to that observed for the dc conductivity (Table 3). The values of the Q_{01} and Q_{02} parameter decrease with the increase of the measurement temperature. Q_{01}, at room temperature, increases with the increase of the TT temperature up to 700 °C. Under these conditions, the parameter Q_{02} has an oscillatory behavior, decreasing from the sample TT600 to TT650, increasing to the sample TT700 and decreasing again to the sample TT800. The parameter n_1 decreases with increasing the measurement temperature, except for the sample TT600 that shows a value of $n_1 = 1$. At 300 K, n_1 parameter value remains substantially constant, with the increase of the TT temperature ($0.83 < n_1 < 0.87$). The parameter n_2 is almost constant for all samples, measuring conditions and TT temperatures. Note that $n_1 > n_2$, in all cases.

Based on the values of the electrical equivalent circuit parameters (Table 4) it was calculated the relaxation time (τ_Z), associated with each relaxation mechanism, and the capacitor value which best approximates the CPE element (C_{CPE}) [18]. The obtained values are in table 5. This table also presents the value of the dielectric constant and dielectric loss for all temperatures of measurement.

It is verified that the τ_{Z1} value is approximately three orders of magnitude higher than τ_{Z2}. However, both parameters decrease with the increase of the temperature of measurement for all samples, i.e., Z'' peak shifts to higher frequencies. The parameter C_{CPE1} presents a behavior similar to Q_{01} (Table 4), in function of the TT temperature, and is approximately two orders of magnitude higher than that C_{CPE2}.

9. Results analysis

The glasses with the molar compositions $92SiO_2$-$4Li_2O$-$4Nb_2O_5$ (92Si) and $88SiO_2$-$6Li_2O$-$6Nb_2O_5$ (88Si) are, due to the high amount of SiO_2, very difficult to prepare by conventional melt quenching method. However, the sol-gel method allows there preparation without major difficulties. The macroscopic aspect of the samples of these compositions depend on the presence, or absence, of inhomogeneities in the glass matrix, the particles size, quantity and refractive index [28]. In the case of the SiO_2:$LiNbO_3$ system, the refractive indices of SiO_2 (~ 1.4 [4]) and $LiNbO_3$ (~ 2.2 [29]) differ considerably. Although the translucent appearance of the as-prepared sample of the 88Si composition, the XRD pattern and SEM micrographs did not reveal the presence of heterogeneities of crystalline or amorphous nature type. The 92Si as-prepared sample is colorless and transparent. However, the results of Raman spectroscopy on the 88Si as-prepared sample showed the presence of bands centered at 240 and 680 cm^{-1}, assigned to vibrations of NbO_6 octahedrons associated with the $LiNbO_3$ structure [30;31;32;33], indicating the probable presence of small $LiNbO_3$ particles dispersed in the glass matrix. These bands are not present in the sample only subjected to the drying treatment at 120 °C, which are also transparent.

Sample	Temp. (K)	τ_{z1} ($x10^{-3}$) [s]	C_{CPE1} ($x10^{-7}$) [F]	τ_{z2} ($x10^{-6}$) [s]	C_{CPE2} ($x10^{-9}$) [F]	ε´	tan δ
	315	0,35	1,20			1873,33	0,47
	310	2,65	1,32			1722,50	0,48
	305	6,92	1,21			1585,02	0,54
	300	26,53	1,06	0,59	1,52	1330,68	0,68
600	295	51,34	1,11	0,72	1,13	1178,92	0,88
	290	75,79	1,18	0,86	0,99	1013,83	1,06
	285			1,45	1,06	859,40	1,33
	280			2,10	1,06	722,08	1,61
	275			3,10	1,08	575,66	1,84
	315	0,50	1,27	0,36	0,82	2265,03	0,89
	310	11,37	2,25	0,63	1,02	1921,36	1,86
	305	26,53	2,27	1,06	1,05	1561,59	2,06
	300	63,66	2,67	1,34	1,01	1266,14	2,36
650	295	132,63	2,42	2,50	1,03	858,17	2,62
	290	198,94	2,83	2,69	1,00	635,53	2,79
	285	454,73	3,01	6,51	1,09	384,23	2,70
	280			8,18	1,05	231,64	2,38
	275			13,84	0,98	155,36	1,93
	315	1,01	4,33	0,19	1,31	4792,50	0,97
	310	9,95	5,82	0,35	1,52	3866,72	1,97
	305	26,53	4,28	0,42	1,25	3363,29	2,59
	300	33,86	3,84	0,57	1,19	3052,55	3,07
700	295	227,36	3,73	0,90	0,89	1668,21	3,11
	290	265,26	2,93	1,56	0,89	1047,35	3,62
	285			2,35	0,86	666,41	3,83
	280			4,08	0,84	329,39	3,61
	275			6,39	0,85	188,13	2,91
	315	1,30	0,62	5,45	1,03	355,43	1,97
	310	10,61	1,24	8,72	1,09	254,31	2,10
	305	22,74	1,39	12,56	1,04	164,21	2,47
	300	61,21	1,42	20,85	1,07	92,67	2,75
800	295	88,42	1,37	29,98	1,04	62,13	2,75
	290	159,15	1,26	45,80	1,07	46,20	2,53
	285			67,78	1,05	36,99	2,20
	280			100,35	1,04	31,24	1,84
	275			147,91	0,96	26,36	1,47

Table 5. Relaxation time (τ_z) and the C_{CPE} capacitors, dielectric constant (ε´) and dielectric loss (tan δ), at 1kHz and 300 K, of the 88 Si samples.

The XRD patterns of the 88Si sample TT at 600 ºC and of the 92Si sample TT at 650 ºC, both translucent, showed that the $LiNbO_3$ crystalline phase is already formed. In the SEM micrographs of those samples it is possible to observe particles with a size that does not exceeds 500 nm, which are associated to those crystallites. The observation of particles in the 88Si sample, treated at temperatures where in the 92Si samples no particles were observed, is justified to the fact that the amount of lithium and niobium ions present in the 88Si sample is greater. As it is known, Li^+ and Nb^{5+} ions, which due to their high field strength, can promote phase separation in the vitreous network [34]. When treated at 800 °C the 88Si sample became opaque. The SEM micrographs of this sample show particles with a maximum size of 200 nm, but by XRD it was detect not only the LiNbO3 phase but also the SiO_2, and $Li_2Si_2O_5$ and Li_3NbO_4 phases. In the samples of the 92Si composition, TT at 800 °C, which are translucent, the $LiNbO_3$ and SiO_2 crystal phases are also present. Thus, the opacity of the 88Si sample, TT800, can be attributed to the presence of particles associated with the crystalline phases of $Li_2Si_2O_5$ and Li_3NbO_4.

The Li_3NbO_4 crystalline phase, present in the 88Si sample TT at 800 °C, compared with the $LiNbO_3$ phase, is richer in Li^+ ions. Thus, taking into account that the molar amount of Li^+ and Nb^{5+} ions present in the sample is equal ([Nb] / [Li] = 1), it can be confirmed the existence, in this sample, of a greater quantity of Nb^{5+} ions inserted structurally in glass matrix, than Li^+ ions. The formation of the $Li_2Si_2O_5$ crystalline phase contributes to the decrease of the number of these Li^+ ions. The absence in the Raman spectra of all the 92Si and 88Si samples, of a band at 800-850 cm^{-1} related to the vibration of the Nb-O-Si bonds, indicates that Nb^{5+} ions are probably embedded in the matrix as network modifiers. The decreasing of the number of lithium and niobium ions, structurally inserted in the glass matrix, with the increase of the TT temperature, is responsible for the decrease of the dc conductivity.

The 88Si sample, TT at 800 ºC, shows the minimum value of σ_{dc}, of all samples of this composition, indicating that this is the one that should have the smallest number of ions structurally inserted in the glass network. However, the σ_{dc} is also affected by the presence of particles, particularly those of $LiNbO_3$, which are characterized by a high resistivity (~ 10^{21} Ω. cm, at 300 K [34])

The particles observed by SEM in the sample TT800 exhibit a morphology and size similar to that observed in the samples TT at 600 and 700 °C. The detection, by Raman spectroscopy, in the samples TT at temperatures above 500 ºC, of vibrations assigned to the $LiNbO_3$ crystal structure (mainly the ones centered at 240 and 680 cm^{-1}) leads us to consider that the particles observed by SEM are $LiNbO_3$ crystallites. Knowing that the particle size is higher in the sample TT 650 than in the sample TT at 700 ºC, and that the intensity of the XRD peaks associated with the $LiNbO_3$ phase is also higher in the sample TT650, it can be assumed that the increase of the TT temperature from 650 to 700 °C promotes a dissolution of particles. Thus, the volume ratio between $LiNbO_3$ particles and the matrix glass is higher in the sample TT650 and the sample TT700 will have a larger number of Li^+ and Nb^{5+} ions structurally inserted in the glass matrix. Thus, it can be justified the observed increase of σ_{dc}

from the sample TT650 to the sample TT700. The decrease of σdc from the as-prepared glass (TT500) to the sample TT650 and from the TT700 sample to the TT800 sample is justify by a decrease in the number of charge carriers.

The 92Si samples present a σ_{dc} much lower than that observed in the 88Si samples. This difference is explained by the existence of a larger number of Li^{+} and Nb^{5+} ions, structurally inserted in the glass matrix, in the 88Si samples network.

In the 92Si samples TET at 650 °C, with amplitude lower than 500 kV/m, it were observed particles in the sample surface by SEM but not detected by XRD. This phenomenon is probably due to the fact that these particles may have a amorphous crystallinity nature. The detection by XRD, in the 650C sample and in the samples TT at temperatures above 650 °C, of $LiNbO_3$ crystal phase, indicates that the particles observed in the 650C sample surface must be associated with $LiNbO_3$.

In both compositions, and in the temperature range of the conductivity measurements, two zones with different activation energies were observed. This behavior was adjusted with an Arrhenius equation and the activation energy calculated. The only exception was the 88Si sample TT at 500 ºC. This atypical behavior can be explained considering that this sample has a microstructure with a high porosity [2]. Kincs et al. [35] found that the non-Arrhenius behavior in glasses disappears with the densification of the glass. Thus, increasing the TT temperature in the 88Si glass composition a structural densification is activated.

In all 88Si and 92Si samples, the σ_{dc} increases with the increase of the measurement temperature. This behavior, typical of a thermally stimulated process, can be attributed to the increase of the charge carriers energy, with the increase of the temperature. Assuming the ionic conductivity model, where conduction is done by the "jumps" of the charge carriers through the potential barriers [25;26;27;28], the increase of the energy of those charge carriers make their movements easier, thus increasing the σ_{dc}.

Samples of the 88Si composition show a decrease in the ac conductivity, with the increase of the TT temperature, indicating that the TT affects the structure of the glass in such way that the number of units responsible for this conduction mechanism (dipoles and/or ions) decreases or their movement, in response to the applied ac field, becomes more difficult. The $E_{a(ac)}$, calculated using the Arrhenius equation, is similar for the samples TT500, TT700 and TT800 (~ 45 kJ/mol). The remaining samples showed different $E_{a(ac)}$ values (TT600 ~ 32 kJ/mol and TT650 ~ 52 kJ/mol). The decrease of the $E_{a(ac)}$ from the sample TT500 to the sample TT600, indicates a decrease in the height of the potential barriers associated with this conduction process, suggesting that the decrease of σ_{ac} is due to a decrease in the number of dipoles, associated with lithium and niobium ions structurally inserted in the glass matrix. Considering that the particles observed in the sample TT600 are $LiNbO_3$ particles, the decreased of σ_{ac} can be explained by the formation of dipoles related with these crystals, which are difficult to depolarization [29;36] in this measuring temperatures. The sample TT650 has the highest value of $E_{a(ac)}$. This is related to the presence of $LiNbO_3$ particles agglomerates, which increases the difficulty of the associated dipoles movements. The decrease of the σ_{ac}, from the sample TT700 to the sample TT800, is justified by the decrease of lithium and niobium ions, since both samples have similar $E_{a(ac)}$ values.

The existence of two zones with different morphology (surface and bulk), in the 92Si composition samples TET, indicate that the dielectric response can be associated to an equivalent electric circuit model as shown in figure 42.

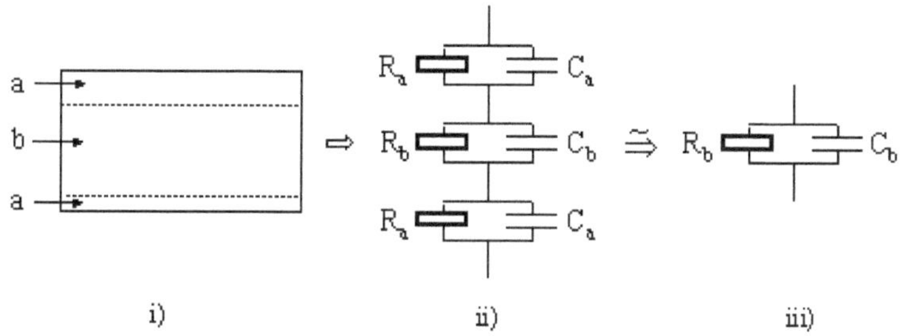

Figure 42. Equivalent electric circuit model: i) sample (a – surface; b – bulk); ii) electric model where R_a and C_a represent the resistance and the capacity related with the sample surface characteristics, R_b and C_b the resistance and capacity related to the bulk characteristics; iii) approximate model.

In the dielectric analysis, the electric circuit comes down to a combination of three capacitors in series: two related with the sample surfaces and the third with the bulk characteristics. Knowing that the thickness of the samples are about 1.0 mm, the thickness of the surface part where the particles are embedded is below 3 µm (sample 700A), the dielectric constant (ε ') of $LiNbO_3$ is higher than 10^3 at 300 K and 1 kHz [29], being much higher than the ε ' of the bulk sample zone, assumed to consist mainly of glass matrix ($\varepsilon'_{SiO_2} \approx 4.0$), it is reasonable to consider that the dielectric behavior can be controlled by the characteristics of the bulk. Briefly, the analysis of the electrical circuit consisting of three capacitors in series shows that the equivalent capacity is

$$C_{eq} = \frac{C_a C_b}{C_a + 2C_b}.$$

If we consider that $C_a >> C_b$ than $C_{eq} \approx C_b$. Therefore the increase of ε' in the samples series treated at 750 ° C, with the increase of the external electric field amplitude may be associated with an increase in the number of dipoles present in the bulk region of the samples. In these samples it was observed a decrease in the $LiNbO_3$ particle size, with the increase of the applied electric field amplitude. This reduction is also associated with the decrease of intensity of the Raman band centered at 630 cm^{-1} related to the presence of $LiNbO_3$ particles [24], indicating a decrease in the number of dipoles associated with those particles. By increasing the electric field amplitude applied during the heat treatment, the σ_{dc} increases. This behavior must be related to the presence of a higher number of Li^+ and Nb^{5+} ions inserted structurally in the glass matrix. Thus it can be assumed an increase in the number of electric dipoles in the bulk zone with the increase of the amplitude of the electric field.

In the samples series TET at 700 °C, the behavior of ε', with the increase of the external electrical field amplitude, is opposite to the one observed in the samples TET at 750 °C. The increase of the particle size in the samples TET at 700 °C, with the increase of the external electrical field amplitude, was observed by SEM and by Raman spectroscopy (increase of the intensity of the band centered at 630 cm^{-1}). This fact will cause a reduction in the number of dipoles in the sample bulk zone, thereby reducing the ε' value. The same occurs in the sample TET at 650 °C. Thus, these results suggest that in the samples TET at 650 °C and 700 ºC the Nb5+ and Li^+ ions migrate from the bulk zone to the surface, contributing to the increase of the particles size.

The dielectric response, using the impedance formalism (Z*), as a function of frequency and temperature for the 92Si composition samples was adjusted to the physical model consisting on the equivalent circuit shown in figure 43 [R_1($RCPE_1$)].

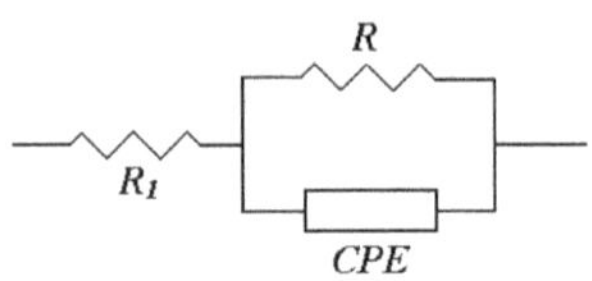

Figure 43. Equivalent electric circuit model.

It is noted that this model fits the experimental data reasonably, showing for all the samples a value of the parameter n very close to 1 (> 0.9), indicating that CPE element behaves very close to a capacitor. The value of R_1 ~0 was considered in all samples.

In the 88Si composition samples it was detected two relaxation mechanisms, which were fitted, using the CNLLS algorithm, to the equivalent circuit model shown in figure 33 ([R_0(R_1CPE_1)(R_2CPE_2)]). From the obtained values, it should be noted that the parameters n_1 and n_2, associated to the elements and CPE_1 and CPE_2, respectively, are always higher than 0.73, which shows that also in this composition, the elements CPE tend to exhibit a behavior similar to a capacitive element. The relaxation mechanism, found in the high frequency region (R_2CPE_2) is assigned to the characteristics of the vitreous matrix ("bulk"), i.e., with the relaxation of dipoles associated with the ions structurally inserted in the glass matrix. The relaxation at the lowest frequencies was associated with surface features, namely those related with the dipoles related with the particles detected at the samples surfaces. The presence of those particles, related with the $LiNbO_3$ crystalline phase, which is characterized by possessing dipoles which electrical depolarization is difficult, justifies the higher relaxation time. The fact that the relaxation time associated with the first depolarization mechanism, observed in the 88Si composition, is of the same order of magnitude to that observed in the 92Si samples composition, suggests that the electrical units responsible for both are the same, i.e., the dipoles associated with lithium and niobium ions inserted structurally in the glass matrix. In the 88Si composition, the relaxation time τ_{Z1}, related with the R_1CPE_1, and associated with the relaxation mechanism at the low frequency zone, is higher (~ 10^{-2} s) than the τ_{Z2}, which also indicates that the units responsible for this relaxation are more difficult to depolarize.

In the 92Si composition, the Z* behavior as a function of frequency, for the TET samples treated at 700 ºC is opposite to that observed in the samples TET at 750 ºC, with the increase of the external electric field amplitude (the series TET at 650 and 700 ºC have a similar behavior). In the samples TET at 750 ºC, the increase of the amplitude of the external electrical field, promote an increase in the values of the R and Y0 parameters. This behavior should be associated with the increase of the number of particles in the sample bulk zone. In the samples TET at 650 and 700 °C, the number of electrical units within the sample (bulk region) decreases with the increase of the amplitude of the external electrical field, and therefore, increases the number of surface particles, justifying the Z*(ω) opposite behavior.

The 92Si composition presents dielectric constant (ε ') and dielectric loss (tan δ = ε'' / ε') values, much lower than those observed in the 88Si composition samples. The justification for this value disparity is the larger number of electric dipoles existing in the 88Si composition sample. In this composition, the decrease of ε', from the sample TT at 600 ºC to the sample TT at 650 ºC, measured at room temperature and 1 kHz, is due to the presence, in the sample TT650, of particle agglomeration and also to high size particles, which promotes the decrease of the number of dipoles associated with lithium and niobium ions structurally inserted in the glass matrix. Furthermore it is likely that the crystal orientation of the particles and agglomerates, do not present a preferential grow direction contributing to the decrease of the dipole moment [36;37].

The parameter tan δ (92Si composition) increases in the samples TET at 750 °C, with the increase of the amplitude of the external electrical field, mainly due to the increase of the ε'' component. This behavior will cause an increase in the conductivity, which is corroborated by the R parameter behavior. The decrease in the value of τz, in the samples TET at 750 °C with the increase of the amplitude of the electric field, suggests that the electrical units, present in these samples, follow the ac electric field more easily. In the TET samples treated at 650 °C, the behavior of tan δ (with increasing the amplitude of the external electric field) is opposite to that observed in the samples TET at 750 °C. This can be attributed to a significant increase in the ac resistivity. Moreover ε' decreases due to the increase of the surface particles amount and thus reducing the amount of ions in the glass bulk zone.

The presence in samples of the 88Si composition of two relaxation peaks, in the Z* spectrum, similar to those already observed in other glass containing $LiNbO_3$ crystals [38;39;40] suggests that the dc conductivity at low temperature and the dielectric relaxation at high-frequency can be assigned to the ion conduction and ionic polarization, respectively. The σ_{dc} behavior at temperatures higher than 300 K and the relaxation process in the low frequency range should be assigned to interfacial electrode-sample polarization [38;39;40].

10. Main conclusions

1. The sol-gel method allows the preparation of glasses and glass ceramics with compositions that by the melt-quenching method are extremely difficult to prepare.

2. The drying process keeps the 92Si composition gel transparent and the 88Si translucent. The heat treatment with or without the presence of the external electrical field, makes the 92Si glass composition translucent. Samples of the 88Si composition, after TT at temperatures above 750 ºC, became opaque.
3. The detection of $LiNbO_3$ crystal phase is observed in the 92Si sample composition TT at temperatures above 650 °C and in the 88Si composition at temperatures above 600 °C. Increasing the annealing temperature it is promoted the appearance of secondary crystalline phases.
4. SEM revealed that crystallization in both 92Si and 88Si compositions are predominantly at the glass surface.
5. Samples of the 92Si composition, TET at temperatures below 650 ºC, promote the formation of $LiNbO_3$ crystallites
6. Increasing the amplitude of the electric field in the 92Si sample series treated at 650 and 700 °C favors the increase in the particle size. The TET at 750 ºC presents the opposite behavior.
7. The decrease in the surface particle size associated with an increase in the number of dipoles within the sample, justifies the maximum value of ε' observed (9,44). Thus, the study of the dielectric constant enables to establish if the crystallization occurs, preferably in the sample surface or in the bulk zone.
8. The high values of ε', measured in the low frequency region (<1 kHz) in the 88Si samples, are due to interfacial polarization. The increase in the amount of surface particles promotes a decrease of ε', indicates that these particles grow without a preferential crystalline orientation.
9. The 88Si composition sample treated at 650 °C shows particle agglomerates that are dissolved, with increasing the thermal treatment temperature. This structural change is substantiated by the SEM results and by the decreasing of the width and increasing of the intensity of the Raman bands. The results of σdc strengthen the hypothesis that the formation of secondary phases ($Li_2Si_2O_5$ and Li_3NbO_4), in the 88Si samples composition comes from the dissolution of the $LiNbO_3$ agglomerated particles.
10. The results of the Raman spectroscopy indicate that the niobium ions are inserted in the glass matrix as network modifiers.
11. The samples of these two compositions present two different temperature zones with different activation energies indicating the presence of two different conduction mechanisms. In the samples of the 88Si composition it was also detected the presence of two dielectric relaxation mechanisms. The $E_{a(dc)}$ associated with the region of low temperatures and the dielectric relaxation mechanism associated with the high frequency region are assigned to ionic conduction and ionic polarization, respectively. The relaxation mechanism observed in the low frequency region and the $E_{a(dc)}$ associated with the region of high temperatures, are due to interfacial polarization phenomena.
12. The CNLLS algorithm associated with an electrical equivalent circuit model was used to adjust the dielectric behavior. The detection of two different relaxation mechanisms in the 88Si samples led to the use of the electrical equivalent circuit shown in figure 41. The results obtained by fitting the experimental data showed that this model may describe the dielectric behavior of these samples.

Author details

Manuel Pedro Fernandes Graça and Manuel Almeida Valente
University of Aveiro / I3N - Physics Department, Portugal

Acknowledgement

The authors gratefully acknowledge the financial support from PEst-C/CTM/LA0025/2011 and from the Portuguese Science and Technology Foundation (FCT) - SFRH/BD/6314/2001.

11. References

[1] Mauritz, www.psrc.usm.edu/~mauritz/sol-gel.html (2002).
[2] I. Schwartz, P. Anderson, H. de Lambilly, L.C. Klein, J. Non-Cryst. Solids, 83, 391-399 (1986).
[3] E.O. Lawrence, Berkeley National Laboratory, "How silica aerogels are made", in http://eande.lbl.gov/ECS/aerogels/saprep.htm .
[4] J.M.F. Navarro, "El Vidrio" (CSIC-Fundación Centro Nacional del Vidrio, Madrid 1991).
[5] S. Sakka, "Treatise on materials science and technology", vol 22, Academic Press, 1982.
[6] F. Mehran, B.A. Scott, Solid State Communications, 11, 15-19 (1972).
[7] M. G. Ferreira da Silva, "Incorporação Estrutural de Iões de Transição em Vidros Preparados pelo Método Sol-Gel", PhD thesis, Universidade de Aveiro, Portugal (1990).
[8] I.M.M. Salvado, "Preparação pelo processo "Sol-Gel" e caracterização de materiais dos sistemas SiO2-ZrO2, SiO2-TiO2, SiO2-Al2O3. Aplicação como revestimentos protectores."- PhD thesis, Universidade de Aveiro, Portugal (1990).
[9] N. Venkatasubramanian, B. Wade, P. Desai, A S. Abhiraman, L.T. Gelbaum, 130, 144-156 (1991).
[10] L.C. Klein, S. Ho, "Glasses for electronic applications", (1991) 221-233.
[11] S.P. Szu, M. Greenblatt, L.C. Klein, Solid State Ionics, 46, 291-297 (1991); S.P. Szu, L.C. Klein, M. Greenblatt, J. Non-Cryst. Solids, 143, 21-30 (1992).
[12] C. Sanchez, G.A.A. Soler-Illia, F. Ribot, D. Grosso, C.R. Chimie 6 (2003), 1131-1151.
[13] K. Chou, J. Non-Cryst. Solids, 110 (1989) 122-124.
[14] M. Prassas, "Silica Glass from Aerogels", in http://www.solgel.com/articles/april01/aerog2.htm .
[15] R. Ota, N. Asagi, J. Fukunaga, N. Yoshida, T. Fujii, J. Mat. Sci., 25, 4259-4265 (1990).
[16] M.P.F. Graça, M.G.F. Silva, M.A. Valente , Journal of Non-Crystalline Solids, 351 (33-36) (2005) 2951-2957.
[17] MPF Graça, MGF Silva and MA Valente, Journal of Sol-Gel Science and Technology, 42 (1) (2007) 1-8.
[18] Manuel Pedro Fernandes Graça & Manuel Almeida Valente, Glass Ceramics with Para, Anti or Ferroelectric Active Phases – book chapter 11 (pp213-292) in: Ceramic Materials", ISBN: 978-953-307-217-3, Intech publisher, 2011.

[19] MPF Graça, MGF Silva, ASB Sombra and MA Valente, Journal of Non-Crystalline Solids, 352 (42-49) (2006) 5199-5204.
[20] T. Cardinal, E. Fargin, G. Le Flem, S. Leboiteux, J. Non-Cryst. Solids 222 (1997) 228-234.
[21] A.C:V de Araujo, I.T. Weber, W.D: Fragoso, C.M. Donegá, J. of Alloys and Compounds 275-277 (1998) 738-741.
[22] Y.D. Juang, S.B. Dai, Y.C. Wang, W.Y. Chou, J.S. Hwang, M.L. Hu, w.S. Tse, Solid State Communications 111 (1999) 723-728.
[23] N. Shibata, M. Horigudhi, T. Edahino, J. Non-Cryst. Solids, 45 (1981) 115-126.
[24] A A Lipovskii, V.D. Petrikov, V.G. Melehin, D.K. Tagantsev, B.V. Tatarintsev, Solid State Comunications, 117, 733-737 (2001).
[25] J.R. Macdonald, "Impedance spectroscopy", John Wiley & Sons, New York, 1987.
[26] M.P.F. Graça, M.A. Valente, M.G. Ferreira da Silva, Journal of Materials Science, 41 (4) (2006) 1137-1141.
[27] A.K. Jonscher, "Dielectric relaxation in solids", Chelsea Dielectrics Press, London, 1983.
[28] H.G. Kim, T. Komatsu, R. Sato and K. Matusita , J. Non-Cryst. Solids, 162, 201-204 (1993).
[29] M.M. Aboulleil, F.J. Leonberger, J. Am. Ceram. Society, 72 (1989) 1311-1320.
[30] R.Claus, G.Borsel, E. Wiesendanger, L.Steffan, Phys. Ver. B, 6(12) (1972) 4878-4879.
[31] A. Jayaraman, A. A. Ballman, J. Appl. Phys., 60 (3) (1986) 1208-1210.
[32] J.G. Scott, S. Mailis, C.L. Sones, R.W. Eason, Appl.Phys. A, 2003.
[33] A. Ridah, P. Bourson, M.D. Fontana, G. Malovichko, J. Phys. Condens. Matter 9 (1997) 9687-9693.
[34] W. Vogel, Glass Chemistry, Springer, 2nd edition, 1994.
[35] J. Kincs, S. W. Martin, Physical Review letters, 76 (1) (1996) 70-73.
[36] MPF Graça, MGF Silva and MA Valente, Journal of Materials Science, 42(8) (2007) 2543-2550.
[37] Glasses with ferroelectric phases, M.A. Valente and M.P.F. Graça, IOP Conf. Series: Materials Science and Engineering 2 (2009) 012012.
[38] E.B. de Araujo, J.A.M. de Abreu, R.S. de Oliveira, J.A.C. de Paiva, A.S.B. Sombra, Canadian Journal of Physics, 75 (1997) 747-758.
[39] C. Hong, D.E. Day, J. Mater. Sci., 14, 2493-2499 (1979).
[40] C. Hong, D.E. Day, J. Am. Ceram. Soc., 64:2, 61-67 (1981).

Chapter 14

Heat Treatment in Molecular Precursor Method for Fabricating Metal Oxide Thin Films

Hiroki Nagai and Mitsunobu Sato

Additional information is available at the end of the chapter

1. Introduction

Industrial processes must minimize heat-treatment to reduce energy consumption and CO_2 emissions during the manufacture of products. In particular, fossil fuels should be conserved for the next generation. However, the use of heat-treatment is essential in the manufacturing processes of many highly functional materials.

In many cases, the materials' functions depend on their surfaces. From this point of view, modification by thin film fabrication on various substrates, as opposed to manufacturing the entire body with the functional material, can generally save resources. The molecular precursor method (MPM) that was developed in our study is a wet chemical process for fabricating metal oxide and phosphate thin films [1-11]. This method requires heat-treatment to eliminate organic ligands from metal complexes involved in spin-coated precursor films and to fabricate thin films of crystallized metal oxides or phosphates. We emphasize the importance of heat-treatment by describing recent results obtained using the MPM, which show great potential for the development of nanoscience and nanotechnology tools and materials.

This chapter focuses on the transparent thin film fabrication of both a visible (Vis)-light-responsive anatase thin film having enhanced UV sensitivity and an unprecedented Vis-responsive rutile thin film on glass substrates. These photoreactive thin films were easily fabricated using the MPM. Heat-treatment under controlled conditions produced these attractive thin films. Thin film fabrication of a highly conductive Ag nanoparticles/titania composite and several metal oxides, including Cu_2O, will be also discussed, illustrating the broad utility of the MPM and the importance of heat-treatment in this novel wet process.

2. Solution-based thin film formation

Novel thin films are an active area of research and are widely used in industry. Most of the thin films have thicknesses ranging from monolayer to nanometer levels up to several micrometers. Due to their relatively high hardness and inertness, ceramic coatings are of particular interest for the protection of substrate materials against corrosion, oxidation, and wear resistance [12-17]. The electronic and optical properties of thin films are used in many electronic and optical devices [18-20]. The wide range of materials, techniques for preparation, and range of applications make this an interdisciplinary field. Many different methods are used to fabricate thin films, including physical techniques and chemical processes. Physical vapor deposition (PVD) and chemical vapor deposition (CVD) are the two most common types of thin film formation methods. PVD methods such as thermal evaporation and sputtering involve atom by atom, molecule by molecule growth, or ion deposition on various materials in a vacuum system [21-23]. CVD and sol–gel methods are less expensive than PVD [24, 25]. The heat-treatment of thin films in these methods is generally important for the formation of crystallized metal oxides.

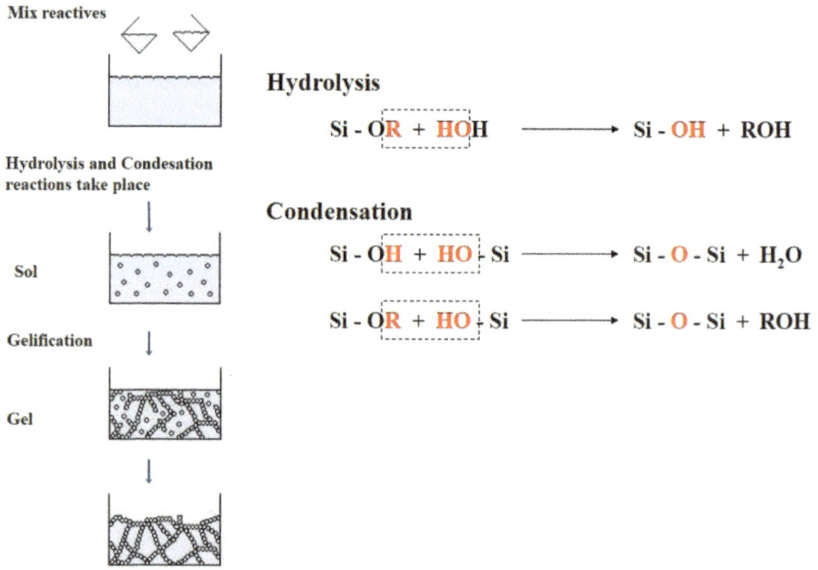

Figure 1. Typical sol-gel process for SiO_2 formation from silicone alkoxide.

The sol–gel method is a versatile technology in which metal/organic polymers are used to produce ceramics and glasses [26-34]. This technology can be used to manufacture thin films in a relatively cheap way compared with PVD. In a typical sol–gel protocol (Figure 1), the process starts with a solution consisting of metal compounds, such as a metal alkoxide, acetylacetonate, carboxylate, and soluble inorganic species as the source of cations in the target oxide. Additional reactants include water as the hydrolysis agent, alcohols as the solvent, and an acid or base as a catalyst. Metal compounds undergo hydrolysis and polycondensation near room temperature, giving rise to a sol in which polymers or colloidal particles are dispersed without precipitation. Further reaction connects the fine particles, solidifying the sol into a wet gel, which still contains water and solvents. Vaporization of solvents and water produces a dry gel. Heating the gel to higher temperatures, where the organic constituents and residues are removed, gives rise to microstructures of inorganic–inorganic composites or hybrids, and glasses and ceramics. In Figure 2, the structural changes to metal oxides from the corresponding hydrolysed polymers by heat treatment of the as-deposited gel. The processes accompanied with dehydration can be categorized into two reactions; (A) the intrachain and (B) the interchain condensation.

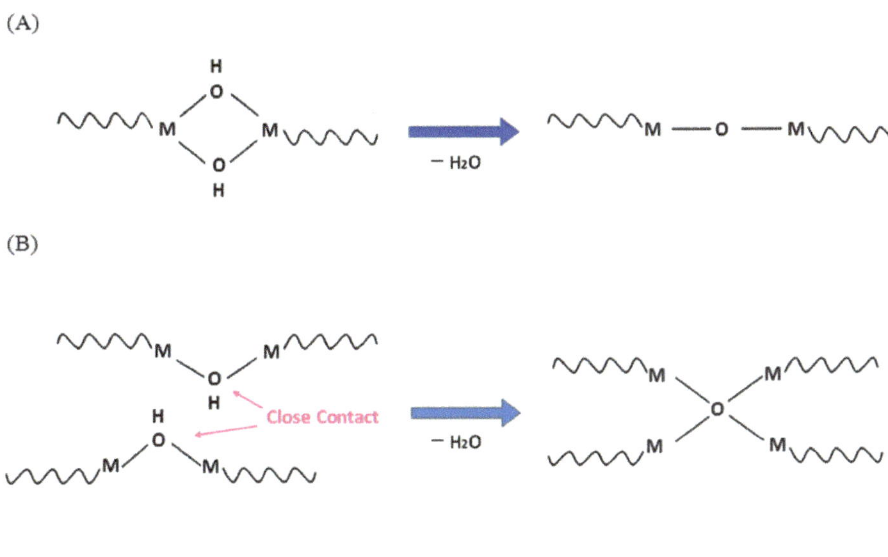

Figure 2. Plausible schematic diagrams for the oxide nucleation process of (A) intra-chain and (B) inter-chain condensation of the metalloxane polymers formed at the early stage in the sol-gel method.

3. Principle of MPM

The MPM is a wet process for the formation of thin films of various metal oxides, including titania or calcium phosphate compounds [1-11]. This method is based on the design of metal complexes in coating solutions with excellent stability, homogeneity, miscibility, coatability, *etc.*, which have many practical advantages. This is because metal complex anions with high stability can be dissolved in volatile solvents by combining them with the appropriate alkylamines. Furthermore, the resultant solutions can form excellent precursor films through various coating procedures. The precursor films involving metal complexes should be amorphous, just as with the metal/organic polymers in the sol–gel processes. If not, it is impossible to obtain the resulting metal oxide thin films spread homogeneously on substrates by subsequent heat-treatment. For this purpose, the alkyl groups in the amines play an important role. Single-crystals of the metal complex can be obtained from the precursor solution in several cases when the alkyl groups in the alkylamines are sufficiently small, *e.g.*, an ethyl group. The model structure of the amorphous precursor films formed on substrates can be examined by means of crystal engineering and based on the crystal structures. Heat-treatment is necessary to fabricate the desired metal oxide films by eliminating the ligand in the metal complex and alkylamine as the counter cation. It is important that densification during heat-treatment occurs only in the vertical direction of the coated substrate.

To the best of our knowledge, the crystallite size of the oxide particles in the resultant thin films fabricated by the MPM is generally smaller than those prepared by the conventional sol–gel method. The smaller size of the crystallites obtained using the MPM may be related to the nucleation process of the crystallized metal oxides. In the nucleation process in the sol–gel method, the polymer chains themselves are rearranged by heat-treatment (Figure 2). The polymer chains should move to produce the core structure of the metal oxide, especially during interchain condensation. In contrast, the nucleation of metal oxides occurs more easily during the MPM. When coupled with elimination of the organic ligands via heat-treatment, a vast number of crystallites can be rapidly formed. It is consequently feasible that the crystallite sizes of metal oxides fabricated using the MPM are smaller than those obtained using the sol–gel method.

4. Heat-treatment for titania thin film fabrication

4.1. Thin film fabrication and basic properties of TiO_2

Titania is a popular industrial material that has been used as a white pigment for paints, cosmetics, and foodstuffs [35]. The photoreactivity of titania is known from observations of phenomenon such as the chalking of white paints containing titania after long-term outdoor exposure. The photoreactivity of titania was first observed by Honda and Fujishima in 1967, and reported in the literature in 1972 [36]. When the surface of a titania electrode (under an appropriate bias between a Pt counter electrode) was irradiated with light of energy shorter than its band gap, 3.0 eV, a photocurrent flowed from the Pt electrode to the titania electrode through the external circuit. The oxidation reaction occurs at the titania electrode, and the reduction reaction occurs at the Pt electrode. This observation showed that water molecules could be split into oxygen and hydrogen using UV light in a sulfuric acid electrolyte. The energy-conversion process that occurs on the titania surface is termed the Honda–Fujishima effect. When titania particles absorb UV radiation, they produce pairs of electrons and holes inside the particles. Because the photoinduced electrons and holes can be incorporated into redox reactions on the titania surface before spontaneous recombination, the surface states of the titania particle are quite important for photoreactivity [37-52].

Titania has three polymorphs: anatase, rutile, and brookite. Anatase thin film deposition on a glass substrate has been achieved using the MPM. A water-resistant coating solution was prepared by the reaction of a neutral [Ti(H_2O)(EDTA)] complex (EDTA is ethylenediamine-*N,N,N',N'*-tetraacetic acid) with dipropylamine in ethanol [1, 2, 4-9]. The anatase phase appears during the heat-treatment of the precursor film at a temperature between 400 and 500°C and is transformed to the rutile one between 500 and 700°C. X-ray diffraction (XRD) analysis of the films prepared by a conventional sol–gel process showed that the anatase phase appeared between 400 and 500°C and was not transformed to the rutile one, even when heat-treated at 900°C. The irreversible phase transformation from anatase to rutile requires heat-treatment. During heat-treatment of anatase, the atoms in the original tetragonal lattice can be rearranged into the rutile tetragonal lattice. The temperature difference between the phase transformation from anatase to rutile in the sol–gel method and the MPM will be discussed in the section **O deficiency in rutile thin films**.

5. Vis-responsive anatase thin film fabricated using the MPM

Many researchers study the fabrication and photoreactivity of Vis-responsive thin films by physical and/or chemical modification of anatase films because of the importance of Vis-photoreactive materials [53-61]. However, there is little information on the enhancement of UV sensitivity of Vis-responsive anatase films.

The implantation of various transition-metal ions such as V^{5+}, Cr^{3+}, and Cu^{2+} into the lattice of Ti^{4+} in anatase thin films was investigated by Anpo *et al.* [62-67]. The photoreactivities of chemically modified anatase thin films decreased under UV irradiation, although those anatase thin films modified with transition-metal ions can behave as photocatalysts under Vis irradiation. Since Asahi *et al.* reported that non-metallic ions such as a substitutional nitride ion at the oxygen sites of anatase are also effective at enabling the thin film to be responsive to Vis light, methods for modifying anatase with tetravalent carbon or hexavalent sulfur cations have also been investigated [68]. Miyauchi *et al.* achieved another chemical fabrication of Vis-responsive anatase thin films, which were modified under NH_3 gas by heat-treating the resulting films formed using a sol–gel method [69]. However, the photoreactivities of those films are lower under UV irradiation than those before modification in all cases. Nevertheless, they can work as photocatalysts under Vis light. Thus, studies on the chemical formation of Vis-responsive anatase thin films with enhanced photoreactivities under UV irradiation are significant from the viewpoint of solar energy efficiency.

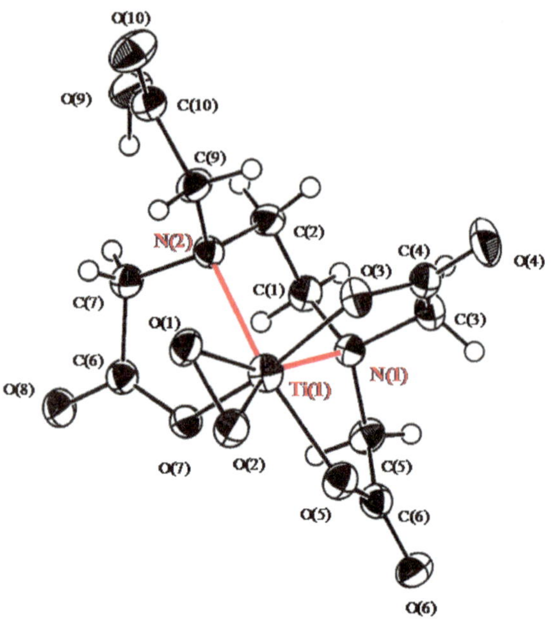

Figure 3. An ORTEP view of the precursor complex having the EDTA and peroxo ligands linked to the central Ti^{4+} ion. The molecular structure was determined by an X-ray single crystal structure analysis of the diethylammonium salt of the complex, although the dibutylammonium salt was employed for the coating solution in order to prevent from the crystallization of the precursor. The single crystals of the identical orange-yellow color could be obtained from a reacted solution of the complex with the diethylamine instead of dibutylamine. The single crystal was $\{(C_2H_5)_2NH_2\}[Ti(O_2)(Hedta)]\cdot 1.5H_2O$; in a monoclinic crystal system, $P2_1/c$ with a = 8.583(1), b = 6.886(1), c = 36.117(2) Å, and β = 92.780(3)°. The full-matrix least-squares refinement on F^2 was based on 3206 observed reflections that were measured at 250 K by using an imaging plate as a detector, and converged with unweighted and weighted agreement factors of R = 0.054 and Rw = 0.061 respectively, and GOF = 1.63. Two Ti–N(edta) bond lengths of 2.307 and 2.285 Å, are slightly longer than the bond length of 2.12 Å in the TiN single crystal. Results indicated that EDTA acts as a pentadentate ligand in the complex, and the peroxo ligand linked to the Ti^{4+} ion has a side-on coordination structure.

The MPM forms transparent titania thin films using the ethanol solution obtained as a coating solution (S_{ED}) by the reaction of alkylamine with a titanium complex of EDTA as the ligand [1, 2, 4-9]. According to single-crystal structural analysis, this Ti complex contains Ti–N bonds (Figure 3). If the Ti-N bonds in this complex can be preserved in the anatase thin film obtained by heat-treatment after coating, a partially nitrided anatase film can be directly formed. However, XRD and X-ray photoelectron spectroscopy (XPS) confirmed that the precursor film formed on a glass substrate using the ethanol solution is transformed into a nitrogen-free anatase thin film through heat-treatment for 30 min in air at a temperature of 450°C or higher. Based on these results, the heat-treatment of molecular precursor films spin-coated with S_{ED} on ITO glass substrates was examined in an Ar gas flow of 0.1 L min^{-1} at 500°C for 30 min. The XRD pattern indicated that the spin-coated precursor film crystallized to anatase through heat-treatment at a temperature of 500°C or higher under atmospheric conditions. Thus, the anatase form was created even if oxygen was not supplied externally to remove organic residues in the metal complex. Furthermore, the chemical bond toward the Ti^{4+} ion from both oxygen and nitrogen atoms can be observed in the XPS spectra of the resultant thin films. The binding energies of Ti $2p_{3/2}$ attributed to Ti–O and Ti–N had typical values of 459.1 and 455.3 eV, respectively (Figure 4) [5, 70-72]. Importantly, the binding energy of N 1s was 396.7 eV, and the existing nitrogen was only in the oxygen-substituted form, not in the chemisorbed form [73, 74]. Thus, the heat-treatment of the precursor films in an Ar gas flow was effective in preserving nitrogen atoms in the complex. However, the photoreactivity of the thin film was not observed after Vis irradiation with a weak fluorescent lamp. Because the partially nitrided film obtained by the MPM alone did not respond to irradiation with a fluorescent lamp, a precursor solution, S_{OX}, with no Ti–N bonds, in complexes whose ligand was oxalic acid (OX), was freshly prepared and the layering of the anatase films was achieved using the two precursor solutions.

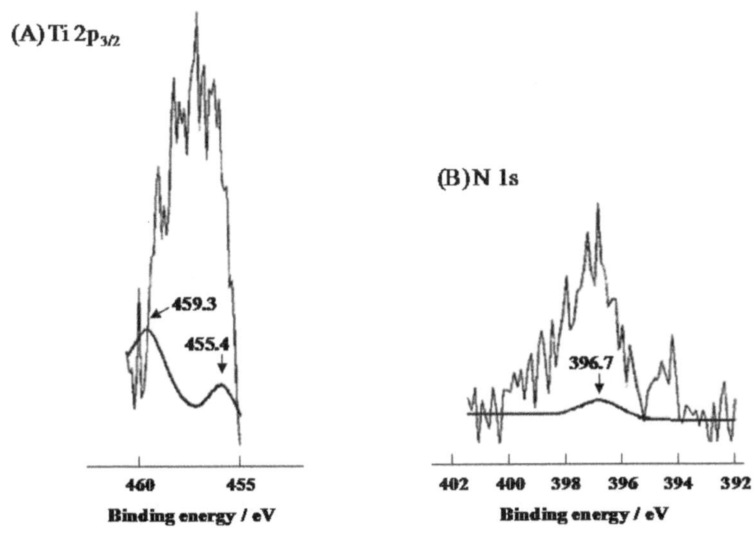

Figure 4. XPS spectra of (A) Ti $2p_{3/2}$ and (B) N 1s of the nitrogen-substituted TiO_2 thin film obtained by heat treatment of the precursor film containing the Ti complex with EDTA ligand on a Na-free glass in an Ar gas flow. The thin solid lines are original data of XPS. The thick solid curves are theoretical Gaussian distribution curves. The dashed curves are theoretically fitted curves by assuming Gaussian distribution [5].

Two types of three-layer film, **OX–OX–OX** and **OX–ED–OX**, were formed on an ITO pre-coated glass substrate by coating and heat-treating the precursor films under an Ar gas flow. A schematic diagram of these structures is shown in Figure 5. The first layer, of thickness 100 nm, was formed by applying **S**ox. The second and third layers were 50 nm thick. The third layer of **OX–OX–OX** was fabricated as a reference by applying only **S**ox, although the corresponding layer of **OX–ED–OX** was fabricated on the second layer using **S**ED, and then the third layer was formed with **S**ox (Figure 6). The overall amount of the heat-treated molecular precursor, energy consumed, film area, and film thickness were kept constant.

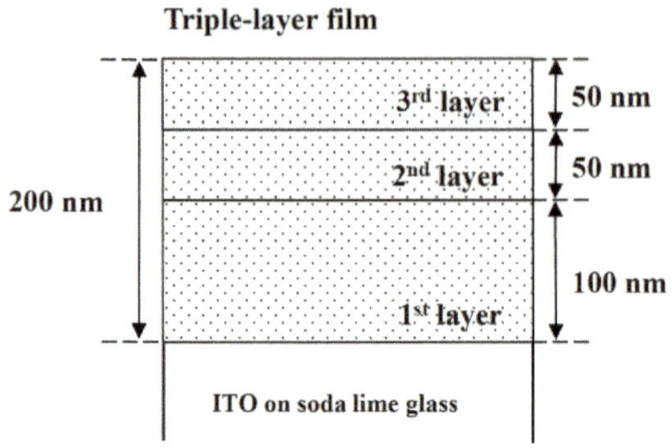

Figure 5. Schematic diagram structures of triple layer thin films. Each precursor film for the first layers was formed by applying **S**ox and heat-treated at 475°C for 30 min. The precursor film of the second layer for **OX-OX-OX** formed by employing the half-diluted solution of **S**ox, and **OX-ED-OX** formed by using the half-diluted solution of **S**ED were heat-treated at 500°C for 30 min. The precursor films of the third layers for **OX-OX-OX** and **OX-ED-OX** formed by employing the half-diluted solution of **S**ox were heat-treated at 475°C for 30 min.

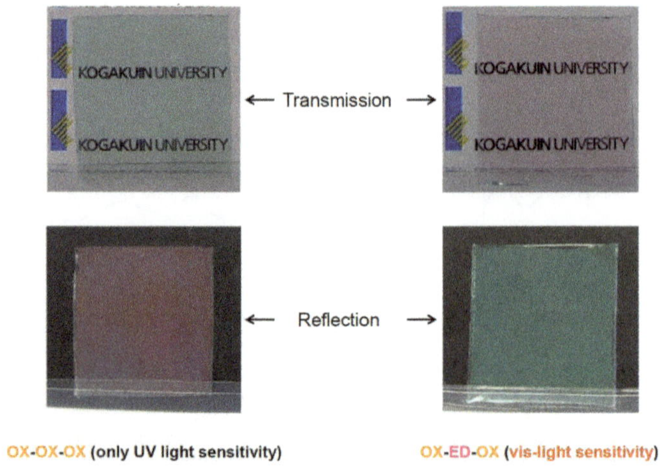

Figure 6. Photographs of the 3-layer **OX-OX-OX** and **OX-ED-OX** films on the ITO glass substrate.

The XRD patterns of both films indicated that anatase was formed. The field-emission scanning electron microscopy (FE-SEM) data of both films show even surfaces without cracks or pinholes. These surfaces were too smooth to detect the roughness by measuring with a stylus profilometer, whose detective limit is ca. 10 nm. The XPS depth profiles of these films are shown in Figure 7. The depth profile for **OX–ED–OX** revealed that nitrogen and carbon were locally distributed in the deep portion corresponding to the second layer. A relative decrease in the amount of oxide ions in the corresponding layers was observed. This suggests that other anions, *e.g.*, nitride ions, compensate for the charge balance. This confirmed that significant amounts of carbon and nitrogen atoms were still present in the second layer, and the substitutional nitrogen atoms were locally distributed in the deep portion corresponding to the second layer. Most nitrogen atoms did not diffuse to other layers, although the amounts of nitrogen and carbon atoms in the other layers could not be neglected. The absorption spectra of **OX–ED–OX** indicate characteristic absorption bands near 480 nm. The **OX–OX–OX** films do not show such absorption bands in the Vis-light region.

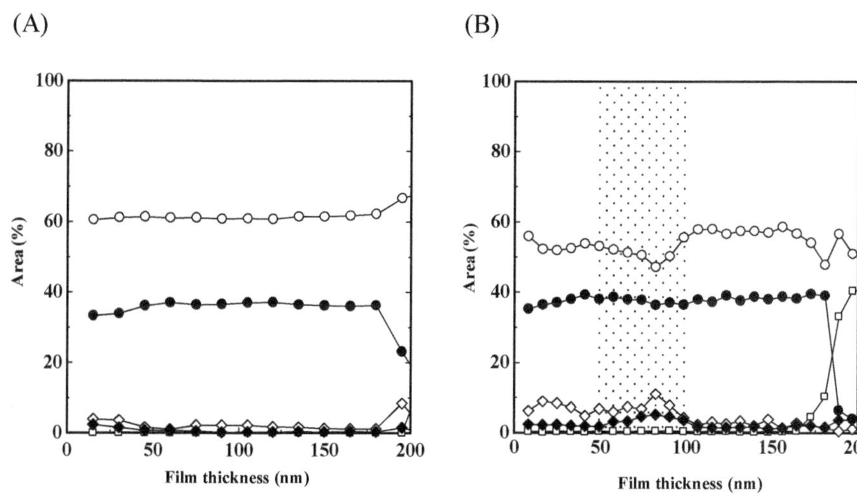

Figure 7. Depth profile of the amount of components in the 3-layer thin film (A) **OX-OX-OX** and (B) **OX-ED-OX**. Notations indicate the energy levels of five atoms in parentheses, —•— (Ti 2p), —○— (O 1s), — ♦ — (N 1s), — ◊ — (C 1s), —□— (In 3d). Net multiplication part is corresponding to the second layer formed by applying the solution S_{ED}.

The photoreactivities of the films were tested using the decolorizing reaction of methylene blue (MB) aqueous solution [5-8, 75-77]. The decoloration rate of 0.01 mol L^{-1} MB solution by the photoreaction with both multilayered thin films under UV or Vis irradiation are summarized in Table 1. The **OX–ED–OX** film has an effective photoreactivity under Vis irradiation. The **OX–OX–OX** film only responded to UV light. It is important that the photoreactivities of the Vis-responsive films are also extremely high under UV irradiation.

The Vis-responsive property of the **OX–ED–OX** film was mainly due to the colored materials that were formed spontaneously during heat-treatment by chemical reactions between the reductant derived from the precursor complex containing OX in the upper layer and the organic residues derived from EDTA ligands in the lower one. Thus, the thermal reactions between the residues derived from the ligand of the precursor complex can afford novel functions such as the Vis-responsive nature of the resultant thin films through heat-treatment of the thin films fabricated by the MPM. The design of metal complexes for the precursor and of the heating program are crucial.

Multilayer film	ν [nmol L^{-1} min^{-1}]	
	under UV irradiation	under VIS irradiation
OX-OX-OX	16(1)	–
OX-ED-OX	21(1)	6(1)

Table 1. The rate ν [nmol L^{-1} min^{-1}] of decoloration rate of 0.01 mol L^{-1} MB solution by the photoreaction with both thin films under UV and VIS irradiation [5]. The rate was measured by the decrease of absorption value at 664 nm of each test solution. Those obtained from the data measured under dark are also indicated. Calculated standard deviations are presented in parentheses.

6. Formation of O-deficient anatase thin film

To clarify the factors for designing an anatase thin film with a higher photoreactivity under UV irradiation, the relationship between the photoreactivity and O deficiency of anatase thin films fabricated with the heat-treated precursor films under regulated conditions was examined. Thin films were formed by heat-treating precursor films spin-coated onto FTO glass substrates with $\mathbf{S_{ED}}$ and $\mathbf{S_{SG}}$ (sol–gel solution) under Ar or air.

The spin-coating method at ambient temperature was used for forming precursor films using a double-step method. The first step used 500 rpm for 5 s and the next step was 2000 rpm for 30 s, in all cases. The precursor films were pre-heated in a drying oven at 70 °C for 10 min and then heat-treated at 500°C for 30 min in a 0.1 L min^{-1} Ar gas flow. A tubular furnace made of quartz was employed for the heat-treatment. Thin films, **ED** and **SG**, were formed by applying the precursor solutions $\mathbf{S_{ED}}$ and $\mathbf{S_{SG}}$, respectively, before annealing in air. The film $\mathbf{ED_{air}}$ was fabricated by firing the precursor film spin-coated with $\mathbf{S_{ED}}$ in air at 500°C for 30 min.

When the concentration of titanium was 0.4 mmol g^{-1} for $\mathbf{S_{ED}}$, the film thickness was 100 nm. An $\mathbf{S_{SG}}$ of 0.5 mmol g^{-1} was stirred for 3 days at ambient temperature to fabricate an anatase film of thickness 100 nm. The post-annealing treatment for the **ED**, $\mathbf{ED_{air}}$, and **SG** thin films was carried out in air at 500°C for 5, 10, 15, 20, and 30 min. The number in the notation used for post-annealed films indicates the annealing time (min). For example, **ED-PA5** indicates an **ED** film post-annealed for 5 min. The photoreactivities of the thin films are presented in Table 2. Each value was calculated as the difference between the decoloration rate under UV-light irradiation and the corresponding value measured for each thin film in the dark.

The maximum photoreactivity of **ED-PA15** produced by the MPM is twice that of **SG-PA10** prepared by a conventional sol–gel procedure.

Annealing Time	v under UV-light irradiation		
(min)	**ED**	**ED$_{air}$**	**SG**
0	2(1)	6(1)	5(1)
5	5(1)	12(1)	6(1)
10	9(1)	12(1)	8(1)
15	16(1)	9(1)	7(1)
20	11(1)	9(1)	7(1)
30	8(1)	5(2)	5(1)

Table 2. The rate v [nmol L^{-1} min^{-1}] of decoloration rate of 0.01 mol L^{-1} MB solution by the photoreaction with each thin film under UV-light irradiation [6]. The rate was measured by the decrease of absorption value at 664 nm of each test solution. Those obtained from the data measured under dark are also indicated. Calculated standard deviations are presented in parentheses.

It is generally accepted that the main factors to consider when designing enhanced photoreactivity of anatase are (1) higher crystallinity, (2) larger surface area, and (3) decreased impurities. The crystallite size is an indicator of crystallinity [78, 79]. Among the crystallite sizes of the three anatase thin films, **ED**, **ED$_{air}$** and **SG**, the **SG** thin film had the largest value and the **ED** film had the smallest (Table 3). These values for the anatase crystallites in **ED$_{air}$** and **SG** thin films were not affected by post-annealing treatment in air. The thin film **ED-PA15** (whose crystallite size was the smallest) showed the highest photoreactivity in the decoloration of an MB aqueous solution among the various thin films formed in this study. The specific surface areas of the thin films were not measured quantitatively because of the difficulties involved. However, the degrees of adsorption of MB molecules in aqueous solution were nearly equal among the thin films, including those formed by the sol–gel method. Therefore, the differences in the photoreactivity among these thin films should be due to other factors than the specific surface area. The XPS spectra suggested that the thin films **SG** and **SG-PA*n*** have higher purities than the other thin films. Therefore, the highest photoreactivity, of **ED-PA15** thin film, cannot be due to its purity. The O/Ti peak area ratio determined from the XPS of the anatase film **ED-PA15** with the highest photoreactivity was extremely small, 1.5. The refractive indices of the thin films **ED$_{air}$** and **SG** increased gradually, depending on the post-annealing time. On the other hand, the refractive index of the **ED** thin film decreased with post-annealing treatment time up to 15 min and then increased with further annealing. The largest index (2.17; **ED** thin film) may be related to the strong and wide absorbance by the above-mentioned impurities. Furthermore, the smallest value (1.99; **ED-PA15)** could be affected by the largest O deficiency in the anatase thin film after purification. The decrease in permittivity of the thin film arose from the lower charge density derived from the O deficiency because the structure of the anatase lattice is rigid. Thus, the O deficiency formed by this method was one of the most important factors for fabricating highly UV-sensitive anatase. This O deficiency was formed during heat-treatment of the precursor metal complexes.

A coordination skeleton of (TiO_4N_2) or (TiO_5N_2) can be assumed in the EDTA complex as a precursor molecule from the structural study of a Ti complex [Ti(H_2O)(EDTA)]·1.5H_2O reported by Fackler *et al.* [80]. In the precursor films, two N and at least four O atoms link to one Ti ion. As a result of heat-treating the precursor complex in an Ar gas flow, neighboring complexes reacted with each other. In this process, several O atoms linked to one Ti ion could be covalently bonded by other Ti ions, and the anatase lattice was gradually created. By eliminating large amounts of C, H, and N atoms with O atoms, oxide ion sites of the anatase lattice were partially occupied by a rather stable nitride ion derived from the coordinated N atom originally belonging to the ligand. As a result, the total negative charge of the N-substituted anatase in the **ED** thin film is ca. 3.6 toward one Ti ion. This value is the summation of 2.8 from the oxide ions and 0.8 from the nitride ions. This charge toward one Ti ion is larger than that of ca. 3.3 by the oxide ions in the **SG** thin film. The substitutional N atoms could be removed from the anatase lattice by post-annealing the **ED** thin film. Consequently, the total negative charge of the **ED-PA15** thin film, whose photoreactivity is the highest, decreased to ca. 3.0. Longer annealing treatment replenished oxide ions in the anatase thin films from their surfaces and the photoreactivity decreased (Figure 8) [6]. Thus, it was elucidated that O deficiency is an important factor to consider when designing anatase photoreactivity. It is also notable that the O-deficient anatase lattice is rather robust because the stoichiometric Ti_2O_3 did not appear at all.

Annealing Time (min)	Crystallite size (nm)		
	ED	ED_{air}	SG
0	—	10	13
5	—	10	13
10	5	11	13
15	4	10	13
20	7	11	13
30	7	12	13

Table 3. The crystallite size of anatase in **ED**, **ED**$_{air}$, **SG** and post-annealed thin films [6]. The crystallite size of anatase was measured with a typical Scherrer-Hall method by employing a peak assignable to only (1 0 1) of anatase, because other peak intensities due to anatase were too low to measure accurately. The crystallite size of anatase in **ED** and **ED-PA5** could not be obtained because the (1 0 1) peak of anatase was also too weak to determine the crystallite size.

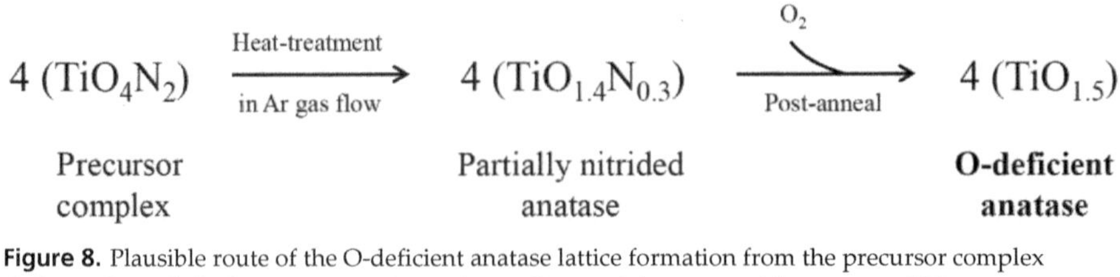

Figure 8. Plausible route of the O-deficient anatase lattice formation from the precursor complex skeleton through the heat treatment in an Ar gas flow and the sequential post-anneal [6].

7. O deficiency in rutile thin film

Rutile is the most stable crystal form of titania. Since Nishimoto *et al.* showed that anatase is more sensitive to UV light than rutile in photoreactions, rutile was believed to be inferior to anatase in terms of photoreactivity [81]. Anatase is important for photocatalysis in pollutant degradation and in the development of photofunctional materials such as films with hydrophilic surfaces under UV-light irradiation. The poor photoreactivity and photosensitivity of rutile is generally believed to be due to its crystal structure. Rutile is primarily known as a useful pigment for white paint, due to its chemical stability [82, 83].

Because the band edge of a rutile single-crystal is 3.0 eV, rutile has the potential to respond to Vis light. Using this knowledge and the results of previous experiments on anatase responses to Vis light, this section describes an attempt to achieve direct fabrication of O-deficient rutile thin films with high photoreactivity using a MPM. The first Vis-light-responsive thin film created from O-deficient rutile is discussed here. This material works without application of an electric potential, due to its unprecedentedly high photosensitivity under UV-light irradiation. The present findings should facilitate widespread practical use of rutile in light-related applications.

The thin films were formed by heat-treating the precursor films after spin-coating onto a quartz glass substrate. **S_{ED}** and **S_{SG}** were applied in an Ar gas flow. The transparent precursor films formed by spin-coating the solutions and pre-heating in a drying oven at 70 °C for 10 min were heat-treated at 700 °C for 30 min in a furnace made from a quartz tube with an Ar gas flow rate of 0.1 L min^{-1}. When **S_{ED}** was used, a transparent rutile thin film **R** was formed. When **S_{SG}** was used, a transparent anatase thin film **A** was formed. The film thickness was 100 nm in both cases.

Each structure was characterized using XRD, Raman spectroscopy, and transmission electron microscopy (TEM). The selectivity was due to the O-vacant sites in the oxide thin films formed at different levels due to the differences between the amounts of oxygen in the two precursors. In this case, the oxygen source required to structure titania was available only in the precursor films when these thin films were fabricated. Therefore, crystallization into rutile, which has many O-vacant sites, and the accompanying rapid elimination of organic residues from the **R** precursor film, occurred because of the heat-treatment.

In contrast, the amount of oxygen available to Ti^{4+} in titanoxane polymers, though significant, was insufficient to develop stoichiometric TiO_2 from **A**. The oxygen defects in an anatase lattice generally lower the temperature of the phase transformation from anatase to rutile [84, 85]. Thus, selective formation occurred according to the differing degrees of O deficiency.

The photoreactivities of the thin films were evaluated by the decoloration rates of MB solutions, which served as a model for organic pollutants in water. The results measured under Vis- and UV-light irradiation are summarized in Table 4, along with those measured under dark conditions (reference values). The data show the effects of adsorption on the samples, vessels, and self-decoloration of MB under each condition. Moreover, the photoreactivity of **R** was extremely high under UV irradiation and higher than the photoreactivity of **A**. This is without precedent.

Notation	v [nmol L^{-1} min^{-1}]	
	under visible light	under UV light
R	9(1)	23(1)
A	—	15(1)

Table 4. The reaction rate v of the decoloration reaction in an aqueous solution containing 0.01 mol L^{-1} of methylene blue under visible- and UV-light irradiation and under dark conditions [7]. Calculated standard deviations are presented in parentheses.

The photosensitivities of **R** and **A** were also examined by measuring the effects of Vis and UV irradiation on the water contact angle for the surfaces of the thin films. The results are shown in Figure 9 [7]. The rutile thin film **R** exhibited Vis-light-induced hydrophilicity with a fluorescent light, even though high-energy light below 400 nm was eliminated. In contrast, Vis light alone did not reduce the contact angle on **A** under the same conditions. Furthermore, a rapid decrease in the water contact angle for **R** was observed with weak UV-light irradiation. The super-hydrophilic property of **R** appeared after only 1 h. When fluorescent light with a UV component was employed, the contact angle on **R** rapidly reduced and the values reached 38° and 10° after irradiation for 1 and 24 h respectively.

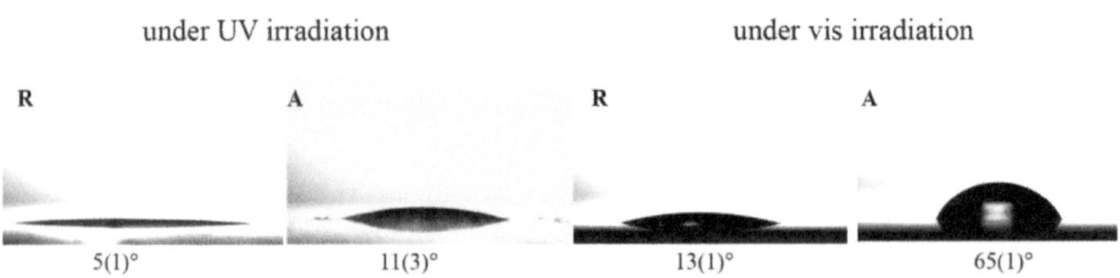

Figure 9. Comparison of the contact angles of a 1.0-μL water droplet on the thin films **R** and **A**. Before the measurement, the thin films were exposed to UV irradiation of 1.3 mW cm^{-2} at 356 nm obtained by a black light (each on the left side), and to visible light without UV light was obtained from a fluorescent light by removing light of wavelengths shorter than 400 nm using a cut-off filter. The Vis light intensity after removing UV components from the fluorescent light, which was estimated by an illuminometer was 0.8 mW cm^{-2} [7].

It is noteworthy that the simple fabrication of a Vis-responsive rutile film with high photoreactivity could be attained. Thus, the O defects in titania are also effective at providing photoreactivity of rutile, which is usually insensitive to both UV and Vis light.

8. Crystal orientation and photoluminescence of rutile in thin film

PL emission has been widely used to investigate the efficiency of charge carrier trapping, migration, and transfer, and to understand the fate of electron–hole pairs in semiconductor particles [86]. It is therefore helpful to examine the position and intensity of the PL bands of semiconductor particles to understand the photoreactivities of the particles [87]. In this section, we report the changes in the PL and photoreactive properties of the Vis-responsive rutile thin film fabricated by the MPM, which are effected by annealing in air at 700°C. The relationship between O deficiency and PL emission was examined to understand the incredibly high photoreactivity of the rutile thin film. Furthermore, the level of crystal orientation of the rutile thin film was quantitatively evaluated on the basis of data from XRD analyses. The amount of oxygen supplied during the annealing process was analyzed by XPS measurements. The growth of crystals and particles was also investigated by crystallite-size measurements and SEM observations. The heat-treatment of the fabricated O-deficient rutile **R** thin film was carried out in air at 700°C for 15, 30, and 60 min. The number in the notation of the post-annealed films indicates the annealing time (min). For example, **R-PA15** indicates that post-annealing treatment of the **R** thin film was carried out for 15 min. The extent orientation factor (*f*) for the (110) plane of the **R-PA*n*** thin films increased with annealing time in air (Table 5).

Film	Crystallite size[a)] / nm	Orientation factor; *f*	O/Ti ratios	
			Surface	Deeper portion
R	15(2)	0.35	1.74	1.75
R-PA15	21(3)	0.67	1.84	1.73
R-PA30	21(3)	0.69	1.89	1.79
R-PA60	20(4)	0.75	1.94	1.85

Table 5. The crystallite size, orientation factor and O/Ti ratio of rutile crystals in the **R** thin film and in the post-annealed **R-PAn** thin films [8]. The crystallite size of rutile was measured with a typical Scherrer-Hall method by employing the peaks assignable to (1 1 0), (1 0 1) and (2 1 1) of rutile. The extent of the orientation was estimated in terms of Logtering orientation factor, *f*, from the XRD peak intensities (*I*). For calculating the orientation factor, the intensity data of non-oriented rutile was cited from the JCPDS card 21-1276. The O/Ti ratios determined by the XPS peak areas of O 1s and Ti $2p_{3/2}$ peaks observed from the surfaces of **R** and post-annealed thin films. The XPS peaks of the thin film surface were measured without bombarding Ar^+ ion beam. The peak area of O 1s and Ti 2p was calculated by FWHM and peak height at the positions 531.0 and 459.0 eV, respectively. The averaged O/Ti ratios determined by the XPS peak areas of O 1s and Ti $2p_{3/2}$ peaks of **R** and post-annealed thin films. The XPS peaks of thin films were measured after bombarding Ar^+ ion beam with 2 kV and 18 μA cm^{-2} for 3 min, in order to remove surface oxides. The peak area of O 1s and Ti 2p was calculated by FWHM and peak height at the positions 531.0 and 459.0 eV, respectively, obtained from each depth profile in Ar^+ ion etching mode.
a) The estimated standard deviations are presented in parentheses

The extent orientation could be estimated from the XRD peak intensity by using the Lotgering method [8]. The terms $I(hkl)_{ideal}$ and $\Sigma I(hkl)_{ideal}$ are defined as the intensity of the peak attributable to the specific plane (*hkl*) and the sum of each intensity obtained for the non-oriented rutile crystals; thus, P_0 can then be expressed as

$$P_0 = I(hkl)_{ideal}/\Sigma I(hkl)_{ideal} \quad (1)$$

In this study, each $I(hkl)_{ideal}$ value was cited from the standard data of the corresponding rutile phase.

The definitions of the terms $I(hkl)_{obsd}$ and $\Sigma I(hkl)_{obsd}$ for the thin films **R** and **R-PA*n*** are identical to those for $\Sigma I(hkl)_{ideal}$ and $I(hkl)_{ideal}$ in equation (1), respectively, and the P_n value can be calculated as

$$P_n = I(hkl)_{obsd}/\Sigma I(hkl)_{obsd} \quad (2)$$

The Lotgering orientation factor f is defined as

$$f = (P_n - P_0)/(1 - P_0) \quad (3)$$

The factor f is defined for P or P_0 values over a certain range of 2θ. As the level of orientation increases, the f value will increase from 0 to 1. The factor f is therefore a measure of the crystal phase orientation. Consequently, the f value of the **R** thin film was the smallest among the thin films fabricated by the MPM. However, the present f value of the **R** thin film was larger than those reported for rutile thin films fabricated by a sol–gel method on several substrates such as quartz glass, alumina, and single-crystals of quartz or silicon [88]. This result may be related to the different mechanisms governing the formation of the rutile lattice. In the sol–gel method, the Ti^{4+} and O^{2-} ions that are tightly linked in titanoxane polymers formed by the condensation of partially hydrolyzed alkoxide might be rearranged to form the rutile lattice at a higher temperature. In the MPM, however, small units composed of Ti^{4+} and O^{2-} ions with higher mobility could be formed when the organics were decomposed and removed by heat-treating the precursor complex. Therefore, a rutile thin film with a higher level of crystal orientation could be formed at these lower temperatures.

After 15 min of heat-treatment in air, the crystallite size of the **R** thin film increased, while those of the **R-PA*n*** thin films remained almost constant (Table 5). These results indicate that crystallite growth at a temperature of 700°C was completed by heat-treatment over a period of 30–45 min. The additional thermal energy was consumed mainly for the process of grain growth after crystallite growth because the grain size gradually increased with annealing time.

Previously, the peak position of the PL emission band obtained for rutile crystals was observed at ca. 450 nm [89]. However, PL emission bands of the **R** and **R-PA*n*** thin films formed by the MPM were not detected at 450 nm (Figure 10) [8].

Nakamura *et al.* reported that in the case of a rutile single-crystal, the PL emission band attributable to the (110) plane can be observed at 810 nm [90]. Taking into account the high levels of crystal orientation with reference to the (110) plane in the **R** and **R-PA*n*** thin films, the PL emission bands observed at around 800 nm in the spectra of the **R-PA*n*** thin films can be attributed to rutile crystals oriented along the (110) plane in these thin films (Figure 11) [8].

For the O-deficient rutile thin film **R** with high photoreactivity, no PL emission was observed in the range 190–850 nm. Thus, as previously suggested, the O-defect sites on the rutile thin film may suppress recombination of the photoinduced electron–hole pairs by electron trapping. In contrast, the PL emission from rutile thin films after annealing in air may be due to the oxide ions that are supplied to the O-defect sites on the film surface, because they function as recombination centers. As a result, the lattice oxygens of titania, especially in rutile thin films, function as recombination centers for the photoinduced electron–hole pairs.

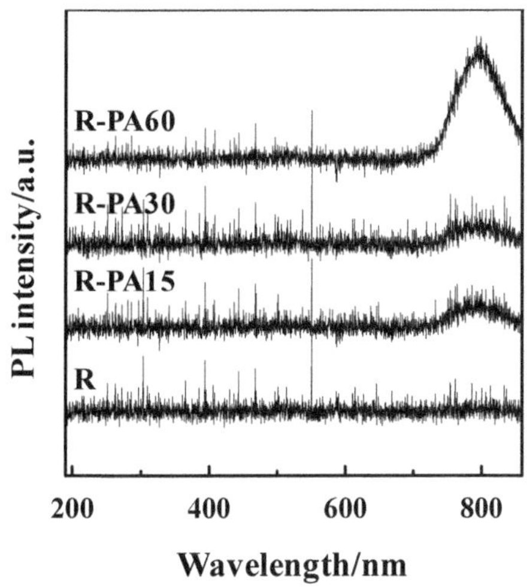

Figure 10. The photoluminescence emission spectra of the thin films, O-deficient rutile **R** and post annealed **R-PA*n*** (*n* = 15, 30, 60). The spectra were measured in the wavelength range 190-850 nm at room temperature [8].

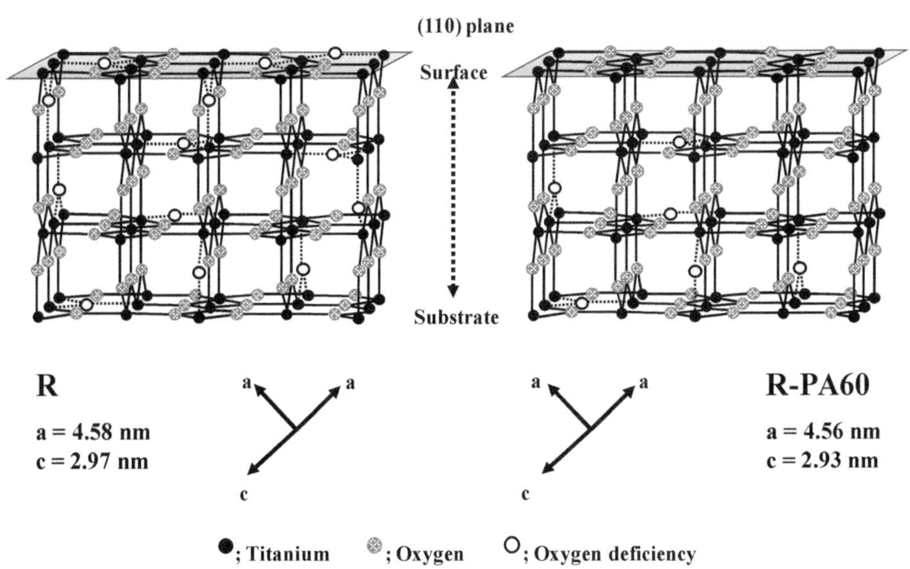

Figure 11. Proposed schematic diagrams for the (110) plane orientated O-deficient rutile **R** (left) and post annealed **R-PA60** (right) thin films are shown. Cell parameters were refined by a least square method [8].

The reduction of rutile surfaces by heated hydrogen activates the photoreactivities of these surfaces [36]. As a result, the formation of an oxygen deficiency provides photoreactivity. The present study revealed that the unprecedentedly high photoreactivity of the **R** thin film is suppressed by oxygen supply during the annealing process. However, the high levels of photoreactivity of these films could be maintained even after oxygen supply to the surface by the post-annealing treatment (Table 5). These results indicate that the enhanced photoreactivity is related not only to the surface but also to the inner part of the thin films as a result of an interparticle electron transfer (IPET) effect, as proposed in our previous paper.

9. Highly conductive Ag-nanoparticle/titania composite thin films

An excellent perovskite-type $SrTiO_3$ thin film was fabricated using a mixed precursor solution from a titania precursor solution containing a Ti complex of EDTA and an SrO precursor solution containing a Sr complex of EDTA [4]. The metal complex ions dissolve independently in each precursor solution and the homogeneity of the mixed solution can be kept at the molecular level. In fact, a mixed precursor solution containing exact amounts of Ti and Sr can be easily prepared due to the excellent miscibility of the solutions. This is the essential difference between the MPM and conventional sol–gel methods in which the hydrolyzed polymers are heterogeneous because of the different rates of hydrolysis of each metal ion. On the basis of this excellent miscibility in the MPM, Ag-nanoparticle/titania (Ag-NP/TiO_2) composite thin films with a wide range of volumetric fractions of Ag in the titania matrix were developed using the titania precursor $\mathbf{S_{ED}}$ [9].

Many researchers have tried to incorporate metal nanoparticles into semiconductor materials to improve the conductivity of the semiconductor. The TiO_2 film's relatively high resistivity of 10^{12} Ω cm at 25°C can be reduced by incorporating metal nanoparticles into the TiO_2 matrix. Electrically conducting particles can be randomly distributed within a semiconductor matrix to form a composite. This composite sample is non-conducting until the volume fraction of the conducting phase reaches the so-called percolation threshold. It has been experimentally and analytically shown that in a conductor/semiconductor composite with the conductor at or above a given volume fraction (ϕ_i), a network of conducting particles is established and thus the composite resistivity suddenly decreases. The most widely applied technique for the preparation of metal/TiO_2 composite materials is the sol–gel process. This has been applied to the fabrication of an Ag-NP/TiO_2 system by Li *et al.* [91]. They prepared a solution for fabricating Ag-NP/TiO_2 composite thin films by mixing a sol–gel solution of titania for thin film fabrication with an 18 mol% silver nitrate solution; a 6 mol% ethanol solution of silver nitrate was also employed. The electrical resistivity of the resultant composite thin film with the highest concentration (18 mol%) of Ag nanoparticles was of the order of 10^3 Ω cm. Because the sol–gel method used in these studies involves metalloxane polymer formation in the medium, and because poor miscibility of each component is inevitable, Li *et al.* reported that it was difficult to obtain a homogeneous solution for silver concentrations above 18 mol%. Therefore, a lower electrical resistivity of the composite thin film may be attained in the event that a solution with an even higher volumetric fraction of Ag nanoparticles can be homogeneously mixed with the titania precursor solution. The electrical conductivity of the resultant films is largely dependent on the volumetric fraction, size, and connectivity of the Ag nanoparticles, and the homogeneity of the dispersed silver in the dielectric titania matrix [92-95].

Using the MPM, mixed precursor solutions for fabricating Ag-NP/TiO_2 composite thin films could be easily prepared. As a result, Ag-NP/TiO_2 composite thin films of Ag volumetric fractions from 0.03 to 0.68 were fabricated with heat-treatment of the mixed precursor films at 600°C in air. To obtain quantitative information about the effects of Ag nanoparticles on the electrical properties, the nanostructures of the films were examined by TEM. The TEM images films with ϕ_{Ag} of 0.26, 0.30, and 0.55 are shown in Figures 12 (A), (B), and (C) respectively [9]. The presence and distribution of Ag nanoparticles (black dots) inside the TiO_2 film can be clearly seen. The percolation threshold of Ag nanoparticles in the titania thin film was found to be ϕ_{Ag} 0.30. It is near the percolation threshold, when Ag particles are still not totally connected, that increasing the ϕ_{Ag} by adding a small amount of Ag nanoparticles helps to build the conductive network and reduce the resistivity of the composite. Therefore, the decrease in resistivity was attributed to a change in the Ag nanoparticles` size, shape, and center-to-center distance between the Ag nanoparticles. As the Ag volumetric fraction increased further from 0.27 to 0.55, the electrical resistivity decreased from 10^{-2} to 10^{-5} Ω cm, respectively. At ϕ_{Ag} 0.61 to 0.68, the resistivity increased from 10^{-5} to 10^{-3} Ω cm due to the inevitable increase in resistivity caused by agglomeration of the Ag particles. This study shows that the MPM, which offers excellent miscibility of the silver and titania precursor solutions, is effective at overcoming the miscibility limitations of the conventional sol–gel method and is necessary for fabricating composite thin films with large ϕ_{Ag} values.

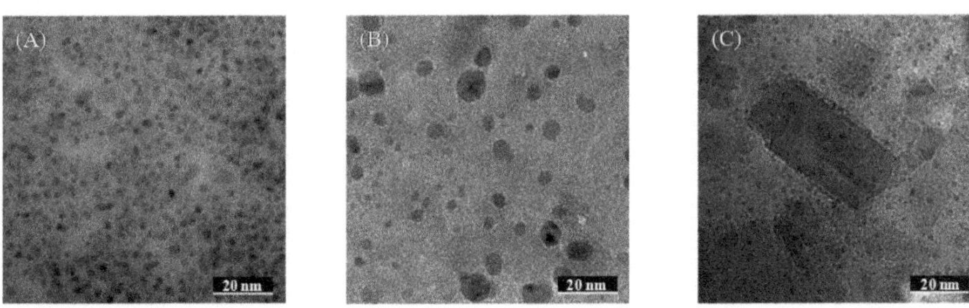

Figure 12. TEM images of the Ag-NP/TiO_2 composite thin films at Ag volumetric fractions, ϕ_{Ag}, of (A) 0.26, (B) 0.30, and (C) 0.55, respectively [9].

The excellent miscibilities of the precursor complexes in the MPM overcame the limitations of the extremely low Ag volumetric fraction in the previous sol–gel process. Therefore, the percolation threshold for the electrical resistivity of the composite film could be examined for a wide range of Ag fractions. Heat-treatment plays an important role in the production of Ag nanoparticles by reducing Ag^+ ions in the precursor film and forming well-dispersed Ag nanoparticles in the titania matrix. This present Ag-nanoparticle/titania composite thin film is useful for fabrication of highly conductive electrodes for devices such as solar cells.

10. Conclusion

The importance of heat-treatment in the MPM was demonstrated through fabrication of thin films of anatase and rutile with unprecedentedly high photoreactivities. This is due to a photoreactive mechanism via O deficiency in the oxide thin films. Based on the excellent miscibilities of the molecular precursors in the $SrTiO_3$ thin film fabrication, heat-treatment was shown to be an essential step. It eliminates organic ligands in the precursor metal complexes and provides important functions to the metal oxides in the chemical fabrication of Ag-nanoparticle/titania composite thin films with high conductivity.

The chemical fabrication of the first p-type Cu_2O transparent thin film was also recently achieved using the MPM, although the heat-treatment of the spin-coated substrates resulted in the deposition of a large amount of powder on the substrates in previous sol–gel studies [96]. The electrical and optical properties of the Cu_2O thin film fabricated using the MPM were consistent with those of similar thin films fabricated using physical procedures. It is very interesting that the charge on copper changes stepwise from +2 to +1 through 0 during heat-treatment of the precursor film involving a Cu–EDTA complex in an Ar gas flow. Based on these results, a transparent dry-type solar cell of area 20 × 20 mm^2 with a combination of Vis-responsive anatase thin films was examined. This film is mentioned in the section **Vis-responsive anatase thin film fabricated using the MPM**. The structure of the solar cell was FTO electrode/n-Vis-responsive anatase/p-Cu_2O/conductive polymer/Ag on a glass substrate, and the photovoltaic nature of the solar cell could be successfully measured under the light from a solar simulator. Thus, the present MPM is useful for fabricating Vis-responsive dry solar cells. The MPM coupled with heat-treatment of various precursor films allows transparent thin films of metal oxides such as Co_3O_4, ZnO, Ga_2O_3, and ZrO_2 *etc.* to be examined and fabricated on glass and/or metallic substrates. The MPM has great potential as a fundamental technology for thin film fabrication by chemical processes.

Author details

Hiroki Nagai and Mitsunobu Sato
Research Institute of Science and Technology, Kogakuin University, Nakano, Hachioji, Tokyo, Japan

Acknowledgement

This work was supported by MEXT*-Supported Program for the Strategic Research Foundation at Private Universities, 2011-2015 (Kogakuin University–Energy Conversion Ecomaterials Center). *Ministry of Education, Culture, Sports, Science and Technology

11. References

[1] Sato M, Hara H, Niside T and Sawada Y (1996) A Water-resistant Precursor in a Wet Process for TiO_2 Thin Film Formation. J. Mater. Chem. 6: 1767-1770.

[2] Nishide T, Sato M, Hara H (2000) Crystal Structure and Optical Property of TiO_2 Gels and Films Prepared from Ti-edta Complexes as Titania Precursors. J. Mater. Sci. 35: 465-469.

[3] Sato M, Hara H, Kuritani M, Nishide T (1997) Novel Route to Co_3O_4 Thin Films on Glass Substrates via N-alkyl Substituted Amine Salt of Co(III)-EDTA Complex. Sol. Energy Mater. Sol. Cells 45: 43-49.

[4] Sato M, Tanji T, Hara H, Nishide T, Sakashita Y (1999) $SrTiO_3$ Film Fabrication and Powder Synthesis from a Non-polymerized Precursor System of a Stable Ti(IV) Complex and Sr(II) Salt of edta. J. Mater. Chem. 9: 1539-1542.

[5] Nagai H, Mochizuki C, Hara H, Takano I, Sato M (2008) Enhanced UV-sensitivity of Vis-responsive Anatase Thin Films Fabricated by Using Precursor Solutions Involving Ti Complexes. Sol. Energy Mater. Sol. Cells 92: 1136-1144.

[6] Nagai H, Hasegawa M, Hara H, Mochizuki C, Takano I, Sato M (2008) An Important Factor Controlling the Photoreactivity of Titania: O-deficiency of Anatase Thin Film. J. Mater. Sci. 43: 6902-6911.

[7] Nagai H, Aoyama S, Hara H, Mochizuki C, Takano I, Baba N, Sato M (2009) Rutile Thin Film Responsive to Visible Light and with High UV Light Sensitivity. J. Mater. Sci. 44: 861-868.

[8] Nagai H, Aoyama S, Hara H, Mochizuki C, Takano I, Honda T, Sato M (2010) Photoluminescence and Photoreactivity Affected by Oxygen Defects in Crystal-oriented Rutile Thin Film Fabricated by Molecular Precursor Method. J. Mater. Sci. 45: 5704-5710.

[9] Likius DS, Nagai H, Aoyama S, Mochizuki C, Hara H, Baba N, Sato M (2012) Percolation Threshold for Electrical Resistivity of Ag-nanoparticle/Titania Composite Thin Films Fabricated Using Molecular Precursor Method. J. Mater. Sci. 47: 3890-3899.

[10] Mochizuki C, Sasaki Y, Hara H, Sato M, Hayakawa T, Yang F, Hu X, Shen H, Wang S (2009) Crystallinity Control of Apatite through Ca-EDTA Complexes and Porous Composites with PLGA. J. Biomed. Mater. Res. Part B 90: 290-301.

[11] Honda T, Oda T, Mashiyama Y, Hara H, Sato M (2010) Fabrication of c-axis Oriented Ga-doped MgZnO-based UV Transparent Electrodes by Molecular Precursor Method. Phys. Status. Solidi. 7: 2471-2473.

[12] Twite RL, Bierwagen GP (1998) Review of Alternatives to Chromate for Corrosion Protection of Aluminum Aerospace Alloys. Prog. Org. Coat. 33: 91-100.

[13] Neto DLP, Atik M, Avaca LA, Aegerter MA (1994) Sol-Gel ZrO_2 Coatings for Chemical Protection of Stainless Steel. J. Sol-Gel Sci. Technol. 1: 177-184.

[14] Metroke TL, Parkhill RL, Knobbe ET (2001) Passivation of Metal Alloys Using Sol-gel-derived Materials-A review. Prog. Org. Coat. 41: 233-238.

[15] Atik M, Neto DLP, Aegerter MA, Avaca LA (1995) Sol-gel TiO_2-SiO_2 Films as Protective Coatings Against Corrosion of 316L Stainless Steel in H_2SO_4 Solutions. J. Appl. Electrochem. 25: 142-148.

[16] Wei Q, Narayan RJ, Narayan J, Sankar J, Sharma AK (1998) Improvement of Wear Resistance of Pulsed Laser Deposited Diamond-like Carbon Films through Incorporation of Metals. Mater. Sci. Eng. B 53: 262-266.
[17] Wei Q, Sharma AK, Sankar J, Narayan J (1999) Mechanical Properties of Diamond-like Carbon Composite Thin Films Prepared by Pulsed Laser Deposition. Comps. Part B: Eng. 30: 675-684.
[18] Nomura K, Ohta H, Takagi A, Kamiya T, Hirano M, Hosono H (2004) Room-temperature Fabrication of Transparent Flexible Thin-film Transistors Using Amorphous Oxide Semiconductors. Nature 432: 488-492.
[19] Hoffman RL, Norris BJ, Wager JF (2003) ZnO-based Transparent Thin-film Transistors. Appl. Phys. Lett. 82: 733-735.
[20] Tsukazaki A, Ohtomo A, Onuma T, Ohtani M, Makino T, Sumiya M, Ohtani K, Chichibu SF, Fuke S, Segawa Y, Ohno H, Koinuma H, Kawasaki M (2005) Repeated Temperature Modulation Epitaxy for p-type Doping and Light-emitting Diode Based on ZnO. Nature Mater. 4: 42-45.
[21] Kim JH, Lee S, Im HS (1999) The Effect of Target Density and Its Morphology on TiO_2 Thin Films Grown on Si (100) by PLD. Appl. Surf. Sci. 151: 6-16.
[22] Walczak M, Oujja M, Marco JF, Sanz M, Castillejo M (2008) Pulsed Laser Deposition of TiO_2: Diagnostic of the Plume and Characterization of Nanostructured Deposits. Appl. Phys. A 93: 735-740.
[23] Tang H, Prasad K, Sanjines R, Schmid PE, Levy F (1994) Electrical and Optical Properties of TiO_2 Anatase Thin Films. J. Appl. Phys. 75: 2042-2047.
[24] Campbell SA, Kim HS, Gilmer DC, He B, Ma T, Gladfelter WL (1999) Titanium Dioxide (TiO_2)-based Gate Insulators. IBM. J. Res. Dev. 43: 383-392.
[25] Wang Z, Helmersson U, Kall PO (2002) Optical Properties of Anatase TiO_2 Thin Films Prepared by Aqueous Sol–gel Process at Low Temperature. Thin Solid Films 405: 50-54.
[26] Sakka S (1994) The Current State of Sol-gel Technology. J. Sol-Gel Sci. Technol. 3: 69-81.
[27] Sakka S (1982) Gel Method for Making Glass. Treatise Mater. Sci. Technol. 22: 129-167.
[28] Sakka S (1988) Science of Sol-Gel Method. Tokyo: Agne-Shofu-Sha. 221 p.
[29] Brinker CJ, Scherer GW (1990) Sol-Gel Science. New York: Academic Press. 908 p.
[30] Hench LL, West JK (1990) The Sol-Gel Process. Chem. Rev. 90: 33-72.
[31] Mazdiyasni KS, Dolloff RT, Smith JS (1969) Preparation of High-Purity Submicron Barium Titanate Powders. J. Am. Ceram. Soc. 52: 523-526.
[32] Sakka S, Kamiya K (1982) The Sol-gel Transition in the Hydrolysis of Metal Alkoxides in Relation to the Formation of Glass Fibers and Films. J. Non-Cryst. Solids 48: 31-46.
[33] Hu L, Yoko T, Kozuka H, Sakka S (1992) Effects of Solvent on Properties of Sol-gel-derived TiO_2 Coating Films. Thin Solid Films 219: 18-23.
[34] Dimitriev Y, Ivanova Y, Iordanova R (2008) History of Sol-gel Science and Technology. J. Univer. Chem. Technol. Merall. 43: 181-192.
[35] Hashimoto K, Irie H, Fujishima A (2007) TiO_2 Photocatalysis: A Historical Overview and Future Prospects. AAAPS. Buliten. 17: 12-28.
[36] Fujishima A, Honda K (1972) Electrochemical Photolysis of Water at a Semiconductor Electrode. Nature 238: 37-38.

[37] Pelizzetti E, Serpone N (Eds.) (1986) Homogeneous and Heterogeneous Photocatalysis. Dordrecht: D. Reidel Publishing Company.752 p.
[38] Serpone N, Pelizzetti E (Eds.) (1989) Photocatalysis, Fundamentals and Applications. New York: John Wiley & Sons. 650 p.
[39] Kamat PV (1993) Photochemistry on Nonreactive and Reactive (Semi-conductor) Surfaces. Chem. Rev. 93: 267-300.
[40] Ollis DF, Al-Ekabi H (Eds.) (1993) Photocatalytic Purification and Treatment of Water and Air. Amsterdam: Elsevier. 820 p.
[41] Heller A (1995) Chemistry and Applications of Photocatalytic Oxidation of Thin Organic Films. Acc. Chem. Res. 28: 503-508.
[42] Hoffmann MR, Martin ST, Choi W, Bahnemann DW (1995) Environmental Applications of Semiconductor Photocatalysis. Chem. Rev. 95: 69-96.
[43] Mills A, Lehunte S (1997) An Overview of Semiconductor Photocatalysis. J. Photochem. Photobiol. A: Chem. 108: 1-35.
[44] Peral J, Domènech X, Ollis DF (1997) Hetrogeneous Photocatalysis for Purification, Decontamination and Deodorization of Air. J. Chem. Tech. Biotech. 70: 117-140.
[45] Fujishima A, Hashimoto K, Watanabe T (1999) TiO_2 Photocatalysis: Fundamentals and Applications. Tokyo: BKC, Inc. 184 p.
[46] Fujishima A, Rao TN, Tryk DA (2000) Titanium Dioxide Photocatalysis. J. Photochem. Photobiol. C 1: 1-21.
[47] Ollis DF (2000) Photocatalytic Purification and Remediation of Contaminated Air and Water. C. R. Acad. Sci. Paris, Serie II C, Chim. 3: 405-411.
[48] Tryk DA, Fujishima A, Honda K (2000) Recent Topics in Photoelectrochemistry: Achievements and Future Prospects. Electrochim. Acta. 45: 2363-2376.
[49] Fujishima A, Tryk DA, Bard AJ, Stratmann M, Licht S (Eds.) (2002) Encyclopedia of Electrochemistry, Vol. 6: Semiconductor Electrodes and Photoelectrochemistry. Wiley-VCH: Weinheim. 608 p.
[50] Hashimoto K, Irie H, Fujishima A (2005) TiO_2 Photocatalysis: A Historical Overview and Future Prospects. Japan. J. Appl. Phys. 44: 8269-8285.
[51] Fujishima A, Zhang X (2006) Titanium Dioxide Photocatalysis: Present Situation and Future Approaches. C. R. Chimie. 9: 750-760.
[52] Fujishima A, Zhang X, Tryk DA (2007) Heterogeneous Photocatalysis: From Water Photolysis to Applications in Environmental Cleanup. Internat. J. Hydrogen Energy. 32: 2664-2672.
[53] Jacoby WA, Maness PC, Wolfrum EJ, Blake DM, Fennel JA (1998) Mineralization of Bacterial Cell Mass on a Photocatalytic Surface in Air. Environ. Sci. Technol. 32: 2650-2653.
[54] Sunada K, Watanabe T, Hashimoto K (2003) Bactericidal Activity of Copper-deposited TiO_2 Thin Film under Weak UV Light Illumination. Environ. Sci. Technol. 37: 4785-4789.
[55] Herrmann JM, Mozzanega MN, Pichat P (1983) Oxidation of Oxalic Acid in Aqueous Suspensions of Semiconductors Illuminated with UV or Visible Light. J. Photochem. 22: 333-343.

[56] Herrmann JM, Disdier J, Pichat P (1984) Effect of Chromium Doping on the Electrical and Catalytic properties of Powder Titania under UV and Visible Illumination. Chem. Phys. Lett. 108: 618-622.

[57] Pichat P, Hermann JM, Disdier J, Mozzanega M, Courbon H (1984) Modification of the TiO_2 Electron Density by Ion Doping or Metal Deposit and Consequences for Photoassisted Reactions. Stud. Surf. Sci. Catal. 19: 319-326.

[58] Serpone N, Lawless D, Disdier J, Herrmann JM (1994) Spectroscopic Photoconductivity, and Photocatalytic Studies of TiO_2 Colloids: Naked and with the Lattice Doped with Cr^{3+}, Fe^{3+}, and V^{5+} Cations. Langmuir. 10: 643-652.

[59] Ikeda S, Sugiyama N, Pal B, Marcí G, Palmisano L, Noguchi H, Uosaki K, Ohtani B (2001) Photocatalytic Activity of Transition-metal-loaded Titanium (IV) Oxide Powders Suspended in Aqueous Solutions: Correlation with Electron–hole Recombination Kinetics. Phys. Chem. Chem. Phys. 3: 267-273.

[60] Karvinen S, Hirva P, Pakkanen TA (2003) Ab Initio Quantum Chemical Studies of Cluster Models for Doped Anatase and Rutile TiO_2. J. Mol. Struct. Theochem. 626: 271-277.

[61] Zhu J, Deng Z, Chen F, Zhang J, Chen H, Anpo M, Huang J, Zhang L (2006) Hydrothermal Doping Method for Preparation of Cr^{3+}-TiO_2 Photocatalysts with Concentration Gradient Distribution of Cr^{3+}. Appl. Catal. B-Eenviron. 62: 329-335.

[62] Yamashita H, Honda M, Harada M, Ichihashi Y, Anpo M, Hatano Y (1998) Preparation of Titanium Oxide Photocatalysts Anchored on Porous Silica Glass by a Metal ion-implantation Method and Their Photocatalytic Reactivities for the Degradation of 2-propanol Diluted in Water. J. Phys. Chem. B. 102: 10707-10711.

[63] Anpo M, Ichihashi Y, Takeuchi M, Yamashita H (1998) Design of Unique Titanium Oxide Photocatalysts by an Advanced Metal Ion-implantation Method and Photocatalytic Reactions under Visible Light Irradiation. Res. Chem. Intermed. 24: 143-149.

[64] Anpo M, Kishiguchi S, Ichihashi Y, Takeuchi M, Yamashita H, Ikeue K, Morin B, Davidson A, Che M (2001) The Design and Development of Second-Generation Titanium Oxide Photocatalysts Able to Operate Under Visible Light Irradiation by Applying a Metal Ion-Implantaion Method. Res. Chem. Intermed. 27: 459-467.

[65] Anpo M (2004) Preparation, Characterization, and Reactivities of Highly Functional Titanium Oxide-based Photocatalysts Able to Operate under UV–Visible Light Irradiation: Approaches in Realizing High Efficiency in the Use of Visible Light. Bull. Chem. Soc. Jpn. 77: 1427-1442.

[66] Anpo M, Takeuchi M, (1998) The Design and Development of Highly Reactive Titanium Oxide Photocatalysts Operating under Visible Light Irradiation. J. Catal. 216: 505-516.

[67] Yamashita H, Harada M, Misaka J. Takeuchi M, Neppolian B, Anpo M (2003) Photocatalytic Degradation of Organic Compounds Diluted in Water Using Visible light-responsive Metal Ion-implanted TiO_2 Catalysts: Fe Ion-implanted TiO_2. Catal. Today. 84: 191-196.

[68] Asahi R, Morikawa T, Ohwaki T, Aoki K, Taga Y (2001) Visible-light Photocatalysis in Nitrogen-doped Titanium Oxides. Science 293: 269-271.

[69] Miyauchi M, Ikezawa A, Tobimatsu H, Irie H, Hashimoto K (2004) Zeta Potential and Photocatalytic Activity of Nitrogen Doped TiO_2 Thin Films. Phys. Chem. Chem. Phys. 6: 865-870.
[70] Valentin C, Finazzi E, Pacchioni G, Selloni A, Livraghi S, Paganini CM and Giamello E (2007) N-doped TiO_2: Theory and Experiment. Chem. Phys. 339: 44-56.
[71] Sathish M, Viswanathan B, Viswanath RP and Gopinath CS (2005) Synthesis, Characterization, Electronic Structure, and Photocatalytic Activity of Nitrogen-Doped TiO_2 Nanocatalyst. Chem. Mater. 17: 6349-6353.
[72] Chen X, Burda B (2004) Photoelectron Spectroscopic Investigation of Nitrogen-Doped Titania Nanoparticles. J. Phys. Chem. 108: 15446-15449.
[73] Saha NC, Tompkins HG, (1992) Titanium Nitride Oxidation Chemistry - An X-ray Photoelectron Spectroscopy. J. Appl. Phys. 72: 3072-3079.
[74] Valentin DC, Pacchioni G, Selloni A, Livraghi S, Giamello E (2005) Characterization of Paramagnetic Species in N-doped TiO_2 powders by EPR Spectroscopy and DFT Calculations. J. Phys. Chem. B 109: 11414-11419.
[75] Li FB, Li XZ (2002) Photocatalytic Properties of Gold/Gold Ion-modified Titanium Dioxide for Wastewater Treatment. Appl. Catal. A 228:15-27.
[76] Tang J, Zou Z, Yin J, Ye J (2003) Photocatalytic Degradation of Methyleneblue on $CaIn_2O_4$ under Visible Light Irradiation. Chem. Phys. Lett. 382: 175-179.
[77] Kwon CH, Kim JH, Jung IS, Shin H, Yoon KH (2003) Preparation and Characterization of TiO_2–SiO_2 Nano-composite Thin Films. Ceram. Int. 29: 851-856.
[78] Yu JC, Yu J, Zhang L, Ho W (2002) Enhancing Effects of Water Content and Ultrasonic Irradiation on the Photocatalytic Activity of Nano-sized TiO_2 Powders. J. Photochem. Photobiol. A 148: 263-271.
[79] Kominami H, Kato J, Murakami S, Kera Y, Inoue M, Inui T, Ohtani B (1999) Synthesis of Titanium (IV) Oxide of Ultra-high Photocatalytic Activity: Hydrolysis of Titanium Alkoxides with Water Liberated Homogeneously from Solvent Alcohols. J. Mol. Catal. A. Chem. 144: 165-171.
[80] Fackler Jr JP, Kristine FJ, Mazany AM, Moyer TJ, Shepherd RE (1985) The Absence of a Titanyl Oxygen in the Titanium(IV)ethylenediaminetetraacetate(4-) Complex: [Ti(edta)(H_2O)]. Inorg. Chem. 24: 1857-1860.
[81] Nishimoto S, Ohtani B, Kajiwara H, Kagiya T (1985) Correlation of the Crystal Structure of Titanium Dioxide Prepared from Titanium Tetra-2-propoxide with the Photocatalytic Activity for Redox Reactions in Aqueous Propanol and Silver Salt Solutions. J. Chem. Soc. Faraday. Trans. 81: 61-68.
[82] Follis D, Pelizzetti E, Serpone N (1991) Photocatalyzed Destruction of Water Contaminants. Environ. Sci. Technol. 25: 1523-1529.
[83] Fox MA, Dulay MT (1993) Heterogeneous Photocatalysis. Chem. Rev. 83: 341-356.
[84] Robert DS, Pask AJ (1965) Kinetics of the Anatase-Rutile Transformation. J. Am. Ceram. Soc. 48: 391–398.
[85] Dorian AH, Hanaor and Charles CS (2011) Review of the Anatase to Rutile Phase Transformation. J. Mater. Sci. 46: 855-874.

[86] Jung KY, Park SB (1999) Anatase-phase Titania: Preparation by Embedding Silica and Photocatalytic Activity for the Decomposition of Trichloroethylene. J. Photochem. Photobiol. A 127: 117-122.

[87] Lotgering FK (1959) Topotactical Reactions with Ferromagnetic Oxides Having Hexagonal Crystal Structures-I. J. Inorg. Nucl. Chem. 9: 113-123.

[88] Nikilić LM, Radonjić L, Srdić VV (2005) Effect of Substrate Type on Nanostructured Titania Sol-gel Coatings for Sensors Applications. Ceram. Inter. 31: 261-266.

[89] Yamashita H, Ichihashi Y, Zhang SG, Matumura Y, Souma Y, Tatsumi T, Anpo M (1997) Photocatalytic Decomposition of NO at 275 K on Titanium Oxide Catalysts Anchored within Zeolite Cavities and Framework. Appl. Surf. Sci. 121/122: 305-309.

[90] Nakamura R, Okamura T, Ohashi N, Imanishi A, Nakato Y (2005) Molecular Mechanisms of Photoinduced Oxygen Evolution, PL Emission, and Surface Roughening at Atomically Smooth (110) and (100) n-TiO_2 (Rutile) Surfaces in Aqueous Acidic Solutions. J. Am. Chem. Soc. 127: 12975-12983.

[91] Li H, Zhao G, Song B, Han G (2008) Effect of Incorporation of Silver on the Electrical Properties of Sol-gel-derived Titania Film. J. Cluster Sci. 19: 667-673.

[92] Ambrozic M, Dakskobler A, Valant M, Kosmac T (2005) Percolation Threshold Model and Its Application to the Electrical Conductivity of Layered $BaTiO_3$–Ni. Mater. Sci-Pol. 23: 535-539.

[93] Lo CT, Chou K, Chin W (2001) Effects of Mixing Procedures on the Volume Fraction of Silver Particles in Conductive Adhesives. J. Adhes. Sci. Technol. 15: 783-792.

[94] Sancaktar E, Bai L (2011) Electrically Conductive Epoxy Adhesives. Polymers 3: 427-466.

[95] Ren P, Fan H, Wang X, Shi J (2011) Effects of Silver Addition on Microstructure and Electrical Properties of Barium Titanate Ceramics. J. Alloy. Compd. 509: 6423-6426.

[96] Armelao L, Barreca D, Bertapelle M, Bottaro G, Sada C, Tondello E (2003) A sol-gel approach to nanophasic copper oxide thin films. Thin Solid Films 442: 48-52.

Chapter 15

Cooling – As a "Heat Treatment" for the Mechanical Behavior of the Bulk Metallic Glass Alloys

E-Wen Huang, Yu-Chieh Lo and Junwei Qiao

Additional information is available at the end of the chapter

1. Introduction

Metals play a significant role in human life since the Bronze Age. Metals' important advantages include higher toughness and predictable fracture behavior in all directions, which are fundamentally essential for engineering applications. In coarse-grain polycrystalline alloys, the plastic deformation is mediated by dislocations within the grains. Micromechanisms of dislocation-based plasticity have been well investigated. Taylor, Polanyi, and Orowan's speculative models and Hirsch and Whelan's experimental results clearly demonstrate that the existence of dislocations in the metals, like a double-edged sword, enhances the ductility, while reducing the theoretical strengths of most of the metallic crystalline systems [1]. However, the toughness depends on the integration of both strength and ductility. Hence, designing of advanced metallic materials to answer the challenging strength-ductility dilemma become an urgent call. There is natural limitation on the conventional polycrystalline metallic alloys. In practical uses, there are always some inherent defects in the crystalline phases, which degrade the alloys properties. Recently, the limitation of the crystalline-material strength was passed when the metallic alloys with amorphous structures were successfully synthesized in many material systems through advanced manufacturing methods[2]. Although most of the metallic elements exiting in the nature are present with crystalline structures which are the most stable structures with the lowest energy state, sometimes they can be made by various ways into metastable amorphous solid forms, such as rapid quenching techniques [3-5], mechanical alloying [6-8], accumulative roll bonding [9-12], and vapor condensation [13]. The characteristics of the mechanical, thermodynamic properties of such category of metallic materials are very similar to ceramic glasses, and thus they are also called as metallic glasses. Moreover, by introducing specific crystalline phases, such as crystalline dendrites, in an amorphous matrix, bulk metallic glass-composite materials demonstrate the improved plasticity and toughness, compared with monolithic amorphous materials [14]. These metallic systems have the capacity of revolutionizing current metal-forming technologies and manufacturing industries.

2. The development of metallic glasses

In 1960, Klement *et al.* [15] developed the first metallic glasses of $Au_{75}Si_{25}$ by the rapid quenching techniques for cooling the metallic liquids at very fast rates of 10^5-10^6 K/s. Their work quickly initiated broad interest among the scientists and engineers because they showed a new "heat treatment" without nucleation and growth of crystalline phase when it cooled fast enough to frozen the liquefied configuration. Later on, the ternary amorphous alloys of Pd-Si-X (X = Ag, Cu or Au) were discovered successfully by Chen and Turnbull [16], and the Pd-T-Si (T = Ni, Co or Fe) ternary amorphous alloys which included the magnetic atoms were also developed in the same period [17]. The maximum size of these metallic glasses could be as large as 1 mm in diameter by using the die casting and roller-quenching method. The effects of the alloy systems, compositions and the existence of a glass transition was demonstrated, it leaded to the first systematic studies in the formation, structure and property investigations of amorphous alloys. Because of their fundamental science interests and engineering application potential, the metallic glasses have attracted great attention since then. The geometry of metallic glasses, however, is limited to thin foils or lines since the formation of glass states needs super-fast cooling rates, which are not easy for industrial mass produciton. How to determine the glass forming ability (GFA) of amorphous alloys and increase the diameter of specimens becomes the important topic in that time. Turnbull and Fisher [18] advanced a criterion to predict the glass forming ability of an alloy. According to their criterion, the reduced glass transition temperature T_{rg}, equal to the glass transition temperature T_g over liquids temperature T_l, or $T_{rg} = T_g/T_l$ is the primary factor. If T_g is larger and T_l smaller, the value of T_{rg} will be higher so that such a liquid can be easily undercooled into a glassy state at a lower cooling rate. Although there are several new criteria proposed following them [19-20], the T_{rg} has been proved to be useful to reflect the GFA of metallic glasses including BMGs.

In 1974, the rods of Pd-Cu-Si alloy about 1~3 mm in diameter, the first bulk metallic glasses were prepared by Chen [17] using simple suction-casting methods. In 1982, Turnbull's group [21-22] pushed the diameter of critical casting thickness of the Pd-Ni-P alloys up to 10 mm by processing the Pd–Ni–P melt in a boron oxide flux and eliminated the heterogeneous nucleation. A series of solid state amophization techniques that are completely different from the mechanism of rapid quenching had been developed during that time. For example, mechanical alloying, strain-induced amorphization in multilayers, ion beam mixing, hydrogen absorption, and inverse melting [23]. The thin films or powders of metallic glasses can be acquired as well as by interdiffusion and interfacial reaction of the temperature just below the glass transition temperature.

In the late 1980s, Inoue's group [24-25] in the Tohoku University, Japan, developed new groups of multicomponent metallic glass systems with lower cooling rates in Mg-, Ln-, Zr-,

Fe-, Pd-, Cu-, Ti- and Ni- based systems. The Inoue group found exceptional glass forming ability in La-Al-Ni and La-Al-Cu ternary alloys system [24]. By casting the alloy melt in water-cooling Cu molds, the cylindrical samples with diameters up to 5 mm or sheets with similar thicknesses were made fully glassy in the $La_{55}Al_{25}Ni_{20}$ alloy. Similarly, the $La_{55}Al_{25}Ni_{10}Cu_{10}$ alloy, fabricated by the same method, was even big with a diameter up to 9 mm.

In the 1990s, the Inoue group further developed a series of multicomponent Zr-based bulk metallic glasses, such as Zr-Cu-Ni, Zr-Cu-Ni-Al, etc. , along with Mg-based, e. g. Mg–Cu–Y and Mg–Ni–Y alloys, all exhibiting a high Glass Forming Ability (GFA) and thermal stability [26-29]. For one of the Zr-based BMGs, $Zr_{65}Al_{7.5}Ni_{10}Cu_{17.5}$, the critical casting thickness was up to 15 mm, and the largest critical casting thickness was 72 mm in the Pd–Cu–Ni–P family [30]. With Inoue's advancement of the aforementioned bulk metallic glass alloys, the BMGs were no longer laboratory curiosity. The possibility of promising engineering applications became reality. One of the examples was that the Zr-based bulk metallic glasses were applied in the industries just three years after it was invented [31]. Subsequently, a set of the very famous empirical rules in order to direct the selection of alloying elements and composition of glass forming alloys have been summarized by Inoue and Johnson as follows [32-33]: (1) Multicomponent alloys with three or more elements; (2) More than 12% atomic radius difference among them; (3) Negative heat of mixing between constituent elements; (4) The deep eutectic rule based on the T_{rg} criterion. These rules concluded critical criteria for the design of the BMGs until 1999. However, the exception was found in the binary systems, such as the Ni-Nb [34], the Ca-Al [35] the Zr-Ni [36], and the Cu-Zr [37-38] alloys. The above systems can also produce BMGs with the size up to several millimeters without the limitations of the eutectics. In summary, the formation mechanism and criteria for the binary BMGs might not follow the traditional multi-component systems. These results suggest that there are many other potential forming systems of the metallic glasses to be discovered.

3. How to describe the mechanical behavior of the bulk metallic glasses

Over the past four decades, considerable research efforts have been made on the BMGs due to their potential opportunity based on the high yield strength, relatively high fracture toughness, low internal friction, high fatigue resistance, as well as better wear and corrosion resistance [31-32, 39]. Although the bulk metallic glasses (BMGs) are one of such species of materials which are considered for future industrial applications, the insufficient plastic deformation at room temperature is still the Achilles' hell for the industrial applications regardless of its highly scientific value. In general, metallic glasses (MGs) are disordered materials which lack the periodicity of long range ordering in the atom packing, but the atomic arrangement in amorphous alloys is not completely random as liquid. In fact, many scholars believe that amorphous structures are composed of short range ordering, such as icosahedra clusters or other packing forms related to the intermetallic compounds that would form in the corresponding equilibrium phase diagram [40-41]. The short range order is identified as a structure consisting of an atom and its nearest neighbors perhaps two or three atom distance. Recently, the study of medium range order is highlighted as a new ordering range between the short range order and the long range order in the amorphous structure [42]. When an amorphous structure is achieved by quenching, it may be composed of icosahedral local short-range ordering, network-forming clusters medium-range ordering, and other unidentified-random local structures [42-43], that is, a complex association in their topology. How to build the atomic structural model in BMGs and how to fill three dimension spaces with these local structural units are still important issues although only limited research has been done so far [44-47]. Due to the difference in the structural systems between metallic glasses and crystalline alloys, it has an unusual performance on the mechanical properties [48-51]. For example, most metallic glasses exhibit evident brittle fracture under an uniaxial tensile loading, but sometimes give very limited plasticity before failure by means of shear-band propagation. Also, BMGs can perform a large global plasticity through the generation of multiple shear bands during unconfined or confined compression test. Activities of shear band are viewed as the main factor on the plastic deformation of BMGs. The more shear bands on the lateral surface of deformed samples, the larger plasticity is obtained. Liu *et al.* [52] show profuse shear bands on a lateral surface of a Zr-based BMG sample after failure. It can be seen the high-density shear bands are distributed corresponding to large plasticity. Despite a wealth of investigations, many questions about shear bands and their microstructures are still unclear so far. For instance, how does a shear band initiate in the MG and develop in to a mature shear band from its embryo, how do shear bands interact with each other, and how would the shear band develop in a composite surroundings such as interactions with embedding crystals? These issues not only depend on the effect of temperature but are also related to the strain rate and other else. The width of a typical shear band is around 10^1 nm[53], and the size scale used to define the structures of metallic glasses is within 1-3 nm.

The studies in the microstructures of metallic glasses in this shortest length scale by laboratory X-Ray, scanning electron microscope (SEM), and Transmission electron microscopy (TEM) are usually excellent and interesting, but they are relatively less in the theoretic models either in the statistic or continuum simulation computing than those in experiments. Through the studies of simulation methods to compare with experimental data, such as electron scattering data and pair distribution functions, it is expected to be able to investigate the shear-band mechanisms and the microstructures of BMGs in depth. Also, a completely theoretic model could be built up due to the combination of computing model and experiments, not only to explain the current experimental phenomena but also to predict the probative behavior in the future.

Molecular dynamics (MD) simulation is one of important simulation methods provided significant insight into material properties under the atomic level. The major advantage of MD simulations is to see a detailed picture of the model under available investigation, and so they have been very instrumental in explaining the connection of macroscopic properties to atomic scale [54]. For instances, MD simulation has been carried out successfully in the studies of various metallic systems such as point defect movement [55], dislocation mechanisms [56-57], and grain-boundary structures in polycrystalline materials [58-61] in recent years. However, a number of limitations in the simulations will also be confronted, while simulations are treated as key insights in the study. Generally, there are three limitations in the current MD simulation, namely the availability of MD potential, time-scale limitations, and the limit on the system size. Two of the later can be alleviated in the promotion of computer efficiency and by adding the parallelization techniques in program, but the former is still challenged on the accuracy of material specificity and on the development for the multicomponent system, especially for the BMGs. Hence, some pioneering attempts [42, 62-64] have shown the possibility of the MD to resolve the mysterious of the BMGs.

4. Microstructures in metallic glasses

In 1959, a structural model of dense random packing of hard spheres is first suggested by Bernal [65] to be a simple model for metal liquids, and subsequently indicated by Cohen and Turnbull [66] that this simple model can be also applied to describe the metallic glasses. In 1979, Wang [67] supposed that the amorphous metal alloys may be a special class of the glassy state whose short-range structure is random Kasper polyhedral close packing of soft atoms similar to those in the crystalline counterparts. This short-range structure is described based on a new type of glassy structure with a high degree of dense randomly packed atomic configurations. The density measurements show that the density difference is in the range 0. 3~1. 0% between bulk metallic glasses and fully crystallized state [68-69]. There is neither splitting of the second peak nor pre-peak at the lower wave vector as seen in the reduced density function curve of the BMGs [68, 70-71]. These results confirm that the multicomponent BMGs has a homogeneously mixed atomic configuration corresponding to a high dense randomly packing. One of the most important topological short range structures developed among glasses and supercooled liquids are the local icosahedral clusters, which are revealed by many simulation studies [72-76]. An icosahedron is the central atom which forms a fivefold symmetry arrangements with each of its 12 neighbors. In contrast is a regular fragment of an FCC order, and the same pair will become an HCP order if the bottom close-packed plane is shifted as the same as the top plane. This fivefold symmetry and icosahedral clusters is also detected from the experiments of liquids and metallic glasses [77-78], even though in an immiscible binary system with positive heat of mixing [74-76]. The binding energy of an icosahedral cluster of 13 Lennard-Jones (LJ) atoms is 8. 4% lower than an FCC or HCP arrangement [79]. The critical size for a transition from icosahedral cluster to icosahedral phase is about 8 nm [25]. Icosahedral packing is a basic structural unit in extended amorphous systems, and the existence of icosahedral clusters offer seeds for the precipitation of the icosahedral phase. The icosahedral quasicrystalline phase will precipitate in the primary crystallization process and then transforms to stable crystalline phases when the amorphous alloys is annealing at higher temperatures [80-82].

In recent years, an order effect, called the medium range order, existing over length scale larger than the short range order but not extends to the long range order as crystalline state, has been detected in the some amorphous alloys [83-85]. Although the icosahedral type model gives a sound description on the structure in the short range order of metallic glasses but fails beyond the nearest-neighbor shell. For instance, how can the medium range order be defined with the local structural unit, and how would the local structural units be connected to full three-dimensional space? Miracle [86] suggested a compelling structural model for metallic glasses based on the dense packing of atomic clusters. An FCC packing of overlapping clusters is taken as the building scheme for medium range order in metallic glasses. Figure 1 illustrate his promoted model of medium range order. A reality check for these previous structural concepts was proposed by Sheng *et al.* [42] with experiments and simulations. They indicate the icosahedral ordering of single-solute-centered quasi-equivalent clusters is an efficient packing scheme, but is not the only type of medium range order. For each one of the metallic glasses, the several types of local coordination polyhedra units are geometrically different, and not identical in their topology and coordination number. They can be considered quasi-equivalent, or cluster-like units for a given glass, supporting the framework of cluster packing. The cluster connection diagrams for the several metallic glasses maybe represents the important question of how the clusters are connected and packed to fill the three-dimensional space, giving rise to the medium range order. It is short range for the packing of clusters but already medium range from the standpoint of atomic correlation beyond one cluster.

A new insight, imperfect ordered packing, which is closely related to the cooling rate, is exposed on the medium range order embedded in the disordered atomic matrix by selected simulation of high-resolution electron microscopy image [87]. It points out that the packing character of medium range ordering structures can be of two types, i. e. icosahedron-like and lattice-like, and indicates that the solidification from melts or crystallization of metallic glasses is controlled by preferential growth of the most stable imperfect ordered packing. On the other hand, Fan *et al.* [88-89] proposed a structural model for bulk amorphous alloys based on the pair distribution functions measured using neutron scattering and reverse Monte Carlo simulations. There are many clusters of imperfect icosahedral and cubic forms. These clusters are randomly distributed, strongly connected, and result in the space between the clusters. The space between the clusters forms free volume, which provides a degree of freedom for the rotation of the clusters under applied load. This cooperative rotation of clusters forms a layer motion (i. e. shear bands), and plastically deforms the amorphous alloys. This model implies that the mechanical properties, e. g. strength and ductility, are dominated by the combination of the bonding characteristics inside and between these clusters. Since the microstructure of the bulk metallic glass decides the properties, it is important to characterize why and how the microstructure changes subjected to fabrication and deformation. The transmission-electron-microscopy (TEM) technique has offered direct observations of local structural characteristics for materials optimization [14]. However, due to the finite sampling space, special care is required to extract the ensemble-average information from the taken image. On the other hand, the neutrons and the high-energy synchrotron x-ray can penetrate the bulk materials, which can give complimentary statistically-sufficient averaged results [14,90-91]. Recently, with the advance of the synchrotron x-ray and neutron facilities, the neutron and synchrotron x-ray diffractometers are equipped with furnaces and other instruments. Hence, the scattering/diffraction measurements become real-time *in-situ* observations. The application of the small-angle neutron scattering (SANS) is one of the complementary approaches to gauge a greater volume without destructive sampling. Prof. William Johnson has shown how to combine the aforementioned TEM and SANS to investigate the microstructure evolution of the bulk amorphous metallic glasses [92]. Dr. Xun-Li Wang demonstrates the *in-situ* synchrotron x-ray study of phase transformation behaviors in the bulk metallic glass by simultaneous synchrotron x-ray diffraction and small angle scattering [93]. The special features of the neutron and synchrotron diffractometers bridges the traditional bulk stress-strain measurements with the microscopic-level understanding [94].

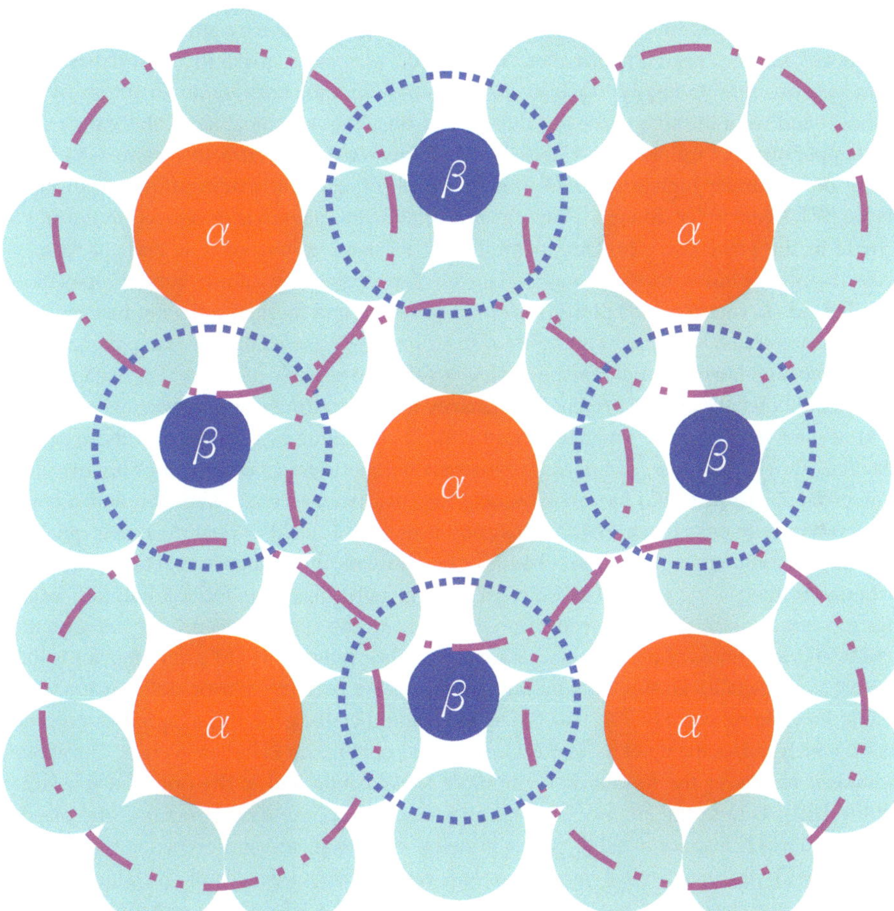

Figure 1. Medium range order model: A two-dimensional representation of a dense cluster-packing structure illustrating the interpenetrating clusters of their efficient atomic packing around each solute α and β.

5. Deformation mechanisms

Unlike the dislocation mechanisms for plastic deformation in crystalline alloys, the amorphous alloys only allow limited atomic displacements to resist deformation as a result of the glassy structure with a high degree of dense randomly packed atomic configurations, when the applied stress is on the amorphous alloys [25,95]. The BMGs have higher tensile fracture strength □ σ_f of 0. 8-2. 5 GPa, Vickers hardness H_v of 200-600, and lower Young's modulus E of 47~102 GPa, than ordinary metallic crystals [25]. It is considered that the significant difference in the mechanical properties is due to the discrepancy in the deformation and fracture mechanisms between bulk metallic glasses and crystalline alloys. It has been widely accepted that shear-band propagation is the major cause affecting the plasticity and toughness of the amorphous alloys. Plastic deformation in metallic glasses is generally associated with inhomogeneous flow in highly localized shear layers with a thickness of about 10 nm. When the shear band went through in the metallic glasses, it is often accompanied with locally rising high temperature to influence the shear flow. From tensile experiments, there was local melting occurs under high strain rate situations such as dynamic loading through unstable fracture [96]. Even under slower loading rates, a veined vein pattern is developed indicating a decrease in the glass viscosity. Due to the highly localized nature of flow and the lack of microstructural features in the metallic glass to distract the flow, shear band propagation typically leads to catastrophic failure. The strain softening and thermal softening mechanisms are closely associated with the localization of shear band [97]. Generally speaking, the metallic glasses have high fracture toughness but brittle as well as negligible plasticity. For instance, the Zr-based bulk metallic glasses present high Charpy impact fracture energies ranging from 110-140 kJ/m^2 and high fracture toughness limit [98]. Their fatigue limit is close to those of the crystalline alloys. However, standard stress–strain fatigue tests show that the Vitreloy alloy (commercial Zr-based BMG) has an extremely low resistance to crack initiation and a crack propagates rapidly once it is formed. If this alloy does start to yield or fracture, it fails quickly. Geometrical confinement of shear bands can dramatically enhance overall plasticity. Furthermore, the plastic yield point of most bulk metallic glasses is located within a small range around σ_y = 2% at room temperature [99]. Composite approach is used to enhance the ductility and toughness of metallic glasses in recent fabricating efforts[100]. Its behavior is like a typical bulk metallic glasses initially but performs as a perfectly plastic deformation after passing the yielding point of 2%. This Pt-Cu-Ni-P bulk metallic glass has a high Poisson ratio of 0. 42, which causes the tip of a shear band to extend rather than to initiate a crack. The above mechanism results in the formation of multiple shear bands and is the origin of the large ductility [101]. Due to the absence of dislocation and grain boundary structures, the plastic deformation mechanism of metallic glasses is well known as shear-band evolution that deeply associates with the mechanical properties and failure behavior in bulk metallic glasses. Nevertheless, the shear band is not the basic defect unit in the deformation mechanisms in the metallic glasses under microscopic scale. In the 1960s, Cohen and Turnbull [102] as well as Spaepen [103] suggested a concept of the free volume which is considered as vacancy-like defect in the metallic glasses, and Argon [104] proposed a theoretical model of plastic flow in metallic glasses, termed shear transformation zone (STZ), which is the fundamental shear unit consisting of a free volume site and its close adjacent atoms in amorphous metals.

In the conceptual framework of free volume, the mechanical coupling is weak to the surrounding of free volumes, and hence the inelastic relaxation becomes possible by local atom rearrangement, without affecting the surroundings significantly [105]. Thus, free volume regions could be the preferred sites where easy caused the glass structure destabilization by either temperature or applied stresses. The concept of free volume is most frequently used in explaining the deformation behaviors and atomic relaxations in the MGs due to its convenient for a measurement (density or enthalpy change), and easy understanding, that is, a necessary open space allowed for a shear process to operate. For instance, a simple relationship $v_f / v_m = \beta . \square \Delta H$ assumes that enthalpy ΔH is proportional to the variation of the average free volume per atom v_f / v_m [106]. Thus, based on the enthalpy recovery measurements, the reduction of free volume difference via structural relaxation, v_f / v_m, was determined. Also, a free volume exhaustion mechanism was proposed by Yang *et al.* [107] to explain the interesting fact that propagation of shear bands in metallic glasses can be retarded, with decreasing temperature and shear strains, in the lack of work hardening mechanisms. It is generally thought that the shear bands could form as a result of the movement and accumulation of free volumes (dilatation expansion). Atomic simulations also show that the local free volumes increase in the BMG provides an open space for the movements of atoms and is associated with the localization of shear band, and the shear softening results from the production of excessive free volume in the shear band [108-110]. Despite the successful description on the strain softening, heterogeneous deformation of MGs, and various mechanical properties of experimental observations, the validity of the free volume theory is questionable, and its atomic basis is still being challenged by atomic simulations. One can easy find the ambiguous characteristics that the free-volume sites may initiate plastic deformation and also can be the result of plastic deformation simultaneously, but not the deformation process itself [111]. Besides, the free-volume model has not made clear motion and rearrangement of constitute atoms within shear bands during plastic flow. To clearly identify the "free volumes" is almost impossible either in experiments or simulations that results in the barrier on the building complete physic model so far. Generally, the concept of free volume is successful as a phenomenology but not as a microscopic theory [105].

6. The model of shear transformation zones

According to Argon's shear transformation zone (STZ) model [112], shear deformation takes place by spontaneous and cooperative reorganization of a small cluster of randomly close-packed atoms [113]. An STZ can supply a small increment of shear strain under the action of an applied shear stress [114], and thus creates a localized distortion of the surroundings to accomplish the shear-band formation. The size of STZ is predicted among the order of 100 atoms from energetic considerations [115-116] and is consistent with the model of molecular dynamics simulation in the investigation of Cu-Ti system [117]. Liu's [118] schematic view on formation of the first STZs in the year of 2012 suggests that a metallic glass system is assumed to be sheared under a constant strain rate. Considering the intrinsic heterogeneity and preexisting flaws resulted from casting, the first STZ generates easily under the action of a stress smaller than macro yielding stress. Since the first STZ alters the initial strain field, the surrounding material is subjected to a self-generated dynamic and thermal noise. With the aid of the heterogeneous strain field and thermal noise, the secondary STZ emerges subsequently in the neighborhood of the first STZ. As a consequence, a shear band nucleus or an embryo of avalanche forms, which induces the yielding of the macroscopic material. STZs have equal chances to propagate forward or backward along the shear plane. For simplicity, here, we only report the case that the STZ starts at one edge and propagates in one direction. Compared with abstract free-volume concept, the STZ mechanism is easy to be studied in the atomic models. A number of MD simulation studies by Falk and Langer summarized the crucial features of STZ mechanisms as follows [111, 119]: (a) once a STZ has transformed and relieved a certain amount of shear stress, it cannot transform again in the same direction. Thus, the system saturates and becomes jammed. (b) STZs can be created and destroyed at rates proportional to the rate of irreversible plastic deformation, and plastic flow can take place only when new zones are being created as fast as existing zones are being transformed. (c) The attempt frequency of the transition is tied to the noise in the system, which is driven by the strain rate. The stochastic nature of these fluctuations is assumed to arise from random motions associated with the disorder in the system. (d) The transition rates between jammed and flowing are strongly sensitive to the applied stress. Recently, they also proposed a criterion (effective temperature) that determines which materials exhibit shear bands based on the initial conditions alone, based on their STZ theory [120-122]. The behavior of the effective temperature as a function of time for a system that localizes and a system that does not localize [120]. Their numerical works show that perturbations to the effective temperature grow due to instability in the transient dynamics, but unstable systems do not always develop shear bands. Nonlinear energy dissipation processes interact with perturbation growth to determine whether a material exhibits strain localization [120]. Argon considered the shear deformation of shear-transformation zones (STZs) as a disk of two-layer atoms around a free-volume site, which had a shear strain of $\gamma_0 \approx 1$ [112]. With a shear stress, τ, the plastic work can be expressed as $\tau\gamma_0\Omega$, where Ω is the volume of the STZ. The deformation of a unit STZ is visualized as an adiabatic process, and the instantaneous temperature increase in the STZ is calculated as [123]: $\Delta T = \frac{\alpha\tau\gamma_0\Omega}{\rho C_p\Omega} = \frac{\alpha\tau\gamma_0}{\rho C_p} \approx \frac{\alpha\tau}{\rho C_p}$, where ρ is the density, C_p is the heat capacity, and $\alpha \approx 0.9$ is the ratio of the plastic work converted to heat. Other simulation studies in the nature of STZ model are keeping on publish [124] and become worthy of investigating this crucial issue in depth.

7. Theory of shear banding and shear band model

According to the experimental observations, the width of a shear band is about 100 nm, which is the same as its offset shear displacement [125-126], and propagation time of shear bands is about 10^{-5} second [127]. Because shear bands are thin, move fast, and are short-lived, to observe the dynamic evolution of the shear bands in the metallic glasses is highly difficult. Building the atomic scale model of shear band such as the development of a shear band inside a binary bulk metallic glass model is very beneficial for studying the shear band mechanism [110]. A simple conceptual quantity excess volume, $v_{excess} = v_{voro} - v_{atom}$, is used to investigate the relation between the free volume changes and shear localization. They suggested that shear banding results from the volume-expansion-induced mechanical softening [109-110, 128-129]. A loop of "local volume increase → local shear softening → large local strain → local volume increase" may be the basic mechanism for deformation and shear banding in MGs. Similar results are observed in the shear-band simulation works of the Mg-Cu systems by Bailey et al. [130], the interactions of the shear bands with the free surfaces as well as with each other result in an initial temperature rise, but the rise of temperature are delayed somewhat with respect to the localization of plastic flow itself. Shimizu *et al.* proposed an aged-rejuvenation-glue-liquid (ARGL) model of shear band in BMGs [131]. That is a more complete theoretical model of shear band than that of others so far. They proposed that the critical condition of initiating a mature shear band (MSB) is not the nucleation of embryonic shear band (ESB), but its propagation. The ESB is easy forming in the MGs. However, to propagate an ESB, the far-field shear stress must exceed the quasi-steady-state glue traction stress of shear-alienated glass until the glass transition temperature is approached internally due to frictional heating, at which point ESB matures as a runway shear crack [131]. In contrast, when applied stress is below the glue traction, the ESB does not propagate, become diffuse, and eventually die. At the same time, an incubation length scale l_{inc} is necessary for this maturation for the BMGs, below which sample size-sale shear localization does not happen. The incubation length $l_{inc} \sim \alpha c_v^2 \left(T_g - T_{env}\right)^2 / \tau_{glue}^2 c_s$, where□ α □ is the thermal diffusivity, c_v is its volumetric specific heat, T_{env} is initial temperature, is the c_s shear wave speed. Through the calculation of this form, the l_{inc} is about 10 nm for Zr-based BMGs [131-132]. Furthermore, it is often questioned whether the shear band mechanisms with regard to metallic glasses is similar to the dislocation mechanisms for crystalline structure, although they are of different definitions. Schuh and Lund [114] found that the plasticity in metallic glasses is consistent with the Mohr-Coulomb criterion by the STZ theory as well as molecular simulation works, and predicted a transition from dislocation-dominated yield processes (following the von Mises criterion) to STZ-dominated yield (following the Mohr–Coulomb criterion) as grain size decreases toward zero for nanocrystalline materials. Ogata et al. [133] simulated the nucleation of local shear transformation zone (STZ) and shear band, under volume-conserving simple shear deformation in molecular dynamics. A significant shear–normal stress coupling which suggests the modified Mohr-Coulomb yield criterion has also been demonstrated. They suggested that the dislocation concept may be applicable to bulk metallic glasses with modifications such as taking into account the structural features of bulk metallic glasses instead of the Burgers vector concept in crystals. The plastic deformation always accompanies the localized heating within shear band that is an important key-point to result in the strain-softening mechanisms and thermal softening on the fracture surface [134]. Understanding the temperature rise in shear bands can also help to improve the ductility and toughness of the metallic glasses. A substantial increase in temperature will correspond with a drop in viscosity governed by the presence of free volume within the metallic glasses [135]. From the calculation of heat conduction theory and STZ modeling, Yang *et al.* [123] demonstrated that the temperature of shear bands at the fracture strength is strikingly similar to their glass transition temperature for a number of BMG systems. This offered a new guideline for the expansion of ultra-high strength bulk metallic glasses from their glass transition temperature, density, and heat capacity values.

8. Conclusions

In summery, the fabrication techniques of the bulk metallic glasses are reviewed chronically. The fundamental concepts of the unique microstructures of the bulk metallic glasses according to different manufacturing are introduced. Moreover, the proposed, but still under debate, various deformation mechanisms are discussed. We wish to draw more attentions from the readers to explore the exciting potential and underneath mechanisms of the bulk metallic glasses.

Author details

E-Wen Huang*
Department of Chemical and Materials Engineering, National Central University, Jhongli, Taiwan (R. O. C.)

Yu-Chieh Lo
Department of Nuclear Science and Engineering, The Massachusetts Institute of Technology, Cambridge, MA, USA

Junwei Qiao
College of Materials Science and Engineering, Taiyuan University of Technology, Taiyuan, China

Acknowledgement

EWH appreciates the support from National Science Council Programs 100-2221-E-008-041 and 99-3113-Y-042-001. JWQ would like to acknowledge the financial support of National Natural Science Foundation of China (No. 51101110).

9. References

[1] Hansen, N., *New discoveries in deformed metals.* Metallurgical and Materials Transactions a-Physical Metallurgy and Materials Science, 2001. 32(12): p. 2917-2935.
[2] Wang, W. H., *Roles of minor additions in formation and properties of bulk metallic glasses.* Progress in Materials Science, 2007. 52(4): p. 540-596.
[3] Gramt, P. S., Prog. Mater. Sci., 1995. 39: p. 497.
[4] Li, B., N. Nordstrom, and E. J. Lavernia, Mater. Sci. Eng. A, 1997. 237: p. 207.

* Corresponding Author

[5] Afonso, C. R. M., et al., *Amorphous phase formation during spray forming of Al84Y3Ni8CO4Zr1 alloy.* J. Non-Cryst. Solids, 2001. 284: p. 134-138.
[6] Koch, C. C., et al., *Preparation of Amorphous Ni60nb40 by Mechanical Alloying.* Applied Physics Letters, 1983. 43(11): p. 1017-1019.
[7] Eckert, J., Mater. Sci. Eng. A, 1997. 226-228: p. 364.
[8] Sherif, M., E. Eskandarany, and A. Inoue, Metall. Mater. Trans. A, 2002. 33: p. 135.
[9] Xing, Z. P., S. B. Kang, and H. W. Kim, *Microstructural evolution and mechanical properties of the AA8011 alloy during the accumulative roll-bonding process.* Metallurgical and Materials Transactions a-Physical Metallurgy and Materials Science, 2002. 33(5): p. 1521-1530.
[10] Yeh, X. L., K. Samwer, and W. L. Johnson, *Formation of an Amorphous Metallic Hydride by Reaction of Hydrogen with Crystalline Intermetallic Compounds - a New Method of Synthesizing Metallic Glasses.* Applied Physics Letters, 1983. 42(3): p. 242-244.
[11] Saito, Y., et al., *Ultra-fine grained bulk aluminum produced by accumulative roll-bonding (ARB) process.* Scripta Materialia, 1998. 39(9): p. 1221-1227.
[12] Schwarz, R. B. and W. L. Johnson, *Formation of an Amorphous Alloy by Solid-State Reaction of the Pure Polycrystalline Metals.* Physical Review Letters, 1983. 51(5): p. 415-418.
[13] Haussler, P., et al., *Experimental-Evidence for a Structure-Induced Minimum of the Density of States at the Fermi Energy in Amorphous-Alloys.* Physical Review Letters, 1983. 51(8): p. 714-717.
[14] Qiao, J. W., et al., *Tensile deformation micromechanisms for bulk metallic glass matrix composites: From work-hardening to softening.* Acta Materialia, 2011. 59(10): p. 4126-4137.
[15] Klement, W., R. H. Willens, and P. Duwez, *Non-crystalline Structure on Solidified Gold-silicon Alloys.* Nature, 1960. 187: p. 869-870.
[16] Chen, H. S. and D. Turnbull, *Formation, stability and structure of palladium-silicon based alloy glasses.* Acta Metallurgica, 1969. 17: p. 1021-1031.
[17] Chen, H. S., *Thermodynamic Considerations on Formation and Stability of Metallic Glasses.* Acta Metallurgica, 1974. 22(12): p. 1505-1511.
[18] Turnbull, D. and J. C. Fisher, *Rate of Nucleation in Condensed Systems.* Journal of Chemical Physics, 1949. 17: p. 71-73.
[19] Lu, Z. P. and C. T. Liu, *A new glass-forming ability criterion for bulk metallic glasses.* Acta Materialia, 2002. 50(13): p. 3501-3512.
[20] Xia, L., et al., *Thermodynamic modeling of glass formation in metallic glasses.* Applied Physics Letters, 2006. 88(17): p. 3.
[21] Drehman, A. J., A. L. Greer, and D. Turnbull, *Bulk Formation of a Metallic-Glass - Pd-40 Ni-40-P-20.* Applied Physics Letters, 1982. 41(8): p. 716-717.
[22] Kui, H. W., A. L. Greer, and D. Turnbull, *Formation of Bulk Metallic-Glass by Fluxing.* Applied Physics Letters, 1984. 45(6): p. 615-616.
[23] Johnson, W. L., *Thermodynamic and Kinetic Aspects of the Crystal to Glass Transformation in Metallic Materials.* Progress in Materials Science, 1986. 30(2): p. 81-134.
[24] Inoue, A., T. Zhang, and T. Masumoto, *Al-La-Ni Amorphous-Alloys with a Wide Supercooled Liquid Region.* Materials Transactions, JIM, 1989. 30(12): p. 965-972.
[25] Inoue, A., *Stabilization of metallic supercooled liquid and bulk amorphous alloys.* Acta Materialia, 2000. 48(1): p. 279-306.
[26] Inoue, A., et al., *Mg-Cu-Y Amorphous-Alloys with High Mechanical Strengths Produced by a Metallic Mold Casting Method.* Materials Transactions, JIM, 1991. 32(7): p. 609-616.

[27] Zhang, T., A. Inoue, and T. Masumoto, *Amorphous Zr-Al-Tm (Tm = Co, Ni, Cu) Alloys with Significant Supercooled Liquid Region of over 100-K.* Materials Transactions, JIM, 1991. 32(11): p. 1005-1010.
[28] Inoue, A., et al., *Preparation of 16 Mm Diameter Rod of Amorphous Zr65al7. 5ni10cu17. 5 Alloy.* Materials Transactions, JIM, 1993. 34(12): p. 1234-1237.
[29] Inoue, A., *High-Strength Bulk Amorphous-Alloys with Low Critical Cooling Rates.* Materials Transactions, JIM, 1995. 36(7): p. 866-875.
[30] Inoue, A., N. Nishiyama, and H. Kimura, *Preparation and thermal stability of bulk amorphous Pd40Cu30Ni10P20 alloy cylinder of 72 mm in diameter.* Materials Transactions, JIM, 1997. 38(2): p. 179-183.
[31] Wang, W. H., C. Dong, and C. H. Shek, *Bulk metallic glasses.* Materials Science & Engineering R-Reports, 2004. 44(2-3): p. 45-89.
[32] Johnson, W. L., *Bulk glass-forming metallic alloys: Science and technology.* Mrs Bulletin, 1999. 24(10): p. 42-56.
[33] Inoue, A., *Bulk amorphous and nanocrystalline alloys with high functional properties.* Materials Science and Engineering A 2001. 304-306: p. 1-10.
[34] Leonhardt, M., W. Loser, and H. G. Lindenkreuz, *Solidification kinetics and phase formation of undercooled eutectic Ni-Nb melts.* Acta Materialia, 1999. 47(10): p. 2961-2968.
[35] Guo, F. Q., S. J. Poon, and G. J. Shiflet, *CaAl-based bulk metallic glasses with high thermal stability.* Applied Physics Letters, 2004. 84(1): p. 37-39.
[36] Meng, W. J., et al., *Solid-State Interdiffusion Reactions in Ni/Ti and Ni/Zr Multilayered Thin-Films.* Applied Physics Letters, 1987. 51(9): p. 661-663.
[37] Xu, D. H., et al., *Bulk metallic glass formation in binary Cu-rich alloy series - Cu100-xZrx (x=34, 36 38. 2, 40 at. %) and mechanical properties of bulk Cu64Zr36 glass.* Acta Materialia, 2004. 52(9): p. 2621-2624.
[38] Hattori, T., et al., *Does bulk metallic glass of elemental Zr and Ti exist?* Physical Review Letters, 2006. 96(25): p. 4.
[39] Greer, A. L., *Metallic Glasses.* Science, 1995. 267(5206): p. 1947-1953.
[40] Doye, J. P. K. and D. J. Wales, *The structure and stability of atomic liquids: From clusters to bulk.* Science, 1996. 271(5248): p. 484-487.
[41] Spaepen, F., *Condensed-matter science - Five-fold symmetry in liquids.* Nature, 2000. 408(6814): p. 781-782.
[42] Sheng, H. W., et al., *Atomic packing and short-to-medium-range order in metallic glasses.* Nature, 2006. 439(7075): p. 419-425.
[43] Tanaka, H., *Two-order-parameter model of the liquid-glass transition. III. Universal patterns of relaxations in glass-forming liquids.* Journal of Non-Crystalline Solids, 2005. 351(43-45): p. 3396-3413.
[44] Hufnagel, T. C., *Amorphous materials: Finding order in disorder.* Nat Mater, 2004. 3(10): p. 666-667.
[45] Fan, C., et al., *Structural model for bulk amorphous alloys.* Applied Physics Letters, 2006. 89(11).
[46] Fan, C., et al., *Atomic migration and bonding characteristics during a glass transition investigated using as-cast Zr-Cu-Al.* Physical Review B, 2011. 83(19): p. 195207.
[47] Fan, C., et al., *Quenched-in quasicrystal medium-range order and pair distribution function study on Zr55Cu35Al10 bulk metallic.* Intermetallics, 2006. 14(8-9): p. 888-892.

[48] Golovin, Y. I., et al., *Serrated plastic flow during nanoindentation of a bulk metallic glass.* Scripta Materialia, 2001. 45(8): p. 947-952.
[49] Murah, P. and U. Ramamurty, *Embrittlement of a bulk metallic glass due to sub-T-g annealing.* Acta Materialia, 2005. 53(5): p. 1467-1478.
[50] Ramamurty, U., et al., *Hardness and plastic deformation in a bulk metallic glass.* Acta Materialia, 2005. 53(3): p. 705-717.
[51] Zhang, G. P., et al., *On rate-dependent serrated flow behavior in amorphous metals during nanoindentation.* Scripta Materialia, 2005. 52(11): p. 1147-1151.
[52] Liu, Y. H., et al., *Super plastic bulk metallic glasses at room temperature.* Science, 2007. 315(5817): p. 1385-1388.
[53] Zhang, Y. and A. L. Greer, *Thickness of shear bands in metallic glasses.* Applied Physics Letters, 2006. 89(7).
[54] Morris, J. R., et al., *Simulating the Effect of Poisson Ratio on Metallic Glass Properties.* International Journal of Modern Physics B, 2009. 23: p. 1229-1234.
[55] Nordlund, K. and R. S. Averback, *Atomic displacement processes in irradiated amorphous and crystalline silicon.* Applied Physics Letters, 1997. 70(23): p. 3101-3103.
[56] Robach, J. S., et al., *Dynamic observations and atomistic simulations of dislocation-defect interactions in rapidly quenched copper and gold.* Acta Materialia, 2006. 54(6): p. 1679-1690.
[57] Van Swygenhoven, H., P. M. Derlet, and A. G. Froseth, *Nucleation and propagation of dislocations in nanocrystalline fcc metals.* Acta Materialia, 2006. 54(7): p. 1975-1983.
[58] Van Swygenhoven, H., D. Farkas, and A. Caro, *Grain-boundary structures in polycrystalline metals at the nanoscale.* Physical Review B, 2000. 62(2): p. 831-838.
[59] Wolf, D., et al., *Deformation of nanocrystalline materials by molecular-dynamics simulation: relationship to experiments?* Acta Materialia, 2005. 53(1): p. 1-40.
[60] Zheng, G. P., Y. M. Wang, and M. Li, *Atomistic simulation studies on deformation mechanism of nanocrystalline cobalt.* Acta Materialia, 2005. 53(14): p. 3893-3901.
[61] Millett, P. C., R. P. Selvam, and A. Saxena, *Molecular dynamics simulations of grain size stabilization in nanocrystalline materials by addition of dopants.* Acta Materialia, 2006. 54(2): p. 297-303.
[62] Lo, Y. C., et al., *Structural relaxation and self-repair behavior in nano-scaled Zr-Cu metallic glass under cyclic loading: Molecular dynamics simulations.* Intermetallics, 2010. 18(5): p. 954-960.
[63] Ikeda, H., et al., *Strain rate induced amorphization in metallic nanowires.* Physical Review Letters, 1999. 82(14): p. 2900-2903.
[64] Schuh, C. A., A. C. Lund, and T. G. Nieh, *New regime of homogeneous flow in the deformation map of metallic glasses: elevated temperature nanoindentation experiments and mechanistic modeling.* Acta Materialia, 2004. 52(20): p. 5879-5891.
[65] Bernal, J. D., Nature, 1959. 183: p. 141.
[66] Cohen, M. H. and D. Turnbull, Nature 1964. 203: p. 964.
[67] Wang, R., *Short-Range Structure for Amorphous Intertransition Metal-Alloys.* Nature, 1979. 278(5706): p. 700-704.
[68] Inoue, A., et al., *High packing density of Zr- and Pd-based bulk amorphous alloys.* Materials Transactions Jim, 1998. 39(2): p. 318-321.
[69] Wang, W. H., et al., *Microstructural transformation in a Zr41Ti14CU12. 5Ni10Be22. 5 bulk metallic glass under high pressure.* Physical Review B, 2000. 62(17): p. 11292-11295.

[70] Wang, W. H., et al., *Microstructure studies of Zr41Ti14Cu12. 5Ni10Be22. 5 bulk amorphous alloy by electron diffraction intensity analysis.* Applied Physics Letters, 1997. 71(8): p. 1053-1055.
[71] Inoue, A., *Bulk Amorphous Alloys.* 1998: Trans Tech Publications.
[72] Honeycutt, J. D. and H. C. Andersen, *Molecular-Dynamics Study of Melting and Freezing of Small Lennard-Jones Clusters.* Journal of Physical Chemistry, 1987. 91(19): p. 4950-4963.
[73] Jonsson, H. and H. C. Andersen, *Icosahedral Ordering in the Lennard-Jones Liquid and Glass.* Physical Review Letters, 1988. 60(22): p. 2295-2298.
[74] He, J. H. and E. Ma, *Nanoscale phase separation and local icosahedral order in amorphous alloys of immiscible elements.* Physical Review B, 2001. 64(14): p. 12.
[75] He, J. H., et al., *Amorphous structures in the immiscible Ag-Ni system.* Physical Review Letters, 2001. 86(13): p. 2826-2829.
[76] Luo, W. K., et al., *Icosahedral short-range order in amorphous alloys.* Physical Review Letters, 2004. 92(14): p. 4.
[77] Li, C. F., et al., *Precipitation of icosahedral quasicrystalline phase in Hf65Al7. 5Ni10Cu12. 5Pd5 metallic glass.* Applied Physics Letters, 2000. 77(4): p. 528-530.
[78] Reichert, H., et al., *Observation of five-fold local symmetry in liquid lead.* Nature, 2000. 408(6814): p. 839-841.
[79] Spaepen, D. R. N. a. F., *Polytetrahedral order in condensed matter.* Solid State Physics, 1989. 42: p. 1.
[80] Saida, J., M. Matsushita, and A. Inoue, *Direct observation of icosahedral cluster in Zr70Pd30 binary glassy alloy.* Applied Physics Letters, 2001. 79(3): p. 412-414.
[81] Wang, W. H., et al., *Phase transformation in a Zr41Ti14Cu12. 5Ni10Be22. 5 bulk amorphous alloy upon crystallization.* Physical Review B, 2002. 66(10): p. 5.
[82] Waniuk, T., J. Schroers, and W. L. Johnson, *Timescales of crystallization and viscous flow of the bulk glass-forming Zr-Ti-Ni-Cu-Be alloys.* Physical Review B, 2003. 67(18): p. 9.
[83] Cowley, J. M., *Electron nanodiffraction methods for measuring medium-range order.* Ultramicroscopy, 2002. 90(2-3): p. 197-206.
[84] Mattern, N., et al., *Short-range order of Zr62-xTixAl10Cu20Ni8 bulk metallic glasses.* Acta Materialia, 2002. 50(2): p. 305-314.
[85] Hufnagel, T. C., *Amorphous materials - Finding order in disorder.* Nature Materials, 2004. 3(10): p. 666-667.
[86] Miracle, D. B., *A structural model for metallic glasses.* Nature Materials, 2004. 3(10): p. 697-702.
[87] Chen, G. L., et al., *Molecular dynamic simulations and atomic structures of amorphous materials.* Applied Physics Letters, 2006. 88(20): p. 3.
[88] Fan, C., et al., *Structures and mechanical behaviors of Zr55Cu35Al10 bulk amorphous alloys at ambient and cryogenic temperatures.* Physical Review B, 2006. 74(1): p. 6.
[89] Fan, C., et al., *Structural model for bulk amorphous alloys.* Applied Physics Letters, 2006. 89(11): p. 3.
[90] Qiao, J. W., et al., *Resolving ensembled microstructural information of bulk-metallic-glass-matrix composites using synchrotron x-ray diffraction.* Applied Physics Letters, 2010. 97(17).
[91] Ott, R. T., et al., *Micromechanics of deformation of metallic-glass-matrix composites from in situ synchrotron strain measurements and finite element modeling.* Acta Materialia, 2005. 53(7): p. 1883-1893.

[92] Loffler, J. F., et al., *Crystallization of bulk amorphous Zr--Ti(Nb)--Cu--Ni--Al.* Applied Physics Letters, 2000. 77(4): p. 525-527.
[93] Wang, X. L., et al., *In situ Synchrotron Study of Phase Transformation Behaviors in Bulk Metallic Glass by Simultaneous Diffraction and Small Angle Scattering.* Physical Review Letters, 2003. 91(26): p. 265501.
[94] Huang, E. W., et al., *Study of nanoprecipitates in a nickel-based superalloy using small-angle neutron scattering and transmission electron microscopy.* Applied Physics Letters, 2008. 93(16).
[95] Takeuchi, A. and A. Inoue, *Size dependence of soft to hard magnetic transition in (Nd, Pr)-Fe-Al bulk amorphous alloys.* Materials Science and Engineering a-Structural Materials Properties Microstructure and Processing 2004. 375: p. 1140-1144.
[96] Flores, K. M. and R. H. Dauskardt, *Mean stress effects on flow localization and failure in a bulk metallic glass.* Acta Materialia, 2001. 49(13): p. 2527-2537.
[97] Flores, K. M. and R. H. Dauskardt, *Local heating associated with crack tip plasticity in Zr-Ti-Ni-Cu-Be bulk amorphous metals.* Journal of Materials Research, 1999. 14(3): p. 638-643.
[98] Inoue, A. and A. Takeuchi, *Recent progress in bulk glassy alloys.* Materials Transactions, 2002. 43(8): p. 1892-1906.
[99] Johnson, W. L., *Bulk amorphous metal - An emerging engineering material.* Jom-Journal of the Minerals Metals & Materials Society, 2002. 54(3): p. 40-43.
[100] Hays, C. C., C. P. Kim, and W. L. Johnson, *Microstructure controlled shear band pattern formation and enhanced plasticity of bulk metallic glasses containing in situ formed ductile phase dendrite dispersions.* Physical Review Letters, 2000. 84(13): p. 2901-2904.
[101] Schroers, J. and W. L. Johnson, *Ductile bulk metallic glass.* Physical Review Letters, 2004. 93(25): p. 4.
[102] Cohen, M. H. and D. Turnbull, *Molecular Transport in Liquids and Glasses.* Journal of Chemical Physics, 1959. 31: p. 1164-1169.
[103] Spaepen, F., *Microscopic Mechanism for Steady-State Inhomogeneous Flow in Metallic Glasses.* Acta Metallurgica, 1977. 25(4): p. 407-415.
[104] Argon, A. S., *Plastic-Deformation in Metallic Glasses.* Acta Metallurgica, 1979. 27(1): p. 47-58.
[105] Michael Miller, P. L., *Bulk Metallic Glasses: An Overview.* Hardcover ed. 2007: Springer
[106] Vandenbeukel, A. and J. Sietsma, *The Glass-Transition as a Free-Volume Related Kinetic Phenomenon.* Acta Metallurgica Et Materialia, 1990. 38(3): p. 383-389.
[107] Yang, B., et al., *Dynamic evolution of nanoscale shear bands in a bulk-metallic glass.* Applied Physics Letters, 2005. 86(14): p. 141904.
[108] Albano, F. and M. L. Falk, *Shear softening and structure in a simulated three-dimensional binary glass.* Journal of Chemical Physics, 2005. 122(15): p. 8.
[109] Li, Q. K. and M. Li, *Atomic scale characterization of shear bands in an amorphous metal.* Applied Physics Letters, 2006. 88(24): p. 3.
[110] Li, Q. K. and M. Li, *Atomistic simulations of correlations between volumetric change and shear softening in amorphous metals.* Physical Review B, 2007. 75(9): p. 5.
[111] Chen, M. W., *Mechanical behavior of metallic glasses: Microscopic understanding of strength and ductility.* Annual Review of Materials Research, 2008. 38: p. 445-469.
[112] Argon, A. S., *Plastic deformation in metallic glasses.* Acta Metallurgica, 1979. 27(1): p. 47-58.
[113] Schuh, C. A., T. C. Hufnagel, and U. Ramamurty, *Overview No. 144 - Mechanical behavior of amorphous alloys.* Acta Materialia, 2007. 55(12): p. 4067-4109.

[114] Schuh, C. A. and A. C. Lund, *Atomistic basis for the plastic yield criterion of metallic glass.* Nature Materials, 2003. 2(7): p. 449-452.
[115] Debenedetti, P. G. and F. H. Stillinger, *Supercooled liquids and the glass transition.* Nature, 2001. 410(6825): p. 259-267.
[116] Johnson, W. L. and K. Samwer, *A universal criterion for plastic yielding of metallic glasses with a (T/T-g)(2/3) temperature dependence.* Physical Review Letters, 2005. 95(19): p. 4.
[117] Zink, M., et al., *Plastic deformation of metallic glasses: Size of shear transformation zones from molecular dynamics simulations.* Physical Review B, 2006. 73(17): p. 3.
[118] Liu, W. D. and K. X. Liu, *Notable internal thermal effect on the yielding of metallic glasses.* Applied Physics Letters, 2012. 100(14): p. 141904.
[119] Falk, M. L. and J. S. Langer, *Dynamics of viscoplastic deformation in amorphous solids.* Physical Review E, 1998. 57(6): p. 7192-7205.
[120] Manning, M. L., J. S. Langer, and J. M. Carlson, *Strain localization in a shear transformation zone model for amorphous solids.* Physical Review E, 2007. 76(5): p. 16.
[121] Shi, Y. F., et al., *Evaluation of the disorder temperature and free-volume formalisms via simulations of shear banding in amorphous solids.* Physical Review Letters, 2007. 98(18): p. 4.
[122] Langer, J. S., *Shear-transformation-zone theory of plastic deformation near the glass transition.* Physical Review E, 2008. 77(2): p. 021502.
[123] Yang, B., et al., *Localized heating and fracture criterion for bulk metallic glasses.* Journal of Materials Research, 2006. 21(4): p. 915-922.
[124] Manning, M. L., et al., *Rate-dependent shear bands in a shear-transformation-zone model of amorphous solids.* Physical Review E, 2009. 79(1): p. 17.
[125] Pekarskaya, E., C. P. Kim, and W. L. Johnson, *In situ transmission electron microscopy studies of shear bands in a bulk metallic glass based composite.* Journal of Materials Research, 2001. 16(9): p. 2513-2518.
[126] Li, J., F. Spaepen, and T. C. Hufnagel, *Nanometre-scale defects in shear bands in a metallic glass.* Philosophical Magazine A, 2002. 82(13): p. 2623-2630.
[127] Hufnagel, T. C., et al., *Deformation and failure of Zr57Ti5CU20Ni8Al10 bulk metallic glass under quasi-static and dynamic compression.* Journal of Materials Research, 2002. 17(6): p. 1441-1445.
[128] Li, Q. K. and M. Li, *Effects of surface imperfections on deformation and failure of amorphous metals.* Applied Physics Letters, 2005. 87(3): p. 031910
[129] Li, Q. K. and M. Li, *Molecular dynamics simulation of intrinsic and extrinsic mechanical properties of amorphous metals.* Intermetallics 2006. 14: p. 1005-1010.
[130] Bailey, N. P., J. Schiotz, and K. W. Jacobsen, *Atomistic simulation study of the shear-band deformation mechanism in Mg-Cu metallic glasses.* Physical Review B, 2006. 73(6): p. 12.
[131] Shimizu, F., S. Ogata, and J. Li, *Yield point of metallic glass.* Acta Materialia, 2006. 54(16): p. 4293-4298.
[132] Shimizu, F., S. Ogata, and J. Li, *Theory of shear banding in metallic glasses and molecular dynamics calculations.* Materials Transactions, 2007. 48(11): p. 2923-2927.
[133] Ogata, S., et al., *Atomistic simulation of shear localization in Cu-Zr bulk metallic glass.* Intermetallics 2006. 14: p. 1033-1037.
[134] Lewandowski, J. J. and A. L. Greer, *Temperature rise at shear bands in metallic glasses.* Nature Materials, 2006. 5(1): p. 15-18.
[135] Spaepen, F., *Metallic glasses: Must shear bands be hot?* Nature Materials, 2006. 5(1): p. 7-8.

Chapter 16

Using "Heat Treatment" Method for Activation of OPC-Slag Mortars

Fathollah Sajedi

Additional information is available at the end of the chapter

1. Introduction

1.1. General

It is well known that a lot of ground granulated blast furnace slag (ggbfs) is produced in the steel-iron industry every year throughout the world. By utilizing this by-product it would help reduce the environmental problems and also provide significant economic benefits. The results of several researches have also shown that the use of replacement materials in mortars and concretes improves durability, which is crucial for structures built in aggressive environments, e.g. in marine structures and structures such as large tunnels and bridges with long life spans. For every ton of Portland cement manufactured, approximately one ton of CO_2, in addition to greenhouse gases, is released into the atmosphere. Therefore, if the part of the Portland cement can be replaced by waste materials, e.g., slag, then the amount of cement needed and hence, the amount of CO_2 released into the atmosphere can be reduced (Lodeiro, Macphee, et al., 2009). Consequently, ggbfs is being widely used as a cement replacement in Portland cement mortar and concrete for improving mechanical and durability properties.

The use of ggbfs has certain advantages because of its excellent cementitious properties over OPC and it is sometimes used due to the technological, economic and environmental benefits. However, the use of slag has been limited because of the disadvantage of its low early strength (Bougara, Lynsdale, et al., 2009). The major factors affecting the early age strength development of mortars and concrete are as follows:

- Mortar mixture proportions including water-binder and sand-binder ratios and the use of supplementary cementing materials such as ggbfs,
- Kind of formwork and size of structural elements, and
- Environmental conditions (Barnett, Soutsos, et al., 2006).

Concretes made with ggbfs have many advantages including improved durability, workability and economic benefits.

Mortar is a workable paste used to bind construction blocks together and fill the gaps between them. The blocks may be stone, brick, cinder blocks, etc. Modern mortars are typically made from a mixture of sand, a binder such as cement or lime, and water. Based on (ACI, 2006) there are four different types of mortars commonly used in building projects namely Type N, Type M, Type S, and Type O. Type N mortar is a medium strength mortar, which means that it is suitable for use both on indoor projects and on outdoor projects that are above grade. Type M mortar is a high strength mortar. Due to the strength of Type M it is usually used in heavy load bearing walls, although it is also sometimes used in other heavy duty applications like masonry that is below grade or that comes in contact with the earth such as retaining walls or foundations. Type S mortar is also a relatively high strength mortar which is suitable for below grade projects and heavier outdoor projects. Type O mortar is the lowest strength mortar and is suitable only for indoor, lightweight applications. The most commonly used mortars for most home improvement projects are Type N and Type S. Type N is chosen for lighter weight or indoor projects, and Type S for projects that require a heavier duty mortar. All of the different types of mortar are made with the same ingredients. The only difference is the proportions of each ingredient in the mix whether for availability considerations or for minimizing the number of different mortar types on the job site. The OPC-slag mortars can be classified into three groups as OPC mortars (OMs), slag mortars (SMs), and OPC-slag mortars (OSMs).

1.2. Research significance

Based on the related literature review there is not much research work regarding activation of OPC-slag mortars, and this is the main purpose of this investigation. In this study, the thermal activation method was used. The compressive strength loss was studied in this research at early and later ages. Strength development of OPC-slag mortars without and with use of activation method was also studied for duration up to 90 days and some regression relationships was suggested. Using the suggested relationships a criterion to forecast the strength behavior of OPC-slag mortars at later ages is established.

1.3. Research objectives

The objectives of this research are as follows:

- To determine the optimum replacement level of slag and the control ordinary Portland cement-slag mortar mix.
- To investigate the effects of thermal activation method on both the early and ultimate compressive strengths and also strength loss of the control ordinary Portland cement-slag mortar.

1.4. Scope of work

The objective of this study is to use higher percentage of replacement slag as possible without any reduction in mechanical properties of mortar such as compressive strength.

Different amounts of slag as replacement for cement were used and the optimum level of replacement was determined. In this investigation only one source of slag was used and the optimum level of replacement was used throughout the study. The optimum level was based on high early strength and lowest strength loss for the mortar mixes.

The focus of this project is to activate OPC-slag mortars using heat treatment (thermal activation method). In this activation method different temperatures within the range of 40 °C to 90 °C were used for activation. The mortars were heated for duration of 2 to 26 hours. The mechanical properties studied in this research are compressive strength at the both early and later ages, strength loss and strength development at later ages of OPC-slag mortars in duration of up to 90 days. The compressive strengths throughout the study were tested for the specimens at 1, 3, 7, 28, 56, and 90 days.

2. Determination of optimum level

2.1. Introduction

Before discussing the activation method, the determination of the optimum level of replacement of slag in OPC-slag mortars is discussed.

2.2. Replacement level

This section reports on the testing of fourteen OPC-slag mortars (OSMs) and two control OPC mortars (OMs) and slag mortars (SMs). The main aim is to determine the level of cement replacement with slag to achieve higher early strength with reasonable flow. The variable is the level of ggbfs in the binder. Graded silica sand was used in all mixes. It was determined that the optimum level of replacement slag is within the range of 40% to 50% of OPC (Ahmed, Ohama, et al., 1999). The optimum level is defined as the replacement level of slag with the highest compressive strength, when used in the mortar while strength loss is the lowest.

2.2.1. Optimum replacement level

It is intended to find the optimum cement at replacement level with slag that gives the highest early strength at 7 days and especially 3 days without the use of any activation method. From Figure 1 it is clear that whenever the level of replacement is more than 40% the early strength at 3 days will be reduced. It can also be seen that although for replacement levels 10%, 20%, and 30% the early strength at 3 days is approximately the same, but, generally with an increase in the slag level from 10% to 40% the early strength increases. However, the early strength at the 40% level is the highest, i.e. 42.4 MPa. Based on the results obtained it can be seen that by increasing the replacement level to more than 40%, the early strengths decreased significantly. This shows that the optimum level of replacement is 40% at 3 days. The same variations for 7-day strengths were observed albeit with slight changes. At 40% optimum level of replacement slag, the strength at 7 days is 55.8

MPa, which is 20% higher than that of OPC mortar. Furthermore, this will continue to gain strength with age.

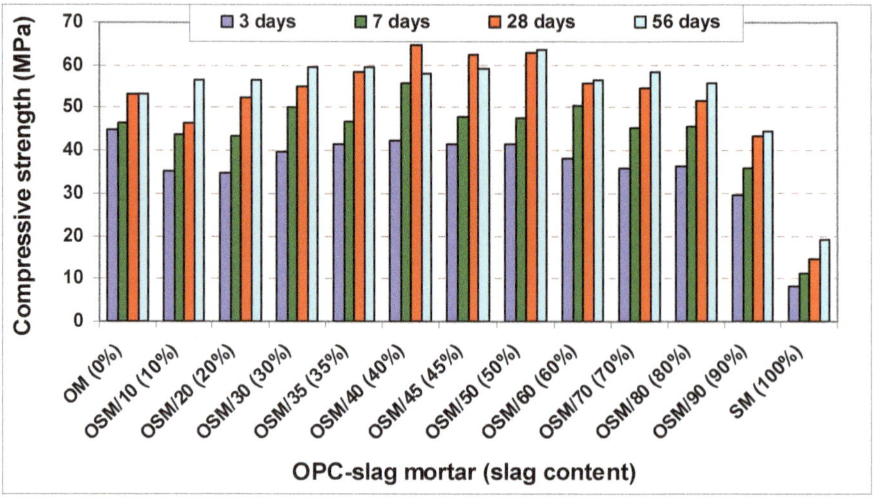

Notes: OSM/i= OPC-slag mortar for i% replacement with slag, OM= OPC mortar, SM= slag mortar.

Figure 1. Variations of compressive strength for different slag contents (%)

The use of different amounts of slag in OPC-slag mortars results in different compressive strengths. Before applying the proposed activation method it is required that several mortars are made with the use of different levels of slag to obtain indication of strengths. Based on this, it was decided to cast samples with 0, 10, 20, 30, 35, 40, 45, 50, 60, 70, 80, 90, and 100% replacement with slag to ascertain the optimum level. The specimens were prepared and hardened samples were tested at 3, 7, 28, and 56 days for compressive strength. Finally, it was revealed that the highest compressive strengths were obtained for samples having replacement level in the range of 40% and 50%. The results showed that there was some strength loss at 56 days compared to 28 days when 40% replacement level was used. In this case, the strength at 28 days was 64.9 MPa, which reduced to 57.9 MPa at 56 days giving a 10.8% loss in strength. In contrast no strength loss was observed at later ages when 50% replacement level was used. A comparison of the results obtained for both percentage levels of slag showed that 50% is the optimum. The strengths for both levels of replacement were very close but, the 50% replacement with slag did not show any subsequent loss in strength. Consequently, in the continuation of the research work this level was taken as the optimum and used for the preparation of mortar samples.

2.3. Summary

In this section it was attempted to determine the best replacement level for slag. Different levels of replacement slag were used to make mixes, i.e. 10, 20, 30, 40, 50, 60, 70, 80, and 90%.

Finally, with comparison of the results obtained for compressive strengths the best level of replacement was determined at 50%. In determination of the best level two factors are considered i.e. early strength at 3 and 7 days and also strength loss that should be minimized. In following parts of the research the best level i.e. 50% was used as the optimum level.

3. Thermal activation method

The objective of this research is to produce a data inventory of the early age mechanical properties, namely the compressive strength of mortars cured at different temperature, as well as the relationship between compressive strength with temperature and the relationship between the compressive strength of specimens cured in air and water at room temperature for 3 and 7 days, for 40% and 50% levels of cement replacement with slag. Thirty-seven mixes of OPC-slag mortars and two OPC mortars were prepared as control. For each mix, two factors are important for consideration. First, using a higher percentage of slag is desirable as it has some economic and environmental advantages and in addition, it helps to improve the durability of the mortars. Secondly, for early strength, it is clear that increasing the level of replacement slag causes early strength to be reduced, as the ggbfs has lower initial heat of hydration than that of OPC. In addition, for early strength the use of a low level of replacement slag is neither economic nor durable. Thus it is desirable to ascertain the optimum temperature and its duration that will give the highest early strength at 3 and 7 days. All the mix proportions made for water-binder and sand-binder ratios of 0.33 and 2.25, respectively, for 40% and 50% replacement level with slag.

3.1. Optimum temperature

In this investigation the effects of different temperatures i.e. 50 ºC, 60 ºC, and 70 ºC were studied on the early strengths at 3 and 7 days of OPC-slag mortars by using 50% replacement with slag. The results are shown in Figure 2. It is clear that 60 ºC provided the most enhancements on early age strength therefore; it is selected as the optimum temperature.

The results obtained in the study for compressive strength based on duration of heat curing are given in Table 1. Based on this, it can be seen that the specimens have higher strengths at 3 and 7 days without use of heat curing and with use of heat curing for duration of 2 hours when they are cured in water compared to curing in air under room temperature.

This has been proven for both OPC-slag mortars with 50% OPC replacement with slag and OPC-slag mortars with 40% OPC replacement with slag. However, as soon as the duration of heat curing is increased to 4 hours and above, the aforesaid statement is reversed. The strength of specimens cured in the air under room temperature is improved compared to those cured in the water. It seems that this is due to the air temperature and high relative humidity of the room's air. As it will be seen in following study, it can be said that both combined effects of temperature and relative humidity are more efficient in strength improvement. Hence, it seems that probably the effect of temperature for duration of at least 4 hours beside the high relative humidity of room's air results in the higher strength for the specimens cured in the air under room temperature after heat curing. This fact is shown in Figure 3 (a) and (b).

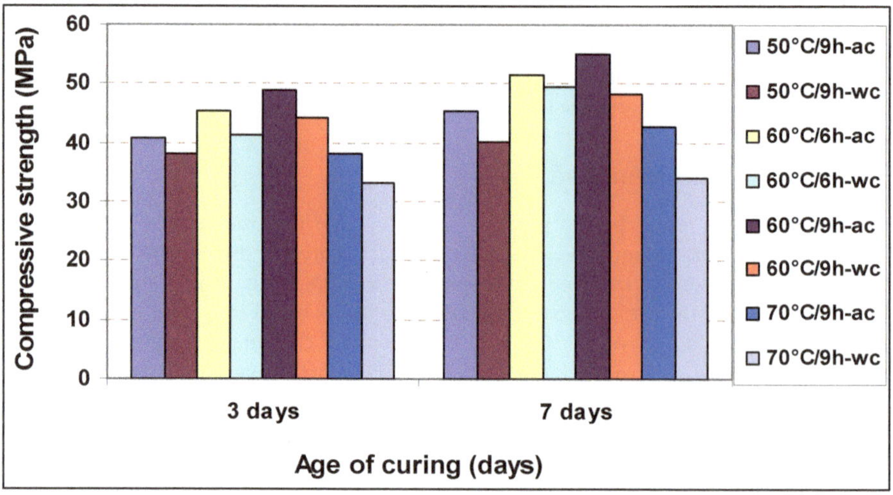

Figure 2. The effects of different temperatures on early age strength of OSMs/50

Duration (hours)	For OSMs/40				For OSMs/50			
	3 days		7 days		3 days		7 days	
	ac	wc	ac	wc	ac	wc	ac	wc
0	33.2	34.5	40.3	47.4	33.6	35.6	37.0	49.6
2	36.6	38.4	44.8	49.8	36.4	37.6	42.5	50.0
4	39.7	35.4	46.2	43.2	42.6	37.7	47.5	47.3
6	45.0	41.2	47.2	44.0	45.5	40.9	51.5	49.8
8	49.6	41.6	52.3	44.8	46.4	43.4	53.1	49.0
10	47.3	40.4	55.6	50.9	50.4	44.0	55.0	48.4
12	49.0	42.5	50.9	46.4	52.6	41.8	57.6	48.3
14	52.6	47.0	56.4	48.5	48.3	41.2	60.0	52.5
16	51.7	45.9	59.0	54.8	51.2	48.4	61.2	53.4
18	55.2	46.1	59.7	50.2	53.5	48.4	59.9	52.4
20	53.1	49.0	61.1	51.2	55.3	49.9	61.6	55.3
22	50.7	43.7	58.8	56.0	50.5	48.4	61.0	56.0
24	54.6	50.1	60.4	56.8	51.5	48.5	62.3	55.6
26	52.5	49.0	57.1	53.7	53.0	46.6	61.2	54.8
For optimum OSM/50 at six ages- air cured under room temperature								
f_1= 15.5	f_3= 55.1		f_7= 61.4		f_{28}= 71.2		f_{56}= 69.6	f_{90}= 73.6
ac= air curing under room temperature; wc= water curing; all strengths are in MPa.								

Table 1. Compressive strength (f) versus duration of heat curing for OSMs/40 and OSMs/50 at 60 ºC

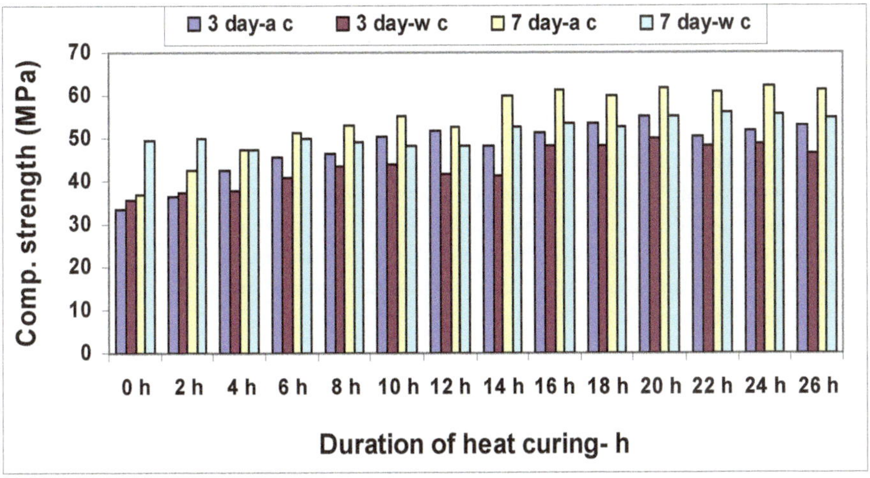

(a) OPC-slag mortar with 50% OPC replacement with slag at 60 °C

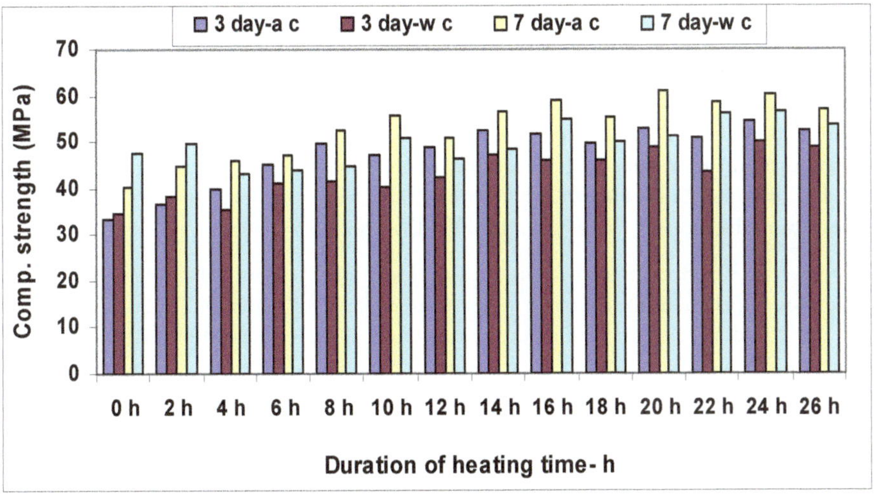

(b) OPC-slag mortar with 40% OPC replacement with slag at 60 °C

Figure 3. Compressive strength versus duration of heat curing of OSMs/50 and OSMs/40 cured in water and air under room temperature.

Based on the results given in Table 1, it can be seen that heat curing at 60 °C for 20 hours, the maximum early strength of 61.6 MPa is attributed to OPC-slag mortar with 50% replacement with slag. It can be seen that the increment percentage is by about 0.62% when compared to a heating time of 16 hours. This shows that if the duration of heat curing is increased over 20 hours, the increase at 7 days strength is not appreciable. Hence, it can be deduced from Figures 3 and 4 that heat curing at 60 °C for 20 hours is the optimum heat curing for the materials used in the study.

3.2. Temperature and humidity effects

From the H- 3 sets mix as shown in Table 2 the effects of ambient temperatures and relative humidity were considered on the strength improvement of the specimens. The specimens were first made and demoulded 24 hours after casting, and then heated at 60 °C for 20 hours and finally, cured in three curing regimes, i.e. at room temperature, in water at 25 ºC - 26 ºC, and in water at 32 ºC. After 3 and 7 days the strength of specimens was determined. The results are given in Table 2.

Based on the given data in Table 2, it is clear that for curing regimes with different temperature and the same relative humidity, higher strengths are attributed to higher temperature regimes. For curing regimes with different relative humidity and the same temperature, higher strengths are attributed to lower relative humidity regimes. From the comparison of the three curing regimes it is seen that the strengths of specimens cured in water at 32 ºC are less than those cured at room temperature. This shows that the increase of early age strength is a function of both temperature and relative humidity effects. In fact, it can be deduced that neither relative humidity nor temperature has the highest strength improvement alone, but rather a combination of both effects are effective on strength development. From the results it is clear that the trend of strength development at 7 days is similar to that at 3 days. It is seen that the strength of specimens at 7 days cured at room temperature and in water at 32 ºC are the same, which shows that the effect of relative humidity over duration of 3 to 7 days is more than that of temperature. It is seen that the highest strengths at 3 and 7 days are attributed to curing at room temperature.

The percentages of strength growth (i) for duration of 3 to 7 days can be obtained as follows: For air cured specimens under room temperature i= 0.97, for water at 25 – 26 ºC i= 0.90, and for water cured at 32 ºC i= 0.89; where i= ratio f_3/f_7. These results show that whenever the specimens are cured at room temperature, about 97% of the strength is achieved at 3 days. This is a major advantage in the precast concrete industry when the specimens are cured at room temperature. For the specimens cured at room temperature, the maximum relative humidity attainable was 85%, while for the specimens cured in water it was 100%.

Based on the results obtained it is evident that the effect of temperature on the strength improvement at 3 and 7 days is higher than that of relative humidity. This is because the strengths of the specimens cured at room temperature and in water at 32 ºC are higher than those in water at 25 ºC at 3 and 7 days by about 14% and 10%, respectively.

It can be seen that with the use of heat curing at 60 ºC for 20 hours the strength at 3 days is, on average, about 97% of the strength at 7 days for specimens cured at room temperature, while, on average, this ratio is about 90% for curing in water at 25 ºC and 32 ºC. This shows that curing at room temperature after heat curing improves the early strength at 3 days extensively, which is very cost effective and applicable in the precast concrete industry. This result also shows that the heat treatment is a useful and efficient method for the activation of ordinary Portland cement-slag mortars and concretes which requires only slat duration and without the use of water to cure the specimens. An elevated curing temperature accelerates the chemical reaction of hydration and increases the early age strength. However, during the initial period of hydration an open and unfilled pore structure of cement paste forms which has a negative effect on the properties of hardened concrete, especially at later ages (Fu, Y., 1996; Neville, A.M. , 2008). Hardened mortars and concretes can reach their maximum strength within several hours through elevated temperature curing. However, the ultimate strength of hardened mortars and concretes has been shown to decrease with curing temperature (Carino, 1984). It was found that by increasing the curing temperature from 20 ºC to 60 ºC and the duration of heat curing to 48 hours causes a continuous increase in compressive strength (Brooks & Al-kaisi, 1990). Studies by (Hanson, 1963; Pfeifer & Marusin, 1991; Shi, 1996) have shown that there is a threshold maximum heat curing temperature value in the range of 60 ºC to 70 ºC, beyond which heat treatment is of little or no benefit to the engineering properties of concrete.

Based on the given data in Table 1 it can be seen that the highest strengths at 3 and 7 days of OPC-slag mortars for 40% replacement with slag and OPC-slag mortars for 50% replacement with slag is attributed to the specimens cured in air under room temperature as:

Curing regime	air curing under room temperature	Water 32 ºC	Water 25 -26 ºC
f_3	58.2	53.3	48.9
f_7	59.9	59.9	54.5
f_3/f_7	0.97	0.89	0.90
f_i are strength in MPa.			

Table 2. Compressive strengths (f) at 3 and 7 days for three curing regimes of H- 3 sets mix

OPC-slag mortars for 40% OPC replacement with slag: f_3= 55.2 at 18 hours and f_7= 61.1 MPa at 20 hours; OPC-slag mortars for 50% replacement with slag: f_3= 55.3 and f_7= 61.6 MPa, the both for 20 hours. The 3 and 7 days strengths of OPC mortars' specimens cured at room temperature and in water are f_3= 45.4, and f_7= 51.4 MPa, and f_3= 43.8, and f_7= 47.8 MPa, respectively. It is noted that the maximum 3 and 7 days strengths of OPC-slag mortars for 40% replacement with slag and OPC-slag mortars for 50% replacement with slag specimens are 21.7% and 19.0% which are 21.8% and 20.0% more than those of OPC mortars' specimens cured at room temperature at the same age, respectively. It is seen that there is strength loss at 56 days compared to 28 days by about 2.2%. This has been previously reported by other researchers (Kosmatka, Panarese, et al., 1991). The main objective of elevated temperature curing is to achieve early strength development. However, it is generally acknowledged that there is also strength loss as a result of heat curing (Bougara, Lynsdale, et al., 2009). Another mix proportion of OPC-slag mortars for 50% replacement with slag was made by using the optimum heat curing at 60 ºC for 20 hours and specimens were tested at ages of 1, 3, 7, 28, 56, and 90 days. To determine the trend of strength development for the mentioned mortar cured at room temperature, the regression technique was used. The equations obtained for this mortar and also OPC mortar cured in water are as below:

$$f_{T-ac} = 5.2128^{*}\ Ln\ (t) + 50.644\ with\ R^2 = 0.9305 \quad (1)$$

For OPC-slag mortar cured in air under room temperature, made using optimum heat curing at 60 ºC for 20 hours, and

$$f_{OM-wc} = 6.1673^{*}\ Ln\ (t) + 35.141\ with\ R^2 = 0.9738 \quad (2)$$

For OPC mortar cured in water; where f is compressive strength in MPa, ac and wc denote air and water curing under room temperature, respectively and t is the age of the specimen in days. The best fit curves are shown in Figure 4.

No	Age (d)	Binomial relationships	Linear relationships	CR
For OPC-slag mortars for 40% replacement with slag, i.e. OSMs/40				
1	3	$f = -0.0455X^2 + 1.865X + 32.921$; $R^2 = 0.9261$	$f = 0.6825\ X + 38.651$ $R^2 = 0.7545$	air
2	3	$f = -0.0131X^2 + 0.8806\ X + 34.825$; $R^2 = 0.8502$	$f = 0.5391\ X + 36.191$ $R^2 = 0.8252$	water
3	7	$f = -0.0347\ X^2\ 1.5959\ X + 40.621$; $R^2 = 0.9107$	$f = 0.6927\ X + 44.234$ $R^2 = 0.8068$	air
4	7	$f = 0.0163\ X^2 - 0.0222\ X + 46.327$; $R^2 = 0.6305$	$f = 0.4011\ X + 44.634$ $R^2 = 0.5815$	water
For OPC-slag mortars for 50% replacement with slag, i.e. OSMs/50				
5	3	$f = -0.0487\ X^2 + 1.9196\ X + 34.298$; $R^2 = 0.9271$	$f = 0.6526\ X + 39.366$ $R^2 = 0.7212$	air
6	3	$f = -0.0184\ X^2 + 0.9857\ X + 35.234$; $R^2 = 0.8492$	$f = 0.5066\ X + 37.154$ $R^2 = 0.7954$	water
7	7	$f = -0.0479\ X^2 + 0.1079\ X + 38.56$; $R^2 = 0.9598$	$f = 0.8628\ X + 43.54$ $R^2 = 0.8291$	air
8	7	$f = 0.0115\ X^2 + 0.0108\ X + 48.789$; $R^2 = 0.7742$	$f = 0.3089\ X + 47.597$ $R^2 = 0.7232$	water
X= heat duration in hours, f is compressive strength in MPa, R^2 is coefficient of determination, CR= curing regime, d= days				

Table 3. Relationship between compressive strength (f) versus heat duration

The relationships between compressive strength and duration of heat curing at room temperature and in water for OPC-slag mortars at 40% OPC replacement with slag and OPC-slag mortars at 50% OPC replacement with slag is shown in Table 3. It can be seen that the best equations are binomial and attributed to the specimens cured at room temperature.

It is also seen that the best fit curve at 3 and 7 days strengths are power equations. According to the results obtained in the study, it can be said that thermal activation is one of the best techniques for the activation of OPC-slag mortars.

No	Age (days)	Power regression relationship	Curing
1	3	$f_{OSM/50} = 1.3991^* x^{0.9147}$; $R^2= 0.8857$	air
2	7	$f_{OSM/50} = 0.4548^* x^{1.2047}$; $R^2= 0.9334$	air
3	3	$f_{OSM/50} = 2.098^* x^{0.8064}$; $R^2= 0.7349$	water
4	7	$f_{OSM/50} = 5.9897^* x^{0.5511}$; $R^2= 0.6897$	water
x= compressive strength of OSM/40 in MPa, R^2= coefficient of determination.			

Table 4. Relationships between compressive strengths (f) of OSMs/50 and OSMs/40

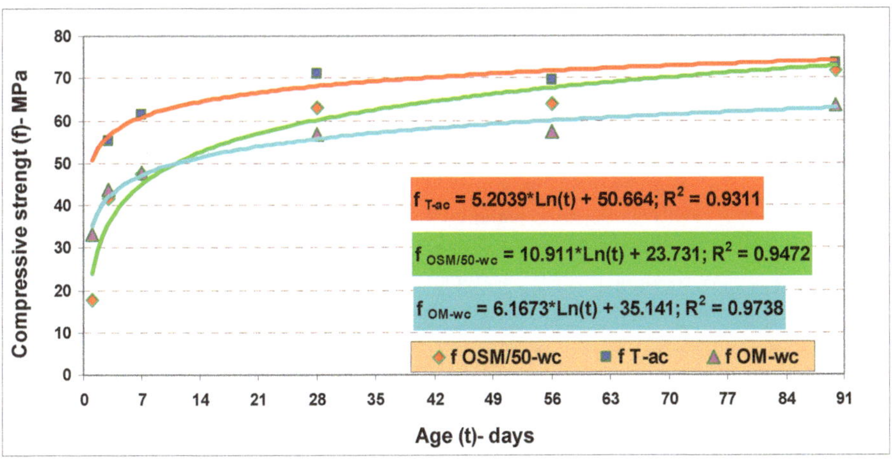

Notes: OM= OPC mortar, T= OPC-slag mortar for 50% OPC replacement with slag made by using the optimum heat curing, f= Compressive strength of the specimens in MPa, ac and wc denote air and water curing under room temperature, respectively.

Figure 4. Strength development for OPC mortar and OPC-slag mortar made by using the optimum heat curing

Based on the results presented in Table 4, it can be seen that there is an acceptable power relationship at 3 and 7 days strengths between OPC-slag mortars for 50% OPC replacement with slag and OPC-slag mortars for 40% OPC replacement with slag for the specimens cured at room temperature.

The given relationships in Table 4 were determined by using the regression technique. Based on the relationships it is seen that the coefficient of determination R^2 of regression for relationships between the strengths at 3 and 7 days of water cured OPC-slag mortars for 50% OPC replacement with slag and OPC-slag mortars for 40% OPC replacement with slag is small. This shows that there is no acceptable relationship between the strengths of water cured specimens. However, there is a proper relationship for those cured at room temperature.

This appears to be due to the behaviour of specimens cured in water, which are still not clear and that is specific for the duration of 7 days. This means that the effect of water on the strength of water cured specimens is different for the durations of 3 and 3 to 7 days.

3.3. Ettringite formation

In this study four extra sets of specimens were tested for SEM images and EDX analyses of H- 4 sets mix given in Table 5 after curing for 7 days. SEM images and EDX analyses were tested on four specimens of the four sets. Each specimen was analyzed to determine if ettringite was present. If ettringite was detected its morphology was noted. It is impossible to quantify ettringite because of the scaling factor and depth of uncertainties of the electron microscope. However, it is significant to note that very small samples measuring approximately 5 mm in diameter yielded large quantities of ettringite. Ettringite was identified visually and from the EDX analysis. The morphologies of the observed ettringite are summarized in Table 5. Samples typically produced ettringite with similar morphologies including lamellar and needles. Ettringite was found in cavities and in the cement matrix.

Set No.	Curing regime	Ettringite formation
Set 1	With heat curing, air cured	Thick and long needles
Set 2	With heat curing, water cured	Needles with lamellar
Set 3	Without heat curing, air cured	Needles with lamellar
Set 4	Without heat curing, water cured	Needles with dense lamellar
Heat treatment was done after demoulding at 60 °C in duration of 20 hours.		

Table 5. Summary of scanning electron microscopy morphology for H- 4 sets mix after 7 days curing

The mechanism of DEF expansion is a highly debated issue. Ettringite Crystal Growth Theory and Uniform Paste Expansion Theory are the two predominant theories. (Shoaib, Balaha, et al., 2000) and (Wang, Pu, et al., 1995) suggested the Ettringite Crystal Growth Theory, which attributes the expansion to pressure exerted by the growing ettringite crystals in the micro cracks between the cement paste and the aggregate.

(Wang, Scrivener, et al., 1994) proposed the Uniform Paste Expansion Theory, which suggests that the concrete expands and then the ettringite forms in the newly created gaps. (Yang, R., Lawrence, et al., 1996) found no evidence to support the Uniform Paste Theory concluding that the ettringite present in the mortar produced the expansion. (Lewis, M. C. & Scrivener, 1996) suggests that both mechanisms are possible and depending on the environmental condition one may be more prevalent. Although other mechanisms have been suggested, expansion most probably results from crystal growth pressure. There are differences of opinion as to whether expansion in mortars or concretes is driven by growth of ettringite crystals at aggregate interfaces or by processes occurring in the paste. If the later expands, gaps will be formed around aggregate particles (Wang, Scrivener, et al., 1994), and ettringite or other phases may recrystallize in them, simultaneously or subsequently. Based on the paste expansion theory the widths of peripheral cracks are proportional to aggregate size; cracks at the interfaces are initially empty. Assuming that expansion occurs through crystal growth pressure, significant growth pressures could not be obtained in relatively large cracks and pastes expand, albeit slowly. Three factors influencing expansion will be considered namely chemistry, paste microstructure, and mortar or concrete microstructure. Proportionality between crack width and aggregate size, which can only be explained by paste expansion, was first reported by (Johansen, Thaulow, et al., 1993). For the H- 4 sets mix 24 hours after casting, the first two sets of the specimens were demoulded and without heat curing but were cured in water and air under room temperature for 7 days, respectively. Another two sets were heated at 60 ºC temperature for 14 hours and then cured the same as the former sets. Ettringite was observed in all of the sets of specimens i.e. with and without heat curing, but with different amounts of crystal size. The results are shown in Figure 5.

The morphology and crystal size of ettringite varies under the different curing conditions the specimens were subjected to. Most of the SEM observation shows that ettringite is normally a slender, needle-like crystal with a prismatic hexagonal cross-section. Its size depends on w/c ratio, that is, the effective space that ettringite is able to occupy (Barnett, Soutsos, et al., 2006). It can grow up to $100\mu m$ long for high w/c ratios. The particle-size of Al-bearing agents is also a main factor affecting the size of ettringite. Large particles of Al-bearing agents form a large amount of small ettringite crystal and the period of expansion can last longer. Small particles will produce large size crystals quickly at any early stage because of the large surface area and fast reaction rate. The ettringite crystals will also be smaller in size in the presence of calcium hydroxide (CH) (Barnett, Soutsos, et al., 2006). The form of ettringite is relevant to studies of the mechanism of expansion. (Lerch, Ashton, et al., 1929) reported that synthetic ettringite consists of long slender needles that often form sphere-like. (Mehta, P.K., 1976) also reported the presence of spheroid ettringite. (Ogawa & Roy, 1982) found that during the hydration, the ettringite formed as very small irregular particles around Al-bearing particles in the early stage, and then changes to long needle-like crystals arranged radially around the Al-bearing particles. This formation was associated with the start of expansion (Barnett, Soutsos, et al., 2006).

Comparison between Figure 5 (a), (b), (c) and (d) show that for each test and in all curing regimes ettringite crystals were detected. For the four sets of tests, the ettringite crystals formed under room temperature curing, i.e. Figure 5 (a), and were more and bigger than those under water curing. Probably, this is the reason for the higher strength improvement of the specimens cured in air under room temperature compared to curing in water. It can be observed that the thickest and the longest ettringite crystals are attributed to the specimens that were cured in the heating process at 60 ºC for 14 hours after casting, and then cured at room temperature for 7 days as shown in Figure 5 part (a). The strength obtained at 7 days for the specimens was about 64 MPa. With the comparison of SEM images and the strengths obtained from the four sets of specimens, it can be deduced that heat curing at 60 ºC for 14 hours increases the rate of ettringite formation and thus the early strength. Whenever there is a greater quantity of ettringite formed it contributes to the higher strength. The details obtained for the compressive strengths at 7 and 28 days, of H- 4 sets mixes are given in Table 6.

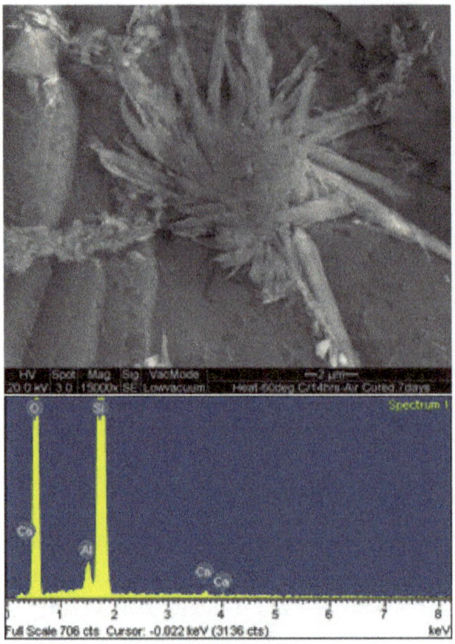

(a) Set No. 1

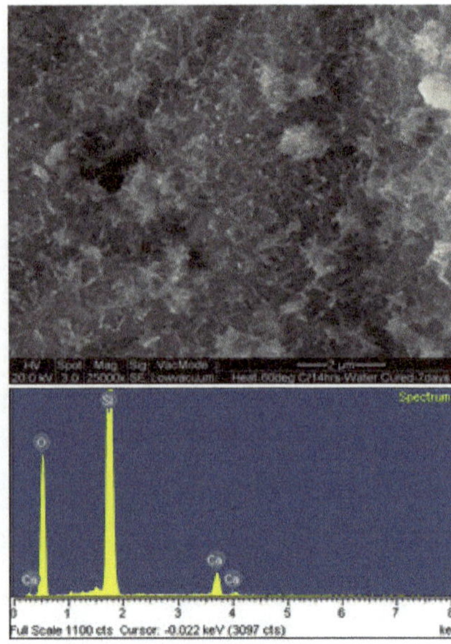

(b) Set No. 2

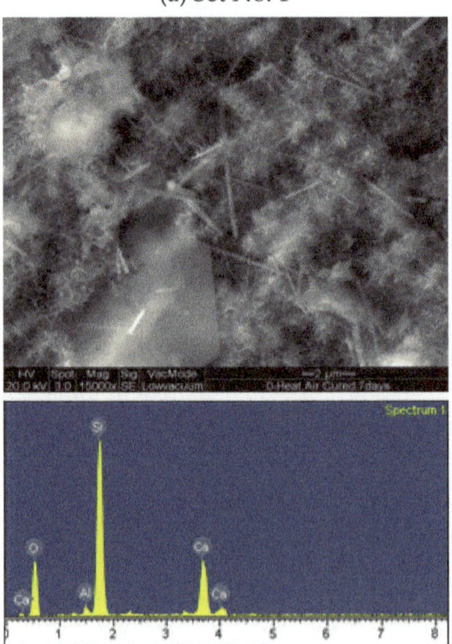

(c) Set No. 3

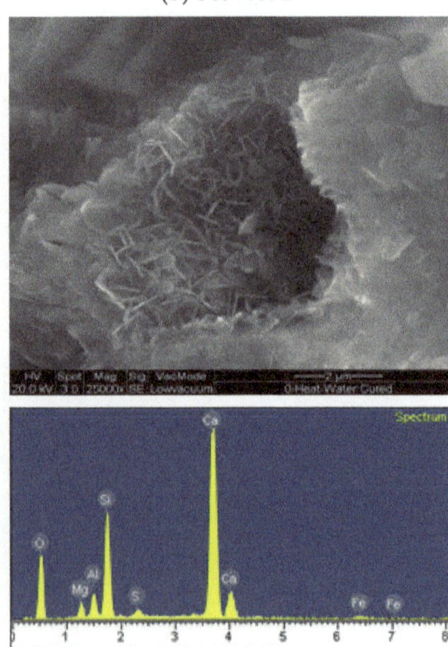

(d) Set No. 4

Figure 5. SEM images and EDX analyses for H-4 set mix specimens

Curing & Strength (MPa)	Type of curing regime			
	Without use of heat curing		Heat curing at 60 ºC for 14 hours	
	Water cured (25 ºC - 26 ºC)	Room temp. Cured (32 ºC)	Water cured (25 ºC - 26 ºC)	Room temp. Cured (32 ºC)
f_7	53.7	38.7	57.6	64.0
f_{28}	61.6	44.5	63.2	64.2
f_7/f_{28}	0.87	0.87	0.91	1.00
f_{28}/f_7	1.15	1.15	1.10	1.00
SEM images	Figure 5, (d)	Figure 5, (c)	Figure 5, (b)	Figure 5, (a)

Table 6. Compressive strength (f) at 7 and 28 days for H- 4 sets mix

Based on the given data in Table 6 it is evident that the highest strength is obtained from air curing at room temperature after heat curing at 60 ºC for 14 hours. Generally speaking, the results obtained can be discussed according to the different curing regimes.

The strength of specimens cured in air at room temperature after subjected to heat curing is the highest. This is followed by the specimens cured in water after heat curing and specimens cured under the same conditions but without heat curing. The lowest strength was obtained for the specimens cured in air at room temperature without heat curing. It was also observed that the thickest and the longest ettringite crystals were formed in the specimens with the highest strength as shown in Figure 5.

Similar to the strength at 7 days, it is seen that among the four curing regimes the strength of specimens at 28 days is the highest for those cured at room temperature after heat curing at 60 °C for 14 hours. From Table 6 it can be seen that the strength growth (i) at 28 to 7 days in different curing regimes is as follows: for without heat and curing in air at room temperature i= 1.15 and for curing in water i= 1.15. Use of heat curing at 60 ºC for 14 hours followed by curing in air at room temperature i= 1.00 and for curing in water; i= 1.10.

From the results observed it is clear that in the cases without use of heat curing, the strength growth of 28 to 7 days is noticeable for both curing at room temperature and in water. The relative strength growth is on average about 1.15. This shows that there is a continuous hydration process progression for duration of 7 to 28 days. It seems that the latent potential of the specimens is gradually released, whilst whenever the specimens are heated, the whole latent potential is suddenly released during the initial days (in duration of the first days) due to the temperature effect. In the case with heat treatment, it is seen that the strength gain is completely different for the specimens cured at room temperature and in water. It is observed that there is no significant strength growth at 28 days compared to 7 days. In fact, it can be said that when the specimens are cured at room temperature after heat curing, the highest strengths are achieved during the first 7 days. This is a major advantage to the precast concrete industry. However, when the specimens are cured in water after heat curing more time is needed to achieve maximum strength. This shows that curing in water is not the best way to cure the specimens after heat treatment from the standpoint of strength gain at early ages. Hence, the comparison of curing regimes at room temperature and in water shows that the best curing regime after heat curing is air curing under room temperature, especially for the precast concrete industry. It is also seen that in the case without heat curing, on average, the strength at 7 days is 87% of the strength at 28 days for both curing at room temperature and in water, and with heat curing this ratio is increased on average to 95%. This means that heat curing improves the strength at 7 days by an average of about 9%.

3.4. Strength loss

The compressive strength loss of mortars and concretes containing supplementary cementitious materials is quite common. In the process of this research, strength loss has been observed several times for some of the mortars prepared. This is because for a variety of reasons; some of the observed reasons in the study are as follows:

Several researchers reported that a high temperature improves strength at early ages (ACI, 2001; Al-Gahtani, 2010; Shariq, Prasad, et al., 2010). At later ages, the important number of formed hydrates had no time to arrange suitably and this caused a loss of ultimate strength. This behaviour has been called the crossover effect (Powers, T.C., 1947). For OPC it appears that the ultimate strength decreases nearly linearly, with curing temperature (Ramezanianpour & Malhotra, 1995).

Generally, the reason for the loss of strength can be due to internal or external factors. The internal factors are those linked to the chemical composition of the reacted products. The most efficient external factors are due to the variability of specimens and testing procedures. Another factor having high importance is the effect of temperature. The initial curing temperature has a significant effect and can reduce or increase strength at later ages. It seems that the main reason for strength loss at later ages is due to lack of inside water in specimens to complete the hydration and pozzolanic process progression. Usually for the duration of 1 to 28 days, the inside water due to the mixing water is available and adequate for the hydration process; however, beyond 28 days it is reduced and then insufficient for the process of hydration and pozzolanic reaction to progress; hence, strength loss occurs.

3.5. Strength development

Customarily, whenever it is wanted to understand the behaviour of a phenomenon, it is accepted to model its behaviour by the use of a diagram or mathematical relationship. Using the relationships can approximately be forecasted the behaviour of the phenomenon at the later ages. In this research based on the results obtained for the OPC-slag mortars with 50% OPC replacement with slag under thermal activation method have been determined to forecast the variations of compressive strengths versus age of curing.

Comparison of all the relationships shows that the most appropriate form of equation to describe the variations of strength versus age of curing is a logarithmic function in the form of $f = a^{*}Ln\ (t) + b$; where R^2 is the coefficient of determination, a and b are constants for a specified mortar, f is compressive strength in MPa and t is the age of specimens in days.

The best fitted curves strength developments are shown in Figure 6.

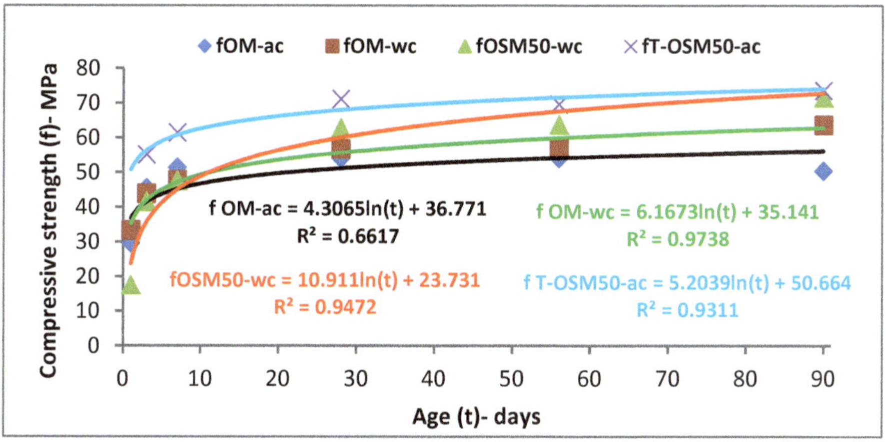

Figure 6. Strength development curve fitting for the optimum OPC-slag mortar activated using thermal activation method and OPC mortars

3.6. Summary

For different duration several temperatures such as 50 °C, 60 °C, and 70 °C were used. Based on the results obtained it can be deduced that heat curing at 60 °C for 20 hours is the optimum. It should be noted that 40% and 50% levels of replacement of OPC with slag were used and the results were compared. In addition to temperature, the effects of relative humidity were also studied. Finally, the results of compressive strength at 3 and 7 days versus heat curing were determined for 40% and 50% levels of replacement. It was recognized that the formation of long and thick crystals of ettringite was the main reason for significant strength improvement at early ages when the specimens were heated and then cured in air under room temperature. The flowchart of thermal activation method is shown in Figure 7.

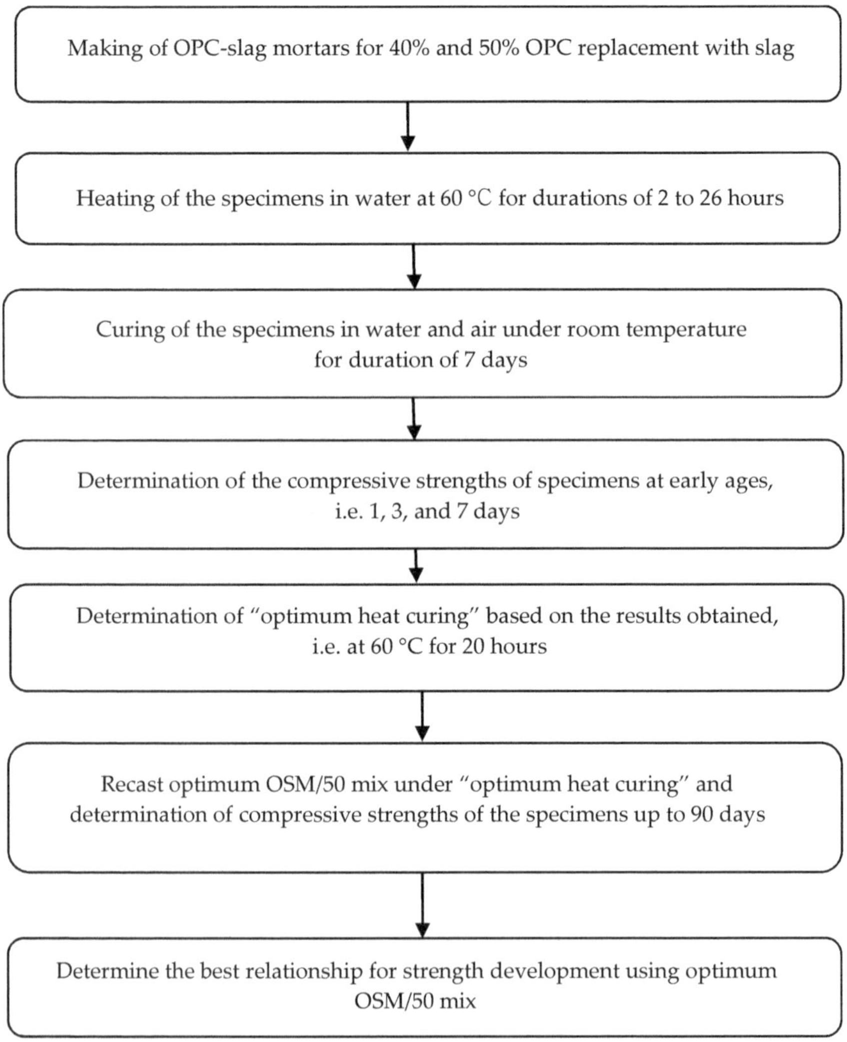

Figure 7. Thermal activation work phase

4. Conclusions and recommendations

4.1. Conclusions

4.1.1. Introduction

In this study 50% level of slag as the optimum was used as replacement for OPC. The thermal activation method (T) was used to activate ordinary Portland cement-slag mortars (OSMs).

4.1.2. Optimum replacement

For the slag used in this investigation, the optimum cement replacement level from viewpoint of high early strength was proven within the range of 40% - 50%. By using the optimum level of replacement slag, i.e. 50%, noticeable strength levels of OPC-slag mortars are achievable without the use of any activation method. In contrary to OPC-slag mortar for 40% replacement with slag, the other mortars made with different levels of replacement slag have shown higher 56-day strength compared to strength at 28-day, but the OPC-slag mortar for 40% OPC replacement with slag gives less 56-day strength compared to strength at 28-day by about 10.8%; namely compressive strength loss about 10.8%. Thus this strength loss phenomenon needs to be further investigated.

The highest growth of 56-day strength compared to 28-day are attributed to slag mortars and OPC-slag mortars for 10% replacement with slag (OSM/10) as 31.5% and 21.7%, respectively. This shows that the strength of mortar including the highest level of replacement slag will be improved at later ages more than others. However, it is well known that the ultimate strength of slag mortars is not significant compared to the strength of the others at the same ages. Therefore, it can be deduced that the slag mortars are the best only from viewpoint of durability.

4.1.3. Thermal activation method

It has been shown that the strengths of specimens cured in water at 3 and 7 days for OPC-slag mortar with 40% and 50% OPC replacement of slag, without and with use of heating for duration of 2 hours, are more than those cured in air under room temperature. However, as soon as the heating duration is increased to 4 hours and more, this effect is reversed. This is a new finding with a major advantage in precast concrete industry and also has many advantages in arid regions for curing of concrete structures.

Based on the experimental results obtained in the study, it can be concluded that there is an optimum temperature for each specific material to obtain high early strength. It was determined that 60 °C is the optimum. Heating duration is also very important for obtaining high early strength. For the slag used in the study, duration of 20 hours is optimum. Usually, as heating time increases towards the optimum, the compressive strength will be increased.

The maximum strengths obtained at 3 and 7 days for OPC-slag mortar with 50% OPC replacement of slag cured in air under room temperature are 55.3 and 61.6 MPa, respectively. It can be seen that these are 21.8% and 20.0% more than those of OPC mortar specimens cured in air under room temperature, and 26.1% and 29.0% more than those of OPC mortar specimens cured in water, respectively.

It was proven that whenever the mortar is heated larger than the optimum duration, it could be seen that this will not lead to an increase in early strength. According to the results of the study and other researches, it can be deduced that the thermal activation is one of the most efficient and applicable techniques for activation of OPC-slag mortars. This is well known specially in precast concrete industry.

The results obtained show that the best relationship of compressive strengths versus heating duration of the specimens cured in air under room temperature and water for OPC-slag mortars at 40% and 50% OPC replacement with slag are power equations. A relationship exists between the compressive strength of the specimens cured in air under room temperature and water for OPC-slag mortar at 40% and 50% OPC replacement with slag at 3 days, but not at 7 days. Generally, comparison of OPC mortars heated in water bath or oven showed that water bath heat treatment gave better results than those of oven heated. It was also observed that the specimens gave higher strengths in air under room temperature compared to water curing after heating in the water bath at 60 ºC for a duration of 20 hours. This is a significant finding with a major advantage in construction and is also economic since water bath heating can be practically implemented. However, oven heated pre-curing, results in higher strengths whenever the specimens are cured in water after heating.

4.2. Recommendations for future works

Among the mixes prepared by using different levels of OPC replacement with slag, mixtures with 50% ggbfs and 50% OPC show the highest strength in the absence of any activation method.

Based on the extra work done in this research by use of "mining sand" instead of "silica sand", it was revealed that using mining sand is preferable to silica sands for activation of slag mortars, thus it is suggested to conduct a new study using mining sand.

In this study a single source of ggbfs was used throughout. It is recognized that other sources may have somewhat different chemical compositions. Thus other sources of the material need to be evaluated in order to determine the influence of the activation methods used in this study to be generalized.

Apendix

Mix proportions

No	Mix name	OPC (g)	Slag (g)	Flow (mm)	SP (g)	Water (g)
1	OM-wc	1200	0	225	40	421.11
2	OSM/10-wc	1080	120	225	68	421.11
3	OSM/20-wc	960	240	220	65	421.11
4	OSM/30-wc	840	360	220	60	421.11
5	OSM/35-wc	780	420	230	40	421.11
6	OSM/40-wc	720	480	210	48	421.11
7	OSM/45-wc	660	540	235	40	421.11
8	OSM/50-wc	600	600	235	40	421.11
9	OSM/60-wc	480	720	220	35	421.11
10	OSM/70-wc	360	840	230	35	421.11
11	OSM/80-wc	240	960	225	33	421.11
12	OSM/90-wc	120	1080	220	27	421.11
13	SM-wc	0	1200	220	30	421.11

Table 7. Mix proportions for determination of optimum level of replacement slag

No	Mix name	OPC (g)	Slag (g)	Water (g)	SP (g)	Flow (mm)
For OPC mortars, room temperature and water cured						
1	OM- ac	1800	-----	631.7	28	230
2	OM- wc	1800	-----	631.7	30	230
For OPC-slag mortars for 40% replacement with slag , cured in water and air under room temperature						
3	H0/0	720	480	421.11	28	225
4	H60/2	1440	960	842.22	82	230
5	H60/4,6	1440	960	842.22	90	230
6	H60/8,10	1440	960	842.22	79	230
7	H60/12,14	1440	960	842.22	79	230
8	H60/16	1440	960	842.22	82	230
9	H60/18,20	1440	960	842.22	73	230
10	H60/22,24,26	2160	1440	1263.33	70	220
For OPC-slag mortars for 50% replacement with slag, cured in water and air under room temperature						
11	H0/0	600	600	421.11	35	230
12	H60/2	1200	1200	842.22	76	235
13	H60/4,6	1200	1200	842.22	91	225
14	H60/8,10	1200	1200	842.22	90	235
15	H60/12,14	1200	1200	842.22	73	235
16	H60/16	1200	1200	842.22	76	235
17	H60/18,20	1200	1200	842.22	62	225
18	H60/22,24,26	1800	1800	1263.33	60	220
OSMs/50 test for three sets of specimens in cured room temperature and in water 25 ºC and 32 ºC after heat treatment 60 ºC for duration 20 hours						
19	H- 3 sets mix	900	900	631.7	35	225
OSMs/50 test for four sets of specimens cured in room temperature and in water after with and without use of heat treatment 60 ºC for duration 14 hours						
20	H- 4 sets mix	1200	1200	842.22	50	225
For optimum OSM/50 at six ages, only room temperature cured						
21	H60/20	900	900	631.7	43	230
H60/i,j,k means 60 ºC temperature with duration i, j, and k hours						

Notes: OSMs/50= OPC-slag mortars for 50% replacement with slag, OMs= OPC mortars, SP= super plasticizer, ac= cured in air under room temperature, wc= water cured.

Table 8. Mix proportions of OPC-slag mortars for thermal activation method

Abbreviations

Abbreviation	Statement
OPC	ordinary Portland cement
ggbfs	ground granulated blast furnace slag
OM	ordinary Portland cement mortar
OSM	ordinary Portland cement-slag mortar
SM	slag mortar
OSM/i	OPC-slag mortar for i% replacement with slag
OSM/50-wc	control mix/ mortar
f_i	compressive strength at i days in MPa
i	relative strength or strength ratio
T	thermal activation method
Wc	water curing
Ac	curing in air under room temperature
SD	standard deviation
R^2	coefficient of determination
XRD	X-ray diffraction
SEM	scanning electron microscopy
EDX	energy dispersive X-ray analysis
XRF	X-ray fluorescence
ASTM	American Society for Testing and Materials
ACI	American Concrete Institute
BSI	British Standards Institute
SCA	Slag Concrete Association
MIA	Mortar Industry Association
CSA	Canadian Standards Association
GSD	grain size distribution
FM	fineness modulus
SSA	specific surface area
SAI	slag activity index
SP	super plasticizer
s/b	sand-binder ratio
w/c	water-cement ratio
w/b	water-binder ratio
DEF	delayed ettringite formation
EEF	early ettringite formation
SEF	secondary ettringite formation
ASR	alkali silica reaction
C-S-H	calcium silicate hydrates
CH	hydroxide calcium
LOI	loss on ignition
RH	relative humidity
AF_m ($C_3A.CaSO_4.12H_2O$)	aluminate-ferrite-monosubstituted or
AF_t ($C_3A.3CaSO_4.32H_2O$)	aluminate-ferrite-trisubstituted or ettringite

Abbreviation	Statement
$CaCO_3$	calcite
$C_3A.3CaSO4.32H_2O$	ettringite
$C_3A.CaSO_4.13H_2O$	imonosulphate
C_2ASH_8	gehlenite hydrate
$M_6.Al_2CO_3\ (OH)_{16}.\ 4H_2O$	hydrotalcite
$Ca_2SiO_4.H_2O$	α- C_2SH

H-3 sets mix is a control mix, which the specimens were first made and demoulded 24 hours after casting, and then heated at 60 °C for 20 hours and finally, cured in three curing regimes, i.e. at room temperature, in water at 25 - 26 °C, and in water at 32 °C.

H-4 sets mix is a control mix, which the specimens were first made and demoulded 24 hours after casting, and then pre-cured without and with the use of heat at 60 °C for 14 hours and finally, each set of specimens was cured in both curing regimes, i.e. at room temperature and in water at 25 - 26 °C.

Author details

Fathollah Sajedi
Islamic Azad University, Ahvaz Branch, Ahvaz, Iran

5. References

ACI. (2001). Guide to Curing Concrete, Reported by ACI Committee 308, *ACI 308R-01*. USA: American Concrete Institute

ACI. (2006). Standard Specification for Mortar for Unit Masonry, Annual Book of Standards, *ASTM C270*. USA: American Concrete Institute

Ahmed, S. F. U., Ohama, Y., & Demura, K. (1999). Comparison of mechanical properties and durability of mortar modified by silica fume and finely ground balst furnace slag. *J Civ Eng, CE 27*(2).

Al-Gahtani, A. S. (2010). Effect of curing methods on the properties of plain and blended cement concretes. *Construction and Building Materials* 24(3), 7.

Barnett, S. J., Soutsos, M. N., Millard, S. G., & Bungey, J. H. (2006). Strength development of mortars containing GGBFS: Effect of curing temperature and determination of apparent

Bougara, A., Lynsdale, C., & Ezziane, K. (2009). Activation of Algerian slag in mortars. *Construction and Building Materials* 23(1), 6.

Brooks, J. J., & Al-kaisi, A. F. (1990). Early strength development of Portland and slag cement concretes cured at elevated temperatures. *ACI Mater J, 87*(5), 5.

Carino, N. J. (1984). The Maturity Method: Theory and Application. *CCA, 6*(2), 13.

Fu, Y. (1996). *Delayed ettringite formation in Portland cement products.* University of Ottawa, Canada.

Hanson, J. A. (1963). Optimum steam curing procedures in precasting plants. *ACI Mater J, 60*(1), 25.

Johansen, V., Thaulow, N., & Skalny, J. (1993). Simultaneous presence of alkali-silica gel and ettringite in concrete. *Advances in Cement Research* 5(17), 7.

Kosmatka, S. H., Panarese, W. C., & Allen, G. E. (1991). *Design and control of concrete mixtures* (Fifth ed.). Ottawa Canadian Portland Cement Association.

Lerch, W., Ashton, F. W., & Bogue, R. H. (1929). Sulfoaluminates of Calcium *Journal of Research of the National Bureau of Standards* 2(4).

Lewis, M. C., & Scrivener, K. L. (1996). *Microchemical effects of elevated temperature curing and delayed ettringite formation. In Mechanisms of Chemical Degradation of Cement-Based Systems*. London

Lodeiro, C. I., Macphee, D. E., Palomo, A., & Fernandez, A. J. (2009). Effect of alkalis on fresh C-S-H gels. FTIR analysis. *Cement and Concrete Research, 39*(3), 7.

Mehta, P. K. (1976). Scanning Electron Micrographic Studies of Ettringite Formation *Cement and Concrete Research, 6*(2), 14.

Neville, A. M. (2008). *Properties of Concrete* (Fourth ed.). Kuala Lumpur, Malaysia: Prentice Hall.

Ogawa, K., & Roy, D. M. (1982). C4AS3 hydration, ettringite formation and its expansion mechanism: III. Effect of CaO, NaOH and NaCl; conclusions *Cement and Concrete Research 12*(2), 10.

Powers, T. C. (1947). A discussion of cement hydration in relation to the curing of concrete. *Proc Highway Res Board 27*, 11.

Pfeifer, D. W., & Marusin, S. (1991). *Energy efficient accelerated curing of concrete, a state-of-the-art review* (No. Technical Report No. 1): Prestressed Concrete Institute.

Ramezanianpour, A. A., & Malhotra, V. M. (1995). Effect of Curing on the Compressive Strength, Resistance to Chloride-Ion Penetration and Porosity of Concretes Incorporated Slag, Fly Ash or Silica Fume. *Cement and Concrete Composites 17*(2), 9.

Shi, C. (1996). Strength, pore structure and permeability of alkali-activated slag mortars. *Cement and Concrete Research, 26*(12), 11.

Shoaib, M. M., Balaha, M. M., & Rahman, A. G. A. (2000). Influence of cement kiln dust substitution on the mechanical properties of concrete. *Cement and Concrete Research, 30*(3), 7.

Shariq, S., Prasad, J., & Masood, A. (2010). Effect of GGBFS on time dependent compressive strength of concrete. *Construction and Building Materials* 24(8), 10.

Wang, S. D., Pu, X. C., Scrivener, K. L., & Pratt, P. L. (1995). Alkali-activated slag cement and concrete: a review of properties and problems. *Adv Cem Res 27*, 10.

Wang, S. D., Scrivener, K. L., & Pratt, P. L. (1994). Factors affecting the strength of alkali-activated slag. *Cement and Concrete Research, 24*(6), 11.

Yang, R., Lawrence, C. D., & Sharp, J. H. (1996). Delayed ettringite formation in 4-year old cement paste. *Cement and Concrete Research, 26*(11), 11.

Permissions

The contributors of this book come from diverse backgrounds, making this book a truly international effort. This book will bring forth new frontiers with its revolutionizing research information and detailed analysis of the nascent developments around the world.

We would like to thank Frank Czerwinski, for lending his expertise to make the book truly unique. He has played a crucial role in the development of this book. Without his invaluable contribution this book wouldn't have been possible. He has made vital efforts to compile up to date information on the varied aspects of this subject to make this book a valuable addition to the collection of many professionals and students.

This book was conceptualized with the vision of imparting up-to-date information and advanced data in this field. To ensure the same, a matchless editorial board was set up. Every individual on the board went through rigorous rounds of assessment to prove their worth. After which they invested a large part of their time researching and compiling the most relevant data for our readers. Conferences and sessions were held from time to time between the editorial board and the contributing authors to present the data in the most comprehensible form. The editorial team has worked tirelessly to provide valuable and valid information to help people across the globe.

Every chapter published in this book has been scrutinized by our experts. Their significance has been extensively debated. The topics covered herein carry significant findings which will fuel the growth of the discipline. They may even be implemented as practical applications or may be referred to as a beginning point for another development.

The editorial board has been involved in producing this book since its inception. They have spent rigorous hours researching and exploring the diverse topics which have resulted in the successful publishing of this book. They have passed on their knowledge of decades through this book. To expedite this challenging task, the publisher supported the team at every step. A small team of assistant editors was also appointed to further simplify the editing procedure and attain best results for the readers.

Our editorial team has been hand-picked from every corner of the world. Their multi-ethnicity adds dynamic inputs to the discussions which result in innovative outcomes. These outcomes are then further discussed with the researchers and contributors who give their valuable feedback and opinion regarding the same. The feedback is then collaborated with the researches and they are edited in a comprehensive manner to aid the understanding of the subject.

Apart from the editorial board, the designing team has also invested a significant amount of their time in understanding the subject and creating the most relevant covers. They scrutinized every image to scout for the most suitable representation of the subject and create an appropriate cover for the book.

The publishing team has been involved in this book since its early stages. They were actively engaged in every process, be it collecting the data, connecting with the contributors or procuring relevant information. The team has been an ardent support to the editorial, designing and production team. Their endless efforts to recruit the best for this project, has resulted in the accomplishment of this book. They are a veteran in the field of academics and their pool of knowledge is as vast as their experience in printing. Their expertise and guidance has proved useful at every step. Their uncompromising quality standards have made this book an exceptional effort. Their encouragement from time to time has been an inspiration for everyone.

The publisher and the editorial board hope that this book will prove to be a valuable piece of knowledge for researchers, students, practitioners and scholars across the globe.

List of Contributors

A.M.A. Mohamed
Materials Technology Unit (MTU), Qatar University, Doha, Qatar
Department of Metallurgical and Materials Engineering, Faculty of Petroleum and Mining Engineering, Suez Canal University, Suez, Egypt

F.H. Samuel
Université du Québec à Chicoutimi, Chicoutimi, Québec, Canada

Grzegorz Golański
Institute of Materials Engineering, Czestochowa University of Technology, Poland

Michal Pohanka and Petr Kotrbáček
Brno University of Technology, Heat transfer and Fluid Flow Laboratory, Brno, Czech Republic

Anton Panda, Jozef Jurko and Iveta Pandová
Technical University of Košice, Slovak Republic

Frank Czerwinski
CanmetMATERIALS, Natural Resources Canada, Hamilton, Ontario, Canada

Anna Biedunkiewicz
Institute of Materials Science and Engineering West Pomeranian University of Technology, Szczecin, Poland

Angel Zumbilev
Technical University of Sofia-Plovdiv Branch, Bulgaria

Sonia Hatsue Tatumi, Alexandre Ventieri, José Francisco Sousa Bitencourt, Katia Alessandra Gonçalves, Juan Carlos Ramirez Mittani, René Rojas Rocca and Shiva do Valle Camargo
Universidade Federal de São Paulo, Universidade de São Paulo, Centro Estadual de Educação Tecnológica Paula Souza, Brazil

Jeffrey C. De Vero, Rusty A. Lopez,Wilson O. Garcia and Roland V. Sarmago
National Institute of Physics, University of the Philippines, Diliman, Quezon City, Philippines

Cao Minh Thi
Ho Chi Minh City University of Technology, Dien-Bien-Phu, Binh-Thach, Ho Chi Minh City, Vietnam

Masayuki Nogami
Department of Materials Science and Engineering, Nagoya Institute of Technology, Gokiso-cho, Showa-ku, Nagoya, Japan

Nguyen Viet Long
Department of Molecular and Material Sciences, Interdisciplinary Graduate School of Engineering Sciences, Kyushu University, 6-1 Kasugakouen, Kasuga, Fukuoka, Japan
Department of Education and Training, Posts and Telecommunications Institute of Technology, Nguyen Trai, Ha Dong, Hanoi, Vietnam
Department of Materials Science and Engineering, Nagoya Institute of Technology, Gokiso-cho, Showa-ku, Nagoya, Japan
Laboratory for Nanotechnology, Ho Chi Minh Vietnam National University, Linh-Trung, Thu-Duc, Ho Chi Minh, Vietnam

Michitaka Ohtaki
Department of Molecular and Material Sciences, Interdisciplinary Graduate School of Engineering Sciences, Kyushu University, 6-1 Kasugakouen, Kasuga, Fukuoka, Japan

Wesley M. Dose and Scott W. Donne
Discipline of Chemistry, University of Newcastle, Callaghan, Australia

Rabah Bensaha and Hanene Bensouyad
Laboratory of ceramics, University of Constantine 1, Constantine, Algeria

Manuel Pedro Fernandes Graça and Manuel Almeida Valente
University of Aveiro / I3N - Physics Department, Portugal

Hiroki Nagai and Mitsunobu Sato
Research Institute of Science and Technology, Kogakuin University, Nakano, Hachioji, Tokyo, Japan

E-Wen Huang
Department of Chemical and Materials Engineering, National Central University, Jhongli, Taiwan, ROC

Yu-Chieh Lo
Department of Nuclear Science and Engineering, The Massachusetts Institute of Technology, Cambridge, MA, USA

Junwei Qiao
College of Materials Science and Engineering, Taiyuan University of Technology, Taiyuan, China

Fathollah Sajedi
Islamic Azad University, Ahvaz Branch, Ahvaz, Iran